D0622314

Canon City Public Library
Canon City, Colorado

CELL OF CELLS

CELL OF CELLS

· · · · · · · · · · · · · · · ·

*The Global Race to Capture
and Control the Stem Cell*

CYNTHIA FOX

W. W. NORTON & COMPANY

NEW YORK   LONDON

Canon City Public Library
Canon City, Colorado

Copyright © 2007 by Cynthia Fox

All rights reserved
Printed in the United States of America
First Edition

For information about permission to reproduce selections from this book, write to
Permissions,W. W. Norton & Company, Inc., 500 Fifth Avenue, New York, N.Y.
10110

Manufacturing by The Courier Companies, Inc.
Book design by Lovedog Studio
Production manager: Amanda Morrison

LIBRARY OF CONGRESS CATALOGING-IN-PUBLICATION DATA

Fox, Cynthia
Cell of cells : the global race to capture and control the stem cell / Cynthia Fox. —
1st ed.
p. cm.
ISBN 13: 978-0-393-05877-2 (hardcover)
ISBN 10: 0-393-05877-8 (hardcover)
1. Stem cells. 2. Stem cell—Therapeutic use. I. Title.
QH588 .S83F69 2006
616' .02774—dc22
2006009917

W. W. Norton & Company, Inc.
500 Fifth Avenue, New York, N.Y. 10110
www.wwnorton.com

W. W. Norton & Company Ltd.
Castle House, 75/76 Wells Street,
London W1T 3QT

1 2 3 4 5 6 7 8 9 0

For my family.

Science is like a stream of water. It finds its way.

—Susan Fisher, stem cell researcher at the
University of California, San Francisco

Contents

Acknowledgments

MANY THANKS TO THE MacDowell and New York Mills artists' colonies for writer's residencies. Thanks to the Japan Foreign Press Center's tireless and gracious organizer, Akira Sugama; its fascinating president, Terusuke Terada; and its formidable managing director, Masahiko Ishizuka, for the fellowship and ten amazing days in Japan stem cell labs. Thanks also to the Japan Society, in particular to Betty Borden.

Thanks to those who read parts of the book, including Dan Bergner, Stephen Campitelli, Kim Clark, Mark Collins, Ann Darby, Jackie Keren, Bill Madison, J. C. Smith, and Dana Tierney. Thanks to *Fortune* international editor Robert Friedman for the exacting and writerly editing job on the article that helped lead to the book. Thanks to Thomas Zweifel for some superb translation and advice re international business. Thanks to my sainted agent, Gail Hochman. Love to the late Robert Towers and the late Robert Bullock, along with J. C. Smith, Stephen Koch, and Richard Locke. Men with gorgeous writers' minds and generous teachers' hearts, they have all inspired people to spend their lives celebrating life by tossing handfuls of words in the air: they have all turned normal people into writers. Love to my family and friends, who put up with me answering every other question with "stem cells" year after year. And much, much gratitude to those at W. W. Norton who put in an unusual and ungodly number of hours on this book.

Then there are the scientists. Stanford University's Irv Weissman says, "Every aspect of biology, cancer, and medicine changes when you think about your problems this way: tissues develop from, and are sustained by, a tiny subset of cells that are stem cells." Stem cells are challenging paradigm after paradigm. The result is scientists who are regularly called upon to serve up a bit of courage with their talent. Harvard adult stem cell researcher Jonathan Tilly has been questioning dogma that was not well proven, yet went unchallenged for fifty years: women are born with a fixed supply of eggs. He caused an uproar when he proposed some egg stem cells persist that might be replicated and frozen, letting women postpone menopause and/or infertility. As of October 2006 the jury was out. Two teams published supportive work; two teams published unsupportive work. But with new data on the way, Tilly continues his pursuit. As he does, Newcastle University has announced the first lab births with germ cells made from mouse ES cells, and Kyoto University has reported that the first mature mammal cells have been cranked backward in a lab to become ES-like sans cloning. That stem cell scientists intent on busting paradigms will prompt some change in fertility seems a good bet.

Then there are researchers facing presidential restrictions on human embryonic stem (hES) cell work. (Due to those restrictions, only $161 million in NIH funds went to hES cells, while $993 million went to adult stem cells, in 2003–2007.) Mahendra Rao was an NIH Stem Cell Task Force member who could have rested on his laurels. But access to hES cells was limited and he wanted to fully understand them. So he moved to private industry, polite yet forthright about the reason. Likewise, Harvard's Doug Melton didn't test political winds. Simply determined to cure his two diabetic kids, he swiftly cordoned off his Harvard lab at great cost, so it wouldn't clash with government restrictions, then created—and distributed free worldwide—new hES cell lines.

"An astounding interest exists for this kind of research," said an October 2006 *Stem Cells* report about the 20-plus nations publishing hES cell work. Thank you to the hundreds of researchers who interviewed for this book in the middle of their often astounding race to capture and control the stem cell.

CELL OF CELLS

Prologue

WITH THEIR GIANT FLIRTATIOUS eyes and permanent smiles, the three camels lounging in the sandy Giza, Egypt, back lot are postcard-perfect. At the approach of Egyptian stem cell researcher Ismail Barrada, all three even duck their heads this way and that, as if posing coyly for a camera.

But when he gets within a few feet of the camels, first one, then all three, begin to roar. The sound alternates, as it grows louder, between the angry bellow of a lion and the anguished, off-key warble of a wounded loon. And just when it seems the animals are done with this otherworldly symphony, one sends his head into a slow dive toward the sand, twisting and untwisting his neck as he goes, as if trying to accommodate the determined ascent of an unpleasant something rising within him. Suddenly, he snaps his shaggy head up toward Barrada with such intensity, it appears he thinks he can shake loose his yellow teeth and send them winging through the air at the hapless scientist. Guests accompanying Barrada jump back, then stare at the yellow grin that remains aimed in their general direction, Cheshire cat–like. After a silence, the guests' comments range from, "These are *our* camels?" to "Are they all right?"

But Barrada, who holds in one hand a plastic bag filled with toilet paper (the Pyramids, his destination, have no bathrooms for his guests), is busily fishing through his shirt pockets with his other hand. "What's that?" he asks distractedly. Still patting at his pockets, he casually swings a leg over the back of one of the camels. As it begins to rise, haunches first, the researcher is pitched forward, and he reaches out to grab the horn on his saddle. But once the beast has clambered onto all fours, roaring away, swiveling its neck in what appears to be an almost successful attempt to tear the foot off his leg, Barrada resumes his one-handed pocket search. "Ah," he says, pulling out a handkerchief. "Warm." He pats his forehead, returns the handkerchief, pats another pocket, pulls out a package of Marlboros, and looks down at his guests as he shakes a cigarette half out of the pack and raises it to his lips. His broad, placid face fills with concern as he slips the pack back into his pocket. "Is something wrong?"

Shamefacedly, Barrada's guests follow his lead, if approaching their camels from behind to avoid those yellow teeth, and muttering, "Whoa there you camel," with every move the animals make. But Barrada notices none of this, now busy chatting with the camels' owner, Ahmed. Ahmed named Barrada's camel "Minnesota" because an American friend of his lives there, and he wants to join her when he has finished his mandatory three-year stint with the army reserve (which pays little, hence the camel business). Barrada is the gynecologist of Jihan Sadat, widow of slain Egyptian president Anwar al-Sadat. He is the husband of an Egyptian movie star and comes from a family of prominent Egyptian doctors. He worked at the University of Minnesota as chief of obstetrical surgery for a decade before moving to Houston, then back to Cairo. He has much to tell Ahmed about the wonders of Minneapolis.

As the guests stumble after Ahmed, Minnesota, and the scientist through the loud and dirty streets of Giza, dodging donkey carts and slow-moving buses, they unhand their saddle horns only to wipe their sweating palms on their pants. After half an hour, the three still have not lost their terrified stiff postures and grim grins. "I don't think . . . maybe we'd better not . . . ," one begins. But then the queasy caravan rounds a crumbling apartment building topped with a satellite dish, and all fall silent.

Rising so high above a grove of nearby palm trees that they look almost fake, like a movie backdrop, the Pyramids are, it seems, impossible in every way. An impossible 4,500 years old—as Ahmed notes—with each limestone and granite block an impossible average 2.5 tons—as Barrada adds—each of their corners is uncannily aligned with the earth's compass points—one, within 1/20th of a degree. Both impossibly familiar and impossibly strange, they are perched above a sweeping valley of sand that begins with no warning: suddenly there is no grass, just desert. Through this landscape, Egyptians are racing horses at an impossibly fast clip, leaving trails of dust behind them and sending explosions of birds into the air. "They are playing Indiana Jones," Ahmed explains.

A *muezzin* starts singing Muslim prayers, as muezzins do all across Egypt five times a day. To the Westerner, his voice is as warbly and off-key as the camels' bellows, though it is more muffled, as if he is singing underwater. A woman in a T-shirt gallops past on a horse, talking on a cell phone. She is followed by an old water peddler who is riding a donkey that is so small, it is barely visible under his *galabaya,* or robe. From far below rises the murmuring of hundreds of tourists spilling out of buses and heading up the hill toward the Pyramids, their sound like that of swarming extras in a Cecil B. De Mille movie on a TV with the volume turned low.

The life that is still swirling around the Pyramids after 4,000 years is so fascinating to Barrada's guests that all three find themselves riding their camels one-handed and/or straight-backed, pointing and craning to see. For a few moments, their fascination has become bigger than their fear.

Barrada, in the meantime, is riding his camel with *no* hands now, pointing upward, looking backward, citing all the impossible facts and numbers associated with the Pyramids, repeating excitedly, "Isn't that something? Isn't that incredible?" He is, it would appear, the postcard-perfect scientist, whose fascination is always like this, every day, bigger than his fear.

But over the coming months, Barrada will need to cling to this quality as tightly as his guests, when the camels begin speeding up, cling to their saddle horns again. For Barrada, inspired by the fact that the United States has backed away from the controversial science of human embryonic stem cells—extraordinary cells that can replicate indefinitely, and form all the cells of the human body—is not only about to launch an embryonic stem cell center in Egypt, which he intends to be the first in the Middle East outside Israel. He is determined to launch it in an international collaboration *with* Israel, which is more experienced, in a new development springing up on the edge of the Sahara called "6th of October City." There is, certainly, poetic justice here. The Pyramids, which Barrada actually passes each day on the way to his lab on the edge of the Sahara Desert, are perhaps the most enduring testaments on Earth to man's determination to live forever. They were built to resemble the "primordial" mounds of earth that appear on the Nile's shores every year after its floods recede. (The first of these mounds gave rise to the eternal Gods, one of whom wept tears that became man.) The monuments are believed to be "Houses of Eternity." Likewise, many scientists believe that embryonic stem cells are the seeds for a potent new regenerative medicine that may prolong more lives than any other single medicine in history. It works, the notion of a lab of "immortal cells" here.

There is one major, modern problem with Barrada's ambitions. The city in which he is building his lab for work he'd like to share with Israel is so named because Egypt still actively, annually, celebrates what it calls its defeat of Israel in a battle that occurred decades ago: the October War of 1973. Egypt and Israel, to put it mildly, don't get along. As a result, Barrada's e-mails to Israel are already being monitored by Egypt's equivalent of the CIA. And his staffing choices have already been questioned by academic brass because some are "white" students from the American University in Cairo.

Furthermore, it is inescapable that Barrada is setting up his stem cell lab in a police state. Brandishing guns, soldiers in white and black are posted everywhere, from malls to the median strips on highways, where they lazily flag down every few cars and search them at will, oblivious to the hours-long traffic jams they cause, apparently unaware of the concept of warrants. Security guards in gray business suits with machine guns strapped to their sides run alongside the cars of government ministers. "It's because of little Bin Ladens" all over, from the streets to government offices, Barrada will casually note on the way to a restaurant on the Nile that boasts two such men posted outside the door. Transported by the awe-inspiring visage of a new science, Barrada has deliberately jumped out of the frying pan of U.S. politics into the raging fire of Middle East war.

More surprisingly, he's not alone.

A few days later, Karl Skorecki stands before a very different wonder of the world, the endless white-blue Mediterranean Sea. The 13-story faculty of medicine at the Technion–Israel Institute of Technology towers over the sea, and Skorecki stands tall within its walls, being the head of its Institute for Research in the Medical Sciences. With other Israeli labs, this institute has pumped out more papers on human embryonic stem cells than have the collected institutes of many other nations combined. His office looks like the quarterdeck of a luxury cruise liner, all windows. And those windows look out on a vast stretch of Mediterranean, a surreal vista broken only occasionally by a seagull, or an empty barge pulling into the port of Haifa somewhere below. It is an end-of-the-world kind of view, not unlike the view—an empty expanse of Sahara—from Barrada's lab.

A quiet-looking man, Skorecki is, like Barrada, anything but, launching into one enthusiastic and articulate sermon about stem cells after another. Names of Israelis appeared on 10 of the first 12 papers describing the existence of the human embryonic stem cell. Israel has since been the first, sometimes with the United States and sometimes with out, to turn the cells into neurons, heart tissue, pancreatic cells, blood vessels. As concerned about scientific ethics as he is excited about the cells, he and others held a series of public meetings with Mufti and rabbis to hash out careful regulations over the course of years, fully legitimizing the research. "The religions were able to come to agreement on this one issue, anyway," he says, smiling wryly.

Skorecki doesn't mention it, but Israel is responsible for another major coup: the first truly successful systemic gene therapy clinical trial in history, which was conducted in the early 2000s by an Israeli scientist using stem cells. There have been 4,000 patients treated via gene therapy by this point in 2003. A reason for the solitary success of Israel's trial, during which Israeli and Italian scientists cured a seven-month-old Palestinian of "Bubble Boy Syndrome": the scientists used stem cells and advanced stem cell techniques. The child is now three, still free of the life-threatening disease.

Skorecki leads a reporter downstairs to a warren of labs, where he pulls a dish out of a room-temperature fridge and pops it into a microscope. At first, it looks like a normal dish of cells: a pencil sketch of a cobblestone floor. Then, as if an unseen translucent jellyfish has been sleeping on the cobblestones and has suddenly awoken, the cells warp and narrow as they lunge in concert at the prying lens, then settle down again. Seconds pass, then the eerie cobblestone creature lunges again, settles again. "An embryonic stem cell that our scientists have turned into human heart cells," Skorecki says, smiling. "They beat, as you can see."

But as he puts the cells away, the window behind Skorecki comes into view, revealing the side of the Rambam Medical Center, the clinical arm of Skorecki's Technion lab. It still holds some of the 50 wounded victims of a suicide bombing that occurred two weeks ago. The bomb killed 21 Israelis and Arabs, including three children and an infant, at a restaurant just down the beach called Makom Maxim. The blast was heard all over Haifa, from the Technion to the Mount Carmel foothills. "It's a new kind of tragedy," says Skorecki, as he glances at the hospital. "Haifa used to be somewhat immune to conflict. It's one of the reasons I'm here. We had many Arab doctors and Palestinian patients here. This has always been the one place in Israel where everyone gets along. In fact, Maxim has been owned jointly by an Arab and an Israeli for 40 years."

But this was precisely the reason this restaurant was blown up, according to press accounts. And since the bombing, Skorecki says, Palestinian patients won't come to Rambam Hospital anymore. "They are not allowed, presumably by the PLO," he continues quietly. "I don't know where they go. We ask them to come, but they won't." Skorecki looks away from the Rambam, where he spends half of his days treating patients, and heads for the door. He doesn't add that by "somewhat immune to conflict" he means there have been only six suicide bombers in Haifa over three years, killing only 74 other people. He also doesn't mention the fact that the Maxim bomb was originally meant for his Rambam.

Back in his office, he glances down at a basketball court that belongs to the military. "The soldiers are always playing down there," he says, his smile returning. "I have to admit, I check. When they're not playing, I know there's something to worry about. When they're playing, all's well."

The courts are empty now simply because it is a Friday, in the hours before Sabbath, Skorecki adds, excusing himself apologetically to head home. But when the writer who has been speaking with him returns to the Hotel Dan, it becomes clear the courts had served as a kind of omen after all. The writer's taxi is blocked by a military barricade. "Uh, oh," the driver says, rolling up his window. A bomb has gone off in the car of a Knesset member who lives near the hotel. The Knesset member is known for his efforts to negotiate peace.

Ejected from the taxi, the writer heads for the hotel on foot. She passes an orthodox rabbi. Old, overweight, and holding his black hat to his chest, he is puffing and his *payes* are swaying as he laboriously kneels, not to pray for Yom Kippur, but to look under his car.

The bomb will only make a couple of news services and will only be discussed formally by the Knesset weeks later. One reason: small bombs are so common now they often aren't mentioned in the local press. From this month, October 2003, until the end of the year, there will be 100 more bombs and 300 more terrorist incidents in Israel: three incidents, one bomb, a day. Like Barrada, Skorecki is pursuing stem cell work against some significant odds.

Then there are the South Korean cloners.

When the gowned Byeong Chun Lee slices tentatively with a scalpel at the belly of a massive, hairy, sleeping pig in a makeshift, tin-roofed operating room one day in July 2005, he exposes an inch of sea-foam-like fat. More arresting, however, is the sight of the pig's four short legs. They had been stoked skyward as if in rigor mortis. Now they are kicking violently at the putrid air. The patient isn't out.

M. Shamin Hossein, who, like Lee, works in the lab of famed Seoul National University (SNU) cloner Woo Suk Hwang, throws his body on the convulsing patient and attempts to replace her anesthesia mask. As he does this, he sweats with an almost cartoonish intensity, rivulets dribbling off his cheeks and joining the rivulets on his arms.

This is unsurprising, and not just because he is wrestling a massive, massively unhappy, ovulating pig on a 92-degree Fahrenheit day. For this is South Korea, which boasts an unusually high humidity, thanks to summer monsoons and mountains that have it 85 percent surrounded, trapping heat and moisture. Visual evidence of the humidity—rambunctious vegetation—is everywhere. The dirt road leading to this Hongseong farm looks like a machete-hacked jungle path. The rice paddy parade between here and Seoul 80 miles north is a relentless electric-green. The mountains seem more like gigantic hills, thickly covered as they are, from bottom to top, with trees ranging from medicinal ginkos to workaday pines. South Korea is just plain overripe this month. Polluted haze like a sullen halo hangs over cities. The smell of market fish and animal flesh percolates in the air. The frailest life-forms persist and/or proliferate robustly.

Including the not-so-frail pig, whose protests continue as surgical lights beam down, unhelpfully radiating extra heat. Spiderwebs cling to the lights like tinsel from an ancient Christmas. Flies circle. Korean *maemi* (cicadas) chatter with an almost metallic twang, sounding like swarms of insect-sized lawnmowers starting and stalling, over and over. The dung scent assaults the nasal passage back at its juncture with the brain, lingering so long these scientists will leave their four-wheel-drive windows open for hours. The odor helps infuse the scene, ironically, with the urgency of a sweltering *M*A*S*H** battlefield operating theater.

Lee swipes again. At last, the patient is still. Lee cuts through fat until he has created a dark hole. He sticks his gloved hand in, pulls out what looks like a pink intestine, and lays it on a cloth covering the pig's belly. It is an oviduct, or fallopian tube, attached to a uterus. Lee slides a straw into the oviduct. A student slides a syringe into the straw, and shoots into the oviduct a substance that looks like water. It isn't. It is fluid containing 150 pig/human embryo clones, who are clones of scores of pig/human fetus clones (growing inside the wombs of seven other pregnant pigs in this barn) and several pig/human adult clones already born, in Seoul.

All these clones possess the same nuclear DNA, which includes an added human immune gene. The point: to make herds of identical pigs whose organs won't be violently rejected by humans because they *are* part human. The goal: to help solve the global organ shortage by letting patients acquire new organs from part-human pigs like these, which have been created via the most efficient germline gentic engineering technology on earth for animals, cloning.

Days earlier, five-month-old fetal clones of the above days-old embryos were born. At first, clonal birth appeared no more attractive than events surrounding clonal gestation. It began with an earthy pink-brown pig hanging in an alien, sterile SNU lab in Seoul. As before, a gowned scientist raised a knife and opened the pig's belly. With a sheet, two other scientists caught a large whitish something that slithered out of the pig. "OK!" the first scientist said jumping back, knife in the air.

The other two swung the bulky sheet into a sink, then pushed it into a small, plastic room-within-a-room: a germ-free "isolator." There, bodiless rubber gloves dangling from holes in the plastic walls suddenly came alive with the inserted arms of scientists. The gloves attacked the wet bulk, which was revealed to be four squirming piglets in what looked like a plastic bag: the placenta. The placenta had been removed in its entirety via the above, unorthodox cesarean section because the birth canal contains pathogens. The idea was to slip the clones from one germ-free world to another so their organs would be as pristine as possible for human transplant. The rubber hands now went after the placenta, cutting it off the squirming piglets.

Next, suddenly: an oddly beautiful ballet. In a larger plastic chamber, four more sets of gloves manhandled the four tiny newborns whose eyes were still shut, washing them and promoting their first breaths. At first the pigs looked like universal fetuses; they could have been any animal. Blind, with flesh the hue and texture of the inside of a seashell, they curled toward their own middles on both ends as if trying to disappear into themselves. They were four identical alien beings that were part human, part pig, engineered partly by nature, partly by man.

Yet soon this scene became striking for the lovely imperfection of it, the nature of it, despite man's role. For as time passed, it became clear that the surreal and sterile man-made womb of the chamber had been overrun with the vissicitudes of writhing, blood-and-seashell-colored life, every example of which was distinct from the next. One bloody piglet was being treated as if she were a human child's nose: a white cloth, which a scientist held between thumb and forefinger, kept squeezing and releasing her as if the scientist were saying, "Blow," over and over. The pig endured this limply. By contrast, in the far corner, another pig turned her face sharply to one shoulder, thrusting her forelegs far from her body as she was swiped. She looked not unlike a distressed diva refusing to be kissed, crying, "Never!" just before the swoon. A third clone lay relaxed, calmly letting blood be wiped off with what looked like a slight smile, as if to the manner born. A fourth was for a while as lifeless as an oven mitt, letting herself be folded, unfolded, then folded again by gloved hands, taking her time before getting around to that first breath.

Breathe she eventually did, and it was on to that ancient ritual: the umbilical cord cutting. A piglet gripped by the neck took on an "I don't want to live" expression, her jaw dropped, as scissors moved toward her lower half. When it was over, she was laid on her side in an aluminum pan that was shot down a plastic tunnel, and slid into the maintenance isolator: a box in a clean facility. In this facility the pig would spend the rest of her days, which would end when she was either autopsied for study, or sacrificed for a dying human in need of a new organ.

Yet seconds after she hit that tray, the just-born clone had righted herself, exercising will the moment the opportunity arose. As she sailed on her small barge into the next part of life, her forehooves were square to the aluminum and her clove-shaped ears were flat to her head: she was prepared. These "identical" clones are alive in so many different ways that the transplant doctor who helped Hwang create them rarely visits. She wants to save her human patients with them, but she also calls them "my babies"; sees their individuality; gets attached. Some initial studies indicate clones can share spookily similar traits. But they can also be quite distinctive, despite scientists' best efforts. Clones can be a potent reminder of all that nature's most powerful creation, the human mind, may never be able to do.

Yet, frolicking in a nearby lab is a potent reminder of all that nature's most powerful creation *can* do. For, minutes away, even older pig/human clones of the Hongseong embryos cavort, at varied ages, in a glassed-in pen. Resembling hairy brown elephants of varying small sizes, they tumble and flump over each other behind a clean-room window like living laundry. Clones of the same age are one thing: they are similar to identical twins. But, genetically speaking, this is a scene of one individual at different stages of her life casually interacting with her own past and future selves. This is new. It is a scene illustrating man's power.

Still, even this is not entirely unnatural. Scientists clone by mixing limited old cells, or just the nuclei of old cells, into potent young eggs, which mysteriously end up reversing, or *dedifferentiating*, the old cells or nuclei back to a young, stem-cell state, so they can grow forward again. In nature, this is similar to the way starfish are believed to reproduce lost limbs, and newts to reproduce lost spinal cords, heart ventricles, tails. They dedifferentiate mature cells back to the embryonic state, then grow them forward again. In fact, newt extract can dedifferentiate old mouse cells. And some scientists think the human body may— naturally, if rarely—react to a dearth of needed cell types by dedifferentiating limited mature cells backward into a potent "young" state, then growing them forward into the needed cells. It is exciting to consider the human lives that are being, and will be, saved by stem cell technologies. It can be dumbfounding to consider the degree to which we are co-opting those technologies from nature itself.

So the Hongseong farm *is* a battlefield. It is part of a lab in which some of the most profound questions of the biotech era have been batted about daily. And it is part of a nation which, like Israel, Egypt, Singapore, Japan, China, Australia, Iran, Taiwan, the U.K., and countless others, has been fighting to answer those questions first, and best, against some sizable odds.

Of course, as international affairs expert Don Oberdorfer warned in 2001, "Hold on to your hats. Korea is a land of surprises." This will prove prophetic: the odds facing South Korea will soon turn out to be unusually large. This is not just because the nation is the size of Indiana, is still transitioning to democracy after an era of military dictatorships ending in 1987, and boasted only one publicly listed biotech in 2000 (23 by 2002). This is because Hwang is about to go into a free fall that will prove to be a global shocker. Right now, in July 2005, he is being lionized for far more than his efforts to reach a midrange goal of regenerative medicine, that is, the above-described creation of cloned part-human animal organs for transplant. He is being revered for steps he claimed to take toward a more critical goal: creating *all-human cloned embryonic stem cells*. Believing this, his nation put him in textbooks and on stamps, and his peers have been echoing Singapore stem cell researcher Vic Nurcombe in branding him "astonishing." But very shortly, toward the end of 2005, it will be found that Hwang and several colleagues committed many fraudulent acts with relation to that all-human cloned embryonic stem cell work. It will turn out to be the biggest act of fraud in science history—in terms of the number of culprits and the extent of public interest—representing a stunning setback for South Korea.

Still, in the months immediately following, as of late spring 2006, the only part of Hwang's work that will be discredited will be his two all-human cloned embryonic stem cell papers. (And Hwang's senior animal cloner, the aforementioned Lee, will not be among the fascinating menagerie of main culprits, as will be detailed.) Hwang's animal work will hold up, including the cloned pigs, the world's first cloned monkey embryo, and the world's first cloned dog, Snuppy—the latter confirmed by two groups in the top journal *Nature*. Indeed, by May, 7 journals will have stood by over 25 Hwang animal articles—and counting. "I have had the 7 papers published in *Molecular Reproduction and Development* examined by 3 experts who concluded there is no indication of fraud," editor Ralph Gwatkin will note in April of the Hwang work—largely pig and cow cloning—published in his journal. And only one company, Virginia's Revivicor, will be substantially ahead of Hwang's group in transgenic animal cloning for organ transplants at that time. (Although others will be more advanced in other cloning areas, as will be seen.) Hwang's crew visited the Hongseong farm at least two days a week for two years, implanting over 100 embryos each time, bringing scores of cloned pig/human babies to life at its isolation facility. This is important to those, like Revivicor CEO David Ayares, who believe such pigs "may change the world."

Tiny South Korea is a pitbull, the world's 11th largest economy and its most wired nation. It has put more than 2,000 people to work on stem cells outside of cloning. And human cell cloning work, with all its bewitching promise, immediately continues elsewhere. Hwang team transgressions will shatter South Korea, but like so many nations in this area of science, it will soldier on. Indeed, the operatic and wholly unprecedented magnitude of "Hwang-gate" will eventually be viewed by some as yet more evidence of the unusual allure of stem cells, of the unprecedented magnificence of their potential.

The United States is the nation with the cash. In the early 2000s, the United States's annual National Institutes of Health budget was some $27 billion, and the annual revenue of its top nine pharmaceuticals was $150 billion. The research budget of the European Commission was about $4 billion, and the rest of the world lagged far behind. Still, far less well-endowed scientists worldwide then, and since, have made huge leaps of faith in their efforts to capture and control the stem cell; to pick up where the United States as a nation left off. (If not where all

U.S. states left off. As will be seen, some key states will rebel, letting some brilliant U.S. scientists enter the race. And when it comes to adult stem cells, the United States continually shines).

In the process, scientists have been soaring to new heights and falling to new depths. Many are creating and regenerating life in ways never seen before. One scientist was trying to create the world's first egg from an embryonic stem cell when he found himself scooped. Midstream he switched tactics, and found himself creating the world's first sperm from embryonic stem cells. Another scientist created an animal embryo from the DNA of two females. Another believes he found adult stem cells that will extend women's childbearing years.

Many have been rushing certain stem cells into clinics too fast, with sometimes catastrophic consequences. Many have been rushing other stem cells into clinics at the proper pace, with sometimes dazzling consequences. Still others are pursuing tantalizing hints, at both bench and bedside, that the stem cell may not just be the body's biggest natural savior, but its biggest natural *killer,* playing a major role in cancer, stroke, and heart disease when it spins out of control.

But the incentive tends to be the same from Seoul to San Francisco, Tehran to Tokyo. Stem cells have scientists in increasingly far reaches of the world believing that the best clues to curing our bodies have essentially been hiding inside our bodies all along. Scientists are reeling over the revolution the cells are bringing to their understanding of how the body works, and of how it can be cycled, subtly, backward in time. Says Lars Arhlund-Richter of Sweden's Karolinska Institute, "This is Nobelworthy stuff, as important as the discovery of DNA." Says Irv Weissman of California's Stanford University: "Every aspect of biology, cancer and medicine changes when you think about your problems this way: tissues develop, and are sustained by, a tiny subset of cells that are stem cells."

Says Barrada simply: "The cells are incredible."

PART ONE

1

UNMADE IN AMERICA

THE TWO GATHERINGS—the first annual adult Stem Cell Wetlab/
Conference: Challenges in the Era of Stem Cell Plasticity in April of
2003 and the first annual Embryonic Stem Cell Biomedicine: The Journey from Mice to Patients wetlab/conference in May of 2003—differed
in focus. The first featured recent work on adult stem cells, rare if
potent cells hidden in remote corners of the adult body that naturally
replicate a limited number of times, but whose purpose is, it is believed,
to help repair the organs in which they are hidden. The aim of the second conference was to present recent work on embryonic stem (ES)
cells, highly potent cells that appear in the human about five days after
conception, can replicate endlessly, and can, it is believed, turn into any
cell or form any organ of the body.

But both these meetings had two qualities in common. They were
both among the first major conferences to receive U.S. federal dollars
not just to discuss recent stem cell advances, but also to give scientists
hands-on experience working with the cells in large wetlabs, an
unusual move that reflected an unusual excitement on the part of the
federal government. And both conferences were set in typical, restless,
all-American cities, cities that had seen better days, worse days,
better—and were determined to see better again, partly with the help
of stem cells.

The first annual adult Stem Cell Wetlab/Conference, for example, was set in Providence, Rhode Island, a city which, for its first 100 years, was better known as a center for principles than products. Founded by Roger Williams, a rebellious minister who left England and was bounced out of Salem, Massachusetts, for his unswerving belief in the separation of church and state, the city was famous from the mid-1600s to the mid-1700s for its "lively experiment" in both the freedom of religion from state persecution and the freedom of the state from religious rule. Eventually the city's position at the tip of Narragansett Bay let it add trade and shipbuilding to its list of accomplishments, and Providence became in the 1800s America's top producer of jewelry and woolen and worsted goods. Because it also owed its success to the slave trade, it was at first reluctant to join the Civil War. But persuaded that the Union—and thus, its economic future—was threatened by war, the city became a Union leader in 1861.

By the 1960s and 1970s, when the textile mills had long since fled south and the area's economy was stewing in recession, it always seemed to be raining in Providence. By the 1990s it would be rescued by service industries and the Information Technology boom, and transform itself into a middle-class paradise of malls and condos remade from old textile mills, a city of brick and wrought-iron elegance. Yet after the crash of IT, it would be in need of reinvention again.

Similarly, the first annual Embryonic Stem Cell Biomedicine conference was set in Pittsburgh, Pennsylvania, another city strategically located for trade, this one at the juncture of three rivers. In the 1800s Pittsburgh was so feverishly devoted to its industries—which included the world's largest steelmakers, and the appliances and condiments of George Westinghouse and Henry Heinz—that its gas lamps were lit 24/7 so its residents could see through its exuberantly polluted air. It boasted a "millionaire's row" of Victorian mansions that played host to President Abe Lincoln; enough wealth that it could afford to toss thousands of coins on its railroad tracks for a memorial flattening when the funeral train of President William McKinley rolled through; enough leisure time that its prominent citizens could afford to spend time planning their own clever tombstones. (The latter included that of an oil wildcatter, which was wired for electricity to keep it from warping in winter, and that of a wealthy physician, on which was engraved a thoughtful list of all the things he never did, including the clubs he failed to join.) Some of those same freewheeling eccentrics joined up with Pittsburgh's vibrant black middle-class community to become engineers on the Underground Railroad, whose tunnels ran beneath the University of Pittsburgh.

Yet sometime after the 1930s, Pittsburgh's many rusting bridges seemed to be the only things holding it together when the steel industry went west. Pittsburgh, like Providence, was rescued by service industries and IT in the 1990s. Also like Providence, it was just able to patch together a new museum and a battalion of inventive yuppie restaurants before, with the IT crash, it too became in need of reincarnation.

Both cities showed in their architecture the torque and tension of what they were, which was typical American cities that lived for change; that *were* the import, export, principle, or flavor of the day; that were less themselves than a reflection of whatever new thing the world hankered after most. So it was fitting that both of these uber-American cities were trying to become a hub for stem cells. The reason this was fitting was not just because stem cells were the technology of the moment. For the writer seeking a metaphor to describe stem cells can find few better candidates than the brilliantly amorphous, strangely egoless, subservient American city. Stem cells, which never stay themselves, which exist to change, and which can form any organ in the body and become whatever the body needs, would appear to have found the perfect home in America.

But the strange fact was, they hadn't. At both these conferences held in the spring of 2003, which were dedicated to opposite ends of the stem cell spectrum, it was clear that this match made in heaven, wasn't. For the first time in its history the U.S. government had effectively given the pass to a hugely promising medical technology. The country that had always scorned other nations for putting the past ahead of the future, that was formed, among other things, in rebellion against one of the most immovable forces in history—state-sponsored religion—had placed religion ahead of all else. "The last thing [the Founders] intended their revolution to produce was a new orthodoxy," wrote historian H. W. Brands in the September 2003 *Atlantic Monthly*. When the Founders wrote the constitution, "they embarked on an audacious and unprecedented challenge to custom and authority."

Yet at the dawn of the twenty-first century, the lead scientist of the great lively experiment of the United States—its 43rd president, George W. Bush—essentially dismissed human ES cells as immoral. He did so based on a belief shared by numerous Christian sects—one not found in the King James Bible—that human life begins when egg meets sperm. Thanks to a new technology called "therapeutic cloning," or "somatic cell nuclear transfer," it was becoming a widely acknowledged scientific fact that life, in principle, can begin when egg meets ear cell, thigh cell, nose cell. Life may even begin with just an unfertilized egg, zapped with electricity, in a process called parthenogenesis. Regardless, it was becoming accepted in science circles that human life may soon begin, like it or not, when humans say it does.

For many U.S. scientists, therefore, the issue by 2003 was quite different. The issue was, When does *meaningful* human life begin? Much of the Western world, in the 1980s, had codified into law the notion that meaningful human life ends when the brain goes. The advent of the ventilator, which could keep a heart pumping for years after the brain was dead, had prompted many Western nations to study the issue for over a decade, and to pass legislation promoting the conclusion that, essentially, "It's about the brain, stupid." Therefore, many scientists by 2003 believed the situation on the other end of life was clear, as well. Many believed that meaningful human life begins the moment the human brain begins to form; to be safe, the moment one cell meets another in the first attempt to form a brain, or approximately 14 days after conception. (The U.K., among other nations, codified this.) That moment occurs long after human ES (hES) cells are formed, around day 5. This, added to the fact that a full 70 percent of embryos never make it to birth *naturally*, should render hES cells "moral" for use, by President Bush's own standards, many scientists believed.

But in August of 2001, almost two years after the hES cell was officially isolated by University of Wisconsin researcher James Thomson (known to cell biology fans by one name: "Jamie"), President Bush declared that only a small number of hES cell lines could be worked on with U.S. federal funds. He still believed the research to be immoral, but he would compromise on lines already in existence on the day he made his speech: August 9. Those lines, which would drop from a claimed 66 in number to 12 within months, would henceforth be referred to with some sarcasm as "the presidential lines." They were, to the despair of many U.S. scientists, few in number and of questionable quality since they were derived before anyone had much experience with the cells. They were also cultured on mouse feeder cells, rendering them undesirable for clinical use, since human cells cultured this way incorporate animal elements that could include animal viruses. (This could occur with animal organs too, of course—another reason animal feeder–free hES cells are considered next-generation therapies.) That there were 400,000 spare in vitro fertilization (IVF) human embryos frozen in U.S. clinics, many of which Bush was essentially consigning to a meaningless death so no federally funded scientist could use the hES cells in them to help save human life, made zero sense to most mainstream scientific and medical groups.

Ever since Bush's August 2001 proclamation, the government had spent far more on researching adult stem cells. In 2002 the National Institutes of Health (NIH) spent $10 million on hES cells, $170 million on adult stem cells; in 2003, it would spend $25 million on hES cells, $190 million on adult stem cells. The result was that, unwittingly or otherwise, President Bush had issued an open invitation for other countries to take the lead in a technology about which former NIH chief Harold Varmus once said: "This research has the potential to revolutionize the practice of medicine and improve the quality and length of life. . . . There is almost no realm of medicine that might not be touched."

The reason for all the excitement: the replicating human ES cell is the most potent cell in existence. Developing soon after fertilization, it forms 210-plus "tissue-specific" adult stem cells, which in turn differentiate into the mature cells that make up all the body's organs and tissues. (It matures, for example, into the neural stem cell, which goes on to form neurons and two other kinds of mature brain cells.) Only small, isolated pockets of those more limited, tissue-specific stem cells remain after birth—commonly referred to, after this stage, as "adult" stem cells. This is one reason why many of our organs can't rejuvenate after serious injury or illness: they're made largely of mature, nonreplicating cells.

Stem cells of all kinds may offer a multifaceted approach to problems that today's drugs, narrowly targeted as they can be, can't touch. For mature cells operate and communicate by flipping thousands of tiny molecules at one another like organic pinballs, or organic words in the cellular language. Most intractable diseases involve the breakdown of many molecular conversations at once. Yet the pharmaceutical industry, with its pills, tends to deal in one molecule, one message, one pinball at a time—enough, say, to pop open clogged blood vessels, staving off heart attacks for a while, but not enough to block all the converging molecular pathologies involved in heart attacks, let alone in other age-related disorders or cancer.

But stem cells are not only entire cells; they are, to one degree or another, the exquisitely sensitive cells of early development. They speak the cellular language fluently, juggling many molecular messages at once, as they do when they're building the body. When transplanted, they appear to respond to molecular cries for help. They can react to heart-attack damage by forming, or prompting the formation of, both blood vessels and cardiac muscle. They react to neural damage either by morphing into and replacing neurons that have died, becoming a seamless part of the brain's conversation with itself, or by issuing molecular instructions, teaching it the language of rejuvenation. Either way, as Human Genome Sciences CEO William Haseltine says, stem cells seem to "remind the body it knows how to heal itself." "Magic seeds," says Harvard neurobiologist Evan Snyder.

Furthermore, not only is this form of "medicine" potentially smarter than we are, knowing its way around the body better than we do, but the body also knows this "medicine" better than it does the conventional pill. So the toxicity that comes with artificial pills, whose molecules fit into the receptors of all kinds of cells—not just diseased cells—causing side-effects, theoretically comes more gently with stem cells. The body is accustomed to regulating the activity of stem cells. It knows how to boss them around, rein them in, keep them in a state of equilibrium called homeostasis. "You can increase the dose of hematopoietic stem cells, for instance, to ridiculous numbers and never get toxicity," says University of Stanford stem cell pioneer Irv Weissman. To repopulate an irradiated mouse bone marrow with new blood cells takes 500 stem cells. "But we've put in 50,000 stem cells, and all you see is that the blood comes back faster. That's it. No toxicity." Thus, says Weissman, echoing many others, "You don't have to understand the disease process" in many instances to get the stem cell to work. "You're just asking the cell to do what it normally does. That's the whole idea."

Press researchers to go into more depth, and the hyperbole soars. Thanks to vaccines, antibiotics, and hygiene, the average life span in the United States and other developed nations rose from 49 to 76 between 1900 and 1997, an age boost equal to that occurring between ancient Rome and 1900 (Romans could expect to live to age 22). Yet, in 100,000 years, the *maximum* human life span of 125 years has stayed the same, suggesting our cells contain a death clock set to go off at a specified time no matter what. Still, even University of San Francisco biologist Leonard Hayflick, who discovered that cellular limit in 1961, believes we may beat the clock, thanks to stem cells. "It is possible," he noted in 2000, "that replacing old body parts with new ones might circumvent aging."

But the issue of which cell to go for—adult or embryonic—became a bitterly debated issue from the day it was announced, in November 1998, that the hES cell had been found. Sylvia Elam's operation in March 1999, which looked like something out of a montage of past and future horror movies and which occurred in one of the aforementioned striving stem cell hubs, illustrated and epitomized many aspects of the debate.

* * *

FIVE MONTHS after Jamie Thomson's groundbreaking hES cell paper was published, Sylvia Elam, a tiny 65-year-old stroke victim and Virginia coal miner's daughter whose husband said she was "too pretty not to be kissed" on their first date, lay on an operating table at the University of Pittsburgh Medical Center. To keep her still, she was wrapped like a mummy, and her head was attached to the sides of a metal box called a stereotactic frame by four screws, two of which were implanted in her temples. Identical men in blue, all eyes and masked mouths, milled about rows of syringes, scalpels, and vials filled with human neurons.

That day in March 1999, as newspapers and science journals worldwide were still roaring over the ethics of newly found hES cells, should have been the scariest of Elam's life, which was once devoted to child-raising and selling Bibles door to door. She was about to become one of the first 12 patients to have neurons created from stem cells planted in her brain. The stem cells came from an adult body, so they were less *politically* controversial than human ES or fetal cells—one reason this trial was approved. But Elam's neurons, the first to be mass-produced ($450 per vial), were far more *medically* controversial.

For they were brewed in a Layton Bioscience lab out of stem cells, culled from an adult body, that had gone awry and had ended up forming a cancer (teratocarcinoma). Scientists had discovered that this odd cancer came from mutated stem cells decades earlier. While Layton's cells came from one of those cancer stem cell lines—and indeed, a particular human line that had been run through a mouse, giving it mouse qualities, according to Peter Andrews of Sheffield University—they for some reason retained some of the talents of normal stem cells. Indeed, Andrews had found in the early 1980s that brain-neuron-like cells could be made out of that particular line of cancerous stem cells. This was, at the time, considered extremely exciting since neurons, being fully mature cells, cannot replicate. There was no supply of replacement human neurons out there, and the normal neural stem cell, which *can* replicate, had not yet been discovered. (The very existence of adult human neural stem cells was not officially verified until November 1998, four months before Elam's operation, thus years from the clinic.) So Layton's tumor-stem-cell–derived "neurons" were well studied as being the closest to normal available for clinical use.

But Layton made a claim that made many uneasy: in the immature stage, before they became neurons, these odd stemlike brain cells replicated so fast that one vial, created in a matter of days, could provide millions of "neurons" to everyone on Earth. When compounds were added, they stopped replicating and went into a "mature" stage, mimicking normal nearly mature neurons that were hopefully safe for transplant. While no one was sure it was true that they were medically safe, one thing was clear: these cells were *politically* safe because they did not come from an embryo.

Yet, all around Elam, "normal" stem cell researchers were gnashing their teeth, believing the FDA had gone too far approving this exploration of a supposedly noncontroversial source. No one knew if the cells, which have some 60 chromosomes instead of the normal 46, could revert. Indeed, some of the substances the company used to make the cancerous cells safe for transplantation—uridine, fluorodeoxyuridine, and arabinosylcytosine—can be reversed or blocked by factors in the body. Before the trial, only one paper on the cells in stroke-induced rats was published—and none in primates. (During the trial, two more papers would come out, if still only describing the cells' impact on rodents.) And the stroke-model rat that was used was not standard, bewildering some. "The passive avoidance task that they see effects on? Has little to do with stroke," said neural stem cell researcher Clive Svendsen, then of Cambridge University.

"It's crazy," concluded another neural stem cell researcher. "What if just one of the cells goes back to a cancer state? You're dead. . . . Those tumor cells grow like rockets. You can get tens of billions of them easily. . . . They're not going to put them in *my* brain."

Still, Elam was the calm at the heart of it, her tranquility a striking indicator of the buzz surrounding stem cells—*any* stem cells. Since a stroke in 1993, she'd been half-paralyzed. She'd watched her life since, unable to live it, wheelchair-bound at her husband's side as he ran their auto-parts shop. She was once talkative, but now many of her thoughts had to be finished by her husband, who picked out the words buried in her garbled speech; who read her eyes in a face that had been stilled, partly by dead neurons, partly by despair. Like 2,000 others who applied for these strange cells, Elam didn't care what they were. They came from a stem cell; they were "magic." "Here's a picture for the family album," she said cheerfully to her husband, as she was moved into a giant humming CT scan.

In the next room, a surgeon and resident huddled over a scan of Elam's brain, which swiftly materialized on a monitor. Her two ventricles resembled a butterfly with one outsized wing, for fluid on the stroke-damaged side had spilled into her basal ganglia after neurons there died, engorging them. Numbers on the monitor—which corresponded to points on the stereotactic frame attached to Elam's head—were jotted down. Blood vessels to be avoided were charted out. Then the resident entered the operating room and sat on a stool by Elam's head. He rested a drill on a shaved square in the middle of her curls; paused to steady it like a soldier with a machine gun; drilled a quarter-sized hole into her skull. A syringe, guided by the stereotactic frame, was pushed through her cortex, and 6 million controversial neuronal cells were shot into her brain.

Suddenly: a soft rumbling. The surgeon's head snapped up. Elam was given only a local and a Versed so he could talk to her; be sure he was avoiding vessels. But Elam was snoring. "She's not supposed to be *that* relaxed," he said. "Let up on that sedative."

Yet Elam would remain so relaxed that two years later, despite a second stroke, she'd request more cells. For half the Layton patients had improved PET scans. The FDA okayed a trial for 36 new patients, to get up to 50 million cells each, starting in 2001.

A few months after her operation, Elam rose out of her wheelchair and walked. It was statistically insignificant. Spontaneous remissions occur. But this trial had many wondering. If a tiny number of stem cells from a *tumor* can prompt activity in a dead area of the brain, what could a large number of *normal* stem cells do? As a proof of principle, the trial hinted that normal stem cells might live up to their billing.

But political storms continued to rage over *all* stem cells in the ensuing years. And the storms reached near-hurricane proportions among those researchers who saw even greater significance in Elam's cells. For oncologists were increasingly looking at cancer and wondering if most cancers actually come from the normal adult stem cells that are tucked away in our adult bodies. The reason for that line of thinking was that cancers behave like, and utilize many of the same mechanisms of, adult stem cells, as many were coming to see. "The question we're discussing [when talking about Layton's neural cells] is a precise question about the origin of cancer," said NIH neural stem cell researcher Ron McKay a month after Elam's operation.

Regardless, the upshot was that the atmosphere at the two pioneering adult and embryonic stem cell/wetlab conferences in the spring of 2003 dramatically reflected the unusually political nature of their nascent field. The science of stem cells advanced in some spectacular ways in the years leading up to 2003, the first year that many Americans began to publish work on hES cells. But the politics of it all advanced as well. American science conferences are generally marked by an air of confidence, quiet and otherwise. The smartest people in the world dart in and out of them like eager fireflies, sure of their own brilliance, yet curious to see if a brief intellectual pairing with another firefly might make them glitter that much more brightly.

But in large part due to the unprecedented intrusion of lay politics into the field, each of these landmark 2003 stem cell conferences possessed an air of uncommon excitement, uncommon unease, uncommon bitterness. The bitterness at one of the conferences was so palpable that one angry scientist would later dub it "a freak show."

.   .   .

ALL WAS at first serene at the Providence conference, which was dedicated to adult stem cells. It was preceded by two days of wetlab, of work, the thing scientists understand and do best. In one lab, a young woman from the Baltimore-based company Osiris slipped the skin off a rat and chattered brightly as she snipped ligaments and muscles off the bones, then crushed the bones in a small white mortar and pestle set. She was showing intense postdocs how to retrieve from the bone marrow adult mesenchymal stem cells (MSCs), which can make bone, fat, and muscle. "Body parts go into the bag!" she called out cheerily to her assistants.

In another room, a fetal surgeon from the University of Pennsylvania, Alan Flake, wearing loupes (jeweler's magnifying glasses) and plastic gloves, was taping an anesthetized pregnant rat to a counter and placing a cocktail-napkin-sized, sterile blue sheet over her, with a tiny hole open over the belly. "Cute, eh?" he said of the miniature operating theater. He proceeded to draw "Ewwwwsss!" from even this hardened crowd when the rat twitched at the first incision he made with his tiny scissors. He paused to wait for an anesthesiologist to put her more firmly to sleep with a toy-sized syringe. After poking the rat a few times, the surgeon then cut open her belly and gently pulled out three or four fetuses in purple, grape-sized yolk sacs, into which he injected blood stem cells that should cure the pups of the so-called "Bubble Boy" or SCID immune deficiency syndrome before they were ever born. Then he returned the fetuses to the rat's womb. Tomorrow he would open up the rat again and be able to see, since the new cells were labeled via gene therapy with a substance called green fluorescent protein (GFP), that the stem cells had been incorporated into the fetuses' bone marrow.

Flake, a thin, quiet man, pulled off this feat once in a boy who should have died in the first two years of life. While the child was still in the womb, Flake had injected his father's stem cells into the boy's bone marrow. Six years later, 90 percent of the boy's myeloid blood cells and 50 percent of his B blood cells were his father's. Even though the two were not a perfect match—something that can foil adult blood stem cell transplants because unmatched cells are rejected—it worked in this case, Flake believed, because in the fetus, our immune systems are not yet formed, so they accept foreign stem cells. Flake hoped to begin larger clinical trials soon, he noted, as he brought a thread high in the air over the rat, looped it, and finished stitching up the tiny wound in an animal that was smaller than either of his deft hands.

Presiding over another room was Morayma Reyes, the University of Minnesota postdoc who had offered evidence of the possible existence of what had been billed as the only "hES-like" adult stem cell: the MAPC (multipotent adult progenitor cell). Reyes believed the MAPC could make any organ, any cell. She was the belle of the ball, partly because her work had reinspired pro-lifers to denounce hES cells, partly because no one had been able to repeat her efforts. Her lab was the fullest, full of curious onlookers, as she popped flasks under a microscope to show all the apparent reason why no one had yet been able to repeat her efforts: her cells divide more slowly than any stem cell yet discovered, embryonic or adult. People must be patient with culturing, she noted. Postdocs leaned in intensely, not just because the claims of Reyes's lab (that of University of Minnesota lead scientist Catherine Verfaillie) were the most dramatic of any in the adult stem cell world, but because Reyes speaks with a heavy accent—in this case, Portuguese —like most postdocs dotting the multicultural, and normally ever-changing, U.S. science landscape.

Yet the air at this conference began to shift when Saul Sharkis, a highly respected blood stem cell expert from Johns Hopkins University, gave a pre-wetlab talk about the flexibility of adult stem cells. Sharkis said that, like others, he had found that so-called tissue-specific adult stem cells can be almost as flexible as ES cells when one does something simple: take stem cells from one area of the body and place them into another. The new area of the body, he hypothesized, can redirect the adult stem cell. "Irv Weissman," he said, referring to a pioneer of the blood stem cell field who has lately been championing hES cells as well, "declared in 1988 that the search for the hematopoietic [blood] stem cell was over. Surprisingly enough, it's still going." Sharkis later followed this opening, good-natured shot with a reference to "big fights" already starting in "other sessions of this conference regarding fusion."

What these seemingly innocuous remarks amounted to were the

week's first missiles aimed from the adult stem cell camp into the hES cell camp, as the cochair of the conference, Roger Williams Medical Center blood stem cell expert Peter Quesenberry, would shortly explain. Making his rounds of the wetlabs the first day, he stopped a writer. "What are you going to write about? You going to write about the political correctness? You going to write about how adult stem cell papers don't get published?" the pleasant, red-faced don of adult blood stem cells asked. He then made an impassioned speech about a rising fear among some adult stem cell researchers: that although adult stem cell work may be better financed by the U.S. government, medical journals may be pro-hES cell, tending to publish negative papers about adult stem cells and positive papers about hES cells. He noted that the debate had heated up significantly thanks to the published claim of Verfaillie in 2002 that she had found an hES-like *adult* cell, the above-mentioned, infamous MAPC.

Before 2002, Quesenberry continued, the hES cell people could claim their cells were younger, more robust, more proliferative and "plastic" (i.e., able to form a far greater variety of cells) than adult stem cells, while the adult stem cell people could only claim that their tissue-specific cells were more controllable in the short run—and less controversial. But a growing number of groups had been claiming that, actually, adult stem cells are not irreversibly tissue-specific, that they, too, can form any number of different kinds of tissues. The claim of Reyes and Verfaillie, made in the prestigious *Nature,* seemed to verify that to many adult stem cell researchers, Quesenberry said.

However, many stem cell researchers—not limited to those studying embryonic stem cells—disagreed. And after two 2003 *Nature* papers were published, offering evidence that adult stem cells may indeed not be functionally "plastic" like hES cells—that they just seemed to be that way because they "fused" with other cells—many pro-hES cell *and* adult stem cell people would begin claiming that this might well be the final answer. Regardless, worldwide, views on either side had been expressed so vocally in recent months that, after a conference in Keystone, Colorado, featuring debates over the existence of the controversial ES-like adult stem cell MAPC, a leading researcher had commented that "she'd never been to a meeting that was so nasty," Quesenberry said. "It's all been politicized—I myself have never seen science so politicized."

The bouts continued into the second day of the Providence wetlab. Over lunch, two PhDs gossiped to another participant that Neil Thiese of New York University and Ulli Weier of Berkeley University had been "very hot under the collar about plasticity," or the "fusion vs. plasticity" debate, that day in a lecture. (Many talks were held simultaneously.)

As the days passed, increasing numbers of adult stem cell speakers talked about how biased they believed the media in general, lay and scientific, to be against their cells. Quesenberry unabashedly devoted part of his formal presentation to the debate, claiming that journals were publishing "weak science" when it came to hES cells and "negative studies" when it came to adult cells. Leiden University researcher Dirk W. Van Bekkum opened his talk with a defiant, "Plasticity is here to stay!" On the other side of the issue, Oregon Health Science Center researcher Marcus Grompe told the assembled room that most of his work indicated adult stem cells don't change fate, but fuse. "And is cell fusion okay? My emphatic answer is, I don't think so! Very few functional hepatocytes (liver cells) emerge from the bone marrow!"

The take-home message of this conference was that, thanks to the politics of it all, even adult stem cell work may not be proceeding as it might in America. HES cell work had received $30 million over the preceding two years, while adult stem cell work had received a sizable $300 million. But for precisely that reason, many of the scientists here claimed that the hES cell camp had risen up to block the publication of adult stem cell papers—and the world's top journals had followed suit.

Over and over, participants either denounced or applauded the two *Nature* papers that found that adult stem cells are not as flexible as hES cells; that, for example, blood stem cells don't turn into liver cells, they fuse with them, becoming strange hybrid cells with too many chromosomes. Over and over, as well, the press's reaction was reviewed and was found wanting either because it involved prematurely declaring the "plasticity" of adult stem cells dead, or very much alive.

A direct panel talk on the topic toward the end of the conference even brought out claws. After Flake referred to fusion as "the F-word," New York University's Thiese said of a fellow speaker that he "has been quite widely quoted in the past few weeks as saying that all plasticity is fusion."

The conference ended on a conciliatory note, however, as all publicly agreed on one thing: the press is bad, be it "*Nature, BusinessWeek* or the

National Enquirer. . . . The contrast between this debate and the media presentation of the issue is striking," Thiese noted. Agreed Grompe: "Some of the reporting that was done on my *Nature* paper implied I said that plasticity is passé, and this was definitely not the issue as we put it." Another fusion paper author, Ed Scott of the University of Florida, added that "to avoid getting killed by you all" he wanted to note that a disparaging comment he made about ES cells "never made it into any press, anywhere."

Science magazine, Quesenberry concluded to the assembled group, made a mistake when it published a recent Irv Weissman paper claiming that certain blood stem cells can't make other tissues. Weissman did not substantially injure tissues before trying his experiment, Quesenberry said. This is key, for many stem cells do not mobilize unless a tissue has been injured, causing it to call out for help. "Very poor editorial judgment," Quesenberry claimed.

Responded the NIH's Donald Orlic: "Oh, he knows very well that injury is something very necessary. That paper has done damage, but I think it will all wash out."

Countered Quesenberry: "As long as we don't wash out first."

*   *   *

A FEW WEEKS LATER, at the NIH-funded Pittsburgh embryonic stem cell conference, the frustration levels and bitterness were strikingly similar, though the gripes there focused on the way hES cells are being underfunded in favor of—and the way the press is biased toward—*adult* stem cells.

Indeed, this particular conference was being hosted by Gerald Schatten, a witty University of Pittsburgh primate-embryo expert who would shortly become partners with then-obscure—but soon world famous—Korean putative human-cell cloner Woo Suk Hwang. Schatten would enter into this unusual partnership in part because of a confused and daunting legal scenario not uncommon in the U.S. states, where organized and thoughtful federal leadership on stem cells was virtually nonexistent. That is, while Pennsylvania was more open to hES cells than many states—possessing the U.S.'s most vocal pro–hES cell and cloning senator (Arlen Specter) and letting scientists work with hES cells—an older law on state books would imprison scientists trying to *create* hES cells. Thus, while Pennsylvania let Schatten conduct research on existing hEs cells, if he wanted to make hES cells by any method that included cloning—and wanted to avoid a turn making license plates—he would have to do it in another state or land.

Regardless, the science talks in Pittsburgh were interspersed with an uncommon number of presentations made by invited patient advocates, politicians, and religious figures, all there to give advice to scientists on something alien to them: how to put an end to the politics of it all.

But politics crept into the discourse, anyway. Igor Lemischka, a Princeton University adult blood stem cell researcher, ended up providing the conference soundbite by referring early on to hES cells as "artifacts" and hES cell conferences "like participating in scientific special olympics." "Only the adult blood stem cell can actually be called a stem cell at this point," he announced.

This prompted NIH stem cell task force member Ron McKay to open his talk with, "Neural stem cells from embryonic cells—that is to say, what I've always *thought* of as the neural stem cells that one can get from what I've always *thought* of as embryonic stem cells. . . ." By the end of the conference, even former University of California researcher Roger Pedersen, who recently escaped the politics by becoming the first scientist to flee the United States for the U.K., was moved to describe the year the hES cell was discovered, not as one of the greatest in medical history, but as "the year politics stole our funding from us."

Still, when the science at each conference wasn't brilliant, it was often exciting. At the Providence conference, University of Pittsburgh researcher Johnny Huard announced that he may have found a robust adult muscle stem cell. And when talk moved away from the supposed plasticity of adult stem cells to their proven ability to repair the tissues in which they reside, it was clear that adult stem cell research was moving along at a fast clip. The ability of adult blood stem cells to rescue immune systems damaged by chemotherapy, and even to attack blood cancers, for example, went unchallenged.

Furthermore, at the Pittsburgh conference, Pedersen talked about steps he had taken to create blood stem cells from hES cells. Hans Scholer of the University of Pennsylvania announced he had created egg cells from mouse ES cells. If he could repeat this in humans, it would mean that someday women may have children in old age. The science at both of the spring 2003 conferences—and the potential exemplified by it—was at times inspiring.

Still, barbs flew throughout the hES cell conference, too. "Ian, was Dolly an embryo?" a priest shouted out to the Roslin Institute's Ian Wilmut, father of Dolly, the cloned sheep. "Yes!" Wilmut responded, causing some laughter, since the hES cell speakers had been accused of avoiding the controversial word "embryo" in their talks. Another scientist admitted that she had been advised not to use the words "therapeutic cloning" but "nuclear transfer" when discussing her work with lay audiences, prompting various scientists to laugh and stumble during their talks, starting to say "cl—that is, nuclear transfer, nuclear transfer . . ."

Linda Lester, an Oregon National Primate Research Center primate ES cell researcher, would conclude later the two stem cell conferences were unusually heated for two reasons. "Since stem cells seemed so unusually promising, everyone had that much more to gain—and that much more to lose," she would note. But it would become clear that scientists weren't the only ones feeling the heat, when, two weeks later—and four and a half years after the discovery of the hES cell—a similar atmosphere of frustration and bitterness pervaded another room, this one in the halls of the U.S. Congress.

*   *   *

STEM CELLS are perceived to be so extraordinary that, even though the most naturally potent ones come from spare IVF-clinic embryos, numerous antiabortion Republican legislators signed on to the cause as early as 1998. The most prominent among those had long been U.S. senator Arlen Specter of Pennsylvania. Between 1998 and the day of the passing of major stem cell advocate Christopher Reeve in October 2004, Specter would hold more than 15 Senate hearings on hES cells, often describing as "scandalous" attempts to squelch work on cells scheduled to be discarded. In early 2003 he wrote to President Bush, asking him to loosen the restrictions that had been placed on hES cell work. The presidential cells, he noted, were all grown on contaminating mouse feeder cells, making them problematic for the clinic.

But in early spring of 2003, two key events occurred. First, Johns Hopkins University researcher Lindzao Cheng devised a clever way to avoid mouse feeder cells (although not all contaminating animal elements). One of the many intriguing aspects of stem cells, he had discovered, is that sometimes they need only one another to grow and change, not a complex recipe of growth factors. He found, in particular, that some adult human blood stem cells and hES cells seem to need largely adult human mesenchymal stem cells (those that form bone and fat) to remain in the stem cell state.

Furthermore, a few weeks later, rumors circulated that some NIH-funded Swedish human embryonic cells had been approved in a "pre-mouse-feeder state." This meant it was possible that some NIH-funded hES cells could soon exist that had never touched mouse feeders, so were ready for the clinic in that respect, should the Swedes use the Johns Hopkins recipe. Specter called a meeting to find out what was going on. It was generally believed that *all* the NIH lines were unsuitable for the clinic—a major reason Specter had been fighting for the federal funding of more hES cell lines.

On May 22, 2003, the hallway outside room 418 in the Senate's Russell Office Building was thick with an overflow of reporters and advocates, the former of whom could be identified by their dark suits and shaved faces, the latter by their T-shirts and beards. As Specter was late, all were given plenty of time to take in the room, which was wedding-cake elaborate, with glistening marble columns, light blue walls, and ceilings crowned and framed by cursive white-and-gold moldings.

Four or five scientists sat stiffly in the front row in straight-backed chairs, looking as uncomfortable in their suits as they would soon be made to feel when Specter burst into the room, apologized for his tardiness, and immediately went into attack mode.

After noting that, time and again through the years, he had been told by scientists that the presidential lines (the 12 or so NIH hES cell lines) should probably not be used in the clinic because they were cultivated on mouse feeder cells, he said he had just been informed that the Swedes might have some presidential lines that had not yet touched mouse feeders, making their NIH-funded stock potentially eligible for the clinic. Calling the news a "shock," he said, "I'm concerned that a politicization of the process has occurred."

NIH chief Elias Zerhouni tried to explain that the NIH had not known about the existence of that particular Swedish stem cell source until recently. (It was later found to be nonexistent.) But Specter, whose Senate seat was being challenged by a Republican who had been battering him mercilessly for his support of hES cells, was relentless—and relentlessly irritated. "It's very upsetting, very upsetting. . . . We're getting contradictory information," he said. Democratic senator Tom Harkin jumped in next: "Is it safe to say the Swedish are ahead of us now?"

As Zerhouni continued to protest that the NIH had not known about the Swedish lines—and that, without publication of their isolation in a major journal, it could not confirm they existed—Specter abandoned one line of irritated questioning for another. Why wasn't the NIH clamoring for more presidential hES cell lines? he asked. "Isn't it important to have genetic diversity? Cells that weren't grown on mouse feeders?"

The exchange that transpired was odd, yet similar to a slew of past governmental exchanges on stem cells. The head of the NIH, a former head of an ES cell lab at Johns Hopkins who had been appointed to his government post by Bush, insisted that the small number of presidential cell lines were enough for U.S. scientists for now even though all the NIH lines were bred in mouse feeder cells. It would take his NIH crew years to completely characterize and understand them, he said. "Having too many lines is not necessarily a good strategy."

The message the Bush-appointed health leader seemed to be sending was that it was Specter who had overplayed his hand in order to score more cells, and that U.S. scientists were satisfied with the cells they had.

Scientists in the audience exchanged looks. This was not what they had been telling the press—and each other—for two years. The NIH had been extremely slow to get the presidential hES cells out there, and those lucky enough to score the cells had been finding that some of the lines were problematic: they flamed out quickly or formed tumors. Mouse ES cell scientists had gone through thousands of different lines before they were able to finally settle on five or six that were stable and usable. HES cell scientists had only a few—at the moment, indeed, only about 12 viable lines—to play with.

"It seems to me," Specter said to Zerhouni, taking off his glasses, "that if there were more lines, more research, we would get there [to the clinic] faster. Is it not fair to say, Dr. Zerhouni, that the vast majority of scientists disagree with your position?" He noted that he had asked the NIH to look for, and invite to this meeting, scientists who could back up Zerhouni's contention that the presidential lines were enough. Not a single one was present. "You cannot even find one non-NIH scientist to come forward and back you up on this issue as to whether there are enough human embryonic stem cell lines?"

Zerhouni indicated that such scientists existed, but were busy at the moment. Yet immediately following him, a series of non-NIH scientists testified that, actually, more hES cells were desperately needed. Said Mark Kessler, a Northwestern University neuroscientist who switched to stem cell work when his daughter's legs were paralyzed in a skiing accident: "In this particular field we are being told we cannot do things in parallel. They have to be done serially. . . . We should be doing them in parallel." He added, "As a scientist it is frustrating to see these hand-cuffs. . . . As a father, it is infuriating." Ray Ogle, a Duke University stem cell researcher, warned that a Chinese postdoc in his lab was in contact with Chinese authorities who had told him that China "is making human ES cell research a cornerstone of its biotechnology industry."

The only actively working stem cell scientist to say he was not unhappy with the presidential cells was the only NIH stem cell task force member who was testifying, Ron McKay. And he spoke slowly and carefully, clearly weighing the possible impact of every word. It was possible that someday the NIH crew would tell Specter it was dissatisfied, he said, but it was not true that McKay could not work with the cells he had available. McKay was a scientist in charge of comparing all the presidential lines, for which academic researchers were charged $5,000 a vial. McKay then added: "We are being very direct, but we are being kind of subtle," drawing laughter from the crowd. ("The NIH has to be apolitical," NIH stem cell task force member Mahendra Rao would explain later.)

Specter, after noting that he quoted McKay "with frequency," tried his question of the day, asking McKay why no non-NIH scientists were there to support Zerhouni's contention that more ES cell lines are not needed.

McKay, looking now as if he wished someone would shoot him, fell uncharacteristically silent for a moment. This was uncharacteristic because McKay was the field's poet laureate, in the way he spoke of the cells, in the way he worked with them. People tended to speak of

McKay's pioneering work coaxing mouse ES cells into neurons as if critiquing a kind of precise, exacting poetry. McKay got to know mouse ES cells so intimately during the 1990s—and became so bemused by their potency—that reporters and scientists alike came to him to get his sense of what the cells were, what kind of a change they represented in our view of how life unfolds, how to translate their beautifully orchestrated molecular messages, how to respond with the same molecular pitch and tone.

If there was anyone in the field who would want everyone working on stem cells, it would be Ron McKay, if only because he of all people would want to be around to witness the great shift in understanding he so vocally believed they heralded. He often used variations on the word "revolution" when discussing stem cells.

McKay loved stem cells. But he was now both an NIH scientist and a kind of NIH official as a member of the stem cell task force. Few NIH officials break ranks and publicly buck the president before Congress. After a spell, he dodged the question, answering that "someday" NIH scientists may well need more hES cells.

"You're a wimp," Specter said.

This was apparently all the opening the clearly relieved McKay needed to signal where he stood. "Well, I'm not a *total* wimp," he said, breaking into a sunny smile.

When the laughter died down, it was Gordon Keller's turn at bat. The prominent Mount Sinai Hospital stem cell researcher said firmly, "I know the field would move faster if scientists had the best tools. Most scientists disagree with [President Bush's policy]."

By the end, Specter was clearly outright angry. "The administration doesn't have the last word," he eventually said. "We can override vetoes. The buck stops right here." He added that he had half a mind to just give up and give Congress's money to "private industry."

But the meeting was adjourned as numerous congressional stem cell meetings had in the past: ambiguously, with no decisions made.

In the elevator minutes later, Kessler said, his face grim: "All scientists know the value of studying more cells, including Dr. Zerhouni. But it was remarkable the way he could avoid saying that, eh?"

Said a young man in the corner: "I think what I learned today was how slimy politics can be."

. . .

"THE BUSH administration is putting only enough money into these cells that history won't be able to say it dropped the ball," said Mahendra Rao matter-of-factly over coffee at a National Institute of Aging cafeteria in Baltimore, Maryland. It was two months later, a glaring hot July 2003 day. Rao was displaying the nonexistent fashion sense of the classic absentminded scientist, that is, he was wearing the same pleasant checkered short-sleeved madras shirt he had worn at the Providence conference two months earlier.

But he was conspicuously lacking the bitter attitude that some stem cell researchers can wear like a shroud. Beyond possessing a pleasant, noncombative personality, there was another good reason for this. As a member of the NIH stem cell task force, and chief of his own hES cell lab at the Institute of Aging, he had little to be bitter about personally, as yet: like McKay, he was one of the handful of U.S. researchers able to work on hES cells—a variety of the presidential lines. Indeed, beyond Jamie Thomson, the University of Wisconsin researcher who isolated the hES cell in November 1998, and Geron, the company that financed Thomson, Rao had put out more hES cell papers than any other American.

But Rao quickly made it clear that it was a trademark of this field that even a man as blessed as he was looking forward with trepidation to the time he might not be so blessed: the immediate future. For everyone's immediate future was rocky in the hES cell field, he said.

"Look in my book," he said, referring to his upcoming *Human Embryonic Stem Cells,* as he ambled across the sweltering Bayview campus. "There is a link to the patent there. It says all you need to know."

The patent represented the second major stumbling block in the progress of the hES cell field, Rao noted. For, in the manner of an increasing number of patents, it was broad—so broad, some believed it could cripple the field for the next 17 years (the amount of time left on the patent). Essentially, the patent declared that the bulk of profits made on most, or the most critical, U.S. human embryonic stem cell work belonged to a single entity, the University of Wisconsin's Wisconsin Alumni Research Fund (WARF) and its main licensee, Geron.

The same patent restriction may also eventually "apply to many countries overseas," where the patent controllers filed, Rao said. Although there is a way out, he added: The European Union may not honor the patent. "It does not like the idea of patenting a human cell." If the EU does balk, might other countries then be emboldened to follow suit, allowing at least the rest of the world to profit substantially from the cells? "Exactly," Rao said. (Indeed, as of March 2006, no other nation will have approved such a broad patent.) And WARF did not apply for a patent in all countries.

He stopped before his building to wrestle with his electronic pass. Seconds ticked by as he focused, unable to open the door. At last. He looked up and smiled absentmindedly as he entered the building. Still, he continued, the fact that WARF did apply for a patent in the EU is worrisome, since traditionally it is the EU that leads the world, after the United States, in biotech. "And while the EU is stubborn about some things, so is the U. S., as we recently saw in Iraq. . . . After you."

It was more than an intriguing idea, that the world could take over for the United States on the hES cell front. It could well be the critical move, said Rao, as he waved at his lab, one of the few in the United States to contain hES cells, and settled into his crowded office. For no matter how much we may learn from adult stem cells, Rao said, odds are good that hES cells are going to be absolutely necessary if the field is ever to go commercial—and become profitable— one way or another.

Rao should know. A neuroscientist, he carved a name for himself in the late 1990s by isolating an even more multipotent mouse neural stem cell (NSC) than the adult NSC isolated by Sam Weiss and Brent Reynolds of the University of Calgary in 1992. He had also done elegant work showing that neural progenitors may be preferable to NSCs for some brain cell replacement therapies (progenitors are stemlike, intermediate cells that are more differentiated—mature and limited—than stem cells, but have not yet reached the rigid, fully mature, nonreplicating cell stage). This was key, he believed then, because it may take a long time to perfectly control the differentiation of hES cells into NSCs, or NSCs into neurons.

However, as the 1990s became the early 2000s, it became clear that adult NSCs and progenitors simply couldn't replicate as robustly as ES cells. And when they could replicate to a good degree (if never to the same degree as ES cells), they often lost their ability to form some kinds of neurons. In other words, Rao said, getting enough functioning adult NSCs for a single transplant, let alone a universal supply, was beginning to look impossible. NSCs are extremely rare in the adult body. Rao began to realize that "the reason adult stem cell biology is so slow is that we just can't get enough of those cells. There are between 5,000 to 30,000 NSCs in the average adult brain. Yet the lowest number of NSCs needed for transplant for a single Parkinson's patient is 100,000 NSCs. The ES cell is the only one that can keep proliferating." (Rao was also not believer in the notion that adult stem cells possess functional "plasticity," or the ability to switch fates to any degree that matters functionally.)

Furthermore, said Rao, "I firmly believed—and believe—that most tumors come from adult stem cells." The reason: adult stem cells are among the few cells in the adult body that can replicate, and each time they do they must recopy the 3 billion letters of their genome. In doing so, they make mistakes. The more mistakes they make, the more likely they are to spin out of control: cancer.

So, beginning in 2002, Rao and a handful of others—including Ron McKay of the NIH—became the first federally funded U.S. scientists to begin work turning far more robust hES cells into neural cells. Among their aims: to figure out how to create such inexhaustible supplies of that most potent cell, the hES cell, that they could weed out whatever cells might go awry in the making, yet still be left with *clinically viable numbers* of neuron-producing cells. Indeed, most of the few U.S. scientists who finally got their hands on the presidential lines in 2002 immediately tried to turn them into either adult blood stem cells or adult neural stem cells, since these were the best known and characterized adult stem cells.

Still, all U.S. hES cell researchers worried then, and on into 2003, about the day they might end up like Richard Garr, who took the closest thing there is to their cells—human *fetal* neural stem cells—out into the marketplace and crashed into a political brick wall.

*   *   *

ONE DAY in February 2001, more than a year after the discovery of the hES cell, NeuralStem CEO Richard Garr entered a College Park, Maryland, lab that was bustling with life in more ways than one. All around him, white-coated technicians dribbled nutrients from pipettes into petri dishes that were stored by the hundreds in dormlike half-fridges. The contents of some dishes looked like the bottom of a fashion model's sink: discarded eyelashes floating in pink water. The contents of others were more complex, resembling smashed dragonfly wings. But the "wings" were mature neurons that had grown out of the "eyelashes," or what Garr called "the world's largest supply of human fetal neural stem cells."

This lab was working with fetal neural stem cells—discarded from family planning clinics, then proliferated in dishes, here—because Garr had determined, like Rao back in the mid 1990s, that neural stem cells from adults were going to be near-impossible to efficiently proliferate to commercial numbers. His fetal neural stem cells were more differentiated, or specialized, than hES cells, so they proliferated less robustly. And they were even *more* controversial than hES cells, since they came from aborted fetuses more than ten weeks old, not from five-day-old blastocysts that hadn't yet begun to form, left over from IVF clinics, as hES cells were.

But Garr's cells already knew how to be neural cells, since they were removed at that later age. They certainly knew how to proliferate more merrily than adult neural stem cells taken from the aging adult human brain. And they represented an appealing opportunity indeed, many doctors felt: to begin trying to cure patients with all manner of brain disorders right away, while hES cell researchers continued to try to control hES cells, and adult stem cell researchers tried to make adult neural stem cells from aging brains more pliable and robust.

It was feeding time, as it almost always was at NeuralStem. Five staffers were engaged in the simple act of feeding the cells all around, which they did for four full hours, every single day. The other four hours, for these five staffers, were spent preparing the cells' "meals." The reason: "The cells are alive, but they don't have fat as we do, so they need constant feeding," Garr noted fondly. Their diet almost sounded like a cereal ad: "all of the 20 amino acids, vitamins, essential minerals, insulin, iron-transporting proteins, and most importantly, Basic Fibroblast Growth Factor, which is the key mitogen that causes the stem cells to divide." Everything a growing cell needs in the formative years.

The atmosphere here was more like that of a nursery than a corporate workplace. Staffers, like parents with wallet pictures, eagerly pulled out dishes to show off some of their 9 billion charges. They traced with pens the loops of increasingly intertwined wings and tentacles of the neurons they had nurtured and watched grow for years. The tentacles and wings drew life from each other in the form of electric currents and excitatory molecules, and gave it back. As Geron CEO Thomas Okarma noted earlier, "Stem cells are living things. They are not inanimate like proteins or pills."

Patrolling it all was the lab head, who spent every day popping 50 dishes under microscopes to "monitor aberrant behavior," as Garr put it. Each neurodegenerative disorder this lab wanted to treat involved the loss of its own unique neurons from its own unique brain region. So each dish contained a slightly different kind of neural stem cell, harvested from a different part of the fetal brain, grown as potential "replacement parts." Here, a dish of midbrain stem cells to replace dopaminergic neurons lost to Parkinson's; here, a dish of spinal cord stem cells to replace motor neurons lost to ALS (Lou Gehrig's disease) or spinal cord injury; here, hippocampal cells for Alzheimer's; here, cortical cells for Tay-Sachs disease.

But these cells were so alive, they could differentiate (or mature) into cells the lab didn't want. The medium in which they basked had to be constantly tweaked. "They are exquisitely sensitive to lots of different things in their environment and can shift in their nature inadvertently," said chief scientific officer Karl Johe, whose hair stood straight up off his head, mad-scientist style. Indeed, one inspiration leading to the launch of NeuralStem—removal of a growth factor to make the cells stop replicating and mature—occurred "because I was with the cells so much I grew to understand them."

These were not words oft heard in the board rooms of the producers of, say, pens. They were not even words oft heard in the board rooms of major drug companies, which deal in bottle-ready small molecules (drugs like aspirin). After giving a brief trial run to adult blood stem cells (the first human stem cells ever officially isolated, back in 1991) and finding that they simply could not be replicated to commercial numbers, U.S. Big Pharma had not thrown itself at *any* stem cells, adult, fetal, or embryonic. But for years, small adventurous labs had popped up across the country anyway. Geron was the largest and most viable of the struggling embryonic stem cell companies. NeuralStem was one of the two largest and most viable of the fetal stem cell companies (the other being Stem Cells Inc.)

But by 2001, both Geron and NeuralStem had been encountering enormous problems getting backing. Since they were companies and thus received no federal NIH funds, both were freer than academics to cultivate whatever cells they wanted. (There are no outright federal bans on the conduction of either kind of research in the United States, just the aforementioned federal funding ban on most hES cell lines and what most scientists consider to be an effective federal funding ban on fetal cells.) But beyond the fact that "pharmaceuticals like pills" and that "cellular transplants don't fit their model—at least not now," as Garr explained, the politics of it all had even the biggest of the Big Pharmas cowering.

Garr wandered into his office and threw himself into a chair. Above him was an article about a child who died of a neurodegenerative disorder. On his desk: a picture of a NeuralStem neuron.

In 1995, three years before the hES cell announcement would be made, Garr explained, he was just a lawyer with a nine-year-old son who had an astrocytoma brain tumor. For years his son had been undergoing chemotherapy; for years, the tumor kept returning. One day he brought his son to Karl Johe's house (their children were schoolmates). He knocked on the door to meet Johe, who was then working in the NIH lab of stem cell biologist Ron McKay, who had published more neural stem cell papers in top science journals than anyone else.

Garr wanted to warn Johe that the tumor affected his son's balance. Johe appeared in a bathrobe, having been in the lab all night. He listened, then said casually, "Someday we'll fix that."

Johe explained to Garr that, in the 1990s (since federal funding for human embryonic cells would not begin until August 2001), McKay's federal lab had been working on adult and fetal rodent neural stem

cells. The adult neural stem cells—originally isolated in 1992 in mice—
ended a century-old belief that the adult brain can't regenerate. For
adult rodent neural stem cells not only regenerated but also formed all
the different mature cells of the brain. And since the brain is at least
partly protected from immune-system attack by the blood/brain barrier,
neural stem cell transplants could conceivably occur without the lifelong regimen of crippling immunosuppressive drugs that would accompany transplants of most other stem cells and organs. The problem, Johe
told Garr: rodents aren't humans. He needed to work on human cells.
Between funding bans and political controversy, only industry was truly
free to do so, and Big Pharma wasn't getting involved.

Garr looked at his son and told Johe not to wait for Big Pharma.
Within months, Garr had drummed up funds and left the law; Johe had
left the NIH. A company was born.

In the years since, Garr continued, Johe had been working with
human fetal neural stem cells. It was rough going at first. The cells were
willful, as neuroscientists everywhere were discovering. Most were
feverishly trying growth factor after growth factor on the cells (growth
factors are proteins that help cells grow) but they couldn't make desired
neurons on demand.

Johe believed the answer to the problem was simple: ease off on
growth factors and just use cells taken from the right parts of the brain,
at the right developmental stage. Take stem cells from the dopaminergic
part of the brain (the midbrain) to make dopaminergic neurons, for
example. And take them at a later stage of development, when they've
already decided what to be. "The right cells do the right things on their
own," Garr said.

At first, Johe failed. While still in McKay's lab, he'd begun culling
cells from different brain regions and tossing them in a dish. Nothing.

But as he toyed with the cells, he realized something. "Stem cells
talk," said Garr, his face rosy with excitement behind his Elvis Costello
glasses. "They say, 'Hey, don't become a neuron. I'm one.'" Perhaps the
final trick was to leave them alone—literally. So Johe separated them to
a certain degree, and removed fibroblast growth factor from the dish.
Voilà. Neurons in a dish. After more tinkering, the company was able to
get them in about the same numbers, "virtually all the time, every time,"
Garr repeated like a mantra.

Pluck the cells at the right time and they even naturally reproduced 30 times, producing over a billion mature brain cells from a single stem cell, said Garr. He pulled out an article, in the non-peer-reviewed, if respected, *NeuroScience News.* The company had published a chart claiming it reliably made seven different kinds of neurons in a dish.

However, partly out of a fear of the political atmosphere surrounding any cells with the words "fetal" or "embryonic" in them, most investors had stayed away, Garr said. The result was that, by 2002, Garr had begun predicting he might have to bring his company abroad. And by July 2003, when his preclinical rat tests of the cells were looking good, he had long been negotiating, actively looking for a foreign company, or one with a foreign branch, with which to merge. Like Geron, the United States's only truly viable human ES cell company, NeuralStem, one of the United States's two viable fetal stem cell companies, needed far more money to test its cells. It just wasn't coming from U.S. sources. Geron had some money in its vaults; it could afford to stay put a while longer. But NeuralStem did not. "The U.S. political and regulatory climate," Garr said, had left him with "no choice."

•   •   •

STILL, DESPITE the pessimism that scientists like Rao and business-men like Garr exhibited over the speed of fetal and embryonic stem cell work in the United States in July 2003, shortly after the congressional hearing that failed again to solve the U.S. hES cell problem, neither man was completely pessimistic about the potential for the work to be rescued abroad, should it come to that. While NeuralStem had just failed to entice a British stem cell company into a merger, it was at the moment being courted by a company with a branch in Singapore.

And as that July 2003 day wound down, the NIH's Rao proceeded to recount, with a quiet if growing excitement, the number of foreign countries he and his NIH stem cell task force had been traveling to in the last year and would be traveling to in upcoming months.

Stem cell task force member Ron McKay was at the moment in Japan, talking to scientists with the Riken Center for Developmental Biology, a $45 million a year stem cell center attached to a hospital. In September of 2003, Rao and another U.S. stem cell task force member would attend an invitation-only stem cell meeting in Sweden. One key reason: to look at those hES cells that might become both available to federally funded U.S. researchers and could qualify for the clinic.

In October, Rao would head for South Korea, where the NIH was co-sponsoring a meeting. South Korea had several hES stem cell lines, Rao noted. And there was a rumor that South Korea was trying to clone human cells.

That same month, both Australia and Singapore were holding their first international stem cell conferences. Singapore's was being billed as history's largest. Nearly all of the United States's top stem cell scientists were planning to go.

All this would be followed up by another NIH symposium in Japan in February 2004 and a major stem cell conference in Beijing that March. The only major country heavily involved in stem cells that some NIH staffer wouldn't be visiting in the upcoming months was Israel—and that only because NIH staffers have visited Israel's labs often in recent years.

Always before it had worked the other way around. The world came to the United States when it wanted to get good science done, let alone funded. Still, these new reverse pilgrimages spelled hope—and not just for those countries. The way was being paved for collaborations between U.S. scientists and scientists worldwide. Indeed, NIH staffers were up front about their desire for this: "To gain access to lines many of which were derived abroad, and to push for collaborative work," Rao said later, by way of explanation for the travel of the NIH staffers. He confirmed that this was unusual. "Not so much need in other advances."

It was an exciting notion, the idea that stem cells could become the basis for a truly global biotechnology. Indeed, a few days earlier, it had become clear that this was already more than a notion. In June of 2003, the world's first international, all-stem-cell society and conference was launched. Called the International Society for Stem Cell Research, or ISSCR, the conference had attracted to Washington, DC, nearly 700 researchers from 21 countries. At this conference, Israeli scientists dominated the hES cell talks. An Egyptian scientist approached Israeli scientists about the start of one of the first Arab hES cell centers and a possible historic Mideast collaboration between Israelis and Arabs.

And from out of the blue, South Korea had materialized. No South Korean group yet had published an hES cell paper in a top Western journal. But more than 28 South Koreans attended this critical first international meeting, representing the largest national coterie outside the United States. They presented 10 posters (that is, written prepublication abstracts) on hES cells, more than any single nation (the United States presented 7 hES cell posters on its own; Australia, 5; Israel, 2; the U.K., 0). Each South Korean poster was presented by more scientists, from more institutes, than is usual, possibly indicating a comparatively huge, coordinated national effort was underway in the country.

The breadth of the international showing at the ISSCR—not to mention the highly coordinated and focused nature of some elements of it—was rare in the history of new professional science societies. It was clear by the summer of 2003 that the globalization of stem cell work had begun. The world was becoming besotted with the window of opportunity that was opened by Bush—and with the cells themselves.

Over the course of the following months, this universal fascination would lead to some highly unexpected developments, from the growth of underground stem cell clinics in third world countries, to scientific breakthroughs in nations where they'd never occurred before, to an unusual case of fraud. A new kind of unifying science nationalism would result in some countries, and an utterly unprecedented fallout, with major potential ramifications, would result between the U.S. federal government and a mounting number of its own states. Pennsylvania, New Jersey, New York, Massachusetts, and California, among others, would begin staging quiet rebellions. Stem cells would not just inspire foreign nations to take on the U.S. government. They would inspire some U.S. states to take on the U.S. government.

And in the critical months following 2003—the year that several nations and U.S. states forged alliances to move the field forward in the absence of full U.S. government support—the groundwork would be laid for a potential source of hES cells that could someday put an end to much of the political controversy (though the controversy generated by the toppling of scientific paradigms, which stem cells were also initiating, was destined to persist for a long time).

"This is a great experiment in democracy," Stanford University stem cell institute chief Irving Weissman would conclude several months later.

"It's breathtaking," NIH stem cell official Ron McKay would add. "It all moved so fast—and it's moving even faster, now."

2

STEM CELLS IN
THE SAHARA

"EGYPT IS A PLACE of contradiction," Ismail Barrada had said in June of 2003, as he sat plumped comfortably in an air-conditioned bus shuttling between the NIH and the Marriott Wardman Park Hotel, the venues of the first International Society for Stem Cell Research conference in Washington, DC.

But "contradiction" is not the word that springs to the mind of the jet-lagged Western traveler landing in Egypt for the first time, especially one first making a pitstop in another stem cell hub wannabe, the United Arab Emirates' Dubai. It is Dubai—a financial center of the Arab world—that leaves the truly contradictory first impression. The planes of its major airline, Air Emirates, are luxurious, with amenities in coach reserved for business class elsewhere: business-class legroom, business-class wide and tapestried seats; business-class facecloths distributed with business-class tongs. TVs with multiple movie channels are sunk into every seat; rolling bars distribute limitless complimentary cocktails; English and Arabic menus offer everything at this time (October 2003) from mousse with raspberry soulis to Tasmanian smoked salmon. Life seems—and smells—great when one is flying Air Emirates.

But travelers exiting all this into Dubai's airport blink like moviegoers stunned by daylight. For Dubai is also a financial hub of the terrorist world. Its banks were launderers of money financing the World Trade Center attacks. Dubai thus makes its travelers wait in long x-ray lines getting both on and *off* its planes. There is a soldier in camouflage and a beret cradling a machine gun every few feet. The air in customs is 9/11 thick, full of people toting suitcases surreptitiously watching people toting guns, surreptitiously watching them.

Yet downstairs, another contradiction: an upscale modern mall, patrolled not by soldiers but concierges, who wear such bright business suits and smiles it appears their sole aim in life is to guide travelers to a safe landing alongside the nearest diamond watch or designer bag sale—of which there are many here.

And the travelers they coo over are the most dramatic studies in contrast. There are backpacking European couples in jeans sleeping, entwined, behind potted plants in the aisles as they await connecting flights. There are Western businessmen working electronic gadgets with one hand as they drag, with the other, rolly laptops like little red wagons behind them, ubiquitous boys with toys. But also haunting these halls are great flocks of eyes-only women, covered entirely in dark *abayas* like ghosts in negative, silently gliding past the spooned sleeping Westerners, or sitting in circles near them, on the floor. Their men lounge in male-only prayer rooms or the other all-male bastion here: smoking rooms.

It is a spooky sight, these segregated flocks. In Saudi Arabia it is against the law for women to drive, or to leave the country without permission from a father or husband. In the UAE custom dictates men can bar women from leaving the country. While no average citizen of either sex in either country is able to vote at this time, women in both countries are profoundly dependent upon their men. In the UAE, men may marry four women at once—of any religion—and divorce is easy to obtain. Women may marry one Muslim man; divorce is difficult to obtain; they can be imprisoned if they bear a child out of wedlock. In Saudi Arabia, women can't board a bus without written consent of their husbands and can be imprisoned for sharing a cab with a man who is not a relative. Many of the men in this airport know many of these women intimately but they instantly left them, to wait for flights separately, as is appropriate form.

Battle gear and business suits, ancient mores in modern malls. Dubai

International Airport offers a fleeting glimpse of a region where extremes of capitalism and coercion uneasily coexist, a place of enormous contradictions.

By contrast, Egypt at first projects a more straightforward image. Travelers unlucky enough to disembark without cash into the dirty Cairo International customs area, with its closed shops and crumbling walls, are stuck in international limbo: stamps must be purchased with cash before the customs line, no credit cards. Next comes the customs cattle call, where the wait can last hours as wealthy Cairenes usher themselves to the front and many poor Cairenes are subjected to interrogations. In the baggage area, taxi drivers disguised as airport officials troll for passengers; outside, undisguised drivers shout offers for rides for 100 pounds Egyptian, 50 pounds over the correct price. There is an ATM, but the numbers on its keys have been rubbed off. Refusal to pay over 50 pounds Egyptian has taxi drivers stomping off in disgust. Travelers are inevitably led down tunnels and into the dark reaches of parking lots, where crowds of men stand by battered vans, announcing the cost is actually 50 pounds *plus* 10 pounds in *baksheesh* (bribes) for each of assembled.

The assault of it all continues. Police in white, with machine guns, gossip along the road leading out of the airport, occasionally raising an arm to pull over cars for random searches or to make foreigners sign red books—a once-quaint custom now used as a tool of friendly intimidation. Out on the freeway, cabs dart about without blinkers; buses whisk past bus stops, leaving passengers hanging out the doors, poised to jump; trucks jammed with squatting workers blast clouds of petrol; donkey carts teeter down the side of the road.

Yet after a few minutes on the highway to central Cairo, Egypt's creeping contradictions start emerging. Out of the darkness, an elaborate white tower looms, a huge limestone scepter of a thing, a Mamluk minaret. It disappears, followed by cement tenements with T-shirts hanging from rotting cement balconies, and another huge white minaret. Darkness and more poverty, then an opulent, domed palace the shape of a cardinal's headdress. None of the looming edifices are dirty or damaged, as such buildings inevitably are in poor areas of U.S. cities. Instead of spraying graffiti all over these monumental reminders of all they do not have, poor Cairenese simply zip in and around them, scrounging for breadcrumbs and *baksheesh*. The word "contradiction" indeed fits, from a purely visual perspective. Egypt's visual assaults are a one-two punch: riches following poverty following riches.

Yet such apparent contradictions, as seen throughout the Middle East, are at least less mysterious to Westerners than they may once have been. They are, many Middle East watchers believe, the result of some direct cause-and-effects the West is slowly coming to understand. There are many reasons it is important for the West to understand the Middle East. Included among these is the somewhat astonishing fact that many Arab nations are making noises about becoming human embryonic stem (hES) cell hubs.

"We believe biotechnology could someday be the new oil of Saudi Arabia," said Sultan Bahabri, board member of the new $100 million Jeddah Biocity, after Jeddah colleague Hammad Al-Omar presented draft *fatwas* to the BIO 2002 conference. Written by the Fiqh Council, the *fatwas* could help give the holy greenlight to fetal—and eventually, embryonic—stem cell work in the Muslim world. (The Congress for Defining Islamic Dogma will offer such a greenlight months later.) Indeed, some 4 percent of Saudis are born via in vitro fertilization, compared to 1 percent in the United States, demonstrating an already more widespread acceptance of embryo-related research there.

Furthermore, University Jihad Institute researchers in Iran in 2003 isolated an hES cell line. Dubbed Royan H1, *royan* being the Farsi word for "embryo," the researchers claimed in September (and would prove in the Western journal *Differentiation* the next June) to have created an hES cell line, and turned it into beating cardiomyocytes. Having started their work mere weeks after the Bush hES cell speech, the researchers now said their next goal was to use the cells to create pancreatic islet cells for diabetics. It all had the Ayatollah Khamenei hailing Iran's new hES cells as if they were weapons in a war. "Attempts made by enemies to humiliate and weaken the Iranian nation have been foiled thanks to the great and praise-worthy efforts of the Iranian scholars and scientists," he said in September. The Supreme Leader also took the opportunity to note that "showcasing its advancements, the Islamic Revolution encourages other nations. Here lies the main reason for the hegemonic powers to form ranks against it." He delivered these accolades personally to the researchers.

But the Middle East's interest does not end there. Jordan is rumored to have set aside $50 million for a stem cell center. In coming months, Dubai will report it is building an adult stem cell center. The UAE is the only government in the world to formally invite Severino Antinori, the eccentric, hugely controversial doctor who is bent on cloning humans, to speak to its medical community.

And in Egypt, where most IVF practices have been cleared by the mufti (donor eggs and sperm are outlawed), a 12 percent infertility rate has prompted the blossoming of more IVF clinics than in Israel (which has the highest number of IVF clinics per capita in the world): over 36, according to Marcia Inhorn's *Local Babies, Global Science.* IVF procedures produce some nine unused frozen embryos per live birth. This means there are a lot of spare embryos—thus, a lot of spare hES cells—sitting in Egyptian freezers. Add all this to the fact that over half of the 50 to 80 million infertile people in the world are Muslims, many of whom are taught it is shameful to be childless, and there are good reasons to believe that if the Middle East wants to become an hES cell hub, it can.

* * *

SINCE BARRADA would like to be at the forefront of hES cell science in the Middle East, it is therefore key that he understand his country's contradictions, or lack thereof, and how to navigate around them. As he relaxed on that soundless bus ride through DC last June, he indicated this was the case. "There are boats that cruise the Nile that are literally floating palaces," he said. "One woman I know has a bathroom as big as my apartment. There were 52 vials of perfume in there, one as big as a bucket. I once saw a Saudi Arabian lose 6 million pounds at an Egyptian table, and just walk away. In medical conference halls, the seats for the doctors are as elaborate and expensive as thrones. Yet somehow . . ." He folded his hands on his stomach. "There is no money for medical treatment."

Barrada had clearly used the word "contradiction" ironically. In 1966, as a medical student, he had left an Old World country in the hands of what appeared to be a great magician, the sunglasses-and-suits wearing Gamal Nasser. Nasser's ability to expel the British from the Suez, collect Soviet funds, and denounce all *kufrs* (unbelievers) while charming the United States indicated to many that Egypt was being whisked into the New World. But Barrada returned 30 years later to a country that had been whisked into a state of relative poverty and bitterness so pervasive some of its streets and newspapers celebrated when two New York buildings, containing thousands of *kufrs,* were destroyed by men led by the son of an Egyptian lawyer—and overseen by a Saudi Arabian whose deputy was an Egyptian doctor.

Egypt isn't deeply mysterious to Barrada. And the What Went Wrong literature that has flooded the Western media since 9/11 has filled in a few gaps for the rest of the world. Egypt is, according to much of this literature, in some ways yet another postcolonial Middle East country whose evolution to democracy was aborted in part by the region's discovery of oil. Oil allowed wealth to remain concentrated in the hands of a very few, who didn't need to distribute civil rights and other privileges to middle classes in exchange for taxes. So they didn't. And when the dwindling middle classes became angry about this, they were told to blame the New World, which was, after all, funding and supporting the status quo. So they did. It is by now an old saw, but one that is accepted by many foreign policy experts. Many of the Middle Eastern states were decolonialized and became independent, at about the same time that oil was discovered. The incentive to build middle classes dispersed in explosive gushers of oil, substantially delaying the move toward the ordered, secular, civil societies many Arab nations aspired to in the early days of Nasser.

The upshot for science, as Barrada notes, has been that it is almost "nonexistent," in some ways, a very good reason to believe Egypt can't become a stem cell hub.

Yet Barrada's aspect in DC that day in June 2003 was not that of a discouraged man. He was in a state of perpetual excitement. After each talk held during the first meeting of the first international stem cell society, which attracted 656 scientists from 22 countries, he would barrel outside for a Marlboro and an excited review of everything that had gone on inside. He used words like "it's all like a science fiction novel" and "with these cells, we may conquer death in the next century." Barrada fully intended to start a modern hES cell center in a country that may have created science but hadn't, as a nation, contributed significantly to it in hundreds of years. Egypt's science is, indeed, so undistinguished that when the Alexandria-born Ahmed Zewail won a 1999 Nobel Prize in science (the first in the Arab world, and the only one, even now), his image was instantly placed on two Egyptian stamps, despite the fact that he was actually an American, who fled Egypt as a lad to get his PhD and never returned.

Yet Barrada believed his goal was reachable, and as the days wore on at that June conference, it was clear he had some good reasons. First, he was Ismail Barrada, son, brother, and nephew of prominent doctors. His name was so established in Cairo medicine that a street was named after his family, and Anwar al-Sadat's wife came to him for her gynecology visits even when he was practicing in the United States. His own wife was a movie star, Lubna Abdel Aziz, who had played opposite the likes of Omar Sharif in the 1960s and remained wildly popular in Egypt. She retained high-level friendships that included Suzanne Mubarak, wife of Egypt's current president. And Barrada had a gynecology MD. If anyone could bring hES cells to Egypt, it was Barrada.

But Barrada would be aided, he believed, by Egypt itself. For while Egypt is like other Middle Eastern states—none of the 22 Arab League nations is at this time an elected democracy—it is also different. It does have a civil society—that is, both public and, increasingly, private institutions, and a rule of law of a kind. It has a per capita income of almost $4,000, giving it an unhappy middle class, but one "within the zone of transition" to successful capitalism, according to *The Future of Freedom* author Fareed Zakaria. Egypt's leaders since the 1950s have consistently made conciliatory gestures to the West, though Sadat paid for that with his life in 1981, leading to a schizophrenic mixture of liberal and repressive policies that has landed Egypt on Freedom House's "not free" list since 1993.

That willingness to look to the West has been due partly to the fact that Egypt's free cash, unlike that of the UAE and Saudi Arabia, has not come directly from oil but more indirectly from the billions in aid (and Suez passage tolls) it collects from the West to buy prime spots *near* oil. This, and its tourist industry, have made Egypt dependent on the West's approval. Furthermore, there has always been a widespread belief that "Egypt is the intellectual soul of the Arab world," as Zakaria puts it; that, while Mecca in Saudia Arabia is Islam's religious heart, Egypt, because of its distinguished past and its status as the largest Arab nation, is its brain.

Regardless, the old joke that Egyptian government employees did 27 minutes of work a day began to change in the 1990s, as current president Hosni Mubarak began a process of decentralization, discarding some unproductive state-run industries and selling others off to private interests. Free speech is stifled; there is effectively only one party; dissenters sit in jail without charge; *baksheesh* and bureaucracy reign. Still, it is believed that Egypt is closer to a state of democracy than its neighbors.

The third reason for Barrada's optimism that week in June had to do with the above-mentioned *fatwa*. The head of the Federation of Islamic Medical Associations may end his professional newsletters with "May Allah (SWT) grant you His Divine guidance and blessings." ("Allah [SWT]" means, essentially, "the pure God.") But a few months earlier it had become clear that in the Middle East, a casual mixing of religion and science may not be a problem for hES cells. For the *fatwa* draft that Saudi Arabia announced (in Toronto, to BIO) in April 2002 was written by the Fiqh Council on Ethics, which governs all Muslims. And that draft proclaimed that Islam accepts, in principle, research on embryos and fetuses because the Koran says the soul doesn't enter the body until 120 days. The *fatwa* draft meant something astounding: from an ethical standpoint, Barrada may have an easier time starting an hES cell clinic on the Nile than he would in New York City.

The last reason for Barrada's confidence was the oddest, if perhaps most interesting: Israel, which has at this point published more papers on hES cells than any country save the United States. At the DC conference, Barrada twice huddled quietly and at length with a charismatic leading Israeli hES cell scientist, Joseph Itskovitz-Eldor. "He's interested in a collaboration," Barrada noted several times, in awe.

The development was, potentially, a coup for Barrada. Elites in Egypt often head abroad for bone marrow stem cell transplants when struck with cancer. Partly for this reason, Barrada's new boss, MISR University chancellor Souad Kafafi, had quickly consented when Barrada, in early 2003, came to her requesting lab space for hES cell work.

Still, he would be starting from ground zero without expert help. Israel could provide that—lots of it. "Joseph said no at first, not because he was against it, but because he didn't want me to get in trouble. But when I talked to him the second time, he asked if I knew people in Jordan. I said yes, I had talked to the minister there. He asked if I had talked to Saudi Arabia, and I said yes again. But, I said, I have to get this lab started, prove it can work, before I can go back to them. He gave me his e-mail. He wants to talk. Isn't this extraordinary?"

There is some precedence for this. In 1989, Israeli blood stem cell specialist Shimon Slavin responded when King Hussein of Jordan requested a bone marrow stem cell transplant for his cancer. And in 2002, Slavin and Italian doctors saved the life of a Palestinian child with an autoimmune disease by giving her transplants of genetically engineered blood stem cells in what turned out to be the first successful gene therapy trial in the world. But aside from clandestine talks between Israeli and Egyptian leaders through the decades to ward off wars, Egypt and Israel have rarely happily worked together on anything. The words "public relations coup" barely described what a stem cell collaboration between Egypt and Israel might mean.

•   •   •

FOUR MONTHS after the ISSCR conference that so transported him, Barrada is sitting in the lush Zamalek garden area of a former royal palace, the salmon pink Cairo Marriott. It is October 16. A *Gulf Today* lead story exults over the fact that President George Bush's numbers are down. The UAE's *Khaleej Times* lead article is an editorial slamming India's decision to buy weapons from Israel. This is followed by a story that is anti-Israel *and* anti-Bush, which is followed by an article on the arrest of 150 peaceful protestors requesting municipal elections at Saudi Arabia's first-ever government reform rally. ("They are a small bunch . . . this won't happen again," an official assures readers.) All day, President Mubarak has been dedicating new private enterprises, from a macaroni factory to the first airport built with funds from a private government incentive funds. He has been doing this as part of anniversary commemorations for the October war, which Egypt generally believes it won against the West, aka Israel and the United States. Egypt has been vigorously celebrating this victory all month, story after story unearthing every detail of the 30-year-old confrontation. It is a fairly ordinary day in Cairo.

But Barrada is uncommonly quiet. He is surrounded by old Egyptian money revitalized by new U.S. cash. Chandeliers beam behind mosque-shaped windows; marble steps wind up to a four-star restaurant and down to a tux-only casino; old palms bend deferentially to a new pool. Barrada's fellow diners are discreetly well-off and have been declared machine-gun-free by a discreet X-ray. The air smells of shish kebab, spiced rice, and *kofta* (spiced sausage). Office lights along the Nile twinkle like fireflies in the LA-thick smog. It is a perfect Cairo night.

Yet Barrada is tapping an unlit cigar absently on an ashtray, distracted. He has spent the last four months taking two steps backward for every step forward, he reports. Most importantly, he says, "They're reading my e-mails."

He explains: For a month after the international conference in DC, he happily chatted up Egyptian academia, sussing out a cellular biology program at the American University in Cairo that could offer him smart science graduates and discussing a collaboration with Cairo's National Cancer Institute to administer blood stem cell transplants to help support future hES cell research. Then, in July, he received an e-mail from Israel's Itskovitz-Eldor, reminding him of their meeting. As their e-mails progressed, Itskovitz-Eldor asked if it would be all right if he contacted the NIH, to ask it to persuade the White House to help Barrada. He also offered to help personally: Come over, he said. I'll show you human-embryonic stem cells. I'll show you what to do with them.

"It was so thrilling," Barrada says. "I was all set to go."

He told Itskovitz-Eldor to wait on the U.S. contacts until he was done clearing his plans, then presented his university chancellor with them. "She said, 'Absolutely not,' at first." Then she relented to some degree, noting that he should check with the National Intelligence Agency, the largest in the Middle East, the equivalent of the CIA.

His first meeting was "great," he says. The officer seemed interested, telling him to write up a report about his intentions and send it along. He did—and waited three weeks. When he couldn't wait anymore, he called, and was told to come in the next day. Meeting him were three officials, three "idiots" Barrada says angrily, who "knew nothing about science, one in full *abaya,* all you could see were the eyes." After listening for ten minutes, they made it clear that they had not bothered to read the report, then told him that his request to start a collaboration with an Israeli scientist, was "of course" impossible. When he began outlining the information in Itskovitz-Eldor's e-mails, one held up his hand. "'We know,' he said to me, and leaned back, grinning. The bastard is reading my e-mails."

It isn't the end of all possible collaborations, Barrada says. Another Egyptian intelligence officer whom he contacted later, a relative of a friend, was more relaxed about it, noting that he might want to consider simply proceeding with the collaboration without telling anyone. "Perhaps I will do this, I don't know. I have to think," he says. "I felt an instant connection with that Joseph. As if he were my brother. It was strange."

His mood improves momentarily when two young ladies join him, his two new American University in Cairo hires, Hala and Miriam. Hala hails from Qatar, a tiny oil-rich nation where her father was a plastics engineer. She now lives in Cairo with her family. Miriam was born in Cairo, grew up in Illinois, and returned to Cairo with her family in time to go to the university. They both majored in biology, graduated in June, and are taking a tissue-culture course to prepare Barrada's lab.

The conversation turns to life in the city. The two girls talk about the dangers of wandering about Cairo alone as women, and the special dangers of wandering it alone without wearing *hijabs* (headdresses), which neither of them do. "My hair gets me into trouble all the time because it's blonde," Miriam notes. "I'm not going to cover it—but I don't wear skirts or shorts or no-sleeved shirts. That would be pushing it." They've become close in part because they live near each other, and so always try to travel together. But that doesn't always solve the problem. This morning, Miriam was waiting for Hala in a café, when men at a nearby table began shouting insults at her. "The owner of the restaurant came up to me and told me to get out, that I shouldn't be sitting alone and was causing trouble."

Barrada has sunk into gloom again. "I don't want to talk about this anymore," he says, handing the girls money to take a cab.

*   *   *

THE NEXT day dawns hot and sunny, as does Barrada's mood. He is tour-guide bright as his driver picks up the two girls and heads for the Pyramids. He chatters about the Nile, how it flows south to north; the roses in the public garden to his right, all imported from—and abandoned by—Britain; the zoo, which looks a little wan, with its carved stone signs bearing both hieroglyphics and giraffes. "There's Barrada Street," he says casually, waving at a short palm-filled boulevard. He points at a rose-colored house. "The house where my father was born." The assistants crane to see.

Then all start talking about McDonald's food, which for the past few years has included the McFalafel. Barrada's mood lifts again and stays lifted throughout the drive to the Pyramids the Egyptian museum. He is clearly in his element. There is much science in Egyptian sightseeing. There is science in the way the Pyramids' blocks were carved so precisely none are separated by more than a few millimeters. There is science in the way the Sphinx mimics the shape of the natural mud lions that form in the Sahara winds. There is science in the bark-brown, linen-wrapped mummies of the pharaohs, whose faces seem caught in an eternal sneeze: eyes near shut, jaws dropped open, hair in an electrified corona around their heads. It was only this month that German scientists announced they'd cracked that mystery. After storing the organs in canopic jars, the Egyptians ran guaiacol from cedar wood through their veins, a chemical that kills bacteria but preserves flesh.

But there is also some Egypt in the science of stem cells. Osiris, the green God of Immortality, whose figure is scattered throughout Cairo museums and bazaars, is the name of a mesenchymal stem company in Baltimore, Maryland. Isis Pharmaceuticals, which traffics in blood stem cells, is named for Osiris's wife, who was responsible for establishing Osiris's immortality. In choosing the names of Egyptian gods to market their cells, researchers have reached over the colossus of modern science to a time of colossal faith, when man was so certain about the prospect of eternal life he built monumental steps to it; packed himself enough gold and food and clothing to get through all of it; staged massive rejuvenation ceremonies to honor it. Egyptian museums overflow with reminders of the Egyptians' confidence they would live forever, packed as they are with treasures excavated from the tombs of Egypt's 100-plus pyramids: alabaster vases, cups, and canopic jars; cloudy Muski glass plates and bowls; gold, copper, lapis lazuli, and turquoise bracelets, tiaras, armlets, and headdresses; cotton slippers, gold sandals, and camel-bone and mother-of-pearl game boards, presumably to combat those moments of colossal boredom that immortality can bring.

Tourist Egypt is one big celebration of eternal life. And while stem cells in the short run are only about modest if unprecedented extensions of life, there is unquestionably something fitting about an embryonic stem cell clinic rising here.

After touring the Egyptian museum, Barrada and his staff hoist a glass of Stella, a state-made beer that has been universally hated for 100 years, at Felfela, a falafel chain. There, while moving on to baba ghanoush (eggplant) and *moosa* (calf knee), they talk about the way films are censored in Egypt. Today, "the simplest kisses are cut out," Miriam says. This was not the case in the forties and fifties. In many ways Egypt is getting more conservative.

Then it is off to one of the world's oldest marketplaces: the Khan al-Khalil souq. Barrada, finally tired, takes a seat beside men who are sitting crosslegged on the floor in Fishawi's Coffeehouse, looking like snake charmers as they take deep breaths out of sinuous *sheesha* (water pipes). "You are fresh as a bird," he says, waving Miriam on.

Miriam, with bright eyes, starts engaging in the ancient art of bargaining. She is consuming in the same stone tunnels and in the same ways—waving her arms, shaking her head, walking away in mock disgust, walking back—that consumers did thousands of years ago. She is surrounded by many of the same artifacts (made of cheaper materials) that were in the museum she just left. She successfully negotiates for a pearl and silver necklace, a bust of Serabis, who is the Panhellenic incarnation of Osiris, the god of eternal life, and an Egyptian T-shirt that makes disparaging remarks about Egypt's eternally awful beer.

*   *   *

ON SATURDAY, Barrada heads for his new lab. Every day, Barrada's driver takes him past the distant forms of the Pyramids to get to work, on the 23rd of July freeway, which travels West out of Cairo into the Sahara. The road is at first lined with palm groves, groomed for their red figs; green rice paddies; sellers hawking live ducks strapped to cages; policemen who stand on the median strip and occasionally pull over drivers for random checks. As the palm groves begin to thin, they are replaced by a vast stretch of desert with hoses running along it, the use for which becomes apparent when a series of luxurious Mediterranean-style houses lead up to a new city rising on the horizon: 6th of October City. The hoses are running water from the Nile miles away to the desert for this city, named after the October 1973 war against the Israelis and built to handle Egypt's overflowing population.

Barrada waves at one of the beautiful huge new houses, which belongs to a film actor colleague of his wife. Still being built, 6th of October City houses the Hollywood of Egypt, Media City, along with marble-making factories, a prefabricated building factory, and Misr University for Science and Technology. The latter is at this moment Egypt's largest private university among the five established since they became legal in 1996 (a phenomenon being echoed in other surprising places, such as Syria) and the home of Barrada's new stem cell lab.

The seven-year-old university is, as its founder and chancellor Souad A. Kafafi will later note, "Pharaonic." Designed by her brother, all the buildings are built of a sunset-colored concrete and gargantuan columns. The library is a massive showcase, with a beautiful inner sanctum that is several stories high, but a minimum of books and desks, and only three computer terminals. When Barrada and his staff arrive outside the university's teaching hospital, they point at a statue of Kafafi. "She had three statues of herself built here," he says, smiling affectionately. All around are students. Fifty percent of them are female. In a reflection of the streets of modern Cairo, nearly all of them wear *hijabs* and some wear the all-concealing black *abaya*.

Inside, the hospital looks somewhat like a modern, marble cathedral. It is large and open, its main atrium open to the ceiling several stories above. Most unhospitalike is the way the ward-rooms all circle around the open space, inviting drugged patients to take a nosedive. Barrada proudly shows off his tiny office and three large labs, which are stocked with impressive equipment not yet used. There are sterilizing laminar flow hoods, a bioreactor for culturing cells, a polymerase chain reaction

(PCR) system to amplify genes, gel electrophoresis equipment to iden-tify and isolate them. A third assistant, a Misr University medical stu-dent, shows off a small flow cytometer. This machine allows scientists to isolate stem cells down to the single cell level, very important to char-acterize their offspring with accuracy. There is much to do, the smallest problem being figuring out how to discourage surgeons from rushing into one of Barrada's labs to use his autoclave to sterilize instruments in the middle of operations.

Out the windows, the unvarnished Sahara is like the backdrop for some ancient painting, waiting to be filled in. It is a constant reminder that, literally, these labs have been set up at one of the ends of the civi-lized world.

Back in his lab, as Barrada and his secretary play tag with the air-conditioning—when she leaves, he turns it up, when he leaves, she turns it down—he talks a bit about Jordan, which had said it would put $50 million into a stem cell center but from which he has not heard since. Ditto re: Saudia Arabia. While Iran will later publish its hES cell work in a respected journal, other Middle East nations may be lagging momentarily. The unstated message: Barrada's lab may be able to score some Arab firsts, if he moves fast enough.

He starts making plans with his assistants. Tomorrow is his first meeting with some top Egyptian scientists to discuss the lab. He out-lines what he will say and mulls again over options for making the lab self-supporting: to start a blood stem cell clinic or a bank for stem cells derived from umbilical cord blood. But all three are worried that Egypt-ian citizens may have problems with cord blood banking, which is ironic given that Muslims theoretically have few qualms about ES cells. Why? "There is a lot of plain stupidity here," Barrada says. A few months ago, he prompted a storm in the Cairo press, and an investigation, when he suggested to one of his patients, who did not have a uterus, that she fer-tilize one of her eggs with her husband's sperm via IVF and inject it into a surrogate's womb. "Scientists—scientists!—argued whether the egg would contain genes from the surrogate," he says, disgusted. And many people called it profane, because Islamic Law forbids a man's sperm to touch the womb of any woman who isn't one of his wives.

And why might there be problems with cord blood? "Well, the tissue of the dead is not supposed to be used, but of course they're not dead, but people apparently think. . . ." He shrugs. "Oh, I don't know. There just may be problems. There always are."

The conversation turns to the effort it had taken to get Hala and Miriam on board. There were problems at the university commission meeting to approve his hires. "At one point, someone waved at me and said, 'Are all your hires going to be white?'" he says. Hala adds, "I walked out. I've never done that before." While Barrada eventually got his way, he knew then that it would be difficult to bring Western scientists and postdocs, a major problem for him. Even the Saudi Arabian doctors planned to give their stem cell center an international staff.

●   ●   ●

THAT NIGHT, outside the Marriott Palace on the Nile, limos pull up preceded and followed by Secret Service–like men in suits, running with radio plugs in their ears and "machine guns tucked in their sides, guarantee it," Barrada says, as his own car pulls away. The reason for all the excitement: a cabinet minister has come to eat.

A few minutes later, at the site of his favorite restaurant complex, Pasha, a docked boat further down the Nile, another series of Secret Service men are posted on every deck, in this case, to accompany the minister of Spain to dinner. Men in ties running alongside cars or posted outside bars, one hand to their ears and machine guns strapped to their sides—these are a common sight in Egypt. The reason: regular assassination attempts on leaders (and the occasional tourist) made by radical Islamists who see the Westward leanings of the government as heresy. Barrada is unfazed by this, gossiping about it briefly with the owner of the restaurant, saying "terrorists," shrugging, then turning the conversation to the radio show his wife does every week, for which he plays the piano.

It is how they met. When he was in his last year of medical school in Cairo, he did a favor for a musician friend and showed up to play the piano for a radio show written and performed by Lubna, who was by then famous, having starred in her first role in what turned out to be the *Gone with the Wind* of Egypt in 1958: *The Empty Pillow*. When she heard his sophisticated improvisational style (a musician with the Cairo Orchestra at age 14, he had been accepted to Paris Conservatory, but went to medical school for his father), "She said, 'Absolutely not!'" Barrada says, grinning. "Then she saw me. . . . We've been together ever since." A school picture of Barrada gives a good indication why: dark hair, angel faced, he was a looker.

When, a few years later, he was accepted to do his postdoc at the prestigious University of Minnesota gynecology program, he convinced Lu to come with him. During his first U.S. surgery, conducted in the impossible atmosphere of a Minneapolis emergency room, he found a rhythm that was "musical. It was music to me." Gynecologic surgery came so easily to him, in fact, that he was made head of the department the same year he finished his training. He also found that he loved the United States, where "there is true justice, the real thing." He stayed for 30 years before returning to Egypt to retire in 1999. But retirement was predictably boring for Barrada. Soon after he read about stem cells, which had been cultivated by gynecologists like him, he knew what he was going to do next.

*　*　*

ON SUNDAY, the beginning of the Egyptian workweek, on his way to his important preliminary meeting with scientists interested in his project, Barrada's mind has wandered back to Israel's Itskovitz-Eldor. "You know, he told me he has sent many ES cell lines to U.S. scientists." This is an interesting revelation. Generally speaking, U.S. scientists receiving NIH funds can't use hES cell lines, outside the presidential lines, unless they set up separate lab spaces. This is one of many indications that there have been quiet undercurrents of rebellion occurring in the world of U.S. science.

Before his meeting, Barrada passes a statue of the university chancellor, who sits in her bronze suit looking over the desert, and enters the campus's main auditorium through a two-story atrium. The auditorium, which is still being built, is vast, with a cathedralesque ceiling, and rife with wrought iron and marble fixtures. "This is where we will hold the first International Egyptian Stem Cell Conference," he says, then leads the way up the stairs to an elaborate, palacelike hallway of glass and marble. Each conference room he passes boasts, as Barrada had noted earlier, tall overstuffed velvet chairs resembling thrones, which are gathered around gigantic oak tables and are each equipped with microphones. "Egyptians like to hear themselves talk," he says. Every conference room and office bears a painting or photo of Mubarak's head.

Barrada enters a well-appointed if messy office, where a secretary immediately offers Coca-Cola and tea. Minutes later the president of the university enters. A former Egyptian envoy to the United Nations, President Mahmoud Naguib has called Barrada twice that day to warn him that he will come to his meeting, but he will not talk, prompting Barrada to twice roar into his cell phone, "You will go. That's an order. *And* I will make you say something."

They are old friends, as becomes clear when the president, a thin gangly man, skips introductions to say, "I mean it. I'll go as long as I don't have to say a word." He turns and explains: "My PhD thesis was on the political playwriting of U.S. writers in the thirties. Cornell University, Tennessee Williams et al. I'm afraid I know nothing about stem cells—although I understand it's the new thing?"

The two are joined by the head of the university's cardiology department, who immediately begins chatting knowledgeably about recent publications indicating that bone marrow stem cells may repair the heart. Then all retire to the university chancellor's office, which is a huge affair, stuffed with gold and purple velvet tapestried chairs and couches, and presided over by another gigantic painting of Mubarak's head.

After six or seven department chiefs and a handful of others have gathered, there is a brief silence. All are seated stiffly in straight-backed gilded chairs that indeed make them look as though they are occupying thrones, as Barrada had earlier noted. Then Barrada says, "And now the president will make a speech." The president shakes his head and laughs. Ice broken, the meeting begins.

The head of neurosurgery notes that stem cells will "revolutionize medicine." A worried looking, slightly disheveled endocrinologist, who had rushed in with folders under her arm, says she's heard that stem cells may help diabetics, important for Egypt since access to insulin is rare. Then the medical school dean, a small gray-haired man, notes there is resistance in Egypt to using cells of the dead, lobbing this bomb with a smile. There is a silence, some raised eyes, then some explanations made in Egyptian. He listens with the same smile, and repeats that, while he is not against using cells of "the dead," many people will be. It begins to become clear: despite the *fatwa* insisting that life in Islam doesn't begin for 120 days, hES cells may be seen by the Egyptian public as verboten as organ donation from the dead.

Barrada, after insisting he can make the Egyptian public see the difference between organ donation and hES cells from five-day-old embryos scheduled to be discarded, recounts his plans. He wants to isolate some hES cells from spare IVF-clinic embryos and work on them for 7 to 12 months, guided by the published work. "Then we'll begin our own experiments." The vice president of the country's pharmaceutical society notes that Big Pharma in Cairo is not going to like this, as it is competition. Barrada notes that actually, in the long run, pharmaceuticals executives will, for "they will want to get in on the action themselves."

Perhaps, the vice president says. What should be done first, he adds, is for Cairo to form a stem cell society. This society will gather funds, hold meetings, and gradually educate both the people and the scientific world. This decision is leapt upon by the assembled with a palpable relief. The Egyptian Stem Cell Society will be formed. Later, decisions will be made.

After the meeting, all gather for catered sandwiches and Nescafé. University chancellor Souad Kafafi, who came into the meeting late, turns out to be a diminutive woman with thick brown hair cut in bangs like a little girl's, powdered white skin, and penciled eyebrows. She speaks broken English, but emits warmth, easily placing a welcoming arm around the waists of visitors. "The Egyptian people will not like this," she says. "Many people will throw pebbles. But I say to Dr. Ismail, 'Just keep going, like this.'" She holds her tiny hands up to the sides of her eyes, like blinders. "When I started this university, many people threw pebbles . . ." because it was—and remains—one of the first five private universities in Egypt, the phenomenon being new to the country. "I just go . . ." she looks for the right word. "Forward." She shrugs, looks at her watch, and smiles. "Have you seen our library? You must be taken there. It is Pharaonic."

After the meeting, the pharmaceutical society vice president drives Barrada home. He is stout, short, and powerful looking. His car is gleaming white. Slung over his rear window mirror are *sebha* (prayer beads) which are used to recite the *tasabih*. "The Pharmaceutical Society will see this as competition," he repeats cheerfully. "But why not? We can try."

*   *   *

THAT NIGHT, Barrada has dinner with his wife, Lubna Abdel-Aziz, at their favorite restaurant complex on the Nile. Lu is wearing a white shirt opened to a leotard underneath, an ornate gold belt, some ornate gold jewelry, and tight-fitting black Capri pants. She is in her sixties but looks in her late thirties with towering high cheekbones and thick black hair. Like the university chancellor, she instantly places her arm around the waists of even first-time acquaintances. By the time the couple hits the top deck, where their restaurant is located, they have been besieged by five fans, all in *hijab,* clamoring for autographs. Lubna signs their napkins, then stands with them for several pictures.

While this is going on, Barrada talks about the meeting. He believes that the society formation is a great idea, but feels that it doesn't offer an immediate solution for his main problem: a way to make his lab self-sustaining. He sighs. "I feel oddly that I've lost a friend in Joseph [Itskovitz-Eldor]," he says. He notes that he clearly cannot e-mail Itskovitz-Eldor anymore. "But if this proceeded with Joseph, we could hit the nail on the head of the bat right away." He stands for his wife.

Lu sits down. An intensely private woman with a wry sense of humor, she has apparently sussed out her husband's mood, for she quietly reminds him that they should contact Suzanne Mubarak and see if she can help with the Israeli/Egypt connection "when things settle down," referring to his National Intelligence Agency problem. Barrada brightens. Several times throughout the past few days he has made calls on his cell phone to his wife, to keep her up to date on all that has happened in the ensuing hour or two. While he may have lost, for the moment, his Israeli collaborator, it is clear he has a lifelong collaborator and powerful support system in the wife who gave up her movie career for him, 30 years ago.

On a parquet stage toward the middle of the restaurant, the kind hauled out in the United States for dancing at weddings, a Tom Jones–like singer launches into "Girl from Ipanema" accompanied by a synthesizer. Barrada begins tapping his fingers in time to it. He likes this song, it is from the "good old days" he notes, the days of early Nasser, when Egypt seemed hell-bent on entering the modern world. "Maybe things will work out," he says.

And over the next several weeks this will appear to be the case. He will hire on several new staffers. He will request and receive two patient wards. He will purchase laser equipment in New Jersey and meet with Wise Young, a top U.S. neuroscientist and codirector of New Jersey's fast-evolving stem cell center. He will arrange for one of his staffers to train with Wise Young and one of Wise Young's staffers to come train with him. Things will begin moving rapidly for Barrada once it hits him that, as a gynecological surgeon, the path of least resistance to a self-supporting hES cell center was fairly obvious all along: he should launch an IVF clinic.

But Barrada's national intelligence agency "problem" will continue apace. "Tell me about the trip you said you were going to take," he says in an e-mail to a writer days later, referring not-so-obliquely to Israel. "But be careful about it." Whether Barrada will be allowed to establish critical and meaningful scientific alliances with Westerners like Wise Young, and with Israelis, will remain to be seen for a while.

3

SCIENTIFIC
PILGRIMAGES

THE JORDANIAN DESERT APPEARS to be on fire. A quarter-mile snake of smoke is slithering across its bland surface. But as the plane circles in closer, it is clear the smoke is actually just a long trail of dust left hanging in the air by a lonely farm truck buzzing through the sand.

Wraiths of real smoke, from cigarettes, haunt the tiny and isolated, air-conditioned Jordan airport, an unusual situation in Middle East airports, where smoking is now only allowed in designated rooms. This gift is eagerly leapt upon by a handful of in-transit passengers, who, on this sunny day in mid October 2003, range from a tired-looking old couple decked in gray *abayas* to a boy in jeans with a backpack to an American with a rolly suitcase. As they smoke, all stare out quietly at a view that is eerily lunar, made more so by the unexpected modernity of this tiny airport, with its clean swath of floor-to-ceiling windows and the characterless stainless-steel gray chairs bolted to the floor, which give the building a spaceshiplike feel.

Getting on a flight to Israel from there is surprisingly tension-free, given the recent increase in Israel's troubles, otherwise known as *matzav,* or "the situation." Airport officials don't even ask to inspect coffee cups today.

Indeed, the only thing distinguishing today's 45-minute flight from

Amman to Tel Aviv are the students. Aged about 15 to 19, they are dressed in a mixture of Turkish and Western garb, T-shirts, exotic shawls, and pasha pants. They gossip as the plane soars over the flat plains of Jordan, the tiny occasional Jordan village, the tiny occasional truck, the frozen oceanlike waves of brown Mo'ab mountains. But when the scene brightens in the reflection from the all-white villages on hills that mark Israel, followed by the white-blue Mediterranean Sea, the kids quiet down. It is a palpable thing, their silence, their noiseless-craning over each other's shoulders to see out the window. It is also a long silence, lasting a good 15 minutes, all the way through the landing. It has the nonstudents looking out as well: What, a bomb? The craning and the silence seem so touristy, the thought is that these kids must be from somewhere else. This must be their first view of Israel.

But after breaking into applause as the plane lands at Ben Gurion Airport, they zip outside and right through the customs desk for Israelis. It is an unusual way to enter a new country: with a plane full of teenagers as unable, or unwilling, to take their home for granted as a plane full of tourists seeing it for the first time.

Regardless, the surprising casualness of Jordanian security, juxta-posed with the unusual intensity of a group of young teens returning home, is a subtle reminder of the shifting soil on which Israeli stem cell scientists stand. In the history of the stem cell world, the city the kids were staring at was once highly significant for being the first place seen by scientists from around the globe when they landed in droves in Israel in the late 1990s and early 2000s to gaze upon the world's first hES cells. For years after the first hES cell paper was published in 1998, sci-entists flooded Israel, home of Joseph Itskovitz-Eldor, Benjamin Reubinoff, Nissim Benvenisty, and Karl Skorecki, all of whom worked freely on hES cells from the late 1990s on. So many scientists came to Israel, in fact, that *Science* magazine actually used the word "pilgrimage" to describe the phenomenon.

In recent months, Tel Aviv has become more historically significant as the last thing Israeli stem cell scientists see before they *leave* Israel, in a reverse pilgrimage, traveling around the world to share their work with scientists now too fearful to visit. Since the intifada started in 2000, there have been 900 violent Israeli deaths, with most of those occurring in 2003, this year. Can Israel retain its lead in the stem cell world under these conditions?

Even when things seem normal, they are not, quite. The phrase "suicide bomber," as regrettable as it is, nevertheless leaps to mind the instant the unwitting foreign traveler fends off the hail of customs questions and enters the narrow battlefields of Israel roads. In Israel, it is quickly apparent, every driver is obliged to drive as if for his or her life: fast and mean, no time for niceties. Drivers are consciously aggressive, often stopping in the middle of the road, blocking all traffic, in order to get out and shake a fist at a driver going too slowly or too uncertainly. Honking is constant. The reason becomes clear quickly enough: all traffic is funneled to the south of the West Bank, chiefly routes 1 or 2, because to wander absentmindedly into a West Bank road is to come to a standstill for hours, thanks to roadblocks. This makes traffic on all Israeli roads nightmarish, a constant monotonous reminder of "the situation." The fact that it's a good idea not to get stopped anywhere here, entangled in a traffic jam near, say, a bus, is a clear factor as well. When traffic jams occur here, many turn around immediately. A primary motivating factor these days in Israel is to get home, fast, whatever the cost, wherever you are.

Yet Tel Aviv is called the yin to Jerusalem's yang, a place of play over passion, and for the traveler waking up to the city, this is apparent, even on October 21, 2003. True, a travelers' warning to U.S. citizens has just been issued by the U.S. State Department due to the unusual murders of three Americans in the Gaza Strip nearly a week ago. And a major terrorist incident in a Haifa restaurant, on October 4, has sparked a new wave of terrorism that by December 26 will be reported as 300 incidents. As a result, Tel Aviv hotel rooms are half-price, and parking lots are bare.

Also true is the fact that guards check under cars all over, and police check pocketbooks of ladies entering the post office to mail a postcard. In malls and stores, it is common to see soldiers on the phone sliding onto the floor to chat, letting their M16s clatter to the ground beside them: the M16 here arouses far less alarm than a backpack or a briefcase.

Still, the air is amiable. There are security guards outside every hotel, but they barely glance at people entering and exiting, busy gabbing and smoking. The Russian and Czech bars on Ha'Yarkon Street are open late and are filled with locals putting back vodka and eating goulash while playing pool and listening to jukebox American music ranging from the sixties on. Motorbikes zip down sidewalks, merrily swerving in and out of trash cans. There is an Internet café on every corner.

And if the beach beyond is only half-filled, it is well filled, with kites punctuating the air; old men and small boys fishing off jetties; lovers tented under towels; dogs dragging leashes and modest matrons wading about in headscarves and old-fashioned black bathing costumes that hang limp to their knees, looking like bemused mourners caught in an unexpected if not unpleasant flood. Half-naked revelers drink up in shack-like beach bars. Hooky-playing lawyers and high school teachers cruise the beach. Beyond, the 4,000-year-old bungalows of Old Yaffa glow bronze. "It is just prejudice," a handball player named Orly says, waving at the empty spots on the beach. "People hear about one incident and decide the whole place is about to blow up. Does this look like it is going to blow up?"

A heavyset solider in olive drab and calf-length boots toddles to the end of a jetty. In his right hand, he clutches the fingers of a woman in a T-shirt and shorts; in his left hand he holds a plastic bag of bread; over his shoulder is slung an M16. To a foreigner, this continually seems ominous. But when Orly, noting that nearly everyone on this beach was once in the reserves themselves, leans back on his elbows in the sun and says drowsily, "See? Safe," it is almost possible to see. It can be, impossibly enough, almost jolly in Tel Aviv.

Indeed, "impossibly jolly" also aptly describes the state of Israeli stem cell science. For, as a tiny country on the other side of both a war and the Western world, it should never have pulled into the lead of the stem cell race, let alone stayed there. B.S., before stems, Israel was already busy as a leader in areas of science related to its weapons work and computers. Between 1996 and 2000, the tiny nation, which is smaller than Rhode Island, put out 1.5 percent of the world's papers in physics, which were cited by other scientists 31 percent more than average; 1 percent of the world's papers on materials science, cited 52 percent more than average; and 2.4 percent of the world's papers in computer science, cited 34 percent more than average. It supports the largest number of hi-tech start-ups per capita in the world. It has attracted venture capital at rates exceeding most European countries.

Between all that, and the political situation, Israel had its hands full, and indeed, with the exception of Teva Pharmaceuticals—which began as a generic drug company distributing drugs on the backs of camels and is now one of the world's top 25 pharmaceutical companies—Israel's biology work was once not widely known.

But in November 1998 and April 2000, the labs of two different Israeli scientists appeared among the names gracing the first two Western papers announcing the discovery of the human embryonic stem cell: Joseph Itskovitz-Eldor of the Technion in Haifa and Benjamin Reubinoff of Hadassah Medical Center in Jerusalem. Israel's biology quickly became known.

*   *   *

IF JERUSALEM is the ultimate city on a hill, Hadassah University Hospital is a city on top of the city on the hill, in more ways than one. The road to Jerusalem from Tel Aviv may be surrounded by rock and sand, undistinguished by much of anything but its history—among other things, it was taken over by Jordan during the 1967 war. Yet Jerusalem is so awe-inspiring it seems almost a caricature of its biblical self, rising suddenly out of the desert, terraces of ancient-looking pines guarding and framing one row of ancient-looking, hugely picturesque white cottages after another.

Then up one goes toward Ein Kerem, a biblical suburb of Jerusalem, and there, on a hill, is Hadassah University. There is a checkpoint out front and a long line for the x-ray at the front door. The university hospital was another site of battles from the 1967 war, but more importantly for science history, it is home to Shimon Slavin, part of the first team to initiate a total cure with gene therapy and adult stem cells, and Benjamin Reubinoff, one of the first two scientists to coax the hES cell into its most astonishing state—that of so-called immortality—thus marking the launch of human embryonic stem cell science.

Reubinoff does not look at all, however, like one of the first few men to gaze into a petri dish and see immortality there. He looks, in fact, at first simply like a guilty man. For the topic of Israel's traffic habits starts the interview, and it is instantly clear that it has not only been brought up before—but that he has been among the "guilty." "Israelis are a little tense lately," he says unhappily. His face squeezes into a grimace. "I know, I see it in America, I always notice this. People don't honk there. And they let you go ahead of them. We don't do this. It is a very different feeling. People are depressed and angry here, and it's there on our roads. I know. I'm sorry."

Adding to the general impression that this is not a man who has freed himself (or at least, some cells) of the mortal coil, he is sitting in a room, located in the genetics wing of the Hadassah Medical Center, that looks more like a converted broom closet than an international HQ of Immortality. It is windowless, tiny, and neat as a pin. Pictures of two children in green pyjamas—patients—peek out from a spot tucked neatly into a bookcase. They represent the room's only bright, expressive touch.

He begins talking about his work slowly, warily, a faint look of dread on his face. But soon he is talking fast, scribbling, jumping up for books and scribbling again. Interviews with scientists often proceed in a pyramidal fashion like this. Unlike others interviewees, scientists are never positive the listener will understand one word of the foreign language they speak, that of their scientific subspecialty. They often start small and contained, then suddenly expand exponentially, either in celebration of the fact they are being understood or too caught up in the beauty of their own language to care.

Indeed, it is in just such a pyramidal fashion that the ES cell story started, with a man in the U.K. and a woman in the United States, then many men and women in many countries racing for similar goals. What distinguishes the ES cell tale from other science stories, however, is the persistently global nature of it. Martin Evans, a British scientist with the University of Cambridge and Gail Martin of the University of California, San Francisco, separately discovered in 1981 a mouse cell that could form all other cells and replicate endlessly: the mouse ES cell.

At first it was just used to vastly improve germline genetic engineering. Until the isolation of the mouse ES cell, which could form the vast majority of tissues in the mouse body, it was profoundly difficult to insert foreign DNA (genetic codes) into a mouse embryo to create a fully "transgenic" mouse, one with a new gene in each cell. This was critical to both create mouse models of human diseases on which to try out new drugs and to, simply enough, figure out what certain genes were for. (Pop an unknown human gene into a mouse, and if it comes out with green eyes, you know the gene codes for eye color.) New genes, or DNA fragments, slipped into mature mouse cells, then popped into mouse embryos, would appear in only a few cells of the resulting new-born mouse. The mouse embryo was not fooled by mature mouse cells: it knew they didn't belong there.

But in the 1980s it was discovered that multicellular embryos recognized mouse *embryonic stem cells* as their own. Mouse embryos were, after all, filled with their own embryonic stemlike cells at very early stages. Scientists discovered that all you essentially had to do was sneak a new DNA fragment into an ES cell, and thus cloaked, coculture the transgenic ES cell with the embryo. Once the multitalented, proliferative, sensitive—and stealthy—ES cell was accepted, it would replicate endlessly in the growing embryo along with its embryonic host brothers, appearing in many or most of the cells of the grown-up mouse—including the germline. In 1982, Richard Palmiter created gigantic mice this way, sneaking into a mouse ES cell a DNA fragment that switched a growth gene permanently on.

Throughout the 1980s, developmental biologists smuggled hordes of new genes or gene fragments into mouse embryos via the clever carrier of the ES cell, creating hordes of new mouse breeds with different human diseases. Drug companies snapped them up and began trying their drugs on them. It wasn't until the early 1990s, however, that some of those scientists began thinking the obvious next thought: Is there an equivalent cell in the human being? Debate that had been raging for decades over the taboo of human embryo experimentation swiftly prompted many of the traditional scientific pioneers (namely, U.S. scientists) to immediately drop that thought. But in countries where such debates weren't raging, countries whose governments were not beholden to a powerful religion (China and Singapore) or whose religion did not hold the moment of conception as sacred (Israel and some Muslim nations), the door was tantalizingly open.

Singapore stepped through first. Throughout the 1980s, a talented IVF clinician with both a veterinarian and developmental biology degree named Ariff Bongso had been busy taking advantage of Singapore's relaxed embryo experimentation environment by scoring, along with a handful of others, some IVF firsts. His approach was intuitive, utilizing his growing familiarity with human and animal germ cells. To increase the chances of a fertilized egg taking hold in the uterus, for example, he examined the procedures of the day and saw a key problem. Doctors were inserting two- to three-day-old human embryos, which they'd created in a dish by mating sperm and unfertilized eggs, into women's uteruses. But in nature, the fertilized egg does not make it down the fallopian tube and into the uterus until about day 5. Reasoning that he should therefore grow his eggs for five days in the petri dish, he tried a number of concoctions on the egg to get it to grow, to no avail. At length, he decided to try to culture the embryos in the environment they would normally inhabit pre-uterus: the fallopian tube. Gathering up fallopian cells from women who had had their tubes cut, he cultured his cells with fallopian cells for five days. It worked. He reported that he ended up boosting the success of IVF in Singapore 25 to 30 percent (although globally the notion is still being tested).

And once he had those five-day-old cells in a dish, it didn't take long for him to recall something he had read in mouse ES cell literature. It is at five days that mouse ES cells appear. Was it possible that there was such a thing as human ES cells? If so, did they appear at five days, as well? He had to know. He asked some of his patients who were done conceiving if he could use some of the spare embryos they were going to discard. Then, using his IVF micromanipulation skills, he followed the recipe for extracting mouse ES cells with spare human embryos. It worked immediately. He isolated some 30 cells that could form the main human tissue categories and could proliferate. They couldn't proliferate endlessly, however, like mouse ES cells. He was cleverly using his fallopian tube cells to feed them, but he was missing some ingredient, he believed Still, he had accomplished more than any human embryologist. "Ariff Bongso was the first person to find hES cells," asserts Harvard University hES cell expert Doug Melton. Bongso submitted his paper to the small, prestigious journal *Human Reproduction,* under the cautious title, "Isolation and culture of inner cell mass cells for human blastocysts." It did not scream "stem cells," and *Human Reproduction* reduced whatever noise it might have made to a peep by burying it in the November 1994 issue, the thirtieth of some forty articles.

The human embryonic stem cell had been discovered, but no one knew it.

Almost no one. IVF specialist Alan Trounson of the Monash Institute of Reproduction and Development outside Melbourne, Australia, produced the world's first pregnancy from a frozen human embryo in 1983. He had heard Bongso speak about his find at a conference. Since he, too, had dual training in animal and human reproduction he instantly understood the potential significance of the find. He invited Bongso to spend a sabbatical with him in Australia to see if they could derive hES cells that were "immortal," or proliferated endlessly. For eight months in 1995 the two IVF pioneers worked on the cells in Australia. Nothing they added to cultures worked. Perhaps the problem was in the extraction, they decided. Since his resident state of Victoria, Australia, outlawed research on human embryos themselves (if not their products), Trounson went to Singapore for a year to work on extraction with Bongso. Still, nothing. "We just couldn't grow the cells beyond a couple of passages," Trounson recalled a few weeks before the interview with Reubinoff.

And the heat was on, at first because "Jamie Thomson came out with his paper finding the first ES cells in monkeys at that time." Then after he had returned to Australia, Trounson heard that a scientist named John Gearhart was working on a very similar hES-like cell, called the human embryonic germ cell. Asked by the American Society for Reproductive Medicine in 1995 to organize a meeting "on something exciting, I said, 'I'll give you some excitement.'" He invited Gearhart to give a speech on his cell. Gearhart described a cell so much like the mouse scientists' pluripotent, wildly pliable mouse ES cell, and so much like Thomson's pliable monkey ES cell—except it was human—that "it knocked me over. It was fantastic."

But it also meant time was running out for the Singaporean/ Australian duo. In 1997, having failed to make hES cells proliferate by adding extra ingredients, Trounson tried to make them proliferate by adding another collaborator: Martin Pera of Oxford University. Well-known for his work on what would eventually come to be understood as cancerous human ES cells—embryonal carcinoma (EC) cells—Pera had always suspected that his weird cancer cells, which contained hair and bones and all manner of human tissue, were actually some kind of mutated human ES cells. He leapt at the chance to take a staff position with Trounson in Australia. But nothing happened, again, until a young gynecologist at Hadassah Medical Center named Ben Reubinoff called Trounson to ask if he could take a sabbatical with him to pick up some extra tips on IVF.

Now acutely aware of the two Americans who were clearly about to pounce on different versions of the hES cell, Trounson asked Reubinoff if he'd like, instead, to figure out if he really had the most powerful cell in existence sitting on a shelf in a freezer in his lab. Reubinoff packed his bags and hit Australia in January of 1998.

After fooling with the frozen cells for a few weeks, and finding they were already dying, Reubinoff headed for Bongso's lab in Singapore to get more and start his analysis. What that turned out to be was a methodical, 24-hour, try-everything approach. He would succeed, but neither he nor Bongso would understand exactly why until years later, when Bongso had repeated his own experiment, bothered by something. Back in 1994, Bongso had followed the mouse recipe very closely, only altering it in the feeder layer: he used those human fallopian tube cells, not mouse embryonic fibroblasts as in the mouse literature, and he tossed in something called LIF, which was used for mouse ES cells. (LIF was a protein produced by cancerous embryonic cells that scientists had long known could, for some reason, keep all kinds of mouse embryonic cells alive.) Reubinoff, on the other hand, stuck with mouse embryonic fibroblasts, and he didn't use LIF. Many people believed that it was the LIF that held Bongso back. But years later it would be clear to Bongso what happened.

The other thing Reubinoff did was refrain from bringing the cells to a low-density level, as was traditionally done with mouse ES cells, whereas Bongso had followed the mouse recipe there. But human ES cells, unlike mouse ES cells, are fascinating in a very particular way, scientists would discover: they differentiated and died when forced to exist alone. "It turns out that human embryonic stem cells are more social than mouse cells," Bongso would note, in an interview in September 2003. "It wasn't the feeder. It was the unusually social nature of our human cells."

The best way to understand this is to think of human ES cells as far more complex, and thus far more in need of close communication with others, than mouse ES cells. The more complex a biological system is, the greater its dependence can be on a variety of communications. The mouse ES cell is, conceivably, less exquisitely sensitive to a huge variety of inputs because it is simpler, needs to get fewer things done. Human ES cells, on the other hand, have a lot to do. They have, among other things, the most complex system in the known universe to create: the human brain. In the same way that human babies remain dependent on their parents longer than most animals, hES cells are turning out to require more preparation before growing up; they are more receptive to, and dependent on, outside influences, longer. They seem to need the exact environment of their "youth" re-created in order to stay undifferentiated and proliferating. When they are together in certain very specific numbers, they are apparently telling each other, "Hey, we're still in the early blastocyst stage, we can't grow up (differentiate) yet." The mouse ES cell, on the other hand, is less dependent on communication from its brothers; it doesn't give and receive the same number of messages, presumably because its tasks are, and will remain, simpler. A mouse ES cell thus turns out to be mousier than the human ES cell.

Whatever the reason, Reubinoff had found a recipe that worked, at last. It was the mouse recipe, minus the LIF, with the cells jammed together, talking, instead of separated. (He also used a different approach to separate colonies.) When he talked about this stage in his life on the phone a few weeks before the October 2003 interview, he did so laughing a lot. It was an "exciting" time he noted, not the least because his approach almost *didn't* work.

When he went to Singapore, "there were a few embryos available there, but not many." After finessing the technique to pull the inner cells out of the blastocyst, and then placing them on mouse feeder cells "we watched for a couple of days. There was some growth, but we were not sure whether this would be the right thing." Then Reubinoff had to get back to Australia, but he didn't want to leave the cells, so "I took the cells back with me. I flew with the cells on my body" that is, on a commercial liner, in his shirt pocket, "and I came to Australia. And here the story is more fascinating because I grew the cells some more in Australia. I was checking these tissues at some point and I was thinking, I don't see anything of interest. I was quite disappointed. I was going to throw them away. And then, at the edge of one of these dishes, I saw something that looked very interesting. I couldn't see it well because it was on the edge of the flask [climbing up the side of it]. I didn't know what it was. I showed it to Martin Pera and he said it looked interesting, but he couldn't see what it was either. It was a very difficult task to get these cells out without losing them, but we have succeeded after a while."

Reubinoff went on vacation to Queensland. But no sooner had he arrived, than he got a call from one of his technicians. "He said, 'You better come back,'" Reubinoff laughed. "I came back and it was really looking interesting. We'd never seen such cells before, they didn't look like mouse ES cells, but they were looking similar to Martin Pera's EC cells [cells that would turn out to be cancerous versions of human ES cells]."

It was exciting, but it was also "stressful," Reubinoff noted. For one thing, they had almost lost the store when the cells started crawling up the side of the dish. To get anything out of a flask one has to remove its cover. The problem was that the cells had climbed right up to the cover—and once it was removed, the medium they were living in would dry out fast, which could easily allow the cells to die. "We knew when we removed the top we'd knock away some cells, and the rest could dry out. So we fire-pooled [created] a very specific capillary tube, then opened the flask and cut gently around this small colony of cells at the edge, then detached what was the one remaining colony and transferred it to another dish. From all those eggs, we had only one single colony of cells. It was everything we had."

The job that remained was a nightmare, letting the colony grow only so much that it remained a colony of ES cells mimicking the blastocyst phase, talking to each other, telling each other to remain young and proliferative. All this while not letting them grow *too* much, because it seemed clear then (and would later turn out to be true) that when there were too many cells piled on top of each other, as if to escape, they either matured into nonproliferating adult cells or died. All this, while keeping the amount of nutrients steady. "So throughout that two weeks, I was flying back and forth between vacation and the job of taking care of the cells," recalled Reubinoff. "One single colony. It was very stressful. We were afraid to lose them, and they were difficult to grow; they tended to differentiate. . . . And we didn't have any back-up. A lot of pressure." But it worked. By the end of September of 2003, the international team had what seemed to be the world's first dish of naturally immortal human cells on their hands.

However, they had barely placed the dish on the table to free their hands to applaud, when they received a rude shock. Theirs wasn't the world's first dish of naturally immortal human cells. For all the noise the United States had made about the ethics of this work, Congress had long ago left a single loophole in its restrictive environment, through which a single U.S. company would slip, and score, reaching Reubinoff's all-important milestone first, if by a nose. For in the United States, while federal funds, at that point, couldn't be spent on *any* work that resulted in the destruction of a human embryo, private funds could. While the vast majority of companies stayed away anyway, afraid of the controversy, a single tiny company, Geron, had been funding a search for the hES cell being done by only two researchers. But those two researchers, James Thomson of the University of Wisconsin and John Gearhart of Johns Hopkins University, were good ones. The upshot was that, mere days before Reubinoff would leave Singapore for Australia with his first real hES cells in his pocket, Thomson submitted a paper to *Science,* reporting that he had isolated the cells in January, nine months earlier. (Gearhart would end up isolating a different cell, the human embryonic germ cell, which was also highly potent.)

The tightness of this race, and the slimness of the margin with which Thomson won it, was breathtaking, especially in light of another irony: four out of five of Thomson's cells were actually from Israel, cultivated with the help of the Haifa lab of Joseph Itskovitz-Eldor, the scientist so admired by Egypt's Barrada. Since 1995, when Thomson published his paper on the monkey ES cell—and when Australia and Singapore scientists were a year into their search for the human version—Itskovitz-Eldor had been pursuing Thomson, hoping he would help him find the human ES cell. In 1997, Thomson had finally agreed, and Itskovitz-Eldor had shipped several Israeli embryos to him in the hands of one of his assistants, who stayed to help. Eventually, only one cell line from Thomson's own U.S. embryos survived to keep proliferating that January, compared to four of Itskovitz-Eldor's. Since Thomson's goal was to create several different lines proliferating for several months before publishing his paper, this means it is very likely he would have submitted his paper much later than August of 1998 if it hadn't been for Israel's Itskovitz-Eldor.

Thus, with the help of one Israeli scientist, the United States would be able to beat by a nose an international team that couldn't get off the ground until it took on an Israeli scientist of its own. Thomson and Itskovitz-Eldor's paper was published in *Science,* making world headlines, in November 1998, just as Reubinoff and Trounson's international team was quietly beginning to write its paper.

In a further irony, *Science* sent the paper to Pera to review before publication, leaving him unable to tell his collaborators that they were racing for a goal they would not meet. So Reubinoff found out about Thomson's paper on CNN, with the rest of the world. "I said, 'OK, what to do?'" He laughed again. "I can't say I was happy, but that's life, and competition."

Still, such was the determination of the international group that it quickly made up for lost time. The Australian, Singaporan, and Israeli team came out with a paper in *Nature Biotechnology* in April 2000 that was not only the first to confirm Thomson's results with a different set of hES cells—proving that his cells were no accident—but also took them one step further. The group created the first neural progenitors (adult brain stemlike cells) and pulsing muscle cell progenitors from hES cells in a dish. The way they pulled it off: they let the hES cells pile up very high, then removed some of their nutrients. The cells began to differentiate. Presumably, this mirrors what goes on in the embryo: when at first hES cells begin to pile up, they tell each other to stay hES cells. But when they reach a certain magic number, perhaps the number required to create an organism, they send each other chemical messages indicating it's time to move and differentiate, to stop nurturing each other and start working together. This is, at any rate, how the international team persuaded their cells to mature.

Indeed, a few months earlier, two Israelis—Benvenisty and Itskovitz-Eldor—had found something similar, in what was the first hES cell paper published after Thomson's. The two placed Thomson's cells (which came from Israelis) in suspension in a bottle, instead of on a flat plate. These formed strange three-dimensional, pulsating embryoid bodies, with cells that differentiate to form the placenta outside and cells that differentiate to form the body inside. The team had thus found something similar to what the international team had found. Nascent stem cells are able to "raise" each other in a number of ways, not just by exchanging chemical messages, but by "touch," or the feel of their place within a three-dimensional crowd either suspended in a bottle or on a plate.

Israel was clearly on a roll. Over the next four years, between Thomson's groundbreaking November 1998 paper and the end of 2003, the tiny nation would put out the vast majority of the world's original lab papers on hES cells, alongside the United States. The scorecard would be the United States 27 and Israel 26, with the U.K. coming in a distant third, with 7. Furthermore, Israel, a David among nations, would score just as many world firsts as the U.S. Goliath—all with *half* as many lead scientists (Israel, 4; United States, about 10). All this is astonishing when you consider that scientific papers with U.S. authorship, according to the *Journal of Translational Medicine,* traditionally attract about half of all citations in Western journals (over 30 million in the last decade alone, followed by England with less than 6 million).

On the phone in August 2003, Reubinoff noted that one way Israel has kept its lead (besides the jump given to it by Bush's restrictive U.S. policy) is to act like a company, as a nation. Israelis split up the job of differentiation, so there would be no wasteful overlap among scientists. Itskovitz-Eldor and Lior Gepstein took heart cells; Benvenisty took gut cells; Skorecki took liver cells; Reubinoff took neural cells. Outside of that, however, they're allowed to work on what they like, which led Reubinoff to become one of the first two scientists in the world to stably transfect (or infect) human ES cells with lentiviral vectors, that is, viruses that could ferry new genes into ES cells accurately. Thomson was the other. (One of the cleverest ways to get a gene into a cell is to load a partly disabled virus—whose job is to invade cells—with that gene, and aim it at the target cell. Like poison on an arrowhead, the gene is ferried into the target by the invading virus.) "When you transfect cells with other viruses you can get relatively low expression, only one copy of the transgene enters the genome of the cell and gets stable expression," Reubinoff noted in September 2003. "With lentiviruses you get a *few* copies into the cell and better expression. May be real advantages."

More importantly, he noted: hES cells will offer more advantages in gene therapy all the way around, for a simple and potent reason. In France, in 2002, doctors announced that the world's first successful gene therapy experiment, wasn't. They had inserted, via a partially disabled virus, genes into stem cells to replace a gene missing in children with defective immune systems. Eight out of ten of their child patients were cured of a formerly lethal disease. But a few months later, two of them got leukemia. (And several months after that, a third child would get leukemia.) In at least two of the cases, the virus had entered the DNA of the kids' cells in the wrong place, too near a proliferation gene. The virus had, in other words, kicked on a proliferation gene that couldn't be kicked off.

But the French used limited adult stem cells, which was one of their problems, Reubinoff conjectures. Since those cells replicate a small amount of time, it's too difficult to winnow out of a petri dish the one cell that may have received the gene in the right place. So the French doctors just had to guess. "But ES cells have a huge advantage because you can infect them with lentiviruses, and then you can take care to characterize the site of insertion," Reubinoff said excitedly. (Translation: you can waste an enormous number of cells looking for the perfect cell because you can always make an enormous amount more.) "With the French, it was more problematic, because they can grow the stems for more limited time. But with ES cells, you can characterize very carefully the insertion site."

Just in case, Benvenisty and Itskovitz-Eldor worked together to put out the first paper showing that hES cells can be transfected using a similar technique—and they added a suicide gene. So should a transfected hES cell *ever* pull a stunt like the French cells, it will die. It's clever: a suicide gene is inserted into the cells which, turned on, kills the cells. But it can only be turned on if a certain antibiotic is given. And again, since hES cells proliferate so robustly, you can run through a huge number to find the ones whose suicide genes slipped into the exact right places in the DNA strand.

But Israel's scientists have not just divvied up duties, they have worked together in a disciplined fashion. Reubinoff worked with Itskovitz-Eldor and Benvenisty on a study that showed that hES-derived cells may not arouse the same immune rejection response as adult stem cells when placed in mice. They may be so immature they sneak in under the radar of the foreigner's immune system. Yet the Israelis also found that injecting overdoses of certain growth factors into the same mice can change the immune-protected quality of the hES cells, causing them to be rejected.

Reubinoff and Benvenisty worked together to show that if a human ES cell is cocultured near a chick embryo, the human cells will show up in most of the tissues of the adult chick. The ramifications were obvious. Beyond the fact that it was now clear that animal/human chimeras could easily be created, the door was opened to curing human embryos of disease before they ever reached the womb.

Indeed, largely because of the collaboration among Israelis (and sometimes among Israelis and Americans), the race that ensued, between November 1998 and the end of 2003, to figure out just what hES cells could do, would turn out breathtaking in the speed, elegance, and sheer deliberative logic of its orchestration. In October of 2000, Israelis and Americans would team up to find the first eight growth factors, or chemical growth ingredients located in all human bodies, that could be used to hasten differentiation of ES cells to early stages of various tissues, from bone to muscle. In November of 2000, a largely different group of Israelis and Americans discovered how to keep an hES cell alive solo, all alone, for a brief while—important so one can exactly reproduce it, and thus guarantee to some degree that all the hES cells that come from it will be just like it (key when placing cells in patients. Pharmaceuticals and patients both like to know exactly what their medications are). And in 2001, Israelis became the first to fully describe in journals how to create primitive insulin-producing and heart cells from hES cells, and, with Australians, substantially improve hES cell freezing.

Furthermore, in 2001, Benvenisty's group was first to describe creating true neurons from hES cells in a dish. Yet another Israeli group discovered a clue to the cells' immortality (a DNA bit that kicks an enzyme called telomerase into gear).

In the United States that same year, Thomson became the first to turn hES cells into primitive blood cells, using an intuitive, all-natural approach that took advantage of the communicative nature of the cells and that had marked the field from the start. Instead of throwing hundreds of molecules at cells, as Big Pharma does to solve many of its problems, to see if one might differentiate hES cells to blood, Thomson cleverly cocultured hES cells *with* adult blood cells, to see if the adult cells would guide hES cells. They did. This was key for many reasons, including the fact that adult blood stem cells are hard to proliferate, so patients rarely get enough (for, say, cancer treatment). (Furthermore, in mice, it is proven that adult blood stem cells are more worn out than fetal blood stem and embryonic blood stem cells. Embryonic blood stem cells repopulate mouse bone marrow seven to nine times over, while fetal (older embryonic) stems do so four to six times, and adult stems, up to three times. And human fetal and umbilical cord stems have much telomerase (an enzyme that prolongs the lifespan of cells) while adult human blood stems have little. University of Illinois oncologist Ronald Hoffman says adult blood stem cells can repopulate an embryonic liver, but retain many of the flaws of their tired adulthood, resulting in an inability to proliferate, and late graft failure.)

The United States's Geron that year became the first to show that neurons can be made from hES cells even after proliferating for six months, and can pass on proper voltage. And Thomson's group tied with Reubinoff's for being first to pop neural cells from hES cells into mice and watch them take root.

In 2002, things slowed down a bit, if each discovery was key. An Israeli and American team, led by Itskovitz-Eldor, made human veins from hES cells sprout in mice. Itskovitz-Eldor's team finessed the creation of heart tissue in a dish from hES cells, luring the cells into becoming two different kinds of beating heart tissue, to provide a model for Big Pharma to try heart drugs on. Geron again reported on longevity: hES cells could be turned into heart cells even after 50 passages (or 260 population doublings) and counting. Geron also got cultures of cardiomyocytes that were 70 percent pure. Thomson created from hES cells cells that form placenta. And two different U.S. teams reported on "stemness" genes shared by all stem cells, adult and embryonic.

In 2003, Thomson and Geron were finally joined by more U.S. teams. By then, delays caused by congressional and presidential deliberations, and the culturing and scale up of the few presidential lines NIH-funded scientists were allowed to use, had finally ebbed. So 2003 saw more than 10 U.S. scientists get into the act, which would result in the United States finally pulling ahead of Israel in terms of published hES cell work: 10 papers to 14, by the end of 2003.

Still, in 2003 Israelis were the first to: pop a suicide gene into hES cells; create vessels in a dish from hES cells; grow hES cells on a scaffold to start shaping organs (the latter with MIT). On the other side of the ocean, Americans would be first to: make liver cells from hES cells; find the Pumilio-2 gene, which helps hES cells become eggs and sperm;

compare genes in mouse and human ES cells; nudge hES cells toward producing skin cells; and compare gene patterns in hES cells and tumors.

Working around different odds, the two nations worked together and apart to lay a solid base for a brand new science. The year 2004 promised to be a big one. If scientists on both sides of the ocean could just identify some of the truly critical growth factors hES cells send each other to differentiate to different states, they would finally be able to start controlling the potent, willful cells. This would mean transplants could begin to be considered.

And that is exactly what would happen.

•   •   •

BUT ON this day in October 2003, a mere two months after his elated phone call, Reubinoff looks troubled. After an hour of talk updating his progress on his latest project—turning hES cells into dopaminergic neurons to fight Parkinson's—and discussing the variety of promising approaches toward neural stem cell work being done by other scientists, he falls as silent as he had been at the outset of the talk. Israel's Ministry of Health has only been able to give its four star scientists $600,000 to $700,000 a year, he eventually explains. Each group has been able to supplement those funds in other ways; in Reubinoff's case, this was funding from ES Cell International (ESI), the stem cell company his international group had founded back in 2000. But ESI has just decided to stop funding its four founding scientists—the two in Australia, along with Reubinoff and Bongso in Singapore—to focus all of its efforts, in-house, on pancreatic islet cells to cure diabetes. It is, indeed, pulling out of Australia altogether and consolidating to save cash in Singapore. Singapore has a lot of money, but it does not have anything close to U.S.-sized reserves, and ESI has decided it has to react by intensely focusing on one area.

Furthermore, it is doubtful at this point that Reubinoff can look to the United States for substantial funds, for it is not *spending* substantial funds on hES cells. While the presidential cell lines are at last being slowly distributed, the government is still only spending some $15 million a year to help scientists study them (compared to $150 million a year it's spending on adult stems). Unless Reubinoff can find some new source of funding, one of the four leading dynamos in the world's attempt to pick up where the United States left off will be slowed to a crawl.

As he takes a turn through his immaculate labs, which are painted white and blue, possess a seafaring décor with shiplike portholes in every door, and are staffed with 18 scientists—"big by Israeli standards"—he expresses hope that the Israeli Ministry of Industry will come to the rescue. But it's clear, he says, that "U.S. big business won't, not in the near future. It is still afraid, I think." It is not just still afraid of the controversy. "It understands this work will take years." Yet of course there is a profound catch-22 here: it will take even more years if Big Pharma doesn't get involved upfront.

It is becoming clear why Itskovitz-Eldor has been courting Egypt's Barrada. The Israelis for years were considered such saviors of ES cell science that they had cloistered lab rats worldwide braving "the situation" to make pilgrimages to their doors. But Israel itself is now in danger of running out of money. The Israeli hES cell saviors may soon need saving themselves.

Despite the spectacular odds, however, almost exactly one year from now, Reubinoff will become the second scientist in the world to prove that hES cells work when he reports in the highly regarded Western journal *Stem Cells* that hES cells can alleviate Parkinson's disease in rats.

And the first scientist in the world to prove hES cells work? Another Israeli, Lior Gepstein, a cardiologist working in the Technion in Haifa.

4

IN THE BEGINNING

ON THE HOSPITAL SIDE of Hadassah University in Ein Kerem, the waiting area outside the office of Shimon Slavin is full and loud with the sound of doctors arguing in a conference room nearby. Clearly, this is not unusual, for none of them jumps up to close the door. At length, a doctor with longish white hair and khakis that drag a bit at his heels comes bursting out. He says distractedly, "Hello, just a minute," to some patients and disappears. He reappears carrying a slide with blood on it, chats with the patients, then says to a waiting reporter, "Hello, just a minute," and disappears again. Some 45 minutes later, his voice comes booming out of his office. "Hello, are you there?" As soon as the reporter jumps in the door, one of his phones rings, and he begins two conversations at once. "Hello? Sit. Not you. You."

The contrast couldn't be more striking between Reubinoff, a young 5-year leader in the uncertain new world of hES cell science, and Slavin, a middle-aged, 30-year leader in the well-established, 50-year-old world of adult stem cell science. It's not just that Slavin, as founder of the bone marrow transplant unit here—Israel's first—has a much bigger office with windows that look out over the hills of Jerusalem. It's also the fact that those windows have, taped to them, x-rays of a patient's chest. To save time, Slavin simply attached them to the window to use the sun as his bulb, instead of bolting out to find some lightbox in a lab.

While Reubinoff's office was utterly tidy, Slavin's office is an organized mess, with pictures of many decades of patients jammed into every bookshelf and counterspace next to piles of journals, books, awards. The walls are alive with bright paintings—some made for him by his patients. With his white hair worn to his collar, and his white lab coat bloodstained and opened to reveal a checkered shirt, Slavin fields alternating phone calls from his cell phone and his desk phone with the air of a man with his feet up on his desk, even though both are (somewhat) firmly planted on the floor. "Da, da," he says into one phone. "What?" he barks into the other. For a moment, he holds both phones to his checkered chest. "Would you like something to drink?"

Shimon Slavin acts like a man who doesn't have to sweat the small stuff, and that is almost exactly what he is. Like Reubinoff, he is a man of many firsts. Unlike Reubinoff, his field is decades old and has thus advanced enough so that most of his firsts are clinical firsts, now. (As University of Minnesota blood stem cell expert Dan Kaufman puts it: "Take two stem cells and call me in the morning? We're not quite there yet. But we *have* been doing stem cell therapy for [over] 30 years—via bone marrow transplants.") Most every experiment Slavin conducts with his hematopoietic (blood) stem cells (HSCs)—the most important cells in bone marrow transplants—carries weight and is taken that way. For the science world understands enough about these particular adult stem cells now that blood stem cell researchers have long been given leave to conduct their experiments on humans, not mice, saving much time. Over 50,000 bone marrow transplants are conducted each year now.

The cumulative effect: Slavin is in an enviable position. He understands his finicky and strange cells so well he has been able to conduct (with an Italian team) the first truly successful systemic gene therapy trial; has improved blood stem cell transplantation immeasurably; is actively bypassing chemo and radiation by using stem cells and lymphocytes (killer T cells, which come from blood stem cells) as targeted cancer killers engaging in "biological rather than chemical warfare" as Slavin likes to say. And, in his most recent work, along with that of a very select handful of other immunologists, he could soon hand to Reubinoff et al. an answer to the biggest problem hES cell scientists face, the only one they're not positive they can solve with money and time: the immune rejection problem.

As noted in the preceding chapter, it was an Israeli team that discovered that hES cells are not special in one critical way, that is, like all cells, when placed in a body different from the one they came from, hES cells will probably almost always will be rejected, eventually. This means that most recipients of hES cells from IVF clinics will have to take crippling, life-shortening immunosuppressive drugs, unless someone can figure out a loophole. Many would like to try to see if that loophole could be closed via therapeutic cloning, the controversial, fascinating approach pursued so eagerly by South Korea, to be discussed in a later chapter. But Slavin is finding that the white horse that may come first to the aid of embryonic stem cells in this regard may be other stem cells: adult stem cells.

In the late 1960s and early 1970s, all this wasn't necessarily in the cards. At that time, Slavin was completing his training in clinical immunology at Stanford University and the Fred Hutchinson Cancer Center—and the adult blood stem cell had not quite technically been discovered. But in actuality, it had, and was being utilized via what was then called bone marrow transplants, the first of which occurred in humans in 1956. One of the most exciting qualities of stem cells are the ever-increasing findings that they have been at the heart of some disease, even several cures, all along, and we didn't know it; that stem cells may be complicit in some of our greatest afflictions and our greatest cures. The story of the HSC—the first stem cell ever discovered and the first to be used in humans—is every inch that story. And it is one increasingly being reviewed by scientists trying to find adult stem cells in other tissues.

Since the age of the Greeks until the 1700s, it was believed that the body was made of four "humours"—phlegm, choler, bile, and blood— with blood considered the paramount humour, or "bearer of life," undoubtedly because it was so clearly the most active part of us. Unlike the brain or the skin, blood was always clearly on the run, always working, always flowing through the body like a river and always self-renewing like a river (making it also obviously chock-full of stem cells, although no one would know that until the twentieth century). All this activity is why, when disease struck, the chief remedy for patients for thousands of years was bleeding. Blood traveled everywhere in the body, as disease did, so somehow disease must be carried by it. Since blood could self-renew, why not just drain everything out and start again?

It would turn out to be true, of course, that blood was the mothering bearer of life, carrying nutrients, hormones, and growth factors to all parts of the body, keeping it alive. But the fact that it was also the body's father-figure, the defender of life—otherwise known as the immune system—would not be known until the 1940s.

People knew the immune system existed. It had been noticed since 430 BC that, when people survived a disease, they often rejected it the next time around, rarely contracting it again. (Wrote Thucydides that year of an Athens plague: "The same man was never attacked twice, at least fatally.") And it had been known since Jenner's 1796 smallpox vaccine that the body can be tricked into rejecting—or becoming immune to—disease by exposure to inactivated forms of it.

But no one knew that it was the churning, "life-giving" blood that was responsible for the magic trick of immunology until the 1940s, when WWII prompted a need for skin grafts. At that time, Oxford University's Peter Medawar noticed that when he tried to engraft skin from one soldier to another but found it was rejected, it was always swarming with blood cells. Medawar moved on to make the Nobel Prize–winning discoveries that blood cells were responsible for rejection of things "foreign" to the body like diseases or skin grafts; that blood cells could be tricked into rejecting disease; and that the body, via the blood, could also be tricked into *accepting* things foreign to it in the fetal or newborn stage (if not, as yet, in the much older adult phase).

Medawar's defining experiment was a fairly simple one. Conjecturing that perhaps the young and pliable immune system would be a *pre*-immune system, and thus would accept as "self" anything that came its way, Medawar injected mixtures of mouse cells into unrelated fetal or newborn mice. When the mice grew older, he gave them skin grafts from the same unrelated mice who had donated the cells. The grafts were accepted.

This discovery, published in 1953, was considered the most important in transplant immunology to date, especially when it was considered in light of a discovery by Ray Owen that fraternal twin cattle sharing a blood system in the womb later tolerate two different kinds of immune cells; that is, they possess two different immune systems. What this all seemed to mean was that we should be able to not just trick our bodies into rejecting disease, but also trick our bodies into *accepting* new foreign cells or organs—and maybe even trick our bodies into rejecting cancer, which is not inherently foreign since it's made of our own cells, thus normally outside the immune system's purview. The immune system had some kind of ever-young, spectacular pliability to it. It was exquisitely trickable.

New generations of immunologists were inspired. Granted, both Medawar and Owen had found that their animals accept only foreign cells when they are in the fetal or just-born phase. But surely a way could be figured out to mimic that early developmental environment in the adult body—to trick the adult immune system into behaving like a fairly naive one.

Israel's Slavin was among the inspired. Another now-prominent figure in immunology, Steve Rosenberg from the National Cancer Institute (NCI), also grew fascinated with the potential power and pliability of the adult immune system. Fascinatingly, while the two would proceed to have similar careers—even sharing some staffers—Slavin would become part of a branch of immunology that would score more successes in part for a very simple reason: it involved stem cells.

As a resident in Boston in the 1960s, the United States's Rosenberg tended a patient who he saw reject his own cancer after a bacterial infection. Deciding that somehow the patient's immune system had not only been aroused by the infection but had also suddenly seen the cancer as foreign since it was present *at the same time* as the foreign infection, he accepted a post as head of the NCI's surgery branch in 1974 and transformed it into the United States's leading solid cancer immunotherapy lab. (Immunotherapy is a beautiful approach that involves co-opting the immune system's own soldiers—often its lymphocytes, T and B cells, which come from blood stem cells—to fight cancer and infection, instead of using artificial chemicals.) Year after year he ran back and forth in one of the biggest bench-to-bedside complexes in America. In one approach, he removed the T cells that were most naturally active against patients' tumors—they could be called Super T Cells—replicated them in a dish, then injected armies of those back in. In another approach, he removed from patients the *tumor* bits that had aroused those Super T Cells, deadened them, then reinjected them accompanied by deadened foreign viruses and other foreign substances, trying to trick patients' immune systems into believing that solid cancers were as foreign as the accompanying substances.

It sometimes worked very well, but rarely for a long time. Sometimes complete remissions could occur, but the cancers almost always returned. This suggested two things. First, cancers are made of all kinds of different mutated cells, so no one breed of T cell was going to be able to erase all the different kinds of solid tumor bits in a cancer. And second: those mature T cells that did work died out after a while, as almost all mature cells do, leaving the cancer free to run wild again.

Over in Israel, Slavin, along with his branch of the field—the blood cancer field—saw far greater success in the same time period. Indeed, blood cancers (leukemias and lymphomas) went from being as incurable as other cancers to highly curable, many up to 85 percent curable, during the course of his career. Slavin, who started Israel's first and largest bone marrow transplantation center in 1978, focused on trying to trick the immune system's blood cells into both accepting organs *and* rejecting blood cancer. Partly because, as it happens, both of these approaches involve highly pliable stem cells, not restricted adult cells, he has had an easier time with the "trickery" required of immunologists.

It was while Slavin was at Stanford University in the early seventies that, inspired by Medawar and Owen, he took a pioneering step. How could one make an adult immune system enough like Medawar or Owen's fairly naive immune systems to blindly accept foreign cells or organs? Since 1961, immunologists had known that the reason blood was so lively and renewable was due to something they would come to call the stem cell, a strangely young and pliable cell that was far more rare in the adult bone marrow than in the fetal bone marrow, but that could reconstitute a mouse if its bone marrow was completely destroyed by radiation. (A famous 1961 experiment by Canadian researchers would strongly hint at the presence of these cells, which were hematopoietic or blood stem cells—if they would not be officially characterized by cell-surface markers until 1988, by Stanford University's Irv Weissman.)

Slavin wondered: Could one create a new immune system in an adult out of some of those pliable blood stem cells? Would the old immune system be accepting of the new if its mature attacking cells were temporarily wiped out, leaving some of its own pliable stem cells behind? He forcefully irradiated adult mouse lymphatic systems, temporarily weakening their ability to produce mature attacking lymphocytes. Then he gave them foreign stem-cell-containing bone marrow transplants—followed by skin patches from the same donors. The mice accepted, and sported, their new skins. It netted Slavin his first paper in *Science,* in 1976. It was too brutal for humans as yet, and worked for some complex reasons. But it indicated that adult stem cells could indeed create new immune systems in other adults.

Meanwhile, in the 1970s, some doctors had been similarly re-creating the immune systems of blood cancer patients whose immune systems were destroyed by chemo and radiation, although doctors tended to use the patients' *own* stem-cell-filled bone marrow, which was removed before the killer therapies, then returned. But when that wasn't possible, experiments like Medawar's and Slavin's gave doctors courage to start trying foreign, if closely matched, bone marrow from siblings. It was toxic, but as time passed, some realized that patients receiving bone marrow transplants from siblings sometimes seemed to fare better than patients getting *their own cells* back. And they sometimes seemed to acquire second, brand-new immune systems, too.

In 1990 a paper came out that explained some things. The University of Wisconsin tallied over 2,000 patients and reported that, fairly consistently, blood cancer patients who had received stem-cell-containing bone marrow grafts from *mis*matched donors survived longer than patients receiving grafts from *exactly* matched donors—when they weren't killed by the treatment. Also, mismatched patients surviving longest were actually those who had gone through, and survived, graft-versus-host disease, where the foreign graft attacked the host. What this, added to other observations, seemed to mean was that, indeed, the foreign stem cells in those bone marrow grafts were building new, secondary immune systems in some patients—which was something mature cells alone could not do. Then the more mature foreign T cells in those grafts began attacking patients' tumors because they saw them as foreign (patients' own T cells could not see their cancer as foreign, since cancer is made from our own cells).

Another 1990 paper showed that this approach could be improved upon by the *deliberate* addition of mature killer T cells from the same donors of the stem cells, after giving the donor stem cells time to settle in. The reason this worked on blood cancer patients was clear enough, at least for Slavin, who had suspected as much a few years earlier and had added T cells to his stem cell grafts in 1987, becoming the first doctor to do so in the clinic. (Although it is widely agreed that the first scientist to officially report in a journal on this approach, called donor lymphocyte infusion [DLI], was Hans Kolb, then of the Fred Hutchinson Cancer Institute, in 1990.)

"It has been known for decades that if I have a cancer, and I try to give it to you, I put a tumor in you, you won't get it. My cancer is killed by your immune system," Slavin says, waving the pointer he uses for presentations in the air as if leading an orchestra. "Our immune systems can't recognize our own tumors as foreign, because they are made of our own cells—*but other people's immune systems can.* When I thought about this, it was such a simple idea I couldn't believe it. I couldn't believe no one had thought of it before. We should use *other people's stem cells and T cells* to cure our cancers."

So three years before the 1990 paper came out, Slavin treated a young boy who had received his sister's stem cells, but relapsed three times. He waited until the child's acute lymphoblastic leukemia was so advanced it caused a bump "like a unicorn's" on his forehead and a growth on his neck. Then he tried the sister's stem cells again, but this time, over the course of six weeks, he infused the boy with his sister's mature lymphocytes as well. "After a few weeks, it worked. The lump grew smaller, the neck mass disappeared. He was cured."

Slavin endured some of the cynicism that is often seen in science. At a conference six months later, he told the story to one of the great leaders in the field, E. Donnall Thomas. Thomas, a University of Washington scientist who would win a 1990 Nobel Prize for pioneering the modern bone marrow transplant, told the assembled about Slavin's discovery. He added that it was only six months out; it could be a fluke; the story "must be taken with a mountain of salt," Slavin says, wincing at the memory. "But a few years later, at a conference at the Waldorf in New York, Donnall Thomas got up on the stage and began his speech with an apology 'to Shimon Slavin, who was right.'"

Slavin beams, brandishing his pointer with a "ta-da" flourish. Slavin's cured boy was described by Johns Hopkins immunologist Ephraim Fuchs in the March/April 2002 issue of *Cancer Control* as "the first patient to receive DLI for a hematologic malignancy in relapse after BMT [bone marrow transplant]. . . . He ultimately obtained a sustained complete remission."

Still, the process of bringing an adult immune system back to a kind of pliable young state in this way—pummeling it with drugs to create within it a fairly unthreatening biological space in which new foreign stem cells could grow—remained brutal. The approach could kill. It was clear that foreign (or "allogeneic") stem-cell-containing bone marrow transplants needed to be gentler. The field focused on that, and by the end of the 1990s, the therapy was made more tolerable by lowered doses of radiation and/or chemo, and gentler immune-system killers. It was called the "nonmyeloablative" approach, the best of which was reported on by Richard Champlin of the MD Anderson Cancer Center in 1997. It allowed many more blood cancer patients to be treated with allogeneic bone marrow transplants.

The bottom line: It was once thought that the biggest advances in oncology in the twentieth century were due to chemo and radiation. But a good percentage of those successes actually were due to the stem cells, and the foreign killer T cell soldiers, that accompanied those therapies, many now believe. Says Slavin, what he and others were discovering was that, "The immune system is the best oncologist"—and that, perhaps, the *foreign* immune system is the best oncologist of all.

Wrote the team of stem cell pioneer Rainer Storb in early 2003: "During the past 50 years, the role of allogeneic hematopoietic cell transplantation [blood stem cells from another person] has changed from a desperate therapeutic maneuver plagued by apparently insurmountable complications to a curative treatment modality for thousands of patients with hematologic diseases. Now, cure rates following [allogeneic transplants] with matched siblings exceed 85% for some otherwise lethal diseases, such as chronic myeloid leukemia, aplastic anemia, or thalassemia." Stem cells were apparently a key reason that many blood cancers became among the most curable cancers.

And the stem cell tactic may soon prompt major successes in the *solid* cancer field, many believe. Taking note of blood cancer successes, including Scripps Research Institute work giving biological space to T cells,

Rosenberg borrowed a part of the approach. Before trying to trick his melanoma patients' own immune systems into attacking their cancers by returning to them their most potent adult killer T cells—as he had been doing for decades—Rosenberg first gave them nonmyeloablative therapy to make "biological space" for the new T cells. It worked. His 2002 *Science* paper reporting on the approach represented one of his greatest successes.

Many in the immunology field believe the *entire* approach should be borrowed for solid tumors like melanoma. The autologous (self) approach of Rosenberg et al., which was about improving the quality or quantity of one's own T cells and returning them, and which dominated solid tumor immunology, has begun to seem unnecessarily laborious. Why try to fight mother nature, trying to trick her into believing her own cancer is foreign? Why not do what the blood people did, which was to run *with* nature, first irradiating that bone marrow to create a new space, then implanting *foreign* stem cells and *foreign* killer T cells that by design will see those tumors as foreign since they come from a different person? The foreign stem cells would settle in to create that dual immune system—either for the short or long run—so that their fellow foreign T cells, added later, wouldn't be rejected by the old immune system. Then the injected T cells could go after those tumors at their leisure.

A number of researchers around the world are trying the foreign blood stem cell/T cell approach now on solid cancers. In 1984 and 1997, Slavin was the first to successfully try the approach on solid tumors in mice and, in 1996, he was the first to observe that this had apparently occurred as a side benefit in a patient with a solid tumor. In 1999, Richard Childs of the NIH pulled off the first deliberate attempt to attack a solid tumor in patients this way, when he threw several kidney cancer patients into remission with the approach. "Blood stem cells may be the most potent cancer immunotherapy," Childs would conclude in January of 2004.

Very recently, blood stem cell researchers have begun adding in some of Rosenberg's tricks. For example, after nonmyeloablative therapy and the administration of foreign stem cell and foreign killer T cells, some researchers wait. Then they *pull back out* those foreign donor T cells that have plunged most deeply into the tumors. They replicate them, and put them back, creating "super specified donor T cell armies."

(Each T cell attacks a different kind of foreign invader, or a different protein on an invader's surface. In a petri dish, you can make perfect cloned armies of protein (or antigen) specific T cells.) Allogeneic stem cell researchers are making headway with these and other Rosenberg-like approaches.

And now that oncologists are, at last, increasingly able to create those dual immune systems with blood stem cells without killing their patients, the eyes of many in the blood transplant field are on the ultimate prize: double transplants in patients. Slavin, once again, performed the first one, in 1985 giving a patient nonmyeloablative therapy, injecting foreign donor stem cells into her to create a dual immune system, then giving her a skin graft from the same donor. It worked.

Massachusetts General Hospital (MGH) followed up on this, in 1999, with the first *functional* double transplant: stem cells plus a kidney from the same donor. Unlike with cancer, it can be critical that few foreign killer T cells creep in when seeking organ tolerance. So MGH came up with a variety of ways to remove contaminating mature T cells. As time went on, MGH's approach began working on patient after patient.

Overall, there is very often still great toxicity associated with the process of using foreign stem cells to mount an attack on cancer or to create dual immune systems. Indeed, while more than 5,000 people in the United States receive allogeneic bone marrow transplants annually for cancer, between 500 to 1,000 die from graft-versus-host-disease. While it's possible that using only tumor-specific super T cells will solve that problem for cancer, this can't solve the graft-versus-host problem in double transplants. MGH is doing astonishingly well. Other institutes are trying. Tweaking is being done worldwide.

It may take some time to work out the kinks, scientists like Slavin believe. But once perfected, this approach could provide a conclusive answer to the most difficult technical challenge in the hES cell field: the problem of hES cell rejection. Once adult stem cell scientists figure out how to create the perfect state of mixed chimerism (dual immune system), all hES cell scientists should need to do is create HSCs from their hES cells, then create dual immune systems out of those HSCs. Ever after, the patient who receives those hES-derived HSCs should accept any organ made out of that same line of hES cells.

But Slavin hasn't been content to just pursue and eradicate cancer cells and plot the creation of dual immune systems for organ transplant patients. Like a kid in a candy store, bemused by the many talents of his cells, he moved on to see how stem cells could improve gene therapy in 2002. Partly because he used the gentle "nonmyeloablative" stem cell approach, his team became the first to conduct a truly successful systemic gene therapy trial, after a run of largely failed trials that began with, ironically, Slavin's doppleganger in the solid tumor field, the NCI's Rosenberg, in 1991.

Throughout the eighties Rosenberg had had some success tricking immune systems into going after their own cancers. But his successes using patients' own mature T cells were, again, impressive but limited. So when French Anderson, a geneticist renowned for his gene therapy work in mice, approached Rosenberg and another clinician, Michael Blaese, to perform the world's first gene therapy trial, Rosenberg jumped at the chance. Perhaps if he could insert the gene for a certain potent anticancer cytokine, or protein, called interferon into his patients' mature T cells, they would more ferociously attack cancer.

In 1991 Rosenberg's team, which included a postdoc from Slavin's lab, was the first to prove that gene therapy could be safe, when they simply inserted a "marker" gene into mature T cells and gave them to patients. A few months later, they inserted the cancer-killing interferon gene. As with most other solid tumor therapies focusing on adult T cells, it worked briefly, then stopped. This experiment was followed up by one in which children with T cells that lacked a key gene (the adenosine deaminase [ADA] gene) had a new copy of the gene inserted into their T cells. This worked for a while and then essentially stopped, as well.

For the next decade, approximately 4,000 patients underwent gene therapy trials—many utilizing mature cells—that, ultimately, largely failed. The reason was a mystery until, in 2001, Slavin had some ADA-deficient baby patients of his own. You can cure this disease if a close blood match is found with stem cell transplants. But these babies had no close blood matches.

"In LA, London, Japan—the whole world had tried again and again to try gene therapy on children like this. But no." Scientists had even advanced to inserting the genes into much more permanent and multipotent cells, blood stem cells, to no avail. But they had never tried the full, modern transplant approach—that is, to issue a certain specific nonmyeloablative therapy to tamp the children's immune systems before administering their stem cells. "Maybe what we needed to do was to create that biological advantage again, to suppress the old abnormal guys, this is what I thought," Slavin says. So, with the aid of Claudio Bordignon, a renowned gene therapist in Italy, "we decided to give the cells the biological advantage in there, the settlers' advantage, just like the settlers in Israel."

Once again, it worked. The fact that it was the first truly successful gene therapy experiment was matched, in reporters' eyes, by the fact that one of the babies was a Palestinian. The article made *Science* magazine in 2002 and attracted worldwide headlines. The trial was "the first complete clinical success with no side effects," European Society of Gene Therapy board member Fulvio Mavilio will note in June 2005. By June 2005, "six patients have been treated, all are cured with no side effects, and the follow-up of the first two patients exceeds four years now." Furthermore, Mavilio will note, a U.K. trial using the same approach will be as successful.

Yet even this isn't enough for Slavin. As he jumps about in his seat, pulling pictures of his patients and cells up on his laptop, waving that academic baton, talking of all the other uses for his magic cells, he looks more and more like an aspiring symphony conductor pretending his stereo is his orchestra. He is trying the approach, he says, on patients with autoimmune diseases whose immune systems turn on their own bodies: multiple sclerosis, rheumatoid arthritis, lupus. He and others are making strides fighting all these diseases with blood stem cells.

And taking his cue from the work of all stem cell scientists who are finding that, in order to get stem cells to do what you want, it often simply takes a return to the pharmacy of the body, Slavin has added mature crushed *bone* cells to his blood stem cells. This way, he has been able to cause bone to grow in the skulls of patients who have had surgery. Bone is notoriously hard to grow. The cells that are capable of doing it are stem cells found in the marrow. Slavin at first tried just inserting bone marrow stem cells into bones with holes in them: nothing. So he thought about the bone marrow and decided to re-create that environment with some crushed bone. "Stem cells are seeds, crushed bone is the soil, this is how I think of it," he says. It has been working so well that Slavin has formed a company around it.

Like others, he is also working on administering blood stem cells to heart attack patients. The approach is highly promising, as will be seen in a later chapter.

Slavin's cell phone, which has been ringing throughout the interview, rings again. "Da, da," he says, then apologizes, saying he must go to his father, who is ill. He slings a bag over his shoulder, then leads the reporter through a number of now-quiet halls, and a completely darkened series of back rooms, to the parking lot. His expertise in maneuvering through these rooms is that of someone who goes home in the dark, after everyone else has gone, all the time.

*   *   *

REUBINOFF HAD advised against venturing into the Old City, the heart of three world religions, alone. "People—including Americans— have been killed in there." But for anyone trying to figure out how or if it's possible for a stable stem cell industry to be established here, let alone retain its leadership here, the Old City would appear to be the center of the storm, the place to go.

This is not, indeed, a thing many tourists seem to be doing lately in October 2003: the Old City's tourist center has been shuttered because of the violence. This will almost immediately become problematic. The way to all the Old City religious sites is through an underground warren of bazaars, filled with men yelling, at first cheerily, then increasingly angrily, for the few customers' attention.

Scores of cavelike warrens offer an often tacky array of goods, from plastic Jesuses and Marys, to T-shirts offering support to post-9/11 America. They also offer an enticing array of ancient cookware, incense, oriental rugs, ornately scrolled boxlets. That the steps leading down into the labyrinth of the Old City have been worn as smooth as a baby's skin is an awe-inspiring testament to the billions who have come here just to stand at the nexus of the world's major religions. But getting lost in this warren can quickly become a frightening experience, as lone customers are followed and jeered at by angry ignored stallmen when entreaties to buy are declined. The insults that fly freely here, where all the world's religions meet, can alternate between "You must be an American" to "You must be a Jew."

The gate leading into the Muslim quarter, by contrast, is full of lines of quiet women wearing *abayas*. Still, Arab soldiers keep all Westerners out: "Not for you." Soldiers in camouflage lounge outside every entrance to the Old City. Men in T-shirts warn visitors, as they enter the church where Jesus' cross was supposed to have stood, the ornate Church of the Holy Sepulchre: "Be careful, you will be robbed." The old man who leads visitors into a cave where Mary was supposed to have been born swears at a few sheckels tip. Children kick balls at visitors' backs and shoot cap guns at them.

And to walk even around the outside walls of the Muslim section is a danger at this time. It is a walk that brings into view the Mount of Olives, the Dome of the Rock, the Jewish Cemetery. Yet cars constantly honk at Westerners wandering about on this side of the Old City, the reason for which soon becomes clear: this is near a spot where Ariel Sharon's wall is being built, the wall that will separate Bethlehem, the place where Jesus was born, from Jerusalem, the place where he died.

It is the Wailing Wall, however, that brings the greatest shocks of all. For every entrance is equipped with an x-ray machine, where soldiers ask questions similar to those asked by customs at the airport. Inside the Wailing Wall courtyard, approximately 12 police cars are lined up; men with rifles stand on walls high above; there is barbed wire fence all around. It looks like a concentration camp.

Yet a closer view of the Wailing Wall brings a more pleasant shock. A rolling table holds piles of brightly colored Torahs, free for anyone to take. Women and men in Western garb stand quietly in front of the wall, nodding at it quietly, as others sit in folding chairs before it, also quietly. The wall itself is all the pale ancient colors of a setting desert sun: white-yellows, rose-whites. Doves sit in its nooks and crannies. A rabbi paces the courtyard in one direction, as a soldier paces in the other. It is impossible to imagine any other part of the world where such extremes of hope and despair coexist.

*   *   *

ON CLEARER days, you can see the Rosh Hanikra grottos near Lebanon from here, says Karl Skorecki. He is sitting looking out over his perfect ocean view in his Technion office in Haifa toward the top of

Israel, watching empty tugboats, like footless slippers, glide past his window.

Skorecki is director of the Rappaport Family Institute for Research in the Medical Sciences, which is located in a 13-story institute flanked by towers that make it look like a man flexing his muscles at the edge of the ocean. He is Itskovitz-Eldor's boss and is in charge of all the hES cell work that goes on here. He is also a Canadian who was lured to the Technion seven years ago after he did some groundbreaking work on disease genetics of Ashkenazi Jews. So when he is asked why the Israelis have done so well, he leaps feetfirst into the subject. He has had plenty of reasons to think about this issue.

All religions here are in agreement that human life is a process, rather than a moment, with "ensoulment" only starting to take root around the fortieth day after conception, he says. As a result, the issue of using hES cells taken from spare embryos on the fifth or sixth day of conception, before they would even implant normally in the womb, is "not trivial." But it is not a major issue. This window helped the various religious groups and governmental authorities build a series of sober regulations to guide the hES cell field early on, even before the publication of that first, famous 1998 U.S./Israel paper. Some of Skorecki's own scientists, who were coauthors on that paper, began lobbying for the government to address the issue as early as the mid 1990s.

But it is more than that, Skorecki says. "Most research in Israel is done in the universities, the currency of science is public. So there is less commercial secrecy here. And all this leads, I think, to more freedom in science to be creative in research."

Skorecki became involved with hES cells in 1997. Joesph Itskovitz-Eldor had picked up his famed micromanipulation skills in the lab of University of Wisconsin's pioneer cattle cloner Neal First in 1985. He called the University of Wisconsin's eventual hES cell pioneer, Jamie Thomson, the minute Thomson published the first monkey ES cell paper in 1995, to ask Thomson if he'd like to help him find the human ES cell. Thomson said no, but Itskovitz-Eldor was persistent. In 1997 he held a scientific meeting to honor First in Rehovot, Israel, and invited Thomson to speak. Thomson did and was persuaded there by Itskovitz-Eldor to collaborate in the search for the hES cell.

"That was when I became involved," says Skorecki. "I had begun studies on telomerase here." Telomerase, which keeps cells replicating endlessly, is expressed most strongly in hES cells. "And as both a nephrologist [kidney specialist] and a physician, I knew there was a shortage of kidneys for transplant. I heard Jamie Thomson speak in Rehovot, and I became fascinated. Perhaps I could generate some in-between kidney cell, something like an islet cell, to produce insulin naturally. I kept my contact with Josi [Itskovitz-Eldor]."

Soon after Thomson and Itskovitz-Eldor's landmark 1998 *Science* paper announcing the isolation and continued proliferation of the hES cell, Itskovitz-Eldor, who is receiving $2.5 million from the United States's NIH, gave Skorecki some of the famous cells. "Josi was very generous in his collaborations early on," Skorecki notes. "I was blown away by how fascinating the cells were, how easy it was to get them to beat in culture."

Since, even though he was made head of the entire Rappaport Family Institute, he has been unable to resist focusing on the hES cells that are taking up a greater part of its research efforts. He and Itskovitz-Eldor became the first to create islet cells that seemed to produce a quantifiable amount of insulin from hES cells in 2001. Skorecki is also studying telomerase levels in hES cells. And in a leading journal, *Proceedings of the National Academy of Sciences,* Skorecki and Itskovitz-Eldor (with MIT tissue engineer Robert Langer) came up with three-dimensional cultures that could be key for both organ creation and tumor studies. They created degradable polymers which, when studded with hES cells, formed liver, neural, cartilage, and blood vessel structures.

Finally, using hES cells, Skorecki recently created human embryoid bodies which, when tumors are placed inside, grow blood vessels the way tumors do in human environments. "Until now, we've had to study human tumors growing in mice microenvironments," Skorecki notes.

These latter two are important studies, for as *Science* noted this year, "the days of monolayer culture are over." Embryonic stem cells should be able to form fantastically complex organs. But to do this, it is becoming clear, they will have to be grown while suspended in fluid, rather than laid flat on petri dishes. They will require the super-communicative three-dimensional culture.

He gives a whirlwind tour of his labs, which includes the beating heart cells made from hES cells seen in the prologue to this book. It will turn out that these cells will soon become quite famous. Lior Gepstein, the head cardiologist in the lab, will make history by turning those cells into "biological pacemakers" that take pigs off of mechanized pacemakers. The paper will come out in *Nature Biotechnology* in October 2004—the same month two other Technion researchers will receive a Nobel Prize for protein work. Gepstein's will make history as the first paper to prove almost unequivocally that hES cells can function.

Gepstein will note a few months later, "It was amazing for us, when we first saw those cells beat in the dish. And they beat for several weeks, they don't stop." Even more amazing: his group can insert a green marker gene into hES cells that turns green just before hES cells become heart cells. "So we can see these cells turn green *before* they are about to beat." That is key for purifying hES cells for clinical trials, to keep out of their biological pacemakers hES cells that may form other tissues. "We're still amazed, every day."

Just before he is to leave for Sabbath, Skorecki pauses a long time when asked about his favorite place in Israel. It is clearly something he has thought about a lot. He moved here with his family, partly for his religion. Yet one major reason to stay has clearly been the hES cells that he talks about with a certain near-religious awe. He would not have been able to study them all these years if he had stayed in Canada, which is at this time still wrestling over how to regulate the cells.

Still, it takes Skorecki a while to answer. Directly beneath him are the navy's empty basketball courts. To his right and far below, in Haifa's port, giant red-and-white cranes swing their necks back and forth, carrying boxes from one place to another. They look like brightly colored, drugged birds, trying vainly to shake huge fishing lures out of their mouths.

After a while, he comes up with a list of places: temples, wineries, scuba diving spots. "But the Golan is the place I would recommend the most," he says quietly. "It is living history."

*   *   *

IT CAN take non-Israelis quite a while to reach the Golan from Haifa. Most roads in Haifa are one-way, and they all seem to circle around and then run ever farther up Mount Carmel, never down it. They take the uninitiated driver past one lovely small white suburban house exploding with brilliant pink flowers after another. It would all look almost like any Western suburb save for the occasional sight of a man in a T-shirt throwing an M16 into the back of a compact car after his luggage, his family waiting around him. On the day of the bombing of a Knesset member's car, the journey is made even more confusing, with roadblocks funneling even local Haifans onto unfamiliar roads.

And the drive up to the Golan Heights along the disengagement zone between Israel and Syria is no picnic. In this desert stretch, the land opens up like a vast blank book, with empty roads that are punctuated with unexpected sights. After a monotonous mile, a bombed-out, graffiti-strewn mosque appears and disappears. There is another monotonous mile, followed by what at first looks like a beige-colored pipe protruding out of rocks, then turns out to be, heart-stoppingly, a beige tank sitting on a hill, abandoned or camouflaged—it's difficult to know. Barbed wire lines the road, punctuated occasionally by yellow signs with Hebrew lettering that say, Undetonated Syrian Land Mines. The desert itself is all the colors of the military, with its scattered olive-drab bushes, cows the varied hues of washed-out khaki, boulders that are rust-colored and gun-gray. Then: a bombed-out stucco building, followed by more military-colored desert, another bombed-out, graffiti-strewn, abandoned mosque.

The only radio station that can be heard up here is Radio Jordan, who's young female disc jockey alternately plays show tunes and Dave Matthews and gives the latest news. Today the news focuses on an Israeli raid on a Jordanian hospital to bring a suicide bomber to justice and a Palestinian town directly south of here, in the West Bank, where two suspected "Israeli spies" are, at this moment, hanging by the neck in the middle of the town square. Occasionally, as the road nears Jordan, an immaculate white UN jeep will fly past, followed by a dirty, open, olive-colored jeep stuffed with soldiers who flick lit cigarettes in the road. Near Lebanon, the monotony is broken by the humps of tents and soldiers sitting on the stoops of prefabricated tin buildings, smoking, guns at their feet. Then: a bombed-out church.

To the ignorant traveler, the entire Golan seems like one giant act of static aggression. While the occasional missile is lobbed into the area from Lebanon, the majority of the bombed-out buildings have been here since the 1967 and 1973 wars. The Golan seems designed to frighten, and it can succeed.

Furthermore, to drive back to Jerusalem the safe way—around the West Bank instead of through it—means a 10-hour drive instead of a 2-hour one. It means hours of driving too fast down too-narrow roads circling too-high cliffs that possess too little fencing, urged on by the honking of weekenders clearly sick of this senseless and endless detour. It means eventually gliding past the whitish carcass of Haifa's recently bombed Makom Makim restaurant, with dead flowers littering a parking lot that was littered with the limbs of dead children just two weeks before. It means hours counting, and losing count of, wailing ambulances; hours of more angry drivers. Sliding into Jerusalem late at night after 10 hours of this, sweating with fright, the first-time traveler to Israel can jump into bed convinced that the nation is on the verge of blowing up both figuratively and literally.

The world's most advanced work on hES cells may be going on here. But it can't possibly continue much longer.

*   *   *

YET TO BE GREETED in the Hebrew University office of hES cell researcher Nissim Benvenisty the next morning is to begin to know again that mysteriously, it can. Benvenisty contracted the flu after spending three weeks traveling from Singapore to the United States and back to Israel to share his work at conferences, since many scientists are too wary to come to Israel lately. With his big ears and wide generous mouth, which breaks into a shy smile, he looks at first like a big, tired kid who just wants to go home. And, after apologizing, as if it were his fault, for the detour around the West Bank he explains quietly that many Israelis make that detour, as he does, because they believe the Jordan River's shores do not belong to them.

And even when the conversation returns to his lab, Benvenisty at first sits slumped in his seat, wary and weary, unsure whether his listener is going to care about his stem cell work and unsure whether he cares. But after a while, the inevitable happens again. As he talks, Benvenisty begins to straighten. He gropes for a pen and paper. And in the amount of time it takes to draw a receptor on a cell, a molecule roaming the bloodstream, and the chain reaction of amazing events that their collision sets in motion, Benvenisty is off. Words tumble over words, and suddenly he's perched at the edge of his seat like an eager driver, zipping in and out of the traffic of his own thoughts.

"You see?" he says, beaming shyly.

The aspect of hES cell work that Benvenisty is most consumed by is genetic engineering. His group was the first to slip new genes into hES cells via a stable approach that involves using chemicals to blast a hole in cell walls. He is also pursuing another answer to the big problem of hES cell immunogenicity. While hES cells do not express markers on their surface (called MHC molecules) that make them targets of immune systems, the addition of certain growth factors to hES cell cultures causes these markers to appear. Thus, in the body, the cells should eventually be rejected. "But I think this is the next breakthrough in my lab," he says. "We are trying to knock out parts of the MHC molecule. We have done it. But I still don't know if it's real or not." If it is, he has potentially created a "universal" hES cell, which would be a huge development.

Benvenisty came to stem cells a bit later than the other top Israelis: the day the famous Thomson/Itskovitz-Eldor paper was published in 1998. "After doing a sabbatical at Harvard where I focused on mouse ES cells, I came back to set up a mouse ES cell lab. But I was waiting for someone to find the hES cell. So the day the Thomson/Itskovitz-Eldor hES cell paper was published, I called Itskovitz-Eldor and invited him to my lab."

They started to work on something that was immediately troubling the entire stem cell field: Thomson's new hES cell lines hadn't formed embryoid bodies when suspended in fluid, as mouse ES cells did. "The worry was that if hES cells can't make embryoid bodies, hES cell research would be limited."

They tried the mouse recipe. In mice, when ES cells are suspended in a three-dimensional medium, not laid out on a plate, a ball of dead cells forms in the middle by day 5, and then out of that, phoenixlike, arise mature differentiated cells. But on day 7, nothing happened with hES cells. Ditto as to days 8, 9, 10. "We knew it would probably take longer in humans," he says. The probable reason: humans live longer than mice. Still, "the minute we finally saw that cystic embryoid body with differentiating cells we were extremely happy to say the least." It took two weeks for the strange bag of human cells to form. "And when it began pulsating, that was really wild. It was clear the field was now open to do studies."

Can Israel retain its lead? He talks about Britain with its "brave" regulations; the United States with its money. "I think in the long run the U.S. will have more to say than it already has. Israel has indeed played an important role. We'll see how long it will last."

•   •   •

OUTSIDE BENVENISTY'S lab, passing a whimsical statue of people holding hands in the shape of a menorah, one of Benvenisty's PhD students, Maya Schuldiner, offers one inadvertent explanation for Israel's extraordinary success in the face of such adversity, when she expresses surprise over yesterday's bombing of a Knesset member's car in Haifa. "I didn't know," she says, shoving her hands in her khaki pockets. "Three years ago everybody talked about every incident. Now, not so much. It's too depressing. It's better just to live. You hear the bombs going off wherever you are, Jerusalem is a small city. But if it's not where you are, you just keep doing research."

She lives and works near the Old City, but she also had no idea the fence was going up between Jerusalem and Bethlehem. "Who knows? I haven't been to the Old City since September 2000. Most people haven't been there since September 2000 [the start of the second intifada]. Most people are sick of talking about the fence. Sick of talking about where it should go up, whether it should go up, just sick. They think, 'Whatever.'"

Whatever the government wants? "Yes."

She talks about the Golan, where she served a compulsory term in the army before moving on to stem cells. The army is there, she says, to guard Israel against neighboring Syria and Lebanon. You learn not to look down, she says, smiling wryly. The reasons: "Those cows you saw? They can't read those signs warning about the mines. They walk into mines all the time." As she talks, it becomes clear why such large swaths of the Golan remain barren. The Golan is being preserved as a bargaining chip for future negotiations with Syria and Lebanon. It is also a buffer zone for Israel.

Schuldiner brightens when the conversation switches back to Benvenisty's lab. It is wildly popular among Israeli students. It is a 15-person lab, but "many, many more students want to come work here, and they can't." This is in sharp contrast to the United States, where NIH officials say scientists are not coming forward to apply for grants for the presidential lines, intimidated by the controversy.

The student looks bewildered when she hears this. Her thoughts are clear. A lot of Israelis would do anything for the controversy in their lives to be limited to a petri dish of cells.

*   *   *

"OH, AND one day San Francisco will sink into the sea."

A few hours later Shimon Slavin is again seated at his desk with an x-ray pinned to the window above him, which leaves the eerie impression that a ghostly chest is hovering over not just his head but all of Jerusalem. As before, Slavin had breezed into the waiting area outside his office a half-hour late, looking oddly like a disheveled angel with his longish white hair and white lab coat flapping behind him. This time, he was holding a vial of blood. "It's your turn," he had said to a transplant patient, a boy wearing a Nike baseball cap and a mask over his mouth. He had disappeared into his office with the boy, his parents, and the vial, to pop his head out a few minutes later. "OK," he had said, "It's your turn."

The San Francisco reference was his immediate response to a query about the traffic. "It is like anyplace. It is like New York," he says, shrugging. The phone rings. "Da," he says, leaning back in his chair. He listens a minute, then smiles broadly. "Hi Michael. . . . You just had a car accident?" Still holding the phone to his ear, he says, looking across the desk, "He just had a car accident. . . . I have an American here who was just telling me what bad drivers we are."

When he gets off the phone, he shrugs again, smiles again. "Maybe we are a little aggressive in cars."

He offers a tour of his labs, which are expansive and possess an astounding view of the serrated green hills of Jerusalem, with its all-white homes and office buildings. There is the cell separator room and a clean room where cells are purified by staffers in spacesuitlike getups and are kept chilled at −18 degrees Celsius. There is a blue-and-white Star of David on every door. In one room, a teenaged boy in a yarmulke is attached by looping tubes to a pheresis machine, which is pulling cells out of his blood. "He is donating his stem cells," Slavin says.

The corner of one hallway, where a woman is eating lunch, looks out over a stunning green valley. "This is where I live," Slavin says, pointing. "See that green roof? It's a French monastery. Filled with nuns. Next to it is the Church of St. John, where Jesus met with Mary Magdalena. Christians make pilgrimages to it three times a year. I live just behind both. You see?"

When asked if his location puts something of a damper on wild parties, he asks innocently, "Why? The nuns? No, no, they're fine with just about everything. They're French."

Back in his office, he moves to the window and pulls down the x-ray. He whips out a Magic Marker and begins drawing cells on the window, over Jerusalem, to make a point about how cell-sorting machines work, which involves coating antibodies in metal powder, letting them seek out and cling to cells of interest, and then weeding out the metal-coated antibodies along with their attached cells. He draws a petri dish on the window, a box for the machine, several dots for cells, several curving arrows, all in different color markers. The result looks like a child's attempt to write hieroglyphics on the sky. Slavin's face, for that, is freckled and round, the kind of face in which you can see the child he was.

Asked why he stays in Israel, he smiles impishly. "I go to the U.S. six times a year. I am director of an institute in Chicago as well as here," he says. "But I'm based here because you have better friends here. Friendships can be surperficial there. If you give a shitty paper in the U.S., they all say, 'Very nice.' Here, they say, 'What a shitty paper.'"

He slows down, thinking. "But I also bump into people on a daily basis here. People here don't just appear at conferences and disappear again. And I served in the military here; I was a doctor in a unit. I am a part of it. I am a part of the history. I am a part of the Bible. I am not orthodox. But I am a part of it."

5

BIOPOLIS

IT IS AGAINST THE LAW for taxi drivers in Singapore to talk politics on the job. They look like men-in-ties sans the ties, dressed in polite, pressed, white oxford shirts and polite, pressed dark pants. They ferry on their roofs scientist Probiotics Yogurts ads, never soiled *Terminator 2* promos. Inside, bulletproof glass does not separate them from their passengers—just signs offering polite health tips.

All this and more should make the Singapore cabbie a very different breed from, say, his Manhattan counterpart. But, just like the latter, the Singapore cabbie's prime avocation is apparently reveling in his city's "quirks." On a hot bright day at the end of October 2003, for example, a driver who's attention has been alternating between his cell phone and his passenger volunteers suddenly that Singapore's airport is rife with x-rays and other scanners not just to look for passengers with a temperature—SARS—but gum. "Gum is smuggled in, especially from Malaysia," he says. He grins viewing the passenger's raised eyebrows in his rearview window.

He adds, suddenly sober: "But think about SARS. Gum brings disease." And he moves on to an analysis of the astronomical cost of the biological science initiative in Singapore ($4 billion), noting that this was the amount the country spent on its highly modern underground in

1978. His patter alone calls attention to another unusual thing about Singapore: at 93 percent, the literacy rate is so high that it is, at the moment, higher than the number of white-collar jobs, forcing many to go blue collar.

The conversation drifts, as it often does here, to Singapore's many cameras. Many drivers proudly point out the cameras on poles all around that regulate traffic and catch speeders, as if they were bizarre species of local wildlife: "There!" "There!" This one is more interested in the "tourists" with cameras around their necks who are actually—he swears—environmental agency agents, hoping to catch in the act litterers, and smokers standing too close to outdoor "No Smoking" signs. He squints out the window to illustrate. No sightings today.

He pulls up to the Raffles Hotel, merrily accepts no tip—as is the Singapore way—and gaily putters off, leaving his passenger standing in front of an all-white throwback to Singapore's past. A leftover from the days of British Colonial rule, which ended in the 1950s, the Raffles' all-white entranceway is lined with white orchid trees, enough to service a hundred proms. Its veranda-studded exteriors are all-white; its doormen wear all-white colonial safari outfits; its lobby has all-white marble floors.

The hotel is legendary for its exotic prices and stories. Joseph Conrad and Jean Harlow stayed here. Rudyard Kipling and Somerset Maugham both wrote about it, Maugham posting himself under a frangipani tree in its Palm Court, the best spot from which to snatch and toss into his stories the rich gossip wafting down from verandas on all sides. Maugham's conclusion was that the Raffles Hotel was the symbol of "all the fables of the exotic east." The island's last tiger was rumored to have been shot beneath—or on—a billiards table here. Raffles bartenders, during WWII, conscious that everything a Raffles employee did was symbolic, threw away all the booze so their Japanese conquerors couldn't have it. Raffles suites, which run into thousands of dollars per night, have parlors and 14-foot ceilings; are doused in the natural perfume of plumeria gardens all around; boast floors made of marble and teak.

But perhaps the most exotic aspect of the Raffles is the activity that would be illegal anywhere else in Singapore, occurring in the hotel's second-floor bar, the Long Bar. Guests here, unlike most every other public place in the city, are allowed to litter. Indeed, guests are encouraged to toss peanut shells on the floor. The Raffles Hotel, in other words, is one of the few spots in Singapore where littering won't score one a minimum $1,000 ticket. It is, as a result, a favorite haunt of slovenly Westerners—as the government means it to be.

"I always wanted to say, 'Meet me at the Raffles for a Singapore Sling,'" says expat Singapore stem cell scientist Victor Nurcombe, looming over a thicket of palms. He throws his long thin body, clad in a short-sleeved polo shirt, into a bamboo chair on the veranda outside the brass and mahogany Billiards Bar. He grins. "Cheers, mate!"

Like Singapore taxi drivers, some Singapore life scientists may have a lot to be cheery about. Nurcombe certainly does. A month earlier, as the Australian-born head of a stem cell unit at the University of Queensland, he had given a talk at Australia's first international stem cell conference. He had ended it with, "I'm on my way to Singapore, where they're building a new complex for me." On the screen beyond, he flashed a picture of Singapore's massive new, 2-million-square-foot, five-building life sciences complex and stem cell hub, Biopolis. The assembled had laughed and applauded.

After his talk that day, Nurcombe had explained that one of Singapore's more endearing new quirks is its deterimation to spend enormous amounts of money on foreign life scientists: "Singapore has offered me five times my current salary. It is also letting me take my entire 11 member staff with me. How do you say no to that?" You don't, he implied, by waving goodbye and shouting out, "Meet you at the Raffles in a month for a Sling!"

The dark-haired, 45-year-old boy wonder, as he's been called in Australian circles, is fresh off his first house-hunting session in Singapore. He orders up a $20 Sling, and once the preposterously large, fruit-stuffed drink arrives, throws himself into shop talk. Nurcombe's attraction to Singapore is almost instantly clear, for his ideas are the scientific equivalent of the extreme spotlessness of the Singapore streets; the extreme lavishness of Singapore's first international stem cell conference, which has been held all this week; the overstuffed Singapore Sling. Nurcombe's ideas are the scientific equivalent of the grand gesture, perfect for a tiny nation-state with a gargantuan chip on its shoulder.

He stretches out his long legs into the veranda hallway and crosses his ankles. Nurcombe's claim to fame is his 1993 *Science* paper report-ing that a sugar on the surface of cells "delivers the bullet to the barrel of the gun." Cells divide, differentiate, or mature only after molecules drifting about the body fit into receptors on their surface, telling them which of the above to do. Those molecules, in some 70 percent of cases, need to bind to Nurcombe's sugar, in its various forms, before they can do any of the above, he has found. "Two-thirds to three-fourths of all the growth or repair molecules out there use heparan sulfate [Nurcombe's sugar] to find the cell surface," he says. It could end up being a short cut of an answer to any number of stem cell problems.

Nurcombe has some very preliminary evidence, for example, that with his sugar molecule, one can make adult neural stem cells prolifer-ate madly—something they cannot do normally and which could help companies create commercial numbers of cells for transplant. He has found that the same is true for MSCs, which form bone and can help bones heal faster. Indeed, one of the biggest problems facing stem cell companies like Singapore's ES Cell International (ESI) is scale-up: how to get hES cells to proliferate faster. His sugar can solve that prob-lem, Nurcombe swears. He just figures out the shape the sugar takes during whatever stage of growth he wants a cell to mimic and creates it for the cell.

But the most exciting potential use for this sugar would be to use it to *de*differentiate mature cells back into the stem cell state, adult or embryonic. "Heroic notion," he admits, but definitely worth looking into. At the opening of this week's stem cell conference, Singapore's first international stem cell conference, Harvard University's Doug Melton, who is trying to find the adult islet (insulin-producing) stem cell in part to restore insulin capacity in his own two diabetic children, announced he believes the islet stem cell doesn't exist. "He writ large one of the problems out there," Nurcombe says. "Not all tissues may have adult stem cells." Perhaps he could create them for scientists, Nurcombe notes, by *de*differentiating mature cells back into adult stem cells or even all the way back into hES cells, without raising the controversy that would come doing this via therapeutic cloning, which must use human eggs to do the same thing.

All this is key also, he says, because the closest adult cell to an hES cell that may exist, the MAPC (multipotent adult progenitor cell), grows very slowly and only in very isolated conditions. So in order to grow enough cells to fix a single kneecap, as things stand now, he notes, "You'd need a culture plate the size of North America. All adult cells grow slowly. Only hES cells grow like wildfire." This matters, Nurcombe says, because a small infarct of the heart "knocks out 100 million cells. A lot of cells are needed for therapy."

On the other hand, Nurcombe continues, there are of course problems with hES cells, too. ESI—the second biggest hES cell company in the world, which was started by the Israeli/Australian/Singaporean team that isolated an hES cell just after Thomson—has many hES cell lines. But like most of the cells approved by the Bush administration, they are "interesting, but difficult to manipulate. There's lots of art in the growth. You can't trypsinize those cells yet, and if you can't trypsinize, why bother? You're working four hours per day just to culture the cells." (Trypsinization is a way of quickly cutting cells out of colonies with a chemical, instead of doing it slowly, with a sharp pipette. It may be key for commercialization.) And the restrictive patent agreements that scientists are forced to make before using the presidential lines have discouraged most scientists, Nurcombe says. "Many people refuse to sign. . . . Everyone is too afraid it will all come down, that the [Bush] presidential lines chalk up to a house of cards."

Thus, here Nurcombe is, in Singapore, where the production of *new* and thus potentially less problematic hES cells is allowed, and where the government has been announcing loudly for years that the world should consider it the number one safe haven for hES cell scientists. At first, Nurcombe worried, since half of Singapore's government is Catholic. He will be working on hES cells, studying his molecule on their surface, to see if he can use that knowledge to bring adult cells back to an hES cell state. "Is this a problem?" he asked Philip Yeo, head of A*Star, the government agency behind Biopolis. "Yeo said to me, 'The Pope doesn't know what he is doing.'" Nurcombe laughs. "Back in Queensland, they don't believe cells are human until implantation. Anglicans believe they are human only after the blastocyst has impanted, the end of the first week. Which is a clever position, because 70 percent of blastocysts do not implant. The bottom line, though—and I say this in my talks to the public—is that Catholics apparently have a problem with God and they don't know it. God produces those blastocysts—and then allows 70 percent to die. Catholics have made God into an abortionist. . . . Hey!"

Nurcombe waves one of his former postdocs, Hiram Chipperfield, into a bamboo chair across from him. The postdoc orders up a Sling. Now a researcher in the lab of the aforementioned Doug Melton at Harvard, Chipperfield chats with Nurcombe about the big bucks Singapore put up for its conference this week: 1.5 million Singapore dollars, or nearly $1 million American. Given that the average U.S. conference costs about $45,000 American, and this conference is only for some 200 people, this is an outrageous figure. Furthermore, the postdoc's boss, Melton, was given a $14,000 plane trip and allowed to take six postdocs with him, after he hesitated about coming. The money being spent made him so nervous, the postdoc reports, that "he was up the entire flight playing with the electronic gadgets."

"In Singapore there is all the money in the world," says Nurcombe, ordering up another Sling. He tells the story of his courtship, which began when the deputy director of Singapore's Institute of Molecular Biology came to a conference attended by Nurcombe. Shortly after talking to the deputy director, Nurcombe discovered he had apparently impressed him, for he was flown out to Singapore along with six of his staff to meet Yeo. "I was dazzled by this guy," Nurcombe says. First of all, Yeo understood Nurcombe's science. Yeo is world-famous for this; armed only with a business degree, he has had experts from all the life sciences visit him on a near-weekly basis to simply educate him.

Second, "When I said to Yeo, 'Look, can you guarantee me a lab budget of $26 million?' Yeo said, 'Why limit yourself?'" Nurcombe was guaranteed a lab with 20 positions and unlimited research funds. He notes that a Russian scientist at that time who had recently relocated to Singapore said, when Nurcombe asked about drawbacks to the move: "'The worst thing is that you have no excuses anymore.' . . . Then I learned that the IMCB [Institute of Molecular and Cell Biology, the division Nurcombe will be working for, which is moving into Biopolis] has never spent out its budget."

Nurcombe said yes to the job.

His home country of Australia, Nurcombe explains, has finally come up with $40 million plus after three years of wrangling for a stem cell institute that will work with hES cells. "But without $400 million you're just pissing in the wind," he says. "It's a very important step for Australia. But if you want to start a spin-off company there, maybe $3 to $4 million can be scraped up for you. Venture capital there only bets on sure things. In Singapore they've set aside $1.4 *billion* for spin-off companies. They've already spent $600 million. . . . And if you want to go on to clinical trials? The $5 to $10 million you'd need is all but unraisable in Australia."

Australia, in general, just isn't plugged in enough to pull into the lead, says Nurcombe. This has happened before. It was an Australian, he notes as an example, who actually discovered penicillin. "[Alexander] Fleming that f——ing f——," he says, reveling in his own audacity, making the postdoc laugh. "He was just quicker with the PR than Howard Florey."

Nurcombe is clearly determined to avoid such mistakes. But he isn't entirely without the occasional—rare—insecurity, especially given the fact that Singapore has hired him on for work that has not been proven yet. After noting that "that Russian guy did have a point about that 'no excuses' business," he leans back in his chair again, falling uncharacteristically silent for a moment. "You know, I started spending months ago," he says slowly, as if checking his own words to make sure they are accurate before moving on.

"They haven't told me to stop, yet."

∗   ∗   ∗

> Please be considerate and hygienic, wash your sh*t stains . . .
> Please be considerate, put your sanitary napkins in . . .
> Please be considerate and don't wet the vanity top . . .

THE SIGNS above line the walls of a bathroom on an executive floor of Singapore's hugely ambitious, brand new, and militantly immaculate $288 million government biotech center, Biopolis. The tiny nation-state's interest in encouraging proper behavior extends to more than the patrons of this particular bathroom, however. Signs are ubiquitous in Singapore, politely informing every traveler and inhabitant what he can and cannot do and in most cases, helpfully informing him of the thousands in fines that shall be levied upon him if he is not polite in the exacting Singaporean Way. Smoking in the wrong outdoor spot, dropping a butt or candy wrapper on the ground, chewing or importing any but the two government-approved brands of medicinal gum, forgetting to wear a seatbelt, and riding an escalator too close to the railing all come with fines ranging from the mid-hundreds into the thousands, according to signs posted everywhere by the obsessive/compulsive tough love, one-party government.

Smuggling drugs into the country comes with an even bigger penalty, according to cards passed out to every passenger on international flights: death. Every citizen knows it is forbidden to discuss religion or race at speaker's corner—Singapore's only forum for public debate. Every reporter and politician knows that Singapore possesses bizarrely strict libel laws. Criticizing the government without—and sometimes even with—backup reams of documents, results in "libel suits" that are generally if not always "won" by the government. When libel laws don't apply, certain government officials freely utilize the option of tax audits, and can breezily inform residents that if they vote for the wrong guy their housing status will never be upgraded. Many domiciles are assigned by the government. Reporters Without Borders (RWB) will name Singapore 144th out of 166 of the world's countries for press freedom in 2005. Calling that ranking, "disastrous," the global watchdog group will note: "Singapore's low ranking was due to the complete absence of independent newspapers, radio stations and TV stations, the application of prison sentences for press offences, media self-censorship and the opposition's lack of access to the state media." In 2004 RWB will note that Singapore's "internal security law allows the authorities to imprison people without judicial approval." It has one of the highest execution rates per capita in the world. In 2005, Jiahao Chen, who, says Index on Censorship, blogged about "the unfairness of the system of state loans for undergraduate studies (and) . . . statistics that showed that Singapore's investment in research produced much paper, but little result in terms of peer review credit" will have to close his blog under threat of government lawsuit. The site will close and link to an apology to A*Star and its chairman Philip Yeo. A cautious few natives use variations on the phrase "benign dictatorship" to describe Singapore; the very few who are less cautious use variations on the word "Orwellian" to describe it.

Into this world of Orwellian benevolence, scores of the world's top stem cell scientists have poured during this last week of October 2003, days before the fifth anniversary of the discovery of the hES cell. They have come with hundreds of staffers. Some 180 researchers have been flown in and are being put up by the government. The government is ostensibly spending the money to get a look at the field it is openly striving to lead, in part by outspending the mere $20.3 million the U.S. NIH will end up spending this year on hES cells (compared to $190.7 million for adult stem cells). But Singapore officials also openly admit they hope the move will persuade some of the scientists to, in effect, refrain from taking advantage of the return ticket. Singapore believes it has the will and the means to become number one in hES cells; it just needs a whole lot more scientists, preferably Western-educated ones.

For their part, many of the attending scientists, some of whom are from countries that either have little money or restrict hES cell work, have come to see if Singapore is serious about hES cells and, if so, what it could mean for them—or what it should mean, given the odd reputation of the "nanny" state. For while Singapore has stern laws regarding just about everything, which can be daunting to Westerners, it also has decided to be unusually laissez-faire in the critical area of hES cells. Its hES cell laws, as they are being drawn up in the parliament, will mimic England's liberal laws, allowing everything except the cloning of human beings. Even therapeutic cloning will be allowed.

Both sides have gathered this week, therefore, to look each other over from opposite sides of a conference room, to decide if they want to cross the boards to ask for a dance. Both sides, it would appear, are serious. U.S. scientists in particular are serious—and are being seriously courted. Anyone trying to contact a top stem cell researcher in America this week is having problems. Just about everyone is here. Some are here out of curiosity. Others are considering mimicking Nurcombe and fleeing permanently to the science haven of Singapore, as European scientists did to the science haven of America during WWII. Only one Western researcher turned Singapore down, and only because of a family emergency.

Singapore is quite determined to win the stem cell race. In addition to the gigantic Biopolis, its 2000–2005 biomedicine budget of over $2 billion is impressive. The government's refusal to tell reporters what portion of that is going to stem cells, however, is not impressive. It is just one more reminder that this is a so-called soft dictatorship.

Still, it is a soft dictatorship—with a smile. And Singapore is clearly spending big bucks, which is just what the hES cell field needs. Israel certainly lacks it, as does the U.K., which is only spending £40 million over the next few years. (Not until a few weeks from now will anyone consider South Korea to be a threat.) Furthermore, Singapore is creating out of its huge new Biopolis, which can hold 2,000 researchers, a multidisciplinary science city where geneticists will work side by side with stem cell specialists, bioinformatics analysts, chemists, nanoparticle experts, et al. Singapore believes a multidisciplinary approach is critical to create a successful biotech market. And the fact is that it is becoming clear to hES cell researchers everywhere that, besides money, "absolutely essential" to the field is going to be "collaborative, stable, long-term multi-disciplinary efforts," NIH stem cell task force member Ron McKay will note later.

Just to pull off his 2002 *Nature* experiment turning mouse ES cells into dopaminergic neurons to combat Parkinson's disease, McKay will note, was "a superhuman task. Every figure in that paper represents 4 to 5 years of work. The figure on behavioral improvement: two guys spent two years doing that. There was a physiology figure—one guy spent three years doing that. The differentiation of the ES cells took three people two years. You can grow ES cells outside the body, but it's not trivial." Recapitulating what the body can do to a stem cell in the petri dish can be done but "it's really quite complicated stuff."

Thus, given that the United States is not spending money to create huge interdisciplinary teams for hES cell work, and other Western nations would like to but can't, the questions of the week for some of the world's top stem cell scientists are: Just how soft *is* this dictatorship? Can hES cells truly grow well in a soft dictatorship? Could they? Singapore *could* end up the field's leader if it attracted hordes of world-class stem cell researchers to its labs. Yet "open" nations "attract more brains," as *New York Times* foreign affairs columnist Thomas L. Friedman asserts repeatedly, echoing many Western analysts. Many brains are thus here this week to do something unusual: weigh the lack of "openness" in Singapore's society against the lack of "openness" in U.S. science.

Of course, the brains are also here to do the thing they all live for nowadays: to see which side in the adult stem cell versus embryonic stem cell debate will win the next round.

*   *   *

THE ROAD to the National University of Singapore conference center is crowded with many breeds of palm trees, trees that excite Westerners in part because they seem to be such alien, prehistoric ancestors of Western trees, their trunks as serpentine and scaly as dinosaurs' necks, their leaves resembling those of ordinary oaks and elms but gigantic, dinosaur-sized.

On the air-conditioned bus to the conference opening, several scientists take off their fogged glasses (the air outside is bathroom-shower humid) wipe them, then look about half in excitement, half in wariness. The endless parade of palms is exotic, but dwarfing them all, like real life behind a movie set, are the skyscrapers of modern Singapore. They were erected by Lee Kuan Yew, the controversial and still omnipresent prime minister who transformed Singapore from a colonial British backwater in the 1960s to a modern metropolis, the cleanest and most organized in Asia, boasting Asia's third-highest per capita income. (Electronics and Information Technology let Lee take Singapore from a third- to a first-world country from the 1970s through the 1990s. But as expected, thanks to China's growing ability to make everything cheaper, electronics now represent only 30 percent of Singapore's manufacturing output, down from 50 percent five years earlier. Hence: stem cells and genetics.)

Furthermore, computerized LED signs regularly peep out over thatches of palms, warning of traffic jams ahead within minutes of their occurrence. They are a constant reminder that these are not your typical meandering jungle roads. Nearly all roads in Singapore are sentient, loaded with optical, video, and load sensors that let traffic be directed— and violators be caught—by a computerized central control. The United States has tried this and has failed, because local and national levels of government can't agree on how and whether to do it and because citizens complain about privacy violations. Singapore was able to do it fast and clean, because the local and federal government are one.

As a result, the chatter in the bus, by the end of the half-hour ride, becomes loud and focused on stem cells. It has become clear that this will not be a wild jungle adventure. This will be work.

Once inside the modern university conference hall, which looks like a hotel lobby with its modern vast expanses of windows, marble floors, and indoor palms, the scientists munch on Singapore dainties, some culled from such exotic sources as seaweed, but most blander to accommodate what are apparently perceived to be uneducated Western tastes. The scientists are then ushered into a wildly modern conference hall by techies dressed all in black and wearing headphones, their job being to open doors and keep food out.

The reason food is so diligently kept out immediately becomes clear: the conference hall is outfitted with expensive high-tech video equipment which must be protected and which, as the conference begins, busily scans the audience. Cameras seek out and rest on the more famous stem cell faces as if they are Oscar nominees waiting for the envelope to be opened, or football coaches at the Super Bowl, and blasts them onto a room-sized screen behind the speakers.

The introductory speaker is Philip Yeo, head of A*Star, the main government body behind the $2 billion life sciences effort that is designed to save Singapore from the ever-growing economic threat of China (and the government body that has threatened to sue a blogger). Yeo utters for the first time a phrase that will be repeated constantly over the next few days: the need for Singapore to stop collecting "industrial capital" and to focus on "human capital." A reporter stationed here will note later that Singapore officials use the term "human capital" all the time: "Scientists aren't people to them, they're 'human capital.'"

The first distinguished speaker of the day is Sir George Radda, the smallish, longish-haired head of Britain's Medical Research Council, the body that drove Britain's successful effort to draw up the most liberal, if careful, hES cell research regulations in the world. Also head of the new International Stem Cell Forum, a group of representatives from 13 nations dedicated to promoting stem cell research worldwide, Sir Radda discusses Britain's new stem cell bank (the first to include hES cells) and the £17.5 billion the European Commission has to spend on research (although hES cell work has yet to be approved).

Up next is Doug Melton, the Harvard researcher pursuing the differentiation of hES cells to pancreatic islet cells. Melton, with his

short gray hair, prominent nose, serious expression, and dull brown
jacket, causes a hush to fall. Melton is a giant among stem cell figures,
in part because of his careful, reasoned denunciations of some of the
stem cell work going on. He recently, for example, dared to call some
NIH researchers to task for their claim that they had turned ES cells
into insulin-producing islet cells. He believes it can be done; however,
he announced that the NIH claimed it generated islet cells because it
saw insulin in its dishes, but insulin is found in the medium the cells
were sitting in. It was a clever coup, to figure out that insulin was in
the proprietary medium used by stem cell researchers everywhere, and
caused a commotion. Some stem cell researchers have been accused
of staying quiet when the science on "their side" of the stem cell
divide isn't great. Melton is famous for his persistent refusal to
do this.

Melton does not disappoint today. After dropping another bomb by
noting that his work indicates there probably *is* no islet stem cell in the
adult human—causing several speakers to scramble back to their hotel
rooms to rework their talks and slides on adult islet stem cells—he
announces that he and Boston IVF have created 17 new hES lines. It is
the first time a U.S. academic has publicly created new hES cell lines in
defiance of the Bush edict. Like the presidential lines, Melton's cells
were cultured on mouse feeder cells, possibly rendering them ineligible
for clinical use—but the announcement is well-received. The new U.S.
hES cells will be characterized, placed in the U.K. stem cell bank, and
distributed for free.

In the United States few will be able to work on them, since most
basic research is done with government funds, and government funds
can't be spent on any but the Bush lines. Melton has helped scientists outside the United States with his move, few inside. Still, it is a
clear and impressive act of defiance, unusual in America, where the
NIH is still finding it difficult to get takers for their few, $5,000 a
pop, presidential lines with their complex intellectual property
agreements.

Melton, who draws chuckles each time he says "presidential lines,"
also notes his cells can be isolated from each other via trypsinization, an
easy chemical approach. So, unlike most of the presidential lines, which
must be grown by the slow "cut and paste" method, letting them double

only every 36 hours and forcing researchers to spend six months growing their cells before experimenting on them, Melton's cells double every 24 hours. His news will instantly make international wires.

Up next pops the elegant Singapore scientist Ariff Bongso, who, as noted in the previous chapter, in 1994 was the first to discover the hES cell. He is a reminder to all in this room of Singapore's place in hES cell history. He is followed by Bing Lim, who was recruited from Harvard to Biopolis by scientist Edison Liu. (And Liu was *himself* recruited from the U.S. National Cancer Institute, where he once served as clinical trials chief, to head up Biopolis's Genome Institute of Singapore [GIS] stem cell division.) The room has now been repeatedly reminded: Singapore is recruiting.

The NIH's perpetually jetlagged and omnipresent Ron Mckay is up at bat. He gives his standard talk about how he can get an adult neural stem cell from the central nervous system to become a peripheral nervous system neuron by adding a single protein. The NIH has been sending him around the world to act as the U.S. government agency's liaison, to help those countries with presidential lines grow their cells—and hopefully, share their results with the United States. McKay was also an author on the hES to insulin-producing paper that Melton recently trounced. He looks exhausted—and is. "I've learned to sleep sitting up," he says, grinning, after his talk.

The room is filled with people not pleased with some of the things Melton had to say. HES cell people will have to scramble to review their hES to insulin-producing cell work, to see if the medium contaminated their cultures, too. (The upshot a few months later will be: some people find McKay's approach works, although indeed, an insulin-containing medium shouldn't be used.) And people here to talk about the adult stem cells that they believe create islet cells are confused. An Oregon University primate embryonic cell researcher will note later that some in the field will be unhappy for a while with Melton's conclusion, which will be published in *Nature*, believing it to be premature.

However, with his announcement about new hES cell lines today, Melton has become the first academic U.S. researcher not only to defy the Bush edict but also to make new lines available for free. (The Bush lines cost $5,000 a vial.) Only those researchers who have access to labs with separate or no federal funding can work on the cells. But it's a bold move and Melton will keep going. Within months, he will have more hES cell lines available for research than the NIH.

In this latest battle in the adult versus embryonic stem cell war, round one seems to have been won by hES cells, in particular, Doug Melton's new hES cell lines.

◆　　◆　　◆

SINGAPORE'S BATTLE to persuade Westerners to find peace on its shores, meanwhile, rages on. That night, the galaxy of stars is lassoed and transported to Sentosa Island, which is ringed with signs sternly instructing all to please refrain from running over the monkeys. Cocktails and more semiexotic Singapore fingerfoods await. The reception is held in a room that is open to the jungle and lined with heavy, gilt-edged mirrors, leaving revelers with the disconcerting impression that they are standing inside an ornately furnished jungle. Anyone without a drink in hand is diligently pursued by cocktail waitresses. Biopolis stem cell head Bing Lim, talking to a reporter, keeps getting wine thrust on him, keeps shaking his head no, please, just a Coke. He is chatting about the fact that he and Biopolis are "aggressively" pursuing scientists from abroad. He has hired five postdocs from the United States one from Yale, one from MIT, three from Harvard. "They tend to be the kind that want adventure," he says.

The finance minister gives a talk that for the first time reveals a crack in Singapore's veneer: "There was a time, during SARS, when we thought this conference wouldn't happen at all," he says. He announces that the U.S. Juvenile Diabetes Association has given Singapore a $3 million joint grant for hES cell work—a blow to U.S. hES cell researchers.

All are in shirtsleeves. Flocks of very large birds scream on all sides. Alan Colman, cocreator of the first adult-cloned mammal, Dolly the sheep, ambles about amiably in his trademark sandals, white socks, and white oxford shirt with the sleeves rolled up. He moved here a year ago to head up ESI.

The head of the biology department at UC San Diego notes he is thinking of focusing on stem cells and the mind in his work, an issue of concern—and fascination—to many stem cell experts. "In what way will the mind, as opposed to the brain, be changed by stem cell operations?"

he muses, stabbing a perfectly round medallion of rare steak with a toothpick. Round and round the caterers come with fingerfood: crab, mushrooms, all kinds of fish and seaweed delicacies. The doors are flung open to the warm moist jungle: one giant bird in particular squawks so loudly all night it halts conversations.

The next day, it is back to the adult-versus-embryonic stem cell wars. Today's target: the so-called ability of adult stem cells to "transdifferentiate" into different tissues like hES cells. Masato Nakafuku of Cincinnati Children's Hospital points out how powerful adult stem cells can be *normally* when they are differentiated the old-fashioned way—to become the tissues they were born to make, not pressured to overcome their limitations via transdifferentiation. He has improved memory with stem cells for the first time by adding adult neural stem cells to a mouse with stroke, he notes. Work on adult stem cells is clearly promising when the attempt is to urge the cells to become what they have been developed in the body to become, no more, no less.

Then Stanford University's Irv Weissman introduces the NIH's Donald Orlic with the following words: "Two years ago Don Orlic shocked a lot of us by saying hematopoietic [blood] stem cells could repair the heart"—and the tension levels rise. Weissman is referring to research that the Providence conference earlier this year fought so hotly over: the 2001 *Nature* paper that claimed that bone marrow stem cells can create both new blood vessels and new heart muscle in injured hearts—the latter against their nature. Functional transdifferentiation of adult stem cells is a lightning rod issue in this field. One reason is no one has proven it exists. Another is that pro-lifers use it to argue that hES cells are not needed. The clearly disappointed Orlic reports that in his follow-up experiment, when he tried a similar approach on monkeys, they saw no improvement.

In response, Weissman asks a pointed question—Did you bring the cells to the single cell level in that original experiment?—clearly aiming to show Orlic that he may have had contaminating cells in his culture. Comments will get even more pointed in quiet corners in the hours following, as people discuss this big new setback. Furthermore, Weissman is about to publish an article in *Nature* that will claim that he repeated Orlic's initial experiment and found it to be wrong: hematopoietic stem cells cannot transdifferentiate into heart muscle.

Weissman's paper, along with another paper making the point even more strongly, will lead a *Nature* editorial to claim that "transdifferentiation" may be a matter of bad technical choices. Many scientists have come to believe that adult stem cells can transdifferentiate, *Nature* will speculate, because of the commonly used technique of "immunofluorescent microscopy, in which specific antibodies bearing tags that fluoresce under laser light are used to track the migration and growth of injected stem cells. Unfortunately, this technique is prone to artifact [creating artificial results]. Antibodies may react with cells other than their intended targets and microscopy is a notoriously subjective business. What's more, some cells fluoresce of their own accord under laser light." Translation: glowing antibodies have been used by many scientists who claim that adult stem cells can transdifferentiate. But antibodies are fallible. The best way to track the differentiation of a stem cell is to insert glowing genes into its genome and so far, when this has been done, transdifferentiation in a significant form has not been found.

The anger and fear that this *Nature* editorial will elicit from scientists on both sides of the issue in the next few months will be real and legitimate. Many headlines in newspapers around the world will get the message mostly wrong and scream that no adult stem cells can help the heart; many others will get it *completely* wrong and scream that no adult stem cells work at all. Pro life groups that had earlier claimed hES cells don't work (pouncing on press articles written about Melton's above-mentioned finding that a single ES cell study was wrong) will now protest that the press is biased against adult cells. The misinformation march will go on for months. In August, even a *New York Times* reporter will casually—and erroneously—write that "so far no one has succeeded" putting adult stem cells "to work to treat diseases." It is a claim that thousands of oncologists and cured bone marrow transplant cancer patients, among others, might take issue with.

It will all have increasingly fed up and fearful scientists in a state of genuine despair in the next few months, grousing off the record about the nightmare of advocacy groups, the press, and one another. One side will accuse the other side of "letting emotions get in the way of science"; the other side will accuse the first of "grandstanding" and being responsible for "a huge amount of very deliberate propaganda in the field." One scientist will claim that he's been having difficulty getting another anti-transdifferentiation paper published because editors are biased; another scientist will decline to go on the record with her belief that some heart regeneration she recently saw was indeed the result of transdifferentiation because "it's not PC—you just can't say that kind of thing anymore."

If this were a normal science field, it is unlikely *any* of this would have occurred. For, all that the *Nature* papers will establish with some credibility is that a certain exact subset of bone marrow cells in Orlic's initial 2001 paper did not transdifferentiate into cardiac muscle. Left open will be the possibility that different purified subsets of bone marrow cells *did* differentiate directly into blood vessels that help those mouse hearts, and that are helping patient hearts in many clinical trials now. It is starting to look highly unlikely that adult stem cells can transdifferentiate in a significant way. But they *can* accomplish neat tricks— the *Nature* papers will not dispute this.

Still, there will be a lot of noise. An extraordinary thing is becoming clear: unlike their peers in other fields, scientists doing the most basic research on stem cells are under an international microscope. In the stem cell field, the smallest study can make front pages and be co-opted by advocacy groups, politicians, bloggers, and religions. And even when their science is perfect, stem cell scientists are inevitably told, by one horde of complete strangers or another, they have made, at the very least, a political or moral mistake.

Doug Melton will later note that the Harvard Stem Cell Institute, which will be established a few months after this conference, will hold public information hearings. "We're quite intent on having forums for public education," he will note. "Our belief is that the public has not enjoyed an informed presentation of the issues. Newspapers make it seem as if people are trying to do something that's unethical."

Ron McKay will go further, wondering if all this is the inevitable result of true paradigm shift. Paradigm shifts in science, as defined by science historian Thomas Kuhn, are essentially those rare moments when some entrenched scientific view of the world is suddenly altered on a grand scale, in "a social way. . . . He believed the paradigm shift was essentially a social construct, where there is a shift in language used by the people. Academics do the work that lead to it, but the transition involves a much broader range of people."

And Weissman will put it thusly: "It's a good thing we're treading on ethical grounds. It means we're getting close to important issues."

*   *   *

THAT NIGHT, the grand opening of Biopolis is held, and with it comes another reminder that Western hES cell scientists are not witnessing the immaculate conception of a new science into a virgin nowhere. They are witnessing a new science being driven and directed by a hand highly experienced at driving and directing—and brokering no opposition: a dictatorship.

When the friendly multicolored air-conditioned Biopolis buses at last pull up to the vaunted center of it all, scientists in each bus are told to leave their bags and briefcases behind. They are then told, by cheerful women who suddenly appear at the top of the aisles, that they will be labeled with little colored stickers that match their buses, ostensibly so all will know where to go at the end, but more clearly so that authorities will be able to separate wheat from chaff when it comes time for dinner. Only the big stem cell stars will be allowed to eat in the same room with the deputy prime minister. The smaller luminaries will eat on another floor. Since the big stem cell stars have been put up in the best hotel, and the smaller stars in a lesser hotel, the sticker-on-the-bus system is certainly a practical way to keep people in their places—if it is something less than humanizing.

It seems a small thing. A deputy prime minister and his minions are attending the opening, so security measures must be taken. But the carelessness with which this is done surprises several of the scientists. They were not informed ahead of time they would have to leave their bags behind. Yet it doesn't appear to occur to the women that any of this might be a problem. Smiling perkily, the women move down the aisles popping stickers on the scientists' chests.

Outside, Biopolis looms. With its vast towers, most of which are all windows and many of which are still under construction, it looks like a series of gigantic empty aquariums. Around Centros, the building that houses A*Star, is a ring of immaculate, plastic-looking trees, grass, wee waterfalls. "Those weren't here last week," an awed administrator says of the flora. Construction workers in blue uniforms are gazing up at Centros, which has spotlights turned on it, movie-premiere style. As the group moves down cement pathway, a PR person points out that the cement curves gently with the slope of the hill. "Feng shui," she says.

Inside, pocketbooks are searched politely, and every scientist is ushered through an x-ray machine. The elevators talk in a seductive female voice and are "smart." If one presses the wrong button, one may just press it again to cancel and try again. Once upstairs, the foreign scientists are placed up front. Local scientists are arranged behind.

The entrance of the deputy prime minister—Dr. Tony Keng Yam, a small gray man surrounded by a coterie of 10 or so more small gray men—is preceded, to the astonishment of some of the assembled, by the sound of coronets. "King music?" someone whispers incredulously. The show that follows is equally astonishing to the conservative scientific assemblage. A Donny-and-Marie look-alike duo jovially patter, trading bad Donny-and-Marie jokes. Two giant monitors once again magnify the faces of those on stage to an unnecessarily gargantuan size. A giant silver pinball with the word "SINGAPORE" on it occupies a pedestal midstage. When the deputy prime minister places his hand inside a giant handprint, there is movie fog and *Star Wars* music. A disco ball flashes yellow lights. The giant silver pinball rises through the fog revealing that it is resting not on a podium but on a giant, twisted strand of DNA.

Biopolis is launched. During the ensuing speeches the words "human capital" are used repeatedly by government officials to describe the most important part of the effort. Talent will be "infused" and "integrated and syngergized." Singapore has decided it is too small to make things (it cannot compete with China's ability to make products cheaply), so it must produce patentable thoughts. Singapore is serious about this, although it is apparently driving home that point by demonstrating how closely it has studied American TV. "Embarassing," mutters an American guest, as the aisles fill with movie fog. "Is this what they think we are?"

Later, a reporter will grouse loudly. The reporter, who is with a Western media outlet but is stationed in Singapore, will only cover the conference for one night, he says, because his editors swear that if he gives them another "Biopolis will rule the world" story they will fire him. It's fine with him, he adds sarcastically, because the stories the government allows him to run have no teeth anyway. Singapore officials rarely offer backup for their figures and reporters have learned not to press for them. The reason? "It's a one-party democracy," he says easily, less bitter than bemused. "No need." Can you publish a story on that? "Are you kidding? I'd be sued for libel. We learn not to ask. No one writes about the fact that this is a one-party democracy, but that's what it is."

*   *   *

BACK TO the salt mines, the last day of the conference. The University of Minnesota's Catherine Verfaillie gives the latest on her "miracle" hES-cell-like adult stem cells, the controversial MAPC (multipotent adult progenitor cell). (As noted earlier, Verfaillie believes that MAPCs are the only adult stem cells that can form any mature cell naturally. No one has yet been able to duplicate her claim, although many of the respected scientist's peers are rooting for her.) The cells do seem to be hES-like in the way they can form many different cells, she says, although they aren't easily coaxed to make heart cells: they don't become "beating cells" as easily as ES cells do. Furthermore, they must be passaged (replicated) at "extremely low density," or after every second doubling, which means to expand (replicate) them to therapeutic numbers quickly at this time is impossible. She has not yet tried them in diseased mice, so it's impossible to know if they are functional.

A thin woman with close-cropped hair and a clipped Belgian accent, there is no joy in her face. Despite her protests that all cells should be studied, Verfaillie's work is constantly co-opted by pro-lifers who say her cells render hES cells unnecessary. Indeed, in June 2004, Jean Peduzzi-Nelson, a University of Alabama pro-life scientist, will be tapped to testify about adult stem cells by an anticloning bill sponsor, Republican senator Sam Brownback, despite the fact that she has published no work on the cells. She will tell a Senate committee (as revealed in the transcript she passed in): "Whether each type of adult stem cells [sic] can make every different cell type in the body is a mute [sic] issue. For example, neurons [nerve cells] can be derived from cells in the adult brain, bone marrow, muscle or skin cells. Also there is evidence from Dr. Verfaillie and colleagues at University of Minnesota that stem cells from adults are able to form any cell type in the body." Peduzzi-Nelson will speak of two "disastrous . . . embryonic cell" trials, never noting they were irrelevant: they were hES cell trials.

In a few months, Verfaillie will announce she is leaving the United States to return to Belgium.

During the break, Peter Droge, a German geneticist with wild blond hair and cornflower blue eyes who has been lured by a big salary to Singapore, notes his group is working on inserting new genes into hES cells. But unlike Australian Vic Nurcombe, a more recent expat, Droge is somewhat disillusioned at the moment. He notes that he is discovering that, contrary to Singapore's vision of one big, happy, well-paid stem cell family leading the world, his group is not yet working with the others. "We're scientists; we do this," he says, making a hoarding motion with his arms. At least at the outset, Biopolis may not yet be inspiring the kind of "synergy" that Singapore is hoping will propel the nation into the lead.

Back at the conference, Stanford University's Irving Weissman, the brilliant and perennially cheerful discoverer of the adult blood stem cell—and devotee of the hES cell—takes the stage for the final talk. During an earlier break, Weissman had cheerfully transmitted his scorn over that week's events worldwide. As the conference was going on, the UN was deliberating, at the United States's urging, whether to ban all forms of cloning. "There's no federal law against it in the U.S. They're trying to get away with a policy that has never been tested in a democracy," he said, smiling sardonically. "The Republican party, supported by evangelistic Christians, wants to extend their moral beliefs outside of the country since they couldn't get it done here." He noted that, as a Jew who was part of a family that came to the United States to avoid this kind of imposition, "I find it particularly offensive that they are trying to impose their religious beliefs on the 60 to 70 percent of Americans who are in favor of this technology."

During the break, Weissman had noted that he was talking collaboration with adult blood stem cell expert Patrick Tan of Singapore because he has collected $11 million in private money for his therapeutic cloning and hES cell center in the United States—necessary because of the Bush restrictions—but needs $40 million. Tan is "doing leading edge clinical research here, he's applying the kinds of things we do at Stanford with adult stem cells. It would be nice to purify stem cells for him."

Singapore in general, Weissman had said, "could well become a center for Asia. Singapore is committed to pluripotent cells, even by nuclear transfer [cloning] if it has to. It's well-to-do, provides a high level of health care, and is surrounded by very large countries. It has a high-class bone marrow transplantation unit. It will be a beachhead for translating progressive stem cell research to third world countries. A number of Chinese nationals born in the People's Republic of China come here or to the U.S. for their education. One of my postdocs has come here. In fact, if I go read my e-mail right now, I'll find nine letters from interested postdocs from China, one from India, one every other year from South Korea. They're out there, ready to be trained."

He had added, about the issue of "transdifferentiation": "Will a blood cell give rise to muscle, brain? No. If you damage a tissue enough, run it over with a train three times a week . . . ," he had pounded the table, laughing, "there will be some fusion in the repair process, my lab has found. But there is no evidence that blood-forming stem cells can turn into other cells."

Yet when it comes time for him to talk before the assembled, Weissman reveals a stance that is far more complicated. For Weissman, who remains the only person to make millions off stem cells so far (the hematopoietic, or blood, stem cell, which he officially characterized in 1988), is making strides with both adult and ES cells. He is a fan of both. He notes that he is working on Slavin's approach, that is, giving allogeneic (foreign) stem cells to breast cancer patients, creating dual immune systems that accept both the foreigner's cells and the host cells, then following that up with killer T cells from the same stem cell donor. Like Slavin, he is very excited by the next step, which he hopes will be the last step necessary: "Adding breast cancer specific T cells."

Yet he is also trying to start a major human ES cell center in California. He is, indeed, working all sides, including the side that believes that many adult stem cells may *cause* cancer. This is a problem for adult stem cells, he notes, but a boon for cancer research. He opens his arms wide, like a preacher. "The isolation of cancer stem cells is going to be an important event to understand."

✳   ✳   ✳

LATER THAT day, reporters attending this conference from abroad are shown around Biopolis. Work has begun again on the complex; trees have been removed again; signs say, *Bahaya Jangam Dekat* (Danger Keep Out). Reporters are brought into Genome, the building that houses the stem cell and genome units. Lobby areas are painted in bright off-colors that are straight out of the 1960s. The furniture is boxy and plain, also straight out of the 1960s. And TVs are set into walls, à la *2001: A Space Odyssey*.

Reporters are shown high-speed DNA sequencers, which look like washing machines and have already captured the gene expression profiles of a number of stem cells. Using machines like these, Singapore scientists have in three years sequenced 70 to 80 percent of the human genome, it is claimed. The reporters are brought past robotic cell colony pickers, which resemble vast honeycombs being systematically raided by scores of disembodied hummingbird beaks. Up on another floor are robotic microarrays with thousands of glass slides, each of which can contain 23,000 oligonucleotide probes that can pick out nucleotide sequences of genes. One of the machines here can analyze multiple slides at once. The SARS virus was decoded here, a scientist says, causing the reporters to step back a bit.

Way in the back of floor 5 sits half of a giant mass spectrometer. It looks like half of a vast steel beer barrel. It isolates and analyzes proteins at the atomic level using x-ray crystallography. It weighs several tons. "I'm just glad that I'm not in the office underneath this thing," a scientist jokes, causing the reporters to step back warily, a second time.

Evidence that everything here is new and actively being installed is reinforced when the reporters get to the elevators, which open to reveal six standing ashtrays, which look like tiny rockets, attended by no one. There is a pause, then the elevator says, sexily, "Down." The ashtrays disappear.

In Matrix, an identical gargantuan aquarium/building next door, the bioinformatics center is being constructed. The group enters a room that would resemble an air traffic controller's headquarters if more than only two of its computers and screens were manned. It is freezing in here, to protect the hardware, a series of large automated sequencers that look like giant air-conditioners blinking with thousands of lights. One of these machines has been running the sequence of the fugu fish for two years, a scientist notes, for Sydney Brenner, the Nobel Prize–winning scientist who has been advising Singapore. (He favors fugu fish for dinner.)

It is all impressive, except for one thing: the press tour of the opening of Biopolis does not include a tour of the stem cell facilities. "The staff is in a meeting," the PR people say, as if that explains everything perfectly for this group that has come halfway around the world to see the stem cell center. The PR people then gaily take the group to Holland Village, the favorite spot of Singapore's considerable and ever-growing expat community, which houses everything from Chinese food restaurants—offering mud carp fish, drunken chicken wings, and jelly-fish—to Starbucks. The PR people fully expect the Western reporters to accept the bizarre omission of the stem cell labs from the tour, in the same way they fully expected scientists and reporters both to comply when they told them to keep their things on the bus. Can a "soft dicta-torship" attract stubbornly independent, competitive, arrogant Western scientists to its labs?

* * *

THE PLEASANT white buildings of the National University of Singa-pore (NUS) hospital are ringed on all sides by exotic trees and armies of signs indicating that littering or smoking anywhere, outdoors or in, will result in a stiff $1,000 fine. Inside, all visitors are met by grim men in hospital masks who direct all visitors into long lines to have their tem-perature taken, thanks to SARS. Once in, the atmosphere is more relaxed: the food courtyard is open and populated by palm trees, crows with long orange legs, and patients hungrily devouring red-bean pan-cakes, pigs roasting on spits, and fish balls.

"Outstanding conference. A galaxy of stars under one roof," says Ariff Bongso, NUS embryologist. Dressed in the uniform here—a long-sleeved white shirt, tie, and dark pants—he lounges back on the couch in his windowless office. For the scientist who discovered the world's first hES cell, in 1994, his office is surprisingly modest—if Bongso himself has the elegant air of someone who has just eaten a rich meal, opened up with arms outstretched on the couch, long legs crossed to the side.

The reason the conference was so successful, Bongso says, was due to the money; the fact that it was invitation only, so the stars would not be bothered by equipment salesmen (and Singapore alone would bene-fit from any collaborations); the fact that each speaker was given 30 to 40 minutes to speak. "I think the researchers felt obliged to come up with new data," he says. He had one disappointment: "I couldn't see the gap being bridged between basic science and the clinic." His team has worked on nothing but that. Among other things, he was one of the first to come up with a mouse-free feeder medium for hES cells. And he and Israel's Reubinoff came up with a faster and more efficient freezing method for the cells.

He moves on to the UN, which is meeting at that moment to decide whether to levy a global ban on all forms of cloning. Last night, at dinner, several of the speakers were discussing ways to lobby the pope, with whom many were meeting soon, he notes. The ubiquitous McKay and others decided they had to offer good science as the best rebuttal to both the pope and the UN. Indeed, Bongso notes, IVF was once a political hot potato; good science won out.

He talks about the early years after 1994, when he first isolated the hES cell but couldn't get it to replicate and survive in the petri dish. In addition to the fact that he had isolated the hES cell batches down to the single cell level, which killed them (because hES cells so often need each other to survive), he now believes that he and the Australian researchers also used the wrong enzyme, too early on, to separate the cells from their blastocysts. "Benjamin Reubinoff came in and said, 'We'll use a different enzyme, dispase,' and it worked. We were on our way. But by that time Jamie Thomson was already doing that at the University of Wisconsin."

He smiles. "Then we went on to show you could make neural stem cells from hES cells in 2000. We were the first to do that. And that was when everyone over here became excited. Alan Trounson [Bongso's lead Australian collaborator at the time] turned to me and said, 'You must get us funds from Singapore.' So we, along with Martin Pera [another collaborator in Australia] went to see the senior minister Lee Kwan Yew, Philip Yeo, and other ministers in the cabinet. Basically everyone wanted to do it, although Philip Yeo said, 'We'll give you $7.5 million but you must set up a spin-off company.' I said to Alan, 'Can you match that 7.5?' He said he'd have to go to the private sector, because the government wouldn't do it. He raised the other $7.5 million from friends. Ben had no money, but he was made one-third partner. We brought a fourth scientist in, Christine Mummery of [the Hubrecht Laboratory in] the Netherlands. She had done good mouse ES cell work. . . . I was the godfather."

The coffers of the brand new company (ESI) contained $17 million at the start of 2000. "At first we were doing everything, looking for cells for Alzheimer's, Parkinson's, etc. I developed some new cell lines. Christine began working on cardiac cells. Benjamin began working on neurons."

But big changes were made this year, 2003, after Alan Colman arrived. Colman, a cocreator of Dolly the sheep, came in 2002 as CSO (later to become CEO) on condition that the company focus its efforts on differentiating its hES cells for one disease. He decided to try and make pancreatic islet cells, the cells that produce insulin for diabetics. "So next year we—the founding scientists—will all have to look for our own funding. ESI is going out on its own."

Is he worried, as Reubinoff clearly is back in Israel? No. Singapore at least appears at this time to be bent on becoming the land of the blank check when it comes to stem cells. "I'm very fortunate. Last night at the dinner I had a discussion with the vice chancellor of the university and he said, 'Just send us a proposal, we'll give you the money you need.' I don't see a problem."

Up next for Bongso is vacation. He has to decide how to spend his blank check. "I'm going to Australia, where my daughter is studying, to reflect. I have a lot of ideas. I'm not a tissue maker. I'm not a stem cell biologist, I do IVF work. My lab is the operating theater, generally. I'm very passionate about the end result: the patient. It's still extraordinary to me that I can have a bundle of cells, and nine months later, a bundle of joy. So I have to think. One thing is clear: if we were still sitting around discussing the ethics of IVF, there would be no IVF."

But for all the talk about IVF, it's the mention of Alan Handyside's work, first brought up earlier in the month at Australia's first international stem cell conference, that propels Bongso out of his seat. Handyside, who invented preimplantation genetic diagnosis in 1990, reported recently that he was able to extract trophectoderm stem cells from mouse embryos without destroying the embryos. If he can pull out similar hES cells from human embryos without destroying them—as Handyside is now trying to do—the controversy surrounding ES cells could someday disappear. Every one of the 1 million babies born by IVF in the United States for example, would have a few hES cells set aside for himself and for hES cell banks. The controversy would be over.

"Yes, yes, we've been talking about this in our lab, too, look," says Bongso, throwing himself into a seat in front of his computer. He calls up pictures of what looks like fried eggs, with yolks of varying sizes. "This one is bad for that. Look, the inner cell mass [yolk] is small. But look at this one." He points to a picture of an embryo with a large inner cell mass. "Very huge. You can get very good cell lines from this. And if we can snip off a small piece of this, and keep the rest intact, we will not be destroying the blastocyst. We're trying to devise a laser pen that will allow us to do this without rupturing anything. In the four-cell stage—the four-blastomere stage—for example. What if we removed just one blastomere? A great challenge." Is it the final answer? "Maybe."

Clearly, for Bongso—who was born in poverty- and strife-ridden Sri Lanka and who is valued by the Singapore government as the man who isolated the first hES cell—the benefits of this soft dictatorship far outweigh the disadvantages.

•   •   •

YET THE disadvantages are not hidden in dark corners in Singapore. Censorship and government accountability problems are as common as weeds in a garden. At the bottom of another British colonial fantasy–hotel of marble, fans, indoor koi ponds, and palms, another group of top international scientists gathers later that day. The hotel is the Fullerton—the $500 to $800 a night hotel where Singapore put up its speakers—and the scientists are the international advisory board that has advised Singapore to take one last step to optimize its chance to become the stem cell capital of the East. The board includes David Hirsch, Columbia University executive vice president for research; Leland Hartwell, Nobel Laureate and director and president of the Fred Hutchinson Cancer Research Center; David Baltimore, Nobel Laureate and president of the California Institute of Technology; John Mendelsohn, president of MD Anderson (absent today); and Hans Wigsell, president of the Karolinska Institute.

Another Nobel Prize winner, Sydney Brenner, calls the group's recommendation "the acclimatization checker." It is explained that the group has recommended that Singapore hire 150 MD/PhDs to guide the "translation" of hES cell work to the clinic. "The culture of clinical medicine is very different from the culture of basic science," Brenner notes. The 150-person group that this committee has successfully advised Singapore to hire would focus on commercializing Biopolis's science, and its first science will be that of stem cells.

But when the assembled local press asks the obvious next question— How much will this cost?—it is greeted with scorn from Singapore officials. "If the committee ends up consisting of one person, it will be small; if it's a large committee, the budget will be large," says Long Hwai Loong, deputy managing director of A*Star. The local press asks the question again, gets a similar response, and apparently just accepts it, the next day not even mentioning the fact that no price tag has been given. In few democracies would officials dare call a press conference announcing a hugely expensive endeavor, funded with taxpayer dollars, and disdain to explain cost.

Still, the charismatic Yeo does add a response free of sarcasm when he adds by way of explanation later. "We are very good at squeezing blood out of stone, here." Throughout, as his well-paid stable of international science advisers talked, he nodded constantly and eagerly, looking down, as if barely able to contain his enthusiasm. After the press conference, when a foreign reporter presses for figures again, another reason for Yeo's likeability becomes clear. He is almost persuasive when he notes that so much is unknown about the stem cell field, yet there is so much potential in it, that it would be pure guesswork and downright wrong to settle upon and release mere numbers. He wants his scientists to feel free to "dream" without feeling constricted, he says. The conference was a part of that initial dreaming period: its purpose was to gather great minds to discuss matters, then sit back and think. When the thinking is done, "Then we'll focus on money," he says, beaming serenely.

But the bottom-line message coming out of this press conference is clear. Any stem cell scientist "defecting" to Singapore has no real idea how big his research pool will be, nor how long it will last.

•   •   •

THE JUNGLE-GREEN Singapore River passes by the Fullerton and winds through the heart of the city. Flat wooden *bum* boats with red lanterns burning in their glassless windows glide up and down, blaring taped, accentless tour-guide voices. The boats are charming, but the tour is surprisingly barren. Most of the talk is about the bridges, clean nondescript structures whose underbellies are as white as the sky above them. Every bridge's builder and dates are duly noted. The good restaurants in the Clarke Quay area are pointed out and represent, it soon becomes clear, much of what Singapore possesses by way of a historical showcase. The quay is quaint, a row of tiny, refurbished pastel gingerbread shophouses boasting restaurants and T-shirt stores on the bottom floors, apartments on the top. Yet they are so squeaky-clean and so dwarfed by the modern cathedrals of commerce behind—skyscrapers—that everything about the quay seems rooted in little but tourism.

The bodiless voice directs attention to the Parliament, perhaps the most ironic building in the city. In keeping with the government's reluctance to share with its citizens its inner workings, the "peoples' house," as the bodiless voice calls it, is a modern gray building, built to mimic but not resemble the neoclassical style, with all its windows tinted black, unseeing, like the windows of a private limosine. The government can see out; the people can't see in. Near it, the voice continues, is a bridge on which old men used to tell stories about ancient China. But there is no one up there now. Just a sign warning of a fine for loitering.

The boat passes the Merlion, or the Water Lion, the city's mascot. It is a huge white half-fish, half-lion, spewing water all day. Singapore got its name from a lion that was supposedly spotted by the first British explorer to the place: "Singapore" means "land of the lion." Yet there never were any lions in Singapore. Singapore's name is based on a fiction.

That evening, a Western reporter stationed in Asia relaxes on another marble veranda outside the Raffles Billiards Room, orders a G and T, and lights into Singapore. Those who run against the state's one party tend to get either audited or sued by the government, he says. Civil rights leaders are still languishing in prison, he adds. He lists the relatives of the former prime minister who run Singapore.

"Man in the street" interviews can never be done in Singapore because "no one talks, no one. They gripe in private—never for the camera. . . . By the way, you should remember that press phone calls in hotels can be tape recorded, and banked." His cell phone rings. He apologizes, chatters a bit, then hangs up. "That was a colleague of mine. She is covering Elite, that famous U.S. modeling agency that is in town." Commercials for its fashion shows have been airing on TVs all week. "Why is Elite here? I'll tell you why. The government throws a lot of money at people when it wants something from them, when it believes it can burnish its image from them. Then when it is done with them, it dumps them. Look at those complaints Alan Colman was making a year ago." Colman, the cocreator of Dolly the sheep, indeed complained to Reuters news service that he had not received the money he had been promised in 2002.

The reporter sits back in his chair, shaking his head. "You know, I asked Philip Yeo once how he expected bright students to stay in Singapore, once they'd experienced the freedom of thought in universities like Harvard and Stanford. He said to me, looking angry, 'What's wrong with Singapore?'"

Still, Singapore, that tiny country with the comparatively huge GNP, is difficult to accurately stereotype. A sign in a cab that evening says, "Beating SARS, it starts with you and me." It is a potent reminder that when Singapore sets out to do something, it often succeeds, leaving its populace happy. Singapore was the first nation to clear itself of SARS. It did this, admittedly, via a few dictatorial governmental moves, but also via some spectacular science.

And that science came out of the government's top genetics and stem cell labs, run by U.S. National Cancer Institute expat Edison Liu and U.S. Harvard expat Bing Lim.

*　　*　　*

BING LIM takes a cab to Biopolis, which has changed even more now that the conference scientists are gone. The waterfall has ceased to fall; an empty hose dribbles onto cement. Patches of grass and plants have been moved, awaiting new homes. There are wooden scaffolds everywhere. Biopolis has gotten back to work.

Lim's office is located in the stem cell section of the Genome building's GIS branch. Office chairs on wheels are still in giant baggies. Workers are climbing up and down outside windows, waving as they skittle past scientists in their offices. Lim, a handsome, diminutive man with a wary face, sits behind his desk and begins talking with his chin in his hand. He starts off with the reason he left a spectacular post at Harvard University to come to a lab in the middle of a tiny tropical nation whose chairs are still in baggies.

"Some cancers start off as stem cells," he says. "Others start as mature cells that begin to turn on the genes of early development— they dedifferentiate back to a proliferative, if uncontrollable, stem-cell-like state." The cells in many cancers that are most virulent, and that can spread from one irradiated mouse to another, are invariably stemlike cells. "We're very interested in stem cells, therefore, as models for cancer."

His lab is already famous for having recently used Singapore's sophisticated DNA equipment, including microarrays, to sequence the genome of stem cells, to find unique "stemness" genes. While two other Western labs did this at the same time, his multidisciplinary group went a step further. It analyzed its own gene sequence along with that of the other two groups and found that there were some 300 genes common to all three groups' ES cells.

Lim's Singapore group went the same extra mile to help sequence the SARS virus. Once again, two other Western groups were able to sequence the SARS genome slightly faster than Singapore. But from the start, Singapore's approach was more thorough, forward-thinking, creative. Lim's group, with Liu's group, decided to decode a *number* of variants of the virus, so they could find the molecular "signature" of each and identify in the future the virus's source. Those efforts immediately came in handy in September, months after SARS had been eradicated. A scientist in a university lab became infected. Normally that would cause a widespread panic and isolation of a large part of the population. But when Lim's lab decoded his virus, they found it was a very specific variant that had come from a very particular lab in Singapore. So Singapore quarantined only a small population of workers, after finding that specific variant in a vial marked "West Nile" that was confined to one lab (not Lim's, nor any other lab at Biopolis).

Both of these encompassing approaches taken by Singapore—to deal with SARS and to find "stemness" stem cell genes—came about because of the close proximity of geneticists to cell biologists, Lim says. The Singapore group thinks with many heads at once, like a Hydra.

Lim shifts restlessly in his chair, glancing at his watch, continuing. Many in his lab are working on adult stem cells, in part so they can later re-create them in a more robust form from hES cells. Biopolis's genetics group is very open to working with stem cell experts worldwide. "We're very interested in people who want to work with us, to take advantage of our genetic technology." Already, they've had interest from scientists at the conference, in particular, from Gordon Keller, an MIT stem cell biologist who has begun to get blood stem cells from hES cells. Keller is also the incoming president of the International Society for Stem Cell Research. This is exactly what Singapore officials were hoping would happen post-conference.

Lim begins pacing with his arms folded. For all of Singapore's commitment, he notes, something does bother him and most other non-U.S. scientists. The United States has a mind-boggling amount of money to spend on biotech, billions of dollars. While the United States is only spending a mere $15 million a year on hES cells now, what will happen to the rest of the world's scientists if all of a sudden the United States changes its mind and kicks a billion in? Much of the work of these scientists, therefore, must be done very fast. "It's about critical mass, hopefully. If you gather enough good people in one place while the stem cell field is young enough, there's plenty of room for one group to get into the competition," he says. "The bottom line is, even though we talk a lot about collaboration, you do want to be the first to get IP rights. The future is going to be very interesting. There will be a big fight over intellectual property."

He stops pacing and throws himself into a chair. "Still, at Harvard, I would have to go out and find collaborators. And there's no motivation for collaboration, unless there's some kind of strong vested interest. Here, the infrastructure for that is in place. It's geared to support shared investigations. Here we have microarry and proteomics facilities [to analyze genes and proteins], people with expertise in amplifying RNA with PCR [gene manipulation technology], easy access to all the biologists in the institute, those who work on cancer biology, infectious disease . . . we have immediate access to these guys. We're experimenting as we go along. There's a possibility here for Singapore to be really strong."

But how about the censorship? He stops shifting and sits still. He must know all the objections. Lim grew up in Malaysia, a next-door dictatorship. "This is not a problem for research," he says quickly. He stops, then continues more slowly, smiling slightly. "And what about the censorship in the U.S.?" He settles back. "It's not complete censorship here, just as it's not complete freedom in the U.S. There's complete freedom here in science, and I think I can speak for my colleagues on that, as well."

The conversation moves on to the adult stem cells that Lim's lab is also working on, largely to serve as a model for making cells from hES cells. One of his scientists, Sai-Kiang Lim, enters and notes she is still working on transdifferentiation, trying to make sure bone marrow cells can't make cardiac cells. She is also working on turning bone marrow (BM) cells into liver cells, something that may end up working.

This may go to clinic here before anything else, because there are no options for extreme hepatitis except liver replacement, Lim notes. Most people have only 18 percent of a liver before they come in with various liver diseases, she says. Groups working with her in Biopolis are calculating the gene expression of liver injury.

Part of the vision of Edison Liu (the GIS head who oversees both Biopolis's stem cell and genetics divisions and the former head of clinical trials at the U.S. National Cancer Institute), Lim says, is "a technology platform called seamless flow." The stem cell group will prep cells. The sequencing group places the sequence of those stem cells into a database, and the bioinformatics group analyzes the database. The latter determine, for example, what group of genes is expressed; their common features; their homologues among different species. Then the biologists try to functionalize all the data. "Biopolis is an assembly line. Everybody does the thing they do best, then passes their data on."

There are four stem cell groups in total in Biopolis. Bing Lim's group; a group led by Larry Stanton, another U.S. expat who was the former director of functional genomics at Geron and is now working on the hES cell genome; a group led by Paul Robson, who is looking at very early embryonic development eggs to the blastocyst stage; and the group led by Sai-Kiang Lim, who is working on adult stem cells for eventual hES cell application.

Sai-Kiang Lim leads a tour of the stem cell labs, which are studded with row upon row of office chairs in plastic bags and opened and unopened boxes. She points out proudly a laser-capture microscope that uses a laser to cut off pieces of cells to analyze. "We are asking things like, 'What genes make ES cells pluripotent?'" She points out a gel apparatus PCR unit, which sequences DNA. "Martha over there is asking, 'What makes adult stem cells different?' Those three students [in front of computers] are ordering reagents. . . . The most senior guy in the lab is looking at endothelial [skin] stem cells. He is correlating gene expression with cell behavior."

She passes a darkened radioactive room, where staffers look at cells that have had a glowing enzyme gene called GFP inserted into them. This helps track the movements of stem cells. Next she moves on to the microarray room, which impresses many of the scientists wandering through. Microarrays can identify the expression of countless genes at once in cells. "We used to be able to do 100 slides at once, but now we can now do 261. We can visualize 19,000 genes at once, in 50 plates. It can work on its own for 24 hours, too. No one has to tend it, anymore." This lab, she notes, also has a $1.5 million spectrometer, which can locate unknown genes.

Lim came on board after running into Bing Lim at a conference. Because they had the same last names, she got the key to the men's room, he got the key to the ladies' room. They liked each other; he hired her on. She, like a couple of other researchers here today, talks a lot about vision. They are here because they like the "vision" of Biopolis, which is being nurtured and shaped largely by American expats.

*   *   *

THE NEXT day at Biopolis, construction work has proceeded. The imported trees that are beginning to line the roads are extraordinary in the sunlight, offering up bouquets of brilliant flowers. The water has been returned again to the waterfall outside the Matrix building, if it is still just flowing out of a hose.

Hwai Loong Kong is executive director of the Biomedical Research Council of A*Star. Dressed casually in a cotton blue jacket, he, like Yeo, is disarmingly chatty, disarmingly friendly. "Did you read the article in *The Scientist* on the UN cloning vote?" he asks immediately. "At this point it's 61 in favor of a total ban, only 22 against. That would be a major disaster."

He looks somewhat cheery as he says this, however, as he well might: before Singapore's parliament right now is a bill that would approve therapeutic cloning. If the UN passes its global ban recommendation, that will make Singapore an even more rarified haven for international scientists. Glancing at his watch, Kong launches into his spiel. A PR person is perched quietly nearby, taping and writing down his every word.

A state developer, Jurong Trun Corp., owns most of Biopolis, Kong notes. The cost of the buildings is 500 million in Singapore dollars. Novartis is one company already in the complex. "Biopolis is a cluster of creative minds," he says, so casually one could almost believe it is not a phrase he has uttered hundreds of times. "When we hired Edison Liu he said, 'As long as I get a new building.' We said, sure—and we built four more while we were at it. We're crazy for smart people."

He easily spins a question concerning the government's reluctance to put price tags on projects. "In the U.S., the government sets the amount, and scientists work with that. We say, 'Whoever has a good idea, we'll match your budget to it.' Our greatest shortage is people, you see, that's all. Biopolis right now is about 50/50 foreigners—and the stem cell leadership is almost entirely foreign. We are very aggressively trying to grow a stem cell program. When we add all of these with the CMM [Center for Molecular Medicine, the group of 150 MD/PhDs working on translating the research from basic science outward] what we should get is a seamless flow of information and technology. The U.S. has been trying to put together something similar in the area of translational research, Ed Holmes at UC San Diego. He's trying to straddle this very rich interdomain area. But we hope it will be Singapore that is the successful test bed, especially since Biopolis is two minutes away from a teaching hospital. Two minutes from the freezer to the ward."

Kong trained as an oncologist at the U.S.'s National Cancer Institute. He worked for seven years there on angiogenesis and gene therapy. He notes that Biopolis's IP rules model those of an American university: the investigators don't own their patents, but they get one-third of the royalties off the commercial end.

Kong emphasizes Singapore's ability to rush clinical trials. The clinical trial phase of all kinds of stem cell research can move faster in Singapore, he says, because Singapore is so small. Drugs go through an

FDA-like examination and through hospital Internal Review Boards, but there are far fewer applications. "So there are seldom major delays." And the process won't be anything at all like the United States's 15 years and $800 million per drug, he says.

So far there have been very few stem cell clinical trials in Singapore, he says. "Hundreds of patients have received blood stem cell transplants, but very few have received MSCs [cells that make bone and cartilage]. Still, some *have* received MSCs to replace cartilage." Many scientists worldwide are working on this exciting approach.

Kong notes, satisfaction in his voice, that he has heard "through the grapevine" there have been a number of collaborations as a result of the conference. "Some of our post docs are right now on their way to U.S. stem cell labs for training. They've gone with our blessing. They'll come back."

Kong's confidence is well-founded when it comes to those students. Postdocs sent abroad with Singapore dollars have to come back, by contract. But the government can't control everything that way. Biopolis has attracted many U.S. companies to it with tax incentives. Can it keep them?

Glaxo, Kong notes, has been in Singapore since the early eighties. Merck is here, as well. "The risk is always there" that such companies will take their tax breaks, then sell off and run. But they haven't so far, he says.

Biopolis, says Kong, is inspired by practical concerns. Furthermore, he says, it is inspired, period. It was conceived in 1999, "when we all met in Philip Yeo's office and he said. 'So what new industry should we go into?' Asia had had a financial crisis from 1997 to 1998." It was decided they should go into the life sciences, the initial push for which had been made by Nobel Prize winner Sydney Brenner, who had come to Singapore in 1983, exactly 20 years before, and told them to build the IMCB, Singapore's first life sciences endeavor, in the early 1980s. The prime minister had at first been reluctant, noting that Singapore was "a nation of technicians." Brenner had said, "And if you don't do this you'll stay a nation of technicians." Like Yeo, Kong says, Brenner's "strength is vision. . . . Philip decided, though, that we would need someone from the NIH in the U.S. He went there in 1999, and three months later, Edison Liu was here. This whole thing was Philip Yeo's brainchild."

·   ·   ·

ESI, THE only hES cell company in the world to *add* staff in 2002 (it grew from 7 to 35 in three years' time), is located in yet another modern building at the other end of town. The company will move into Biopolis soon. The roads to ESI move through wooded areas, past hundreds of three-story terrace houses crammed tightly if neatly into the side of every road, beer gardens, a small factory or two, something called the Pink Peppercorn Club.

Alan Colman is wearing the casual duds he favors at conferences— white shirt, khakis, sandals, white socks. The Dolly the sheep cocreator is slightly more tense than usual, for he is engaged in numerous huddles to discuss what the conference meant for Singapore. But, in stark contrast to his Singapore bosses, there is little shtick in Colman. His shtick, if anything, is the fact that he remains at all times confidently, and a tad recklessly, himself. He immediately demonstrates this when asked how he feels about the announcement that was trumpeted at the conference that the U.S. Juvenile Diabetes Association had bypassed the United States to provide Singapore with $3 million to work on its hES cells.

"Actually, it's not that much," he says, his face placid. "It's only about $3 million. And in fact, I'm going to New York next week to make sure that I'm not a part of that, that this company should remain eligible for other, better funds."

Colman taps a pencil on the table. He was more relaxed a month ago, at Australia's first global hES cell conference, during which he skipped lectures to hit a soccer match with his son. He is about to find out just how much a part of the blank-check CMM he will become. The plan is for him to become one of the group of 150 MD/PhDs to be funded for "translational medicine" in Singapore, which will mean a lot more money and minds for his company. Colman envisions the CMM as something like the Institute for Molecular Medicine at Oxford, run by David Weatherall, a famous hematologist. "It is for up and coming scientists to do what they want for five years. A kind of research hotel, where people can do research they want without having to worry about teaching."

The very notion of it seems to soothe him into Total Colman Mode: hands behind head, sandals crossed. "I know they haven't set a figure for this [the CMM] yet. But Singapore has a history of delivering. Of course, I still don't know where Singapore gets its money. . . ." He laughs. "Singapore Airlines, for example, pays for its planes in cash. You can't write that off."

With a Boston company, Curis, from which it bought its diabetes franchise, Colman's ESI is examining insulin-producing cells from cadaver pancreatic ducts. Colman is hoping to find the related adult pancreatic stem cell. If he can find it, and can next persuade an hES cell to form it, he will have two cures for diabetes in his hands. "We don't know which cells are the stem cells," he admits easily. "We want to mark the different types of cells in the ducts genetically. . . . Then we'll do our wonderstuff: where do these cells come from? I have put some of my own people in there at Curis."

His people are learning about adult beta cells in Boston, so they can transfer that knowledge to hES cells here. "I'd like ultimately to get their people over here, but Singapore is difficult for many. It's halfway around the world from home, difficult to get green cards. And there are other difficulties. For example, there aren't enough dead people here." He smiles, clearly enjoying the fact that he is unraveling a lot of Singaporean spin here. The United States is the best place for adult stem cell work from a simple perspective, Colman continues: there are a lot more people there to die and collect adult stem cells from. "It can be a matter of economies of scale. The U.S. is by far the best arena for doing adult stem cell work of this kind."

Indeed, Colman is even relaxed as he admits that Singapore has still so far only ponied up 17 million in Singapore dollars for his company—a figure that has remained unchanged from the start. The money is coming, he says benignly. "And I'm going to be good at spending it."

Colman has perpetually smiling eyes and rustled red hair. Behind the conference table is the window of his office. Posted on the windowsill: a toy giraffe and a tiny red car. "I'm very grateful for the opportunities Singapore is giving to us. Its investors have deep pockets and a long term view." Still, he is blunt where his bosses are bombastic at every turn. There were some venture capitalists at the conference, he notes, but "a lot of it was just dog and pony show stuff. Singapore invests in some of their venture funds, so quid pro quo, some of them come here to listen. There have been no decisions. They tend to like molecules, an easier sell. But one is apparently excited about all this, Stellus Apodopolis VC, which put up half of the original $17 million."

Colman is also blunt about ESI's prospects. He is starting collaborations with two U.S. stem cell stars, Melton and Ari Brianvalou of Rockefeller University. "While we've scaled things up, we still have problems converting the cells to islet cells. Still, it's a big market." Another argument for attacking diabetes first is that, since it is not an immediately life-threatening disease, "regulatory hurdles may be less."

Unlike his bosses, he doesn't see clinical trials being faster in Singapore. "We foresee doing our clinical trials in the U.S. or some European country. Drugs can be fast-tracked in the U.S.—Gleevec [a new cancer drug] went to clinical trial in less than 15 years."

He's even forthright about his potential to pull ahead of Geron, his chief commercial rival. "Maybe someday," he says. "But Geron just raised $60 million in the last few days."

He does refuse to confirm, however, the rumor that ESI will have to look elsewhere to derive new hES cell lines. The word is the company will have to go to Europe for the new lines because it can't get enough embryos here. The reason: professional jealousy in the IVF clinics. Fiefdoms. It is a complaint being voiced from Egypt to China.

And a recent development showed for Colman that his instincts have carried him to the right place. PPL Therapeutics recently tried to persuade ESI to buy its stem cell work before it went under. "Why bother?" he says, shrugging. There's irony in this. Colman was at PPL when he and Ian Wilmut's team at Roslin in Scotland cloned Dolly. When PPL indicated shortly thereafter that it was more interested in transgenic animal work than stem cells, he had tried to buy its stem cell division. PPL, basking in the international limelight, wasn't interested then, even though it ended up focusing on transgenic animal work anyway. All that was the reason he went to ESI in the first place, some two years ago. Now PPL is gone, and Colman is CSO of the second-largest hES cell company in the world.

"Unusually pleasant divorce though," he muses. "My company didn't have to pay severance." Because Singapore made him a nice personal offer? "You assume right. And income tax is fantastic here. If I offer 100,000 in Singapore dollars to a postdoc, he pays 4,000 Singapore dollars in taxes."

It is in the end, therefore, not altogether unsurprising that he claims that he was misquoted in a Reuters article, shortly after he arrived in Singapore, that indicated he was unhappy here, "I was misquoted, because I was simply saying I needed at least $70 million and it looked like I wasn't going to get that," he says, not looking entirely comfortable. "It was taken as a criticism of the government, and it shouldn't have been. SARS has showed the power of this government."

And the censorship? "I don't notice it," he says more easily. Besides, he adds, "This is a very benign, complacent population. And in many ways rightly so, because it has seen a great improvement in quality of life in the last 20 years. Western values of liberty are simply not so important here. Eighty percent of the population lives in government-subsidized condos, but there are no ghettos. They are mixed racially, the condos, which I think is very smart. . . . The worst part about living here is the humidity, the best part is the cleanliness. I've done a lot of traveling being here. Indonesia, Malaysia, Sri Lanka, Thailand . . . And I am training to scuba dive. You have to adapt to your environment." He grins.

"Dolly was very good to me."

*   *   *

DESPITE THE problems Colman has encountered, it is true that he is standing in a very good spot for stem cell research at this moment. Singapore has some obvious flaws, from a Western perspective, that bothered more than one scientist. "Disturbing," one of the speakers will call the city-state. "Amazing attempt at social engineering." This is a view that many Asians who live in more "controlled" democracies than those in the West view with "contempt," Asia watcher James Fallows has written. "Thirty years ago, they say, their countries were poor and dependent. Now they are prosperous and confident. Americans can consider this a mistake all they want." Be that as it may, when one's goal is to attract Westerners to one's shores, it is a view that matters.

Yet Singapore's emphasis on collaboration is spot-on for the complex stem cell field, many scientists believe, as is its apparent willingness to spend a lot of cash. "They are investing heavily and they have a national plan," notes University of Toronto public health services professor Abdallah Daar. Furthermore, history has shown that, at least in the short run, "heavy-handed" government can sometimes be an asset. "Don't under-estimate the value of the ability to focus, know where you are going as a nation, have a strategic plan." Daar adds that Singapore may end up changing to meet the needs of the very biotech industry it seeks to domi-nate. Rockefeller University cloning expert Peter Mombaerts agrees. Notes hES cell expert, the University of Sheffield's Peter Andrews, "I think they will have important insights and relevant people."

Still, Singapore will not succeed over the next couple of years to attract *any* of this conference's brighter lights to become permanent fix-tures in Biopolis's firmament. And such is the unpredictable world of the stem cell that Singapore will, in a few months, be squarely knocked out of the spotlight for several months by an obscure group of scientists in South Korea. That group, upon announcing that they have cloned the first human cells, will appear to prove that Dolly can be very, very good to scientists—not to mention to entire struggling nations—indeed.

6

WORLD CLONING WARS

*We say, "Let's move the sky and see if it
happens."*

—*Curie Ahn,
member of the South Korean
cloning team of Woo Suk Hwang*

AMY, AMERICA'S FIRST ADULT CLONED COW, appeared from a
bucolic distance to be your standard, noncelebrity Every Cow. At the
end of a lovely September 2003 day in Storrs, Connecticut, on an
electric-green slope between Horsebarn Hill and Swan Lake, the black
beast with the white triangular birthmark on her forehead seemed to be
doing everything anyone could ask of a cow, and she was doing it to per-
fection: standing still, staring at and doing nothing, in the company of
some 40 other cows who were similarly absorbed in their excellent pursuit of nothing.

But things were different in the red barn behind Amy, which possessed a distinctly churchlike air—if a standard barn smell. Through a
row of windows, a heavenly parade of dusty sunbeams fell onto the hay-
strewn aisle to the milking station. To enter was to startle flocks of spar-
rows, who instantly shot to the four corners of the barn making the
sound of robes, swooshing (or spirits, rousing). As one walked, one was
uncomfortably aware of the stares of many utterly still, quietly judgmen-
tal, normal farm cows, who grew restless as it became clear that you
were, traitorously, not food. They seemed not unlike pious parishioners
waiting for the service, annoyed to see that not only were you not the
pastor but that you were wearing jeans in the presence of the Lord.

It was the perfect build-up to the Blessed Event to come immediately after, through a door beyond: the machine-driven milking of two cloned Holsteins named Betty and Cathy, caught face-to-rump behind aluminum bars. Both were born, like Amy, without egg ever meeting sperm; both were genetically near-identical to Amy. Both, like Amy, were black with white triangles on their foreheads, and they gazed up at visitors with the exact same (one could swear) look of unusually bright curiosity in their cartoonishly large, Milk Duds brown, artfully lashed eyes.

But what genuinely made this barn spooky was entering it with the knowledge that if you thought you might see the cloned bovines not just look alike but *act* alike, you might be right. For in a few weeks, their creator, one of the United States's top cloners, the University of Connecticut's Jerry Yang, would publish a paper noting that these clones were behaviorally different from the other cows here, and behaviorally similar to each other. Like 16-year-old Aspen (who was an ancient, postmenopausal matron—with a white triangle on her forehead—when her then-13-year-old ear cells were removed for the cloning), her holy trinity was more high-strung, expectant, and demanding than the other cows around them. They were all prima donna cows like their mother/sister/master copy.

Indeed, old notes taken of this grand dame when she was a young champion milking cow indicated that years ago she was unusually serious, not playful, highly curious, very aggressive sexually and during confrontation, and a meticulous groomer. Thirteen of her 14 naturally conceived calves possessed none of these qualities. But *all four* of her clones (a fourth, named Daisy, would die of a bovine virus contracted a few years after birth, and shortly after the study) possessed all of the above qualities, after being examined according to rigorous, standard animal observation guidelines, during a two-month period when they were young and kept in a pen area with four matched controls.

The four identical animals also sought each other out, constantly, somehow instantly knowing each other, despite the fact that all had been conceived in a fifth, unrelated, surrogate cow's womb.

Yang's article would conclude: "Behavioral trends were . . . observed in the clones that indicated that they exhibited higher levels of curiosity, more grooming activities and were more aggressive and dominant than controls. Furthermore, these four clones preferred each other or the donor as companions, which may indicate genetic kin recognition."

Moving away from the cloned twosome, who were still staring up at her expectantly, a Yang assistant put it thusly: "Nobody much likes the clones around here. You talk to the people who milk them: they're a pain. Aspen was rambunctious and so are they."

The ramifications, if borne out in other studies, may have some significance. Behavior may be driven by genes to a startling degree. Those grief-stricken pet owners out there who have been banging down pet-cloning company doors to get Bowser back may not be as delusional as has been thought.

True, it may well be discovered this does not hold for all cloned animals (the Hongseong pigs frolicking at the outset of his book, for example). And CC, the first cloned cat, made by Texas A&M University researchers, has a personality that is distinctive from her original— although *The Scientist* jokes that "felines do things as they see fit, even accessing their genomes." But it is also quite possible that some aspects of personality are "clonable," given the number of genes found yearly that scientists argue do or do not effect personality. The notion brings home the kind of impact on both research and therapy that cloning technology could bring, if allowed to proceed unfettered. To make a fairly exact genetic copy of an old individual's cells—except rejuvenated, returned to an embryonic state—is a mind-boggling feat (clones *do* possess different mitochondria, tiny energy packs: these derive from the different eggs used to create them). That this approach, because it creates hES cells immunologically identical to the donor, could also lead to an end run around hES cell rejection problems (since hES cells as they exist today come from spare IVF-clinic embryos, each of which match only 1 in 40,000 patients) seems almost extra-credit, in the face of the knowledge to be gained. The prospect still seems bizarre, even years after Dolly. Unfertilized eggs turn back the clock on the nuclei of old cells? "Still seems like a miracle to me," says Rockefeller University cloner Peter Mombaerts, who notes it is still possible not *all* cells can be turned backward this way—although the list of cells that *can* grows daily. If we can figure out how the egg does this, we can theoretically, someday, just swallow a cocktail of growth factors as we grow old to keep churning our cells backward.

The above is the sci-fi future that most cloners simply dream of. It is the outer limits of stem cell research, not least because there is an effective federal funding ban on human cell cloning in the United States. But by September 2003, 46-year-old Jerry Yang had been staring hard at that future for years. With his long-time agricultural cloning work with embryonic cells, which started back in 1983, Yang had helped establish some of the procedures used in 1997 by the Roslin Institute's Ian Wilmut and Keith Campbell to create Dolly, the first mammal cloned from an adult cell. (One of those procedures was the "combination activation" approach, which involves mimicking the activation that a normal sperm initiates in a normal egg by both adding chemicals [calcium] like those that woosh into the egg after the sperm penetrates it and a jolt of electricity à la Frankenstein. Yang developed the approach, if for the less historic cloning of embryonic mammalian cells, in the late 1980s.)

Prominent University of Michigan animal cloner Jose Cibelli noted in a documentary on Yang made early in 2003, "Jerry prepared the stage for Dolly." And at the First New England Symposium on Regenerative Biology and Medicine thrown by Yang this month, another top animal cloner, the University of Pittsburgh's Gerry Schatten, would tell the room, which included most of the top diplomats from the Chinese Consulate in New York: "If Jerry and Cindy [Yang's wife, molecular biologist Cindy Tian] hadn't started cloning cattle, we wouldn't have cloning."

Some cloning history in a nutshell: until the landmark year 1997, a giant year in science history, scientists could only clone embryonic mammal cells. Cambridge University's Steen Willadsen was the first to do this in farm animals in 1986, when he published a *Nature* report revealing that he had placed a sheep embryo cell in contact with an enucleated egg from another sheep and zapped them with electricity. The cells' membranes had fused "like the skins of merging soap bubbles," as Wilmut likes to put it, and the donor embryo's nucleus—the DNA-fueled control room of any cell—went to work in its new cytoplasmic home. Shortly after, the first cloned mammal via nuclear transfer was born.

But while this was interesting for agriculture, it was of less interest to humans. The real coup would be to take a cell from an old broken-down sheep, fuse *that* with an unfertilized egg, and get cells that were exact, nonrejectable copies of the old broken-down sheep—except, again, completely rejuvenated, reverted to their *embryonic* state. Most developmental biologists, since the beginning of developmental biology, believed firmly this was impossible. The hands of the cellular clock, it was believed, simply cannot be turned that far back.

The few attempts made failed, reinforcing the dogma. Then, in the mid 1990s, Wilmut's team had one of those obvious-in-hindsight thoughts. Perhaps one should fuse the adult cell only when it is in a state most resembling that of the egg; that is, the quiescent G0 or its similar G1 state, when it is not busy with the chores of being an adult cell: duplicating chromosomes, pumping out proteins. The Scottish team found that adult cells starved of growth factors fell into a quiescent state like hibernating bears. When the team then fused hundreds of quiescent adult cells with enucleated sheep eggs that were in metaphase II (the phase before release from the ovary), they ended up with Dolly.

In the years since, Yang, among others, had discovered that adult cells and oocytes (unfertilized eggs) can be successfully cloned in various cell-cycle stages if forced, and if other conditions are altered. Success rates are hugely variable. Still, Yang says, the main reason no one cloned an adult cell before Wilmut was "a matter of the mind. No one really thought it could be done. I didn't." Regardless, Wilmut and his chief collaborator, Keith Campbell, had the kind of minds that mattered: gifted, stubborn, curious to a fault. Their work electrified the world of developmental biology. "When Ian   a friend from Cambridge—told me about Dolly before his paper, I thought, 'Well, I have to go rewrite all my lecture notes,'" says Alan Trounson, Monash University hES cell pioneer. "It made me know: nothing's impossible. Nothing."

In the six years following Dolly, scientists all around the world scored a number of smaller, yet still significant, adult cloning coups. In addition to the above, Yang cloned the United States's first cow from an adult cell, the world's first mammal from an adult skin cell, and the world's first male animal from an adult cell. He was first to prove that clones from postmenopausal large animals can give birth normally and was among the first to confirm that cloned telomeres (which help determine age) are not prematurely shortened in large mammals. He was also first to prove that clones can be derived from adult cells cultured for two to three months, and would in June 2004 be first to prove that large mammals can be cloned from clones (serial large mammal cloning). He was a cloning player.

Furthermore, by September 2003, Yang's new $20 million Institute for Regenerative Medicine had been operating for a year and was holding its opening ceremony at the University of Connecticut to coincide with its first regional regenerative medicine conference. Its staff of six distinguished scientists, which would include scientists from Ian Wilmut's lab, the Jackson Lab in Bar Harbor, Maine, and Rudy Jaenisch's cloning lab at MIT, had been focusing on differentiating animal ES cells to tissue cells and attempting to dedifferentiate (as the egg does) mature adult cells back to ES cells. "We're all working around the theme of therapeutic cloning and stem cells," Yang said a few months before the opening. "I am very excited."

But, by September 2003, it was beginning to seem unlikely that Yang could keep his position toward the head of the cloning pack if he stayed in the United States. He had accepted the above offer from the University of Connecticut in 2000, a year before George Bush won office, when the pro-stem-cell Bill Clinton was still president, and all things seemed possible. Ever since Bush took office, it seemed more and more unlikely that Yang's efforts would be allowed to culminate in the pursuit of one of the biggest goals in this field: therapeutic cloning of human cells via the human oocyte. For Bush almost instantly signaled he would support any bill that called for an outright cloning ban in academia and industry alike, and never wavered.

This was a catch-22 for Yang. Given the stifling political environment, U.S. researchers, more than most, needed to study how therapeutic cloning might work via human oocytes, to see if they could do it in a noncontroversial way acceptable in the United States, that is, via growth factors (not the human egg). Yet other countries that allowed unfettered hES cell and human therapeutic cloning research were the only ones that could do the research to establish an eggless, noncontroversial, path to therapeutic cloning (aka somatic nuclear cell transfer) in the first place. And therapeutic cloning was looking important. As Kyoto University's Norio Nakatsuji, Japan hES cell pioneer, would note in 2004: "Somatic cell nuclear transfer is the shortest route to the clinic for hES cells, as things stand now."

Thus, the crowning future of Yang's field might not lie due north of Horsebarn Hill nor anywhere in the United States. There certainly was potential money to be made in animal cloning. Quality and quantity of product from normal beef cattle and dairy cows is variable. In the $30 to $35 billion U.S. beef industry, in which 40 million head of cattle are slaughtered every year, ranchers might gain an extra $200 per head, or $8 billion, if they could produce healthy clones. Similarly, in the $18 to $20 billion U.S. dairy industry, there are cows that produce 25,000 pounds of milk a year, far over the national average of 16,000. Output could theoretically be doubled by cloning the best animals.

But for cloners with higher ambitions, like Yang, the future of therapeutic cloning seemed to lie elsewhere. The future belonged, possibly, to the U.K., where laws allowed and encouraged human cell therapeutic cloning (although big, U.S.-sized bucks had not yet been thrown at it). The future belonged possibly to China, where big bucks were being thrown at it *and* the law allowed it. The future belonged possibly to Japan, whose scientists had scored an outsized number of animal cloning firsts; where the law would also shortly allow cloning work on human cells; and to which the United States's top cloner, Teru Wakayama, fled in 2002.

Furthermore, China had been repeatedly trying to lure China-born Yang back, as it was successfully—more successfully than any other nation—luring back scores of regenerative medicine postdocs and mature scientists.

The main reason Yang had not yet seriously contemplated returning was the metastatic salivary gland cancer he contracted a few years back. The United States was still the best country in which to have cancer and he had already undergone five surgeries on his face and lungs at a top East Coast cancer center. He was at the moment in remission (it was a slow-moving cancer) with only two small nodules in his lungs, but this was not expected to last. (It would return in early 2004, and reach both lungs by September 2005.)

So Yang had a unique problem. If he stayed here, odds looked good that he would not be able to push his work to the outer limit. But if he responded to numerous requests to return to China, there was no guarantee that he would get anywhere close to reaching that outer limit, no matter how much money—and freedom—was offered to him. China by no means, as yet, possessed the proper infrastructure for a booming biotech industry. And as it lacked the United States's cutting-edge oncologists, Yang could die before he got anything done.

Further complicating things was the fact that Yang was a hero in China. He was born in 1957 in Liasanjian, a poor farming village in North Central China, during a period of famine when 20,000 Chinese peasants died of starvation and the ensuing cultural revolution ensured few would go on to college. The diminutive Yang had not loomed large on the massive flat horizon, which looked in season like a green, agricultural superhighway, crisscrossed with crops of sweet potatoes and corn that were yet not enough for the even more massive populations of the surrounding villages. A catastrophic 1966 flood seemed to seal the deal. Yang was destined to lead the life of a swineherd. "People ask if I ever dreamed of becoming a scientist," he told *The Hartford Courant's* William Hathaway in 2000. "In China where I grew up, a dream was having a full meal."

Then in the late 1977, a window of opportunity: the state was allowing teens to take a test for college. He crammed for a month; got into the Beijing Agricultural University; went. He later ended up at Cornell University. Even though he arrived speaking no English, he received his PhD in 1990, and became head of Cornell's Department of Animal Science two years later. He started cloning work on animals in 1983 (if only with embryonic cells, not believing a Dolly was possible). To do so, he had to scrounge up funds from companies because Cornell did not quite get the significance of the work.

But scrounging was what his life had been about. He found the funds and time to send for, and marry, his college sweetheart, molecular biologist Cindy Tian, along the way. A few years later, he found one of the few U.S. schools that did "get" cloning, the University of Connecticut. The school relieved him of the need to scrounge by giving him a funded lab in 1996 and making him a tenured professor four years later, the shortest tenure track in the history of the school.

Throughout, Yang was also starting multi-million-dollar organizations to build bridges between U.S. and Chinese scientists. He started an IVF program that allowed small milk-barren Chinese cows to give birth to huge, prolifically milk-producing U.S. cows. All over China, to this day, bony beige Chinese cows are giving birth to massive, brightly colored Americans, causing shocked Chinese mother cows to do double-takes before resignedly accepting their strange children—gratis Yang.

China wanted Yang back—badly.

Yang, of all people, understood the limitations of his own work at this stage. His face had once been a perfect heart shape. Now there was a split in his pointed chin, thanks to one surgery. There was an uneven indent on the left side of his face that made him look, perennially, like someone who had just been punched by some great, gnarled fist, thanks to another surgery. His chest was scarred, thanks to a lung surgery. His back was scarred, thanks to a surgery that transferred fat from there to his ravaged face. He went home for lunch every day because he'd had so much of his jaw removed that his lips couldn't close around the rim of a cup, or to conceal his eating, and this embarrassed him. He knew he couldn't save himself via his own research—there was no time, no matter which country he chose.

But, as an embryologist by hard-won training, and a nuclear transfer pioneer by hard-won design, he was in the unique position to at least pursue a rudimentary understanding of both the state of birth, and, incredibly, that of rebirth, before he died. How much could he possibly contribute to that understanding if he stayed? How quickly would he die if he left?

Just before his first regeneration conference, he would say only: "In the area of therapeutic cloning, we can only focus on animal models here. Obviously China is in a position to be way more advanced. They can work on human cells. They have money, and they have thousands of monkeys available for research, which we don't have here [due to animal rights restrictions]. A huge center in China is always inviting me to go there. Many of my postdocs have gone. I always want to go back."

He would conclude with, "We'll see."

But soon enough, Yang would be reconsidering his position again. For in five months, a stunning advance would appear to come from cloning's far-left field—South Korea—and jack up the stakes for scientists. For South Korean researchers would claim in February 2004 that they had cloned the first human cells. The news would come a mere 10 months after prominent University of Pittsburgh cloner Gerry Schatten had predicted in *Science*, in his report on his hundreds of failed attempts to clone a monkey, that cloning of human cells may be impossible. It was as if someone "drew a sharp line between old world primates—including people—and other animals, saying, 'I'll let you clone cattle, mice, sheep, even rabbits and cats, but monkeys and humans require something more,'" he told *Science* in an article accompanying his April 2003 paper. *Science* concluded, "Nature already seems to have imposed its own limits on cloning." Ten months later, *Science* would do a 180, publishing the South Korean report—with, it would turn out, the help of an excited, chastened Schatten.

Ironically, it all seemed to mirror what had occurred in the embryonic animal cloning field: in 1984 prominent Wistar developmental biologist Davor Solter declared all mammalian cloning "biologically impossible"; in 1986, Willadsen published in *Nature* the fact that he had just cloned the first sheep with embryonic cells.

Regardless, the South Korean paper would seem to conclusively prove what U.S. stem cell scientists had feared for so long: biotech was no longer solely the province of the West. The advance would alternately depress, agitate, and exhilarate U.S. scientists. Despite the contention of many that this supposed first South Korean coup was simply a logical next step that any decent cloner could have pulled off in the right permissive environment—and with the right bounty of human eggs—the behind-closed-doors response in many of those same corners was, Holy Cloned Cow. *It apparently works.*

Johns Hopkins (and former Christopher Reeve) neurologist John McDonald would note in October 2004, "We thought it was impossible to clone human cells. Why? Because some major Western groups tried, and failed. Then this group from South Korea came along and proved you can. It was really nifty to see the other groups admit that they missed it, and why. Fascinating."

Still, five months before the South Korean announcement, it was business as usual on the political end of the worldwide cloning debate, as well. That is, it was a roller-coaster ride.

*   *   *

THE LAST MONTH in October 2003, at the Singapore conference, things at first seemed to be looking up for cloners. Indeed, the biggest and last laugh at the lavish conference concerned cloning. It came during the concluding speech by Alan Colman, who, after noting the success of the conference, said he'd found the answer to the field's woes when it came to getting the world to accept therapeutic cloning. "Clone

Philip," Colman said, referring to the dynamic Singapore official behind the conference, Philip Yeo. On the screen above, suddenly, there was one Philip Yeo, two Philip Yeos, three . . .

The roar of laughter that ensued wasn't unlike that of battle-weary soldiers who, if they weren't yet home, believed they could at least see it on the horizon. There were reasons to indulge in the idea that therapeutic cloning, the approach that might steer hES cells into the clinic most quickly, might be accepted worldwide. Singapore, of course, was demonstrating a powerful commitment to regenerative medicine, as were many other Asian countries, the U.K., and Israel. True, the Dickey-Wicker amendment of 1996 had forbidden the use of U.S. federal funds to harm an embryo. And the U.S. House in July 2001, after a mere three hours of debate and a lot of name-calling ("mad scientists" being one), had passed a bill that would ban all forms of human cloning and impose jail sentences of up to 10 years and fines of $1 million. ("It's amazing how incompetent the House was," MIT Whitehead Institute cloner Rudolf Jaenisch would comment.) But that bill had died conclusively in the Senate, and nothing similar had been passed since. It remained legal in the United States to clone cells in the private sector.

Furthermore, a team of American and Chinese scientists had just caused a storm by announcing at an October 13 conference that they'd achieved a pregnancy by using nuclear transfer to inject the nucleus of an older woman's fertilized egg into the cytoplasm of a younger woman's enucleated, fertilized egg. But it was a mini-storm, incomparable to the hurricane caused by the contested claim of Advanced Cell Technology (ACT) in November 2001 that it had cloned a handful of human cells. And 66 of the world's top scientific academies, from seven continents, had just come out in support of therapeutic cloning. In the uncertain realm of regenerative medicine, these were coups.

Yet the moment of levity at the international Singapore conference proved merely a time-out. For before and after Colman's speech that day at the end of October 2003, some of the same merrymakers were huddling in corners, worrying over the biggest threat to therapeutic cloning yet. As noted earlier, that week, the United Nations General Assembly legal committee was considering a resolution draft that the United States, Costa Rica, and 42 other nations had cosponsored. The resolution supported a total ban on all human cloning, both reproductive and therapeutic, worldwide. While nonbinding, it would send a powerful message: therapeutic cloning, perhaps the best way yet known to make hES cells clinically viable, was *bad*.

Instead of being sanguine, therefore, many in the room were nearing sanguinary. Another series of countermoves was being evaluated, including an upcoming meeting with the pope to be conducted by three of the United States's top scientists.

Still, by end of 2003, things would settle down. In November, the Organization of Islamic Conference (OIC) pushed and won a motion in the UN legal committee postponing the cloning vote for another two years. That this compromise came from the OIC was no surprise to those familiar with Islamic traditions—the Koranic sources and law—which say life does not begin at conception. (Depending on their sect or country, Muslims see the soul as entering the body between 40 and 120 days. This opens the door to hES cells, which are isolated in the blastocyst stage, from all sources.)

The upshot: on November 6, 2003, by an 80–79 margin (with 15 abstentions), the cloning vote was delayed until 2005. The stunningly close margin had the UN committee frozen like an audience of Pompeii victims for a long moment, swiveled in their seats, staring up at the numbers: both sides had come *that* close to a win/loss on the merits? (The undecided, and most of those against a total ban, voted for the delay. Most total banners voted not to delay.)

There had been an aftershock. Encouraged by the slim margin, the Costa Rica coalition, with the support of the United States, began quietly lobbying General Assembly members. Instead of rubber-stamping the legal committee's vote, as is the norm, Costa Rica suggested the Assembly take the unusual step of rejecting it and voting on the merits. On December 9, after patient advocacy groups and scientists had done a mass double-take of their own, then pelted the UN with another round of letters, the General Assembly rejected the legal committee's resolution. But because the "total ban" group still appeared to lack a majority, members agreed, without a vote, to simply postpone the matter for one year, not two.

The British delegation was "profoundly unhappy. . . . It is clear there is no consensus in respect to therapeutic cloning research," British Deputy Ambassador Adam Thomson said in a widely distributed notice to the press. Supporters of the Costa Rican resolution have effectively destroyed the possibility of action on the important area on which we are all agreed: a ban on reproductive cloning." Egypt's envoy signaled her alarm over the precedent set by reversing a legal committee vote.

Still, it was done. The future of therapeutic cloning remained, for the moment, just as it had always been: up for grabs.

This wasn't true for reproductive cloning. Even the UN's milder competing cloning resolution, offered by therapeutic cloning advocates Belgium and Britain, advocated a reproductive cloning ban. Many major international bodies have been quick to condemn reproductive cloning. And the Indian delegate to the UN legal committee noted in November that, judging from statements made by UN members, "every member state" is against it. "There is virtual consensus out there on reproductive cloning," agreed Dr. Thomas Murray, head of the Hastings Center for Bioethics, in early October. "The consensus is that it's a bad idea." Beyond the danger it poses to the child due to unsolved technical challenges, reproductive cloning elicits a universally visceral response, Murray said, most likely because it challenges the institution at the core of most religions and cultures: the family. "It upsets all kinds of relationships, particularly that of parents and off-spring." Therapeutic cloning elicits a far greater variety of responses. "For the foreseeable future," Murray had predicted, "therapeutic cloning laws will be all over the map."

Did therapeutic cloning stand a chance of being accepted globally? Thanks to some fascinating UN and local debates, clear themes had at least emerged throughout the pivotal year of 2003, and on into 2004, in arguments pro and con. In many developing nations, for example, there had been more than a fear that commercial use of the technique would create a demand for the eggs of poverty-stricken women; there had been an outright conviction. Nigerian envoy Felix E. Awanbor told the UN committee after the November 6 vote that Nigeria supported a total reproductive and therapeutic cloning ban because women from developing nations, "particularly Nigeria, are most likely to be at risk as easy targets to source the billions of embryos required for scientific experimentation." Uganda's envoy echoed that conviction to help explain why her nation made the "difficult" decision to part ways with the Organization of Islamic Conference, of which Uganda is a member, to vote against a delay of the total ban it supported so "wholeheartedly."

At a UN debate two weeks prior, three other UN representatives defended their support of a total ban for the same reason. Fiji's Asenaca Uluiviti, Sierra Leone's Allieu Ibrahim Kanu, and Portugal's Sebastao Jose Povoas all stated that banning all forms of cloning, not just reproductive cloning, would protect poor women. All three voted not to delay the vote. Whether a growing international trend—banning profiteering from egg sales—would reverse the convictions of such nations, remained to be seen.

Also clear during the UN debates was the fact that many lobbying for total bans apparently still believed that therapeutic and reproductive cloning were one and the same; that the technology to create a neuron was the same as that to create a human. Sierra Leone's Kanu noted at a final cloning debate that if the UN called for a ban on reproductive cloning only, there was no guarantee that unscrupulous scientists wouldn't just use therapeutic cloning to accomplish it. Anacleto Rei A. Lacanilao of the Philippines said his nation supported a total ban because it believed cloning was a single process; there was not one kind of cloning for research and one for reproduction; the science was the same. If research cloning was allowed, he said, it would be only a matter of time before an embryo was implanted in a womb.

Among those favoring a total ban, some also clearly believed that therapeutic cloning offered nothing new. During debates, Chilean envoy Pedro Ortuzar said Chile would vote for a full ban because he believed other research tools could perfectly well substitute for therapeutic cloning.

And a few times during UN deliberations it was noted, without irony, that scientists would be better off trying to cure disease—as if therapeutic cloning was about something else, entirely—or poverty. United Republic of Tanzania envoy Andy Mwandembwa explained his country embraced a total ban because therapeutic cloning would send the wrong signal by authorizing the creation of human embryos for experimentation. The substantial resources that were expended on cloning, he said, should be diverted to real crises, like HIV/AIDS. Envoys from Fiji and Uganda made the same point.

Education and promises for strict regulation could ameliorate such concerns, many scientists believed. But many others believed that, no matter how often these were voiced, they weren't the reason the UN decision was tabled. Said University of Pennsylvania ethicist Arthur Caplan, who made a presentation to the UN committee, in November 2003: "These were concerns, but I don't think they are what swung this vote. I think the whole vote keyed on the status of the embryo. It was a battle about the metaphysical and moral standing of the embryo."

Indeed, many Catholic countries voted not to delay a vote. Ireland, Argentina, and Peru, to name a few, had already codified the Catholic belief that the right to life of the "unborn child" was equal to that of the mother from conception on, making it difficult to justify creating a cloned embryo only to destroy it.

Yet, fascinatingly, it was becoming clear that not all Catholic countries were walking in lockstep with the pope. After much deliberation, the Spanish government would by the spring of 2004 be expected to allow limited hES cell research, even announcing it would form a bank for its hES cells. Spain supported a total UN cloning ban in 2003, but its new government, elected in March 2004 after the Madrid bombing, would eventually approve limited human therapeutic cloning. Brazil's Chamber of Deputies would pass a confusing bill in March 2004 that would prohibit creating embryos for research but allow therapeutic cloning. It would be variously interpreted. But the government had earlier earmarked $2 million for stem cell work that the science minister had said should include therapeutic cloning. And Brazil's UN delegation in 2003 recognized "the potential of therapeutic cloning in the alleviation of suffering" and voted only for a UN reproductive cloning ban. Many Brazilian scientists believed, as of early 2004, that therapeutic cloning would be allowed. "Let's hope," said Mayana Zatz, director of Brazil's Human Genome Research Center. Within a few months, that hope would be granted: a measure would pass allowing hES cells to be derived from spare IVF-clinic embryos that have been frozen for three years.

While the Mexican parliament passed a ban on all cloning of human cells in December of 2003, Mexico was not one of the 44 cosponsors of the total UN ban motion. It chose to delay a vote for two years; its envoy, Alfonso Ascencio, noted his nation needed time to weigh both "human dignity" and "freedom of scientific research."

Furthermore, at a final 2003 UN cloning debate, Cuban envoy Bruno Rodriguez had said he felt reproductive cloning was ethically unacceptable. But, he said, science should not be restricted, as the approach may help target cures for degenerative brain disease. Cuba was one of 13 co-sponsors of the resolution draft calling for a reproductive cloning ban only.

Because Cuba had a solid if small biotech industry, one could conjecture that economics played a role in its stance. But this wasn't true of other Catholic nations willing to consider therapeutic cloning. And ethicists were observing that a small if growing number of Catholics outside the biotech world were rethinking the issue. An earlier, "centuries-old" Catholic belief was that "a certain amount of development (beyond conception) is necessary in order for a conceptus to warrant personal status," Yale University theologian Margaret Foley wrote in 2003. Since this mirrored the fact that "embryological studies now show that fertilization is itself a 'process' not a 'moment' . . . a growing number of Catholic moral theologians do not consider the human embryo in its earliest stages . . . to constitute an individualized human entity."

Indeed, like Islam, many Protestant sects also saw the start of human life as a "process." And Orthodox Jewish law "is very clear," said Technion stem cell scientist Karl Skorecki in the fall of 2003. "The full sanctity of life begins at birth." Outside the womb the embryo has little status, Skorecki said; in vivo, it gains in status as it grows, starting to be taken seriously at about 40 days, as is true for some Islamic sects, when ensoulment occurs. This clarity—and early public education sessions—was one reason human embryonic research generated comparatively little controversy in Israel, he said.

Still, some nations were unlikely to budge on the issue, and not all of those were Catholic. Other Christian denominations possessed similar objections. Norway is a largely Lutheran, rural nation in which religious leaders have a certain sway. It was one of the few countries that had already banned it all by the end of 2003. A law banning reproductive cloning and therapeutic cloning "as a method to produce ES cells for medical research" went into effect in January 2003. The Norwegian Biotechnology Advisory Board tried to at least save other sources of ES cells. "The board made two statements suggesting that research on left-over IVF-eggs should be allowed," said Norwegian Biotechnology Advisory Board senior adviser Ole Johan Borge. But "neither the Government nor the Parliament followed the board's advice. The law is final. The board has no authority to change it or lift the ban by it ourselves." The nation also rejected the two-year UN delay option.

Indeed, despite the exceptions, it was tempting to view the therapeutic cloning roadblock as a purely Western one on a shallow level and a purely Western fundamentalist Christian one at a deeper level. To a certain degree, this was true. Therapeutic cloning advocates Belgium, Sweden, and the U.K. may have been Protestant countries that had to varying degrees adopted more liberal laws than Norway, but their governments were more secular than Norway's. Similarly, there was strong support for therapeutic cloning in China, where religion "plays almost no role in government, it's an almost scientistic nation," as Caplan noted. China was a cosponsor of the UN resolution draft to ban reproductive cloning only, as was Singapore, with its largely Chinese and Muslim population. India, a Hindu and Muslim nation, began hES cell work years ago. It banned reproductive cloning in 1997 but one of its UN envoys, Dr. Manimuthu Ghandi, told the UN in 2002 that it supported therapeutic cloning, if with tight oversight. Scientists have speculated this may have been due, in part, to a belief in reincarnation, shared in Hinduism and Buddhism, that places value in the notion of "recycling" life. This is simplistic, but indeed, India voted to delay the UN vote for two years.

And Russia, long a comparatively secular nation, also voted to delay a UN vote for two years. It was, in 2003, working off a five-year moratorium on human cloning alone.

Furthermore, the eleventh-hour entry of the Organization of Islamic Conference, which offered the UN debate's first compromise, certainly seemed to reinforce the view that Western Christian fundamentalism was all that stood between therapeutic cloning and world approval, Caplan suggested in December of 2003. If it had come to a vote on the merits, while Caplan did not believe the United States had the 100 votes it claimed, it might have won "by a very small margin, maybe." But, he said, "the opposition to a full ban was strong" not least because "it was, politically, going to be very difficult for the U.S. to line up support. Many third world countries were saying, 'You're accepting a Western moral framework if you oppose the right of China or India or Singapore to do this research.' It was starting to look like moral imperialism."

Still, "moral" and "metaphysical" reservations clearly did not belong to the West alone. It may have been telling that the OIC did not cosponsor the reproductive cloning-only UN ban proposal. "The Islamic organization may do that, eventually," Caplan said that December. "But I think they wanted time to think, to talk to their religious leaders. And religious leaders go slow, they see questions in terms of eternity, not next month's vote."

One thing was clear: nations that saw the start of human life as a "process, not a moment" tended to have an easier time with the notion of therapeutic cloning. Indeed, as noted earlier, some countries were coming to view the start of human life the way the West now generally views its end: as a matter of the mind. Since most Western nations define human life as ending with the brain, the reasoning went, shouldn't human life be defined as starting with the brain, the period where cells first begin to differentiate, somewhere along day 14 (long after the stage—day 5—where hES cells are formed, and must be isolated)? The additional, inescapable fact that therapeutic cloning had muddied our definition of conception—no longer just egg and sperm, it was (theoretically) egg and any cell—bolstered the conviction of countries like the U.K., Belgium, Singapore, and China that work on embryos up to day 14 of development was ethically sound.

Regardless, the glass remained half-full, or half-empty, no matter what your views, at the start of 2004. At that point, only 19 nations had either passed laws allowing therapeutic cloning or exhibited a strong interest in doing so. A November 2003 Eos Gallup Europe poll of 15,000 Europeans from all 30 of the current and future EU states found that the majority supported therapeutic cloning, but the majority was slim in most of those countries. The world was neatly divided.

And that situation would undoubtedly not change, Murray predicted, "until ES cell research begins to show therapeutic potential." Caplan agreed. Certainly, in the near future, he said, "no shift will happen in the U.S.A. position unless there is a new administration here, and that is a very remote possibility. The same allies will line up for the total ban. I think the UN will stalemate." Still, he said, by the next vote, "there may be persuasive medical evidence that stem cell research will work, or that it won't. If that happens, the UN debate would change accordingly. Evidence of strong success in animals could tip the debate toward allowing human research. Setbacks, or no progress, could tip the debate the other way."

Enter South Korea.

*   *   *

THREE SOUTH KOREAN institutions were among the handful that had created their own hES cell lines by the time of the Bush cutoff for federal hES cell funding (August 2001). But in comparison to countries like the United States, U.K., Australia, and Israel, the nation was utterly silent on the topic of its hES cells for over two years. South Korea's stem cell lines were discounted by some as among the many dubious-to-nonexistent lines that the Bush administration had used to pad its final tally of lines eligible for federal funding.

Then, in June of 2003, just after Pittsburgh and Providence had thrown their contentious opening ES cell/adult stem cell conferences, respectively, and just before Yang would throw his opening New England cloning conference, there was a surprise. More than 28 South Koreans came from out of nowhere to attend the first annual meeting of the largest global stem cell conference, the June 2003 International Society for Stem Cell Research meeting in Washington, DC. As noted in an earlier chapter, Koreans represented, by far, the largest national coterie present outside of the United States. Koreans presented 10 posters on hES cells, more than any single nation (the United States presented 7 posters on its own; Australia, 5; Israel, 2; the U.K., 0). Furthermore, each Korean poster was presented by more scientists, from more institutes, than is usual.

Still, this caught the eye of almost no one back in June of 2003. Alan Trounson, head of Australia's stem cell effort, had heard a rumor throughout 2003 that the South Koreans had $80 million to spend on hES cells. But few noticed June's sudden bombardment, except for the NIH's Mahendra Rao, who had been asked to give a talk at what was to be South Korea's inaugural stem cell conference in Seoul that upcoming October. That conference had yet to be announced, but Rao thought something was going on, he noted that June. He just didn't know what it was. As a representative of the NIH, he would attend the South Korean conference to find out.

At the Seoul conference, South Korean researchers had presented preliminary evidence that they had learned how to weed out hES cells destined to become pancreatic cells (a step toward combating diabetes), growing hES cells on human feeder layers, and mass-producing hES cells—three of the most pressing concerns on the minds of many hES cell researchers. Still, the conference was not announced in any journals. It received zero press. And the South Koreans had not published hES cell work in top Western journals. After the conference, Rao knew they were also working on cloning human cells. He did not yet know if they had succeeded.

But in December 2003, some key events began to attract attention. South Koreans presented four posters on hES cells to the prestigious American Society for Cell Biology in San Francisco—the conference's *only* posters on hES cells. One was chosen as a "hot pick" by conference brass. In addition, several South Korean groups published four papers on hES cell and human cell cloning work in Western journals. One of those described the first use of hES cells to successfully combat some Parkinson's symptoms in mice—if it was not published in a high-profile stem cell journal.

At about the same time, an obscure animal cloning specialist at Seoul National University named Woo Suk Hwang made world headlines, including a story in *Nature's* news section, with the claim that he had cloned "mad cow disease resistant" cows. Hwang also detailed the production of the first cloned human/cow cells, the world's first paper on that phenomenon. And he shortly thereafter announced he had cloned some pigs with human genes—a claim backed up by published work. Said South Korean President Moo Hyun Roh after visiting Hwang's lab, "I am electrified at this. The government will not spare anything to support further advancement of the study."

Apparently, the president wasn't kidding, for that very month the Korean National Assembly announced that it had approved a licensing procedure for human therapeutic cloning. Seeing as there was no organized public cloning debate in South Korea preceding that government announcement—and no cloned human cells anywhere that the world knew of—the latter at first seemed the oddest development of all.

But the timing of the government's move would soon become clear, if incredibly, no news leaked to the West that anything highly unusual might be occurring in South Korea until 3:59 P.M. on Monday, February 9, 2004. At that time, Sean Tipton, head of the stem cell advocacy group CAMR (Coalition for the Advancement of Medical Research), sent out an excited e-mail to some prominent stem cell figures. "Big week for science" the header read, followed by, "Hey folks, this week's SCIENCE will have an article from a Korean lab that has gotten an ES line from cloned human embryos. [The lab] did SCNT [somatic cell nuclear transfer, or cloning] of a woman's own cell (a cumulus cell) and her own egg took the embryo to the blastocyst stage and got a robust set of ES cell lines from it. This is a very legit paper in a high prestige journal. We are working on a CAMR statement praising it. This is embargoed until Thursday afternoon, though we are trying to get something ready in case the embargo is broken."

A silence followed on Monday, Tuesday, and Wednesday morning. Then, at around 4:00 P.M. on Wednesday the 11th, 21 hours before the press embargo was to lift, *The New York Times*'s Gina Kolata posted a story under the headline, "Scientists Create Human Embryos Through Cloning." Noting that the news embargo was broken earlier by a South Korean newspaper (*JoongAng Ilbo*), the article's first quote came from Robert Lanza, scientific head of ACT, the only U.S. company to have persistently tried to clone human cells in the past. "You now have the cookbook (for human therapeutic cloning)," he said flatly. The second quote came from Richard Rawlins, head of assisted reproduction laboratories at Rush University Medical Center. "My reaction is basically, 'Wow' . . . It's a landmark paper."

The *New Zealand Herald* chimed in at 9:00 P.M. with the head, "First cloned human stem cells reported by scientists." *The Wall Street Journal* followed at 11:38 P.M. with a story headlined, "Human Embryo Successfully Cloned in South Korea." There was another brief lull. Then, an explosion.

As if being fed onto a tickertape, from the wee hours of 2:00 A.M. U.S. EST to 4:21 A.M. U.S. EST on February 12, more than 60 articles from around the world silently clocked onto the wires. The stories were appearing in publications ranging from the U.K.'s *Bath Chronicle* to the *South African News*, *The Missoulian* in Montana, Australia's *ABC Online*, *Trade Arabia*, *Novinite* in Bulgaria, *RTE Interactive* in Ireland, *US Newswire*, *The Tallahassee Democrat*, Canada's *Globe and Mail*, Russia's *Itar-Tass*, and the *Wyoming News*. By midmorning, stories were flying onto the wires and the Internet in dense flocks. It was just the start. By month's end several hundred publications would post articles about the apparent coup; by the end of four months, thousands of publica tions would mention the achievement. The University of Pennsylvania's Hans Scholer, listed by *Science* as an expert consultant willing to comment, would still be receiving 100 e-mailed inquiries a day as late as February 29, over two weeks later.

However, the immediate upshot was that, by the time the day's cloning press conference commenced at 2:00 P.M.—when the embargo was officially lifted—journalists worldwide were scrambling for their phones, fighting busy signals to score a coveted conference-call slot. So instead of a handful of local reporters, the world was waiting as two small South Korean men in meticulous dark suits and bright white shirts walked into the press room. They were introduced as the aforementioned Woo Suk Hwang and South Korean Stem Cell Center head Shin Yong Moon. They took their seats. After some initial bungling prompted laughter (the press and the Koreans both had to be sternly advised that their cell phones had to be turned off), Donald Kennedy, editor of *Science,* made the announcement official.

"This represents an extraordinary series of technical accomplishments," Kennedy said. "The authors will probably modestly avoid telling you the extent to which they applied great experimental ingenuity and care into the culture medium, the fusion and activation process, and the process of somatic cell nuclear transfer itself which is difficult and challenging, and has brought trouble in some other hands. The result is a terrific accomplishment . . . that has very important potential, long range therapeutic impact." He noted the study's possible limitations: of 242 human eggs, only 1 cloned hES cell line was generated; and the team couldn't clone male cells or any cells coming from individuals not the egg donors. Regardless, Kennedy continued, the Koreans' work represented a landmark step toward the routine creation of hES cells that will not be rejected by the body.

The two scientists spoke briefly and quietly. Their voices hesitant, they briefly ran over some key details of their paper and called upon world governments to ban reproductive cloning. They were followed by an ethicist who declared that stem cell research was now officially "international."

With that came press questions, and the honeymoon showed signs of being over before it began. Few if any reporters seemed interested in the science. A British reporter immediately asked The Question, even though it had already been answered: could the South Koreans' technique lead to reproductive human cloning? Perhaps, the Koreans quietly repeated, adding that nations should work on laws banning such a step immediately. Another reporter asked if the cells would be made publicly available. The Koreans noted they would be happy to work with others. Were they going to commercialize their cells? The scientists noted that Seoul National University had applied for a worldwide patent, but that they themselves did not stand to gain. Was there any way the Koreans could guarantee the method would not be used to create human clones? The scientists repeated their earlier answer.

There was a moment of humor. When asked, "How easy will it be to imitate what you've done?" Hwang noted that it may not be easy because "the human oocyte is very sticky." Added Moon, "Like oriental rice, very sticky."

But by the end of the press conference, it was clear that this was going to be yet another stem cell first that might generate equal parts controversy and praise.

Still, as the days passed, while the honeymoon was not Hollywood-perfect, it was among the most pleasant to be enjoyed by any hES cell researchers to date. Since 1997, the year of Dolly, geeks, politicians, and gurus alike had been fretting over the day that the first human cells would be cloned. But perhaps in part due to the rigorous and careful science of the Koreans' work, in part due to their emphasis on its therapeutic ramifications, during the first weeks after their paper it rained valentines on the two scientists—who were, as it turned out, humorous, highly quotable, and a great story all on their own.

Hwang was a 51-year-old workaholic born to a poor family of six children. Fatherless since age five, he had helped his mother raise cows to support the family, whose home was heated with cow pats. Sometimes the family ate grass to survive. The constant contact with cows gave him affection for—and he insisted again and again, grinning—understanding of, the cows he would eventually clone. The first in his village of 50 to graduate elementary school, he earned a full scholarship to Seoul National University.

Over and over he said, seriously, that a majority of South Koreans tolerated cloning of human cells perhaps because at the heart of

Buddhism is reincarnation, the "recycling" of life. Over and over he told the press, less seriously, that part of his teams' success could be attributed to Korea's metal chopsticks. The metal makes them difficult to grip, so Koreans have a developed sense of hand-eye coordination, he claimed.

Moon, for his part, revealed himself to be "a beer-drinking Methodist" and 56-year-old IVF specialist who was more middle class than Hwang in the years after the Korean War, but still poor. ("My first words in English were, 'Hello soldier, give me a chocolate!'" he repeated over and over, also with a smile.) He deferred graciously and constantly to Hwang, who was the one who had pulled off the cloning feat. (Moon, who isolated Korea's first hES cells, took over at the hES cell stage.) "He works morning, noon and night over there: it is a cloning academy," Moon repeated proudly.

When contacted by cell phone after the initial press conference, the two were in a loud and crowded Seattle Starbucks, celebrating with the many staffers they had brought. Laughing and happy, both insisted on passing the phone around to their underlings to give them proper credit. And while sipping Cokes in front of the Space Needle, *The New York Times* reported, the two took pictures of the reporters taking pictures of *them,* in part for the humor; in part as a gesture of thanks for the attention. Indeed, such was the near-boyish enthusiasm of the two that when they returned to Korea with the words "We have placed a Korean flag at the summit of global bioengineering research," the statement seemed less a boast than an effort to share their overflowing excitement with everyone they saw.

But also becoming clear in the weeks after the *Science* paper was the fact that the achievements of tiny South Korea might soon seem to range far beyond one experiment; that South Korea wanted to place more than one flag on that summit. Indeed, it was looking as if South Korea, of all the nations involved in the stem cell race, may have been making, from the start, some of the smartest choices of all.

For South Korea's supposed cloning coup may have been no accident, said NIH stem cell task force member Mahendra Rao a few days after the announcement. Three years earlier, he noted, South Korea organized a highly efficient $80 million private/public hES cell and cloning initiative that allowed no effort to be replicated, and no resource wasted. The result was labs that "rival any in the world, and I have said so in a report to the NIH. The labs there are five times better than mine at the NIH in terms of square footage, supporting infrastructure equipment." In agreement was Franco Marincola, NIH Immunogenetics Lab director, after speaking at an adult stem cell conference in Seoul. "Definitely Korea is very aggressive in getting things done in the most efficient way," he said in an e-mail in February. "Their labs and hospitals are very competitive and definitely comparable to the U.S."

Furthermore, said NIH stem cell task force head James Battey, "That cloning paper involved a huge team of investigators—the South Koreans are very good at assembling large teams of scientists." This, combined with the fact that South Korea had been making basic-science investments representing a greater percentage of GDP than the United States since the 1990s—although little of that until recently went to life sciences—had apparently resulted in a bioscience culture that had helped catapult South Korea to the forefront of the stem cell effort for the moment.

Rao was particularly enthused. "Three years ago the country's three main IVF clinics—the Pochon University clinic and the MizMedi clinic, which were private, with the government's Seoul National University clinic—decided to join forces rather than compete. Moon was responsible for it. He formed an intriguing public/private stem cell consortium out of them, and convinced the government to fund most of it with a promise of $80 million over ten years."

Rao continued: "It sounds like a huge stretch but, when the U.S. decided to build the atomic bomb, they had all these scientists all over the U.S. and they had great facilities. But the most critical decision was to get them all in one place for the Manhattan Project. The reason you do that is that you get a huge synergy between people being able to talk to each other and be able to share one little reagent or one skill or one technique. That's key, all of us realize it, and that's why the NIH did the same thing—pooled resources—for the human genome project. And that's the kind of synergy they have in Korea."

Such intensive collaboration is becoming the rule of thumb in any number of fields, he added. "A lot of the funding agencies and other groups in science are coming to the realization that large-scale science works better than small. It used to be, for scientists, 'I will do my own thing and I will surprise the world with a major breakthrough.' But if you look at the Nobel prizes in the last five to ten years, it has by and large been about large-scale science. The development of the MRI, the sequencing of the human genome, homologous recombination work, that zebrafish mutation project, and so on. All of them are the result of large-scale work, because it became clear it would not be possible for a single lab in this day and age to do it."

Furthermore, said Rao, Korea's collaborative ties bind beyond country lines. "Moon went to school—Virginia's Jones Reproductive Institute—with a prominent Chinese embryonic stem cell researcher. They've been helping each other out; they're drinking buddies." That researcher, Cheng Guihan of Beijing's Third Medical University, is an IVF pioneer in China.

Beyond the unusually collaborative nature of Korea's endeavors, Rao said, the nation has been wise with its spending. Unlike Singapore, "they're not going out and building new buildings." Old buildings are simply being stocked with new equipment donated in part by Samsung. Hwang's lab may be in a prefab building on the outskirts of the SNU campus, but "it's state of the art, inside. . . . There is a 'vision thing' going on. Korea's scientists have got it."

U.S. scientists were not alone in their admiration. A month after the coup, traditional rival Japan would have kind words. "So skillful, so successful," Shin-ichi Nishikawa, Riken stem cell leader, would note. "I was surprised when I went there," pioneering mouse cloner Teru Wakayama would volunteer. "They are so smart and very motivated." The University of Kyoto's Norio Nakatsuji, who created Japan's first three hES cell lines, would add, "The government is pushing very hard over there."

So by the time the two scientists returned to their country, their government and much of the citizenry appeared to have "gotten it," as well. The sense of nationalism the coup had stirred up surpassed any such feeling generated by the country's hugely successful car company, Hyundai. It was almost on a par with the Ayatollah's announcement that Iran's isolation of an hES cell line proved his country's superiority in the war on the West. By the end of February, as a direct result of the cloning paper, the government announced it would nearly double Moon's stem cell network funds from $80 million to $130 million and would create a "medical valley" dedicated to transplantation and stem cells.

In making the former announcement, Health Minister Hwa Jung

Kim said: "If Professor Hwang's achievement is adapted for practical use, then Korea will become the center of the world in somatic treatment and organ transplants. . . . If we are the first to commercialize this technology, then it will be inevitable for patients with incurable disease to come to Korea." A few days later *The Korea Times* quoted him getting even more specific: "We need to create a complex to monitor all related research on cloned embryos. The complex will cover research, treatment and practical application."

Furthermore, the minister of science announced he would form an association to fund stem cell research for Hwang, and made the first of several government announcements promoting a Nobel for Hwang. This was not likely—U.S. scientists noted the Koreans' accomplishment at that time appeared to be one largely of great effort, and opportunity. There was no completely surprising twist in their work.

Still the Korean press went wild, one day denouncing the government for having allowed Hwang to stay in a $50 a night motel while he had been presenting his paper in a swanky U.S. hotel, the next decrying his salary (50 million won a year, or $40,000). Hwang was given a Korean science prize totaling $250,000. Among the people, the nationalist element and the cheerful anti-U.S. sentiment was so vocal that it became clear that, although another ingredient may not be necessary for stem cell science, it was helpful: the desire to trounce the United States's arrogant behind. It was an element that had also clearly been influential in the Middle East's decision to cause the UN cloning vote delay.

Whatever the ingredients that had gone into the South Koreans' apparent success, one thing was clear: it was a success by local standards and international standards. In March, the private Korean Maria Biotech published a paper on hES cells in the prestigious British journal *Human Reproduction* for the first time. An SNU report appeared in another prestigious British journal, *The Lancet,* finding that blood stem cells injected into the coronary arteries may make it beat more strongly, but may also contribute to atherosclerosis. Hwang received scout offers—and offers to speak—from 70 stem cell labs all over the world. South Korea, almost overnight, was being considered an authority on stem cells.

All this was not surprising to some Korea watchers outside the stem cell field. Business consultants Frost & Sullivan in November 2003 had reported, "Government initiatives such as introducing favorable policies and promoting start-ups with support for R&D activities have placed the South Korean biotechnology industry among the fastest growing biotech hubs." Biotech was new to the country, but the country was clearly determined to catch up.

But if some U.S. stem cell researchers and biotech analysts weren't completely surprised, many were exceedingly unhappy. In March 2004, a U.S. NIH official announced that of the 78 hES cell presidential lines, only 23 colonies would *ever* be viable. That same month the Pentagon announced it was funding an hES cell project in Sweden. Said Harvard/MIT stem cell researcher George Daley, echoing the sentiments of many, in a widely quoted statement to the press: "Federally funded scientists have to drive Model T's while Korean scientists get to drive around in the newest Porsche. It's crazy." "South Korea" was becoming the catchphrase that defined everything that was wrong with U.S. science.

Things remained almost supernaturally upbeat for the South Korean team. Then, in May, *Nature* reported that two of Hwang's anonymous egg donors may have been on his staff and that the hospital board overseeing ethics for his project was not in compliance with Korean Food and Drug Administration rules because it only included one nonscience member. Few publications bothered to run with the accusations. But the Koreans' return to the United States on June 1, 2004, four months after the publication of their paper, would not be quite as triumphal as it might have been. An air of uncertainty would hang over them for the first time, despite the fact that they had been invited to the United States as the world's leading human therapeutic cloning experts, to literally set the world straight.

*   *   *

ON ALL SIDES of the coffin-narrow United Nations pressroom on this gloomy June 2004 day are photos of important men in ties—and the cultural equivalent thereof—shaking hands. Diplomats shake the hands of presidents who shake the hands of sheiks above the line of sight, while battered encyclopedias, dictionaries, and UN guides tip this way and that in rickety bookcases below. It is as if these books have been blown about so relentlessly through the years by the force of all that important handshaking, all around, that their librarian long ago gave up trying to keep them tidy or up-to-date.

On a tiny, supernaturally bright, makeshift stage sits a clutch of cloning and stem cell specialists looking uncomfortable. The event was organized by the Genetics Policy Institute (GPI) to educate the press about therapeutic cloning, the day before GPI will educate 150 UN staffers. But once again, few questions have focused on science (most have focused on whether therapeutic cloning will lead to human cloning) and the scientists' faces run the gamut of emotions from bored to bothered.

Later, outside the room in a cooler and larger hallway the scientists look happier. They loosen their ties and separate and come together in the more natural formations of scientists in their native conference-break habitat: small gaggles of two, sometimes three, huddling in corners, comparing notes.

The exception is Woo Suk Hwang. The meticulously dressed, pleasantly smiling scientist is late breaking into the hallway because the minute the press conference ended, he was surrounded by reporters. Within seconds of his escape, he is again encircled by reporters, numerous boom muffs hanging over his head, numerous tape recorders at his face. As he talks, his expression alternates between amusement and alarm, as if he feels cornered by a horde of vaguely menacing marionettes and sock puppets.

Talking nonstop at the side of the hallway is Bernie Siegel. Famous for launching a lawsuit again the Raelians, the UFO group that claimed to have cloned a baby, to force them to provide documents proving their claims, Siegel is the event's organizer. Siegel formed the GPI, he explains, after he organized a petition asking the World Congress to rule human reproductive cloning an obscenity in 2003. His work during that time with diplomats led to a tip that the Bush administration had been quietly making UN rounds for months, trying to engineer a full UN cloning ban.

He instantly set up the institute and organized a letter-writing campaign, becoming a key force in the 2003 UN cloning vote postponement. "We were late getting around to the issue that last time," he says, as he surveys a room of stem cell stars. "We won't be late, the next time—the vote in October 2004. Hopefully the gathering today and the panel tomorrow will change the outcome of the UN debate. At the very least, it will show the UN that the stakeholders on the therapeutic cloning side will not go silently off into the night."

Indeed, later that night, at a reception held at a roped-off section of the elegant Metrazur restaurant in Grand Central Station, the scientists are not silent. Clutching their wines, they gather in large groups and talk loudly. Hwang looks more comfortable, able now to huddle like his peers, with fewer reporters around.

But when a *Nature* staffer begins talking quietly in a corner, there is a reminder that while the painted stars on the vast vaulted pool-green ceiling above are fixed and immutable, the human stars beneath them are not. "The belief in the office is that the charges against Hwang are true, that some of the eggs he used in the experiment came from his researchers," the staffer says. "And frankly, we're quite surprised this hasn't yet been followed up in the press."

If this turns out to be true, it will not impact the Koreans' apparent main accomplishment: the world's first cloning of human cells. But it could hurt their reputation, because it has the "whiff of coercion to it," as *Nature* had quoted University of Pennsylvania ethicist Arthur Caplan saying. This could make it more difficult for South Korean stem cell scientists to publish in major Western journals; collect funds from backers; be accepted. The Koreans' star could be dimmed, is the thought at this time.

Behind the *Nature* staffer, a small crowd of reporters has been gathering around John Wagner. Word has gotten out that a clinical trial that the University of Minnesota professor may head—giving Geron's hES cells to spinal cord patients—may be historic for two reasons. First, it should be the world's first clinical trial with real hES cells conducted by reputable scientists. And second, Geron hopes to go ahead with the trial next year.

It is big news, and the look on Wagner's face begins to alternate between amused and alarmed, as the press gravitates toward and around him.

*   *   *

THE NEXT DAY, the cloning universe remains a bit tilted, as the two South Koreans share a stage with Western cloners, instead of dominat-

ing it. The seminar for the UN delegates proceeds as such public seminars normally do: a bit hostilely, with the audience asking at times
pointed questions based on ethics, not science. The pro-cloning scientists, who include Ian Wilmut, father of Dolly the sheep, are asked how
to tell the difference between cloned cells that could be turned into a
human being and cells to be used as ES cells ("Intent," says MIT's Rudy
Jaenisch); whether there have been misleading facts circulated (the fact
that cloned cells actually could create a healthy human being, says
Wilmut); whether egg trafficking may become a problem in third world
countries (delegates "worry too much," says Moon; the UN should
police that, says Gerald Schatten of the University of Pittsburgh). Hack-
les are raised; one scientist walks away from a pro-life reporter who
keeps asking insistent questions.

But the cloning world rights itself, placing the two South Koreans at
its heart, at the lunch afterward, when Hwang is again overcome by a
sea of reporters.

In the lobby of his small Avalon hotel (he moved out of the elaborate
digs the UN provided in order to stay with his son, who is going to
school in the United States) Hwang later proves to be as boyishly enthusiastic and as relaxed as he appears. He's meticulously dressed in a dark
suit and bright white shirt, and is flanked by somber marble pillars—but
he reaches out, scientist-style, to draw enthusiastically all over the wall-
paper with his hands when he needs to describe his work.

The team's success was due to four modifications in the standard
approach, he says. They slightly shifted two ingredients in the culture
after the nuclear transfer. But more importantly: "The human egg has
a very unique character because it is very sticky . . ."—he touches a
thumb and fingertip together as if making a shadow puppet for a
child—"and very fragile. It is extremely vulnerable to rupture during
micromanipulation. So we squeezed the egg very gently" to get the
nucleus out, to make way for the adult nucleus. "We have 10 years of
animal cloning experience, pigs and cows, so we knew. Some people
use squeezing for bovines. We modified that. We made a slit in the
outside of the egg, and squeezed it between a micro-needle and
pipette."

(Johns Hopkins neurologist John McDonald will note a few months
later that it was this advance that many believed truly seemed to set the

South Korean effort apart. "I really think it was how they enucleate the egg. . . . There are key things in the cytoplasm that are important, and the old-fashioned way of sucking the nucleus out, sucked out too many other pieces. It was that simple. There were three major things, but that was the major advance that everyone points to.")

Hwang wanders a bit, enthusiastically: "We are now trying to clone a dog that way. No one has done this yet. We have gotten to the morula stage." One of his students helped clone the first cat, at Texas A&M. "I suggested the cat, I told her that dog was very, very hard. . . ." He laughs when asked if Korean chopstick prowess helped. "I think so. Koreans use steel chopsticks, and they are much more difficult to use than wooden chopsticks." It is impossible to tell if he truly means this.

He returns to the human work. Also key was the reprogramming time, the time that the egg and its new nucleus are allowed to get adjusted to each other before they are zapped into a state of activation, or fusion—something the sperm can prompt automatically after it penetrates the egg but which must be done artificially here. "We tried 10 minutes, 15 minutes, one hour, two hours, four hours, 10 hours," Hwang says, the list an enviable thing for U.S. scientists: each of those attempts involved human eggs, a scarce, expensive commodity here. "Eventually we found two hours was best." It allowed 25 percent of the eggs with their new/old nuclei to survive. "There is no two-hour reprogramming time needed for other species. For pig, reprogramming and fusion should occur at the same time, for example. The dog is close though. We are now studying why. One of my PhD students is studying this reprogramming."

Next they tried substances to kick off that fusion. They used a substance that other investigators have used to get unmated human eggs to develop a bit, using a concentration they arrived at, again, only after an enviable period of trial and error.

Hwang says the result was that 25 percent of the original 242 eggs were "fertilized," a result comparable to what occurs in cattle or pigs. A total of 30 blastocysts were obtained, and from that they obtained a single hES cell line, he says. The fact that the old mature cell came from cumulus cells, the autologous old mature protective coating their donor had around her egg, means there is a very slight chance that cloning didn't occur: that the group generated the first hES cells from an egg, period. But that chance is slight, *Science* had noted. The group received the help of the University of Michigan's Jose Cibelli, an expert in that phenomenon (parthenogenesis).

How do Koreans look at cloning? Hwang notes that some Korean bioethicists are up in arms, so he has temporarily halted his human research to let rules be established for the technique. But for many, the idea of "recycling is natural and normal because of Buddhism. A very famous leader in Buddhism came to talk to me. He explained the similarities of Buddhism and cloning. He agreed with my research. Almost every Buddhist leader supports this in Korea."

When the rules are established and he is free to get to work again, "we will be able to make ten new hES cell lines in a single year. We have the optimal facts. I think we can establish a single line out of only a few dozen eggs this time."

It was early 2003 when the team lured the first of their 30 fused eggs into the eight-cell morula stage, just before the blastocyst stage, that he knew they succeeded. "I knew we would get the hES cell then." Done by the fall of 2003—just as Singaporeans were ripping the plastic off the chairs in their multi-million-dollar complex—the Korean team submitted the paper. Then they told almost no one about it until February 12, 2004, when *Science* made the announcement, he says. (Although later it will be revealed that *Science* turned it down once. Then Hwang told Schatten about it—and Schatten helped Hwang rewrite it, and resubmit it to *Science*.)

Has life changed for him? "Well, I traveled here this time on a business-class ticket," he laughs. "I've been recently to Hong Kong, the UN, Pittsburgh . . ." He laughs again. "I speak today to 500 Korean Americans. I have received 30 invitations already to speak this year, and have accepted 10. . . ."

His demeanor changes dramatically, however, over the claim made by *Nature* that two of the women who donated eggs were on his staff. "They made some very important factual mistakes in their news," he says, his face suddenly hard. "They apologized this week in their new edition, but I am not satisfied." The apology will be for a less important matter. *Nature* will not retract its ethics claim, but it will, in upcoming months, publish some very flattering words about Hwang.

Hwang relaxes and smiles when when the subject is changed to what he is most looking forward to. "I want to start making new human cell lines with this technology," he says.

Many others want him to as well. In the four months since the announcement of his accomplishment was made, he has received 3,000 e-mails asking for help. Some 100 a day—and counting.

•   •   •

EARLIER THAT DAY, Hwang's boss, Shin Yong Moon, relaxes in the lobby of the UN Millennium. The lobby is all dark mirrors and potted plants. As relaxed and happy as Hwang, Moon confirms that the team's apparent success came from a certain highly organized sense of intent, and perhaps nationalism: "We are all in Seoul, so it is easy for us to communicate and visit each other's labs. We arranged it so that no one was—or is—repeating each other's work. And we are establishing a single hES cell (general hES, not cloning) research protocol for all Korean researchers to follow. Every month each group confers." This organized effort enabled Korea to become one of the few nations to contribute stem cells to the Bush collection. Moon was made head of the entire South Korean stem cell effort in 2001.

The reason was simple: He was the first doctor in South Korea to successfully produce IVF babies, being a good friend of Ariff Bongso's in Singapore (the scientist who created the first hES cells) and having attended a famous IVF clinic, the Jones Reproductive Fertility Clinic, in 1983. That year, he finished his gynecology training in Korea and wanted to study chromosomal analysis, cytogenetics, and cell culture. He wrote a letter to Howard Jones. When he got no response, he decided to visit the clinic anyway. Arriving there, he found Jones was in Italy for two weeks. He waited. Eventually he met Jones, who said there were no openings. Moon told Jones that when he gave a talk in Korea once, Moon was the one who had translated his manuscript for him. Impressed by his persistence, Jones created a job for him.

When Moon returned to Korea, his first 40 attempts at IVF were unsuccessful, so he returned to Jones with his data. Jones told him he was too ambitious, trying two drugs together at once up front. "'Don't play the viola and violin together at first,' he told me." So Moon went back and tried them first separately then together. It worked. "After that I played viola and violin together, no problem." He beams.

Still, it took a long time to establish Korea's first normal hES cell lines—those generated before the cloned hES cell lines. "There were a lot of failures, before we finally succeeded," he says. "HES cells only grow in groups; it was difficult."

Life has changed for him, he says, but the culture in Korea should keep him grounded. "We have a dislike for what we call 'airport professors' who travel all the time, giving canned speeches, repeating themselves all the time," he says. "I've refused a lot of invitations. We should just work."

With that, he ambles off to find a Starbucks.

*   *   *

MOON MIGHT as well amble while he is in the United States. He and Hwang have voluntarily halted their human cloning work to allow the government to catch up and establish methodical regulations. They have promised not to work on human therapeutic cloning for a few months. But they will easily retain their apparent lead during that time by improving their animal cloning work. And Hwang in particular will find himself showered with the money to do it. The science-loving country that boasts that 75 percent of its households have broadband, the highest concentration in the world (compared to 20 percent in the United States), will reward Hwang, big-time. In 2003, the government had given him only $474,000 for his lab (private investors filled in from there). In 2004, however, the government will name him the nation's top scientist, which comes with an award of an extra $300,000. Various local institutes will talk to him about clinical trials and millions more.

Furthermore, in mid 2004, the newly reelected South Korean president will announce he is making science his foremost concern, anointing his science minister as one of the nation's top three vice prime ministers. South Korea until 2004 awarded a higher percentage of degrees in science and engineering than any other country, but it ranked 15th in the amount spent per student. That, the president vowed to change.

For Hwang's part, he will reportedly turn down, in the summer of 2004, an $866 million offer from the United States. The Korean government, which *Nature Biotechnology* will report in December 2004 has already set aside a whopping $4.4 billion for biotech from 2000 to 2007, will establish an unusual pan-governmental task force to support him. In September of 2004 Hwang will be declared a governmental VIP and issued a 24/7 bodyguard and a decade of free flights on Korean Air.

And Hwang's work will appear to continue in such a dynamic fashion that a U.K. science delegation visiting in the fall of 2004 will be besotted. "Hwang is the God of Cloning, there is no doubt about it," U.K. delegation member and King's College stem cell researcher Stephen Minger will gush after his trip. Minger was the first, with Austin Smith of Edinburgh University, to acquire a U.K. license to derive new hES cell lines. (As of March 2005, that number will swell to 10, and 5 to 10 more will be licensed to work with the cells.)

Moon's national hES cell network is impressive enough, Minger will note: by March 2005, his 300 researchers will have created 36 hES cell lines and counting. But then, there's Hwang. "Hwang's lab does 1,200 nuclear transfers of pig and bovine oocytes a day. It's an amazing place. They're going to dominate this field. Large numbers of very enthusiastic, very committed young researchers. Amazing to walk in and watch them work, watch them interact with Hwang."

Hwang's extended lab boasts an astounding 125 people who work only on cloning. There are 10 to 12 cloning micromanipulators (the main cloning instruments), which generally cost $50,000 each. It's likely no lab anywhere has as many. "I watched a guy enucleating oocytes—one every 30 seconds he was popping them out, one after another." The labs themselves are not "super stellar fantastic whizzy wow, but they are very functional, filled with top-of-the-line micromanipulators and microscopes. Really as good as anything I've seen. . . . There was literally a table with 8–10 guys and women sitting around mounds of ovaries. . . . He's got the biz. Enucleating a cell every 30 seconds, one after another, just pop, pop, pop . . . He's got very, very highly skilled people."

Indeed, in 2004 Hwang's team will personally help naysayer Schatten develop the first cloned monkey blastocysts. Minger will conclude: "Boom! After so many years, cloned primate embryos." Schatten will become Hwang's partner, visiting the clinic three times and leaving ecstatic. "We could have been struggling for decades," Schatten will tell reporters. "Now our work is taking off fabulously. I think the whole world owes the Republic of Korea a debt of gratitude." Hwang will also begin collaborating in 2004 with Johns Hopkins neuroscientist (and former Christopher Reeve doc) John McDonald on fusing hES cells and somatic adult cells "to see if we can dedifferentiate the cells without using an egg," according to McDonald.

Finally, in early 2005: an extraordinary request. Shortly after receiving his U.K. license to clone human cells, Ian Wilmut, the Father of Dolly, will ask Hwang to collaborate with him on his first human cell cloning project: creating cells to replace those lost to Lou Gehrig's disease. Hwang will visit Wilmut's lab in May "to check the potential of Wilmut's research," *The Korea Times* will report.

The lab even has a motto, the beguiling if awkward *hanul eul gamdong shiyeara* (moving the heart of the sky), *Nature* will breathlessly report. Clearly, South Korea has captured the heart of the stem cell world.

Can Korea become a leader in that world?

Looked at from a strictly historical perspective, the answer at this point is a firm "maybe." South Korea was called an "economic miracle" for decades. While it was geographically small, its economy, with its world-class car (Hyundai) and electronics (Samsung) companies, was ranked 11th globally for years. This was surprising not just because of the nation's size. Until 1987 South Korea was run by military dictatorships, and until the economic crisis of 1997, it was run by a top-down, still heavy-handed "developmental state." Its banks were cozy with its government, which was cozy with its *chaebols* (large business conglomerates). Excessive central planning like this can ultimately fail in part because, by definition, it doesn't provide fertile ground for new ideas; in part because it can breed corruption.

But, as Michael Breen has written in *The Koreans*, while South Korea saw much corruption, it "could be described in our century of failed experiments as one of the world's most successful centrally planned economies." One reason for this, many Korea watchers believe, is the very nationalism that has helped float Hwang's boat so successfully thus far. "Unlike Western companies, whose raison d'être is to increase the wealth of their shareholders, Korean firms existed initially for nation building,"

Breen noted. Agreed Harvard University Korea expert Carter Eckert, "Nationalism has often been cited as a cultural factor in economic growth, especially in late-developing countries, where it can function as an ideology of popular mobilization and legitimacy during the hardships and socialization of rapid economic growth." Absence of nationalism can present "a serious obstacle to economic growth . . . in Korea there has been no such obstacle."

There has been no such obstacle, many believe, because Korea, located between China and Japan, has been invaded some 900 times in 2,000 years, the last time from 1905 until WWII by the Japanese. The *han* (bitterness) that has built up over this is massive and was "consciously and effectively harnessed in the service of economic growth by South Korea's developmental state," contended Eckert. South Koreans have been willing to sacrifice a lot, and tolerate a lot (including dictatorship), for the best revenge—success.

Furthermore, South Korea has long subscribed to Confucianism's reverence for learning. One result: the nation possessed the highest level of education in the world vis-à-vis wages by the late 1970s. On the day of national college entrance exams, mothers hit temples before dawn to release onto rivers good-luck boats made of newspaper. Commuters leave the house an hour early to clear roads for test-takers. Police are on alert to rush late test-takers to sites.

It all added up to a certain outsized economic prowess for decades. "South Korea's success has come from a combination of profit and pride," says Abdallah Daar, University of Toronto public health policy analyst. "The thought is, 'When Samsung is up there, we're going to be respected.' People are incredibly hardworking, on the job until 7 or 8 at night. Kids go to school in shifts so they can work."

In 1997 the economy crashed, in part the result of rampant state corruption. This "severely dented the country's growth prospects and led to foreign investors selling their assets," according to Chiltern International's Faiz Kermani. But some subsequent developments seem to bode well for the future. The people rebelled and elected a left-wing outsider who cleaned up many branches of government. International Monetary Fund (IMF) loans were tendered only after Korea agreed to increase state transparency. The ruling *chaebol* lost some of its power. All this resulted in a general loosening of ties between bank, business, and government—although all along Korea's autocracies may have been more open than others, according to Asia commentator James Fallows. In the 1980s, 30 of the 36 members of the state's most influential think tank had U.S. PhDs. South Korea "is the first [Asian nation] to challenge the idea that 'Asian-style democracy' might be enduringly different from that of the West," he has contended.

Most importantly for scientists, all this democratization has been accompanied by a loosening, since 1994, of the central state's control over biotech plans. Changes have included divvying up power over biotech funds between many state entities, and increased collaboration between state and private industry. These changes may have come about because, throughout the 1990s, a "fractious but effective democracy" was growing, as longtime Korea watcher Don Oberdorfer has put it. Alternatively, they may have been the natural consequence of the demands of the new globalism, says University of Toronto political scientist Joseph Wong. "Advanced Asian economies such as South Korea's have had to shift their industrial focus away from conventional manufacturing sectors toward postindustrial sectors including biotechnology," Wong wrote in 2004. This has required a shift from "the industrial learning paradigm to a new knowledge creation paradigm where technology innovation, rather than technology borrowing, is key. . . . The state has *had* to reposition itself."

Regardless, the general consensus at this time is that South Korea is serious about biotech, and could become a force in the stem cell world should more coups occur, legitimizing Hwang's work, many are speculating excitedly at this juncture. In the meantime, however, competition will come from a source right next door: Japan.

*   *   *

IN AUGUST, the Siberian tundra surrounding the Maksunuokha River looks like Kansas covered in moss. This impossibly remote area, to which a handful of Japanese cloning scientists traveled often in the summers of the late twentieth and early twenty-first centuries, is impossibly flat, yet impossibly cushy, because it is indeed simply hundreds of miles of flat green moss. On closer inspection, it is also dotted with hundreds of small freezing lakes with no banks. The vaguely surreal,

Dali-esque view all around is of fields of moss that roll along and disappear from time to time, falling into the lakes.

But the view is also occasionally interrupted by ugly, mottled-gray gashes of mud, dug by human hand. For this tundra in northeast Siberia, 500 miles north of the arctic circle, was once rife with live mammoths, massive hairy creatures the size of several elephants, with huge tusks that curled back toward their heads as if to allow them to carry and relocate entire Taiga forests, larch by Siberian larch. It is now rife with extinct mammoth remains, many of which were preserved in their entirety here because the clumsy creatures would often get trapped in mud ravines and, unable to generate warmth by movement, die within minutes.

Russian scientists have been dog sledding across this tundra, poking about, since the creation of the USSR Mammoth Committee in 1948. And in the late 1990s, they were joined by French and Japanese scientists, including a prominent Japanese geneticist named Akira Iritani, a famous IVF specialist who heard about the Russian expeditions, heard about cloning work, and became inspired to try to put these together to do something outrageous: help geologists restore the tundra to its steppe-grassland Pleistocene state in part by bringing back to life mammoths once so numerous, and weighty, moss was literally unable to grow under their feet.

The less difficult aspects of this plan to create "Pleistocene Park" began in 1989 with the introduction of two dozen Yakutian horses. Their hooves did indeed break up the moss and allow the beginnings of grasslands to sprout. The near-term plan, which took form in the early 1990s and is shared by numerous Russian, Canadian, and Japanese groups, is to also bring in musk oxen, bison, and moose.

But the long-term hope is that in a few years, once-extinct cloned mammoths will be added to the crew, followed perhaps by once-extinct cloned Siberian tigers, steppe lions, giant deer, and ancient foxes, all recycled from DNA gleaned from frozen specimens found on the spot. Earlier, scientists had tried many artificial approaches to restore the steppe-grasslands, then they realized: once-native animals may take care of it. The notion was not unlike the difference between using chemical drugs to cure patients and luring patients' bodies back to a healthier, younger state via stem cells.

Regardless, the belief has grown that, together, cloned extinct animals and modern animals who have survived for tens of thousands of years may help restore a swath of Siberia twice the size of Manhattan to the state it was in during the Pleistocene era 10,000–100,000 years ago (when glaciers covered most of the earth except for Siberia, Alaska, and the landmass that then connected the two). As fanciful as all this sounds, it represents what some consider the ultimate practical use for cloning outside human therapy: the preservation/resurrection of endangered/extinct species.

Wakayama City, Japan, home of Akira Iritani's Kinki University, is no Siberia. Located one hour from Osaka on the ocean, it is bordered to the north by the typical Japanese blink-and-you'll-miss-it mountain range, in this case, the lovely wee Izumi mountain range, which is covered in bamboo trees that look like giant dried ferns. Taxi drivers, wearing blue blazers and white gloves, sit in lace-draped seats. The warm current flows closest to Honshu (Japan's mainland) at the southernmost part of Wakayama, so the fishing—and the eating—is the best in Japan. There are plenty of ancient temples; plenty of oddly shaped rocks meant to represent "eternal" animals from turtles to cranes; plenty of hot springs, precious pilgrimage paths, sea air. Wakayama City's somewhat seedy downtown, with its rundown shops and leagues of neon 7–Elevens, has that past-its-prime, old seaside-resort feel. But it, too, is precious: hand cut bonsai trees and handmade shrines grace nearly every tiny garden, which every tiny, slate-topped house possesses.

Still, science is all about men and women leaving the world in which they live to explore other worlds. Usually molecular biologists leave this world in their minds, sinking to its molecular level, from within the safety of their labs. Akira Iritani would literally leave his lab for Siberia, since he was in search, not just of the molecular basis of life, but the molecular basis of death returned to life, an endeavor requiring a special effort.

In 1986, Iritani was part of the first team to produce a mammalian live birth from immobile sperm, via the intracytoplasmic sperm injection (ICSI) technique. He would soon after hook up with a scientist named Kazufumi Goto, who would in 1990 achieve the first mammalian birth using ICSI and *dead* sperm.

"We found we could use sperm boiled to 19 degrees centigrade," says the septegenarian Iritani over a cup of tea in his University of Kinki office. It is March 2004, a month after the first apparent Korean human cell cloning coup. With his flowered tie and aviator glasses, he looks far younger than his years—indeed he looks as if he has stepped out of a 1970s documentary. "Sperm DNA, we discovered, is very strong. The approach began to be used in clinics worldwide, for immotile and low numbers of sperm."

Then, at an early 1990s reproduction conference, *Newsday* reporter Bob Cooke fancifully suggested that the crew use the technique on Ice Age man remains, which got Goto and Iritani thinking. Such an experiment on a man would be too dangerous, but what about bringing back some extinct animal, such as the popular mammoth, by mating frozen mammoth sperm with an elephant—and then mating the off-spring? A Canadian and Russian group was already planning a Pleistocene Park and had discovered that indigenous animals can go a long way to restoring environment. Mammoths could conceivably go an even longer way by capturing the attention—and the *funds*—of the world. "We were able to create live births from the mating of fresh eggs and pig and bull sperm that had been frozen at –20 degrees centigrade. . . . We decided we could create a mammoth/elephant if mammoth sperm was frozen at –20 degrees centigrade."

For months, Iritani brooded. At length, he realized that even if he could mate unfrozen mammoth sperm DNA with a fresh elephant egg, it would take generations of breeding to get a pure mammoth. "We gave up that project—but within months, Ian Wilmut produced the first (adult) mammalian clone."

Immediately, Japanese researchers began work on what would become the second and third cloning coups in 1998: Teru Wakayama would clone the first mice from adult cells and colleagues of Iritani's at Kinki University would clone the world's first cows from adult cells. Japanese researchers would also work with the University of Connecticut's Jerry Yang to produce the first cloned male from an adult cell (a prize bull) in 1999. And in 2000, researchers at Tskuba University would become the first to publish that they had cloned pigs from fetal skin cells. Japan, more than most other nations, realized the potential of cloning right away, and got to work proving it right away.

For his part, Iritani immediately understood the implications for his Siberian dream: by fusing the nucleus of a mammoth cell with an enucleated fresh elephant egg, he might get a living, embryonic, exact copy of a mammoth, right away. "This is when we knew how we would do it. So later in 1997, we went to Siberia, to look for intact frozen mammoth DNA." They went with a Russian crew led by Petr Lazarev, the geologist head of the Mammoth Museum in Yakutzk.

It was a grueling trip to an arctic circle spot near the Kolyma River, part of the area destined to become Pleistocene Park, one where many mammoths were found. It was August, when the permafrost (ice-water subsoil) melts four meters, so land and river were messy, boggy, frightening. After traveling by jet to Yakutsk, and by battered charter plane to Cherskii, "We made our way down the river on a timber transporting boat," Iritani says. "It was a swamp all around, very dangerous." They finally made it via a rubber boat to a set of cliffs where mammoth remains were found before—Duvannyi Yar. They clambered over gray, claylike cliffs that often crumbled beneath their feet. For a landscape harboring so many dead remains, it seemed remarkably alive as it fell apart all around, and under, the scientists.

They found 30 meters of skin and returned triumphantly to Japan only to discover that it was the skin of a 30,000-year-old rhinoceros. The rhino was a big hit with the International Embryo Transfer Society at its meeting in 1998, but Iritani was disheartened. He and Goto had formed the Mammoth Creation Society in Japan that year, but its funds were almost exhausted by the expedition. So after he made another failed expedition in 1998, "We told Dr. Lazarev that we couldn't afford to come with him again until he found some good mammoth remains."

Four years later, Iritani got a call. Lazarev had been told of a mammoth foreleg along the River Maksuonokha. Iritani, by then 71, decided to send his student, Hiromi Kato. In 2002, the international team carefully extracted the leg, then had to leave it behind and wait for over a year, while the Russian government dithered over how to release the remains to the Japanese. Among other problems, a competing group had appeared in Japan. Determined to display a full mammoth at the 2005 World Expo, they were offering huge sums for mammoth remains. The upshot: in 2003, Lazarev arrived at Iritani's door with a tiny liquid nitrogen container of a piece of mammoth foreleg the size of a napkin. The section, by then, was not viable—the DNA was incomplete, broken.

So Iritani's group waited some more. In the meantime, Iritani inserted a spinach gene into pigs, which may result in healthier pork. It is the first time a vegetable gene has been introduced to meat. It will be published in an April 2004 *Proceedings of the National Academy of Sciences*, he says.

Iritani discusses this a bit, hands folded on his stomach. "It is interesting," he agrees amiably. "Perhaps we can make meat much more healthy this way." But the 76-year-old fairly leaps forward again when the subject changes to "frozen zoos" in Cincinnati, San Diego, and Omaha, where the DNA of endangered species is being preserved in anticipation of the day they might be cloned. He discusses the endangered species that have been cloned thus far, which include a gaur and perhaps a Siberian ibex (a claim made by China). South Korea's Hwang is working on the Korean Baekdu tiger; China, the panda; Australia, the wombat. All of these projects are fascinating; none are as ambitious as the project of Iritani and his international crew.

"I do think it is possible to bring frozen mammoths back to life again," he says, eagerly. This summer, he has been invited to a frozen zoo to discuss it.

In the spring of 2005, his team will announce it will soon head for Siberia again.

But Iritani is not the only Japanese researcher out to stun the world with a stem cell or cloning coup. In May 2002, the Japanese government opened a $53 million stem cell complex, the Riken Center for Developmental Biology (CDB), the world's largest bench-to-bedside stem cell complex. The annual budget of the Kobe, Japan, institute is larger than that of MIT's famed Media Lab: $45 million. There are more than 500 staffers, 243 of those scientists. And à la Western science, Riken gives unprecedented authority to young group leaders to design their own projects. Normally only older established scientists possess this freedom in Japan. But here, all team leaders (there are 30 of them) are between the ages of 28 and 40, according to Riken stem cell chief Shin-ichi Nishikawa. Riken emphasizes production, another Western approach, asking researchers to justify new contracts every five years. And it actively courts Western scientists. "This is an exciting experiment to establish a new institute merging basic and medical developmental biology, as well as creating an environment that has been generally difficult to find in our universities," CDB director Masatoshi Takeichi told

Nature in 2002. "We will give young researchers a degree of independence and use of facilities that is unknown in Japan," agreed Yoshiki Sasai, a CDB group director.

Furthermore, after the United States, Canada, and Germany, in the 1990s Japan put out more mouse ES cell papers than any country in the world (at 500). Given that the United States and Germany have placed a legal stranglehold on federally funded hES cell work, and Canada has been dithering, Japan is positioned to score big with regard to both straight hES cell and cloning work.

Additionally, Japan is establishing other key global alliances. Both Kyoto University and Riken's CDB have struck a partnership with a top stem cell institute at the University of Edinburgh. In August of 2002, Sosei and Stem Cell Sciences of Australia, Singapore, and Great Britain formed Japan's first stem cell company, Stem Cell Sciences KK, in Kobe. In 2003, Japan helped form the International Stem Cell Forum, an alliance of 12 countries whose goal is to accelerate the pace of global research. Schering AG, a huge pharmaceutical company, has moved a branch with 200 staffers near Riken in part to collaborate with its bench-to-bedside stem cell facility. One-third of its work will focus on stem cells, unusual for a Big Pharma company.

Unfortunately for scientists, Japan has gotten off to a slow start in the area of human ES cells. The government deliberated for years before finally setting up its hES cell rules and directives, allowing the first three lines to be distributed to Riken in early 2004 and more to be distributed shortly after. But Japan has moved comparatively faster in the area of therapeutic cloning. For, among other things, in 2002, Riken lured the then second most famous therapeutic cloning team in the world to its labs: Teruhiko (Teru) Wakayama and Anthony Perry.

In 1998, this duo had cloned the world's first mouse, Cumulina, from an adult cumulus cell in the ovary, which as noted earlier, a world-famous developmental biologist, Davor Solter, had once declared "biologically impossible." (Solter discovered cell markers that led to the isolation of the hES cell.) The mouse oocyte is highly fragile and its development time (and thus, presumably, its redevelopment time) is very fast. Even Yang told the press it was "impossible" to clone mice. He hadn't yet realized, he would tell a conference in 2003, that in cloning, "'Never say never.' It was the stupidest comment I ever made." Just as the work of Yang and others had informed Wilmut's to a certain degree, Wilmut's work had informed Wakayama's.

Now, in March 2004, Wakayama's lab is a very well-funded, very gung-ho, cloning factory.

•   •   •

RIKEN IS not inviting, unlike its main academic partner, the nearby University of Kyoto, located near the heart of Kyoto—which is in turn the cultural heart of Japan. Kyoto is virtually exploding with 2,000 beautiful temples and shrines boasting everything from singing "nightingale" floors to rice paper walls. The Allies carefully avoided these in WWII: too much beauty. The city, which feels like a large village, is crisscrossed by shallow, massive rivers with wide yellow banks where lovers meet, their bicycles lying beside them. Everywhere are artisan shops tucked into alleys; rickshaws drawn by bored T-shirted boys; eel, octopus, and squid so fresh they wriggle in their cellophane on market tables.

The last week of March 2004, Kyoto is especially beautiful, for the vaunted *sakura*, or cherry trees, are in full bloom, as has been predicted by TV weather ladies. (Like ES cells, which must be captured early in an embryo's life or they will vanish, *sakura* only bloom for a week at a time in each Japanese city.) The trees bend this way and that over streets tossing pink and white petals on passersby like giant bridesmaids throwing rice.

Ironically, Kyoto—where Japan's first hES cell lines were derived the previous year, at the University of Kyoto—is also rife with shrines to the unborn. Sheds lined with stone statues of babies often wearing knitted red caps were erected centuries ago for young women whose lives would be ruined if they had had babies out of wedlock, so procured abortions. They are still actively visited today by modern women who have had abortions. Japan is Buddhist and Shinto. As Hwang noted, Buddhism is about reincarnation, life folding itself into life. However, both Buddhism and Shintoism also heavily emphasize the sanctity of the body. This explains the shrines to the unborn children and the fact that it took a long time for Japan to accept organ transplantation, and to establish rules for hES cells and therapeutic cloning. All life is recyclable; all life is sacred.

Another dichotomy: Japan's reverence for both past and future. At this temple, a monk in a brown robe with a shaved head is chatting on a phone. ("Talking to God," a Japanese woman whispers, smiling.) At a nearby temple, sick people wait patiently to scoop ancient healing waters out of a pool with a cup on a string. Then they pop the cup into a microwave that zaps the germs. "They believe the waters come from the gods and will cure them," says a native Kyotian. "But they also believe in the logic of ultraviolet waves. That's Japan." On TV this week, there is a movie that has an evil Commodor Perry bringing corrupting commercial Western ways to nineteenth-century Japan. It is interspersed with ads for "Ambitious Japan" depicting Japanese men in business suits bolting through an airport à la O. J. Simpson.

Japan has a mind-numbingly lengthy past, one that has been, in many ways, preserved intact more successfully than that of many other nations. But, as its leading role in technology has proved, it can embrace the future more easily than many other nations, as well. It is a matter of perspective as to whether it was amazing that Japan only spent a few years devising careful regulations for hES cell work before approving it, or amazing that it spent such a long time doing so.

But approve it all—with regulations—Japan recently did, and Riken's CDB will be the prime beneficiary. Built on a man-made island off Kobe, it is the business face of ancient Kyoto, Japan's other face, the one staring hard at the future.

It is accessed first by a long moving sidewalk that eases the com muter slowly past one ultramodern, all-gray building after another, many of which are still being built. Elevator music plays. At the end of this seemingly endless walk, a new highway lifts off the land over the water and ends in midair. There are no cars on the road and not a frivolous *sakura* in sight. Yet every other parking lot is jammed with cars.

Riken itself is a massive gray building surrounded by armies of tiny exotic tree-lets kept in place with strings. A sign out front bears an outline of a man, with a red slash through it. This huge biotechnology center is not complete, but the parts that are, are humming. Riken is grimly all-business already.

The offices outside the lab of Teru Wakayama, however, bustle with friendly noise and friendly T-shirted young scientists. It is one of three labs at Riken (along with more at the University of Kyoto) focused on therapeutic cloning; that is, on the ways the oocyte is able to dedifferentiate adult cells. The lab beyond is ultramodern, equipped with five $50,000 micromanipulators that allow Wakayama's increasingly experienced crew to enucleate 150 oocytes an hour. Wakayama's goal is to increase the success rate of nuclear transfer in mice—critical because there are more mouse disease models than any other kind. (Right now, on average, only 2 percent of mouse eggs turn into live mice and 20 percent yield ES cells.)

It is 3:30 and the lab is empty, but not because anyone has gone home. His crew got in before 9:00 A.M. and are at lunch, Wakayama explains. They will be back in an hour so they can work until 10:00 P.M.

He shows a film of his work. One pipette holds an egg. Another pipette pierces the egg, sucks out the nucleus. Yet another pipette pierces the egg again; sends a mouse cell nucleus into the dark center; pulls out. Three hours after this nucleus is injected, chromosomes will start to form. Four days later, cloned blastocysts will form. The film proves what they say about Wakayama: he has "magic" hands, "the fastest hands in the business." This is key partly because mouse eggs mature—and dedifferentiate—faster than those of other mammals. It is also key because cloning's success rate is so low it must be done fast, for any work to get done at all.

Wakayama is a good-natured, impatient kid in a T-shirt: the kind who throws himself into chairs. He, like South Korea's Hwang and Scotland's Wilmut, is the center of attention wherever he goes, and tomorrow's Riken stem cell conference will be no exception. But he is as casual about his accomplishments as he is about his dress—at last, able to afford to be. His boss Shin-ichi Nishikawa has high hopes for him. Nishikawa is the group director of the Laboratory for Stem Cell Biology at Riken's CDB and a star in his own right. (He was the first to get a hematopoietic [blood] stem cell from a hemangioblast—the progenitor of both hematopoietic and endothelial [vessel forming] stem cells—and the first to publish the discovery of the hair follicle stem cell.) His hope is that Wakayama will "bring up a generation of young scientists with his superior cloning skills."

Indeed Wakayama has trained a cloner whose work analyzing genetic and epigenetic changes occurring during cloning is constantly cited today: another one of Riken's prized group leaders, heading up another ambitious cloning team, Atsuo Ogura.

Still, Wakayama's surroundings have not always matched his talents. As gifted as he is, he found it difficult to find a permanent home for himself in the United States. One reason was uncertainty over rights to the cloning work, according to a source close to Wakayama and Perry. But another may have been controversy over the human applications of the cloning approach, which kept funding down even for animal work.

"We only improved on the [Wilmut] technology a little for the mice," he begins, talking about his 1998 mouse-cloning coup. "We just used a new micromanipulation machine [the piezo input pipette drive unit] which made the nuclear transfer process happen in a shorter time, so it was less traumatic." And since mice develop so quickly, theoretically, this demands that the opposite be true—that is, that the injected adult nucleus have its clock reset quickly by the enucleated oocyte.

Another innovation Wakayama brought to Wilmut's approach, some note: after placing the nucleus of the adult cell into the enucleated egg, he refrained from zapping the egg with electricity, instead relying on chemicals to gently activate the new egg.

Despite Wakayama's casualness, it is clear he never would have pulled off his cloning coup without youthful determination, aka rebellion. He and Anthony Perry were on fellowship in the University of Hawaii lab of Ryuzo Yanagimachi back in 1996. (Yanagimachi did key work fertilizing eggs with dead, freeze-dried sperm.) Wakayama had earlier cloned mouse ES cells. "I didn't believe it was possible to clone adult cells," he says. "Then Dolly happened. I wanted to try it right away."

But when he asked Yanagimachi, he met with resistence. "It was not a nuclear transfer lab, it was a fertilization lab. He said, 'You cannot do it.' But he was 75 years old. He didn't come to the lab much. So . . ." Wakayama pauses, then grins again. "I did it, after hours. It took three days total. Three experiments. It worked on the third try."

Davor Solter, the developmental biologist who had declared mammal cloning "biologically impossible" in 1984, had tried for years to pull off what young Wakayama did in three after-hours days.

To Yanagimachi's credit, "When I opened up a pregnant mouse, and showed to him the little cloned embryos, with their beating hearts, he changed his mind. He was happy. He encouraged me then. He was a very gentle man; in Japan, they'd be angry."

So in July of 1997, with the help of Perry's English, Wakayama wrote the paper and sent it to *Science*. The magazine rejected it. In October, they submitted it to *Nature,* "and then spent a lot of time battling with the referees [reviewers]." It took a long time for the skeptical *Nature* reviewers to come around, but in July of 1998 Wakayama's paper declaring the cloning of 31 female mice from adult mouse cumulus cells was published. Wakayama and Perry had scored a 2 percent success rate in mice (which Wakayama would later get up to 5 to 10 percent in some instances). The paper was published alongside a mea culpa from Solter, who declared that it was now clear that cloning of adult mammalian cells was real.

But by the end of 1999, Wakayama and Perry had left Hawaii. Despite the fact that Wakayama had signed away his rights to his work at the request of Yanagimachi, it is generally known that fights were raging among the university, a venture capital company, and Perry's attorney over who owned them. Furthermore, Rockefeller University neuroscientist Peter Mombaerts was so excited by Wakayama's work that he flew to Hawaii several times to observe—as did many scientist "pilgrims" at the time—then helped lure the pair to New York.

At Rockefeller University, Wakayama in particular would score more coups. He would create six generations of clones of mice (clones of clones of clones . . . etc.). He would establish that the telomeres of cells in cloned mice (caps on chromosomes that signal the age of the cells) are not shorter than in normal mice. At Rockefeller, he also would be the first to get a high success rate acquiring mouse ES cells via cloning: 20 percent. And he, with Mombaerts, would be the first to establish the feasibility of therapeutic cloning with his April 2001 *Science* paper that showed ES cells from cloned embryos could create neurons.

As spectacular as the work was, Wakayama and Perry had some initial difficulty acquiring NIH funding for it. "There is no culture of federal funding for this kind of thing—nuclear transfer, IVF—in general," Mombaerts would later note. And patent problems remained. But Riken offered money, as did the controversial cloning company, ACT. So in 2001 the pair moved on to ACT, in Worcester, Massachusetts, for a temporary stint until Riken was ready—bolstered by an NIH grant they belatedly scored. While there, Wakayama would "help, just a little bit," the ACT crew with their attempts to clone human cells, which led to what many called an infamous premature paper that ACT published in late 2001 in a new online journal. ACT claimed it had created the first cloned human cells. The cells replicated a bit, then died. It was equivalent to crying success when a rocket gets six feet in the air then crashes. Many staffers did not want to publish. Success in the cloning of human cells, they and most other scientists believed, would only come when cloned cells reached the blastocyst stage, allowing hES cells to be culled, as the South Koreans would appear to do in 2004.

But the CEO, Michael D. West, insisted. Because the experiment was not a success, it drew criticism worldwide, and the field came down hard on ACT. So did the U.S. government: ACT was audited shortly after. Some ACT staffers believed the intent was to intimidate the company into closing.

By then, Wakayama was on his way to Riken, with Perry not far behind.

As Wakayama will note at a conference, once he improves the success rate with nuclear transfer (that is, once he is able to persuade far more than 2 percent of his blastocysts to regularly form living mice) the work will get even harder. So far he has found that, even among those 2 percent that make it, after one year, cloned mice tend to weigh twice as much as normal mice. Their placentas are also larger. There are other genetic abnormalities, as well. (MIT's Rudy Jaenisch contends: "There are no normal clones.") Wakayama will have to improve the *quality* of his success, too.

By the same token, he notes, his first cloned mouse, Cumulina, lived two years and seven months—which is *longer* than the average mouse. Telomeres, which help determine the age of a cell, are often *longer* in cloned mice than normal mice. And abnormalities of the clone do not appear to be passed on to the (naturally born) offspring of the clone. So success is within reach, Wakayama believes.

He glances at his watch, then rises, noting that he has to work on his speech, in English, for the conference. "English is harder than cloning," he says, apologetically.

Wakayama's partner in cloning crime, Anthony Perry, sits in the office of his brand new lab on the other end of Riken, looking even more relaxed. Also T-shirted, he has a staff of 10, much fancy equipment, a very nice budget. On his desk is a cartoon of mice talking about recent developments in human science.

"Clinical trials are easier here," he says, folding his hands behind his head. "There is a guy here doing a trial using artificial joints coated with [adult] stem cells. He didn't even have to check with an IRB [institutional review board.] He just did it."

Perry looks satisfied, as indeed he should be. He also looks as if he feels safe, as he is, comparatively speaking. "Looking back, there were times throughout our spell in America I felt traumatized and scared," he says, musing. But within minutes, he begins launching into a diatribe that few, if any, U.S. scientists would dare: he points out that U.S. scientists have been far too harsh in their total condemnation of human reproductive cloning purely on the basis of the abnormalities that so often occur. U.S. scientists do this, undoubtedly, to throw a bone to the conservative right. The problem is, Perry says, these florid condemnations cast aspersions, by association, on the ES cells derived from cloned embryos and could ultimately damage the progress of the therapeutic cloning field. Those abnormalities could well be reversed someday, as scientists well know, he says.

Perry feels so safe here that in November of 2004 he will publish in *Nature Biotechnology* an article chastising scientists for caving in to the right this way. Improvements are already being made, he will note, pointing to an exciting article by Michele Boiani, one of Wakayama's former trainees, now at the University of Pennsylvania. Her group, led by Hans Scholer, reported in the October 2003 *EMBO (European Molecular Biology Association) Journal* that they were able to improve the birthrate of mouse clones *eight times over*—in a clever way. Boiani had noticed cloned embryos often possess fewer cells than normal embryos and that the size of the deficit is often in direct proportion to success rates. (There are 9 percent fewer cells in bovine cloning, which has the highest success rates; 19 percent fewer in porcine cloning, which has slightly worse success rates; 43 percent fewer in difficult rabbits; and 55 percent fewer in very difficult mice.) Boiani's group aggregated two or three cloned mouse embryos at the four-cell stage, then placed them in wombs.

The result was more ES cells expressing the quintessential ES cell gene—called Oct4—and eight times more births. If it holds up in other labs, this means that, as so often is the case, what the confused stem cells in rapidly reprogramming cloned embryos need is not ham-handed human intervention, just communication with other stem cells.

"Progress can be expected," Perry will conclude in the article, writing in his office nestled in the safe concrete harbor of Riken.

•   •   •

THE LAST, best Asian refuge for ambitious therapeutic cloning advocates, China, did not fare as well as other Asian nations seemed to throughout 2003. Despite the steady backflow of Chinese-born postdocs from the United States to their homeland to stem cell and cloning centers, the mounting yuan being spent, and breathless articles on China's potential in the area of regenerative medicine in publications from *Wired* to *Nature,* only a single notable stem cell paper from the nation was published in 2003. And that paper, about ES cells from so-called cloned rabbit/human embryos, was rejected by a slew of top journals, eventually landing in a third-tier Chinese publication.

But while the buzz about China will continue throughout 2004—indeed, it will get even stronger for some astonishing reasons that will be outlined in a later chapter—the year will also offer scores of Western scientists like Jerry Yang, who were perennially facing the stay-or-go conundrum, a whole new series of confusing options. Fed-up California scientists, their tempers brought to a boil by the apparent South Korean coup and the prospect of four more years of the "anti–stem cell president," will begin hatching a bold plan. That plan: to bypass the U.S. federal government altogether and take their plea for funds for more extensive hES cell research—$3 billion worth—straight to the people, in a historic state referendum vote to coincide with the November 2004 presidential election.

This will threaten to create not only another East/West divide—this one between the east and west coasts of America—but an unprecedented divide between several U.S. states and their own federal government.

Meanwhile, just to confuse everyone and stretch their loyalties to the hilt, work on the far-better-funded *adult* stem cell field will begin making some truly exciting strides from 2004 through 2006. These advances will occur in the lab and in some of the world's top hospitals.

It will all have talented U.S. scientists like Jerry Yang swiveling their heads like the possessed girl in the classic horror movie *The Exorcist.*

PART TWO

7

Double-Edged Sword: Stems Cells as a Cancer Cure and Cause

Someday perhaps it will turn out to be one of the ironies of nature that cancer, responsible for so many deaths, should be so indissolubly connected with life.
—Pioneering oncologist Charles Oberling, 1946

KOTARO YOSHIMURA IS SITTING at a table that is casually decorated with grinning human skulls in a University of Tokyo basement room. It is March 2004. The room is Sovietlike, with dented and tearing linoleum and peeling walls. There is an old curtain behind a video screen; an old TV; old gray filing cabinets. The hall outside is lined with old fridges. All in all, like Benjamin Reubinoff's office in Israel—which is new, but windowless and small—this room couldn't look less like a site where a kind of stem cell history is being made. It more resembles a place where bad people are sent when they are being punished.

And indeed, it is both these things. For this is where Yoshimura shows visitors the results of his stem cell research. While his work is potentially important in both a medical and business sense, it is wildly un-Japanese in the traditional cultural sense, for it is about breast enlargement, a Western—if increasingly global—preoccupation. But Yoshimura is doing it with stem cells, an approach that could someday meet a serious need—reconstruction for breast cancer patients. So Yoshimura has become the world's first doctor with a highly reputable university to enhance or restore women's breasts with stem cell mixtures—

although at this stage, at least some of his work is consigned to the basement.

Many Japanese people nod a lot or chirp "Hi! Hi!" (yes, yes) a lot. Yoshimura does both constantly, which can make him seem—to the newly arrived English speaker—like a man with short-term memory loss. He also is among a minority of Japanese researchers (many of whom were schooled in the United States) to mix up his English *r*'s and *l*'s. But while Yoshimura's aspect may seem hesitant, like his workplace, his demeanor is deceptive. The year before this, in 2003, an Australian scientist named Wayne Morrison announced he'd been able to coax stem cells in a pig to form breasts by inserting a capsule covered with stem-cell-attracting growth factors into its chest, then looping the pig's femoral artery into the capsule, to give it a blood supply. His goal was to bring this breast apparatus to patients by 2005. And this month, MIT is coming out with a paper announcing it has created functional human breasts in mice by injecting, into the chest area, young cells from human breast tissue, which formed treelike mammary ducts that lactated when the mice became pregnant.

But Yoshimura is ahead of them both in one key sense: he has moved into the clinic. With a stem cell mixture he has created larger breasts in two female human patients since January, and by April 2005 he will have augmented the breasts of 13 women and restored a man's deformed face. Given that 1 in 8 women in the Western world get breast cancer, that Asia is fast catching up, and that there are scores of cancers, surgeries, and diseases that result in tissue deformity, Yoshimura's approach could become quite a winner worldwide, and he knows it.

"Ten years ago the American Society of Plastic Surgeons recommended that liposuction injections into the breast should be stopped," he says, referring to the transplantation of fat directly from the belly or thigh to the breast. The reason: calcification, fibrosis, and cell death generally occur when fat from one part of the body is transferred into another part, because the fat dislikes being detached from its original blood supply. But Yoshimura continued the work through the years, finessing it to where this calcification more rarely occurs, he says, partly because of his injection technique—and his cells: sensitive adult stem cells.

Yoshimura fiddles with a silver laptop, trying to get images up onto the screen on the wall. After several minutes, he fails, sighs, and turns his laptop around. A film pops up on the computer screen. It is a film of a small white human breast being injected with fat in multiple areas.

A courtly 71-year-old Japanese guide, here with a visiting writer, stares in horror a moment, then excuses himself and bolts out of the room.

One reason U.S. surgeons couldn't get fat transplants to work, Yoshimura explains (as the breast on the screen is squeezed in a gloved hand and poked with a fat-containing syringe) was simple: they transplanted fat blobs that were too big. "The outside surface of the fat was OK, it could attach to the living tissue in there," he says. "But the inner surface died. In our case, we began to simply inject very small amounts, diffusely, across the breast. When the fat tissue was small enough, it could just hook up to blood vessels. It was mostly surface. So it was fine."

The problem with the approach was that, since only small injections could be done, only a total of about 100 cc of fat could be transferred. "And even Japanese women want 300–400 cc of augmentation. The approach was just not good enough." And while the United States turned to artificial implants, those, famously, weren't good enough either. "In the U.S. every year 250,000 patients receive breast augmentation," says Yoshimura. "But every year, 50,000 are removed." Often human tissue will harden up around silicone or saline implants. The implants can leak. And even when artificial implants last, they rarely do so more than 10 years.

He starts another film, explaining that we are about to see his final answer: stem cells. A surgeon is shoving a thick hose into the thigh of an unconscious patient in a fluorescent OR. At the other end, the hose snakes away from the patient and leaks a yellowish-red river into a jar. When the surgeon is finished, the jar looks like a blender filled with a thick raspberry and lemon sorbet shake. The enzyme collagenase is sprayed into the jar "to loosen things up," or to dissolve the connective tissue in the fat, Yoshimura explains. A gowned assistant places the blender in a centrifuge at the corner of the operating room and turns it on. The "blender" spins furiously.

After this process of "density gradient centrifugation" has separated the substances based on weight, the centrifuge is turned off. Mature red blood cells that have settled to the bottom are thrown away, leaving a reddish layer of blood cells that includes stem cells that make blood and blood vessels, and a layer of fat that contains fat progenitor cells. "Marc

Hedrick, an LA plastic surgeon, showed that some cells in this fat layer can form multiple lineages: bone, cartilage, connective tissue, cardiac muscle, fat," Yoshimura says. "Later he reported they also contained endothelial cells. In other words, there are mesenchymal stem cells [MSCs] in there." It is unclear at this time whether MSCs can form functional versions of the above tissues. This is one thing Yoshimura is testing now. The main stem cell he is sure of, he says, are endothelial progenitor cells, long proven to form blood vessels. These should vascularize the fat grafts, keep them alive.

Indeed, over the past three months, Yoshimura has found that injecting large doses of these fairly purified populations of potent cells, along with small doses of mature fat, does the job—at least thus far. He continues rolling the film, which shows a surgeon again injecting a needle into breasts; this time, with aspirates of both fat and stem cell mixtures. (The courtly Japanese guide, who reentered only moments before, murmurs "*Sumemasen*" [excuse me] and again bolts out of the room.) The film flashes to a month later. Before and after photos indicate the breasts have grown visually. Hopefully, they are becoming vascularized, which would mean that both the injected stem cells and native breast stem cells have been galvanized to action by the inflammation caused by the injection of the syringe, and the "blending." The fat, and vessels, should continually be replenished thereafter by the stem cells.

In April 2005, Yoshimura will report via e-mail that he has injected a total of 13 patients, and all are doing well. No major complications. The average patient's endowment has been enhanced by two cup sizes and happily has stayed there (not decreasing or, God forbid, continually growing).

If all continues to go well, Yoshimura's stem cell soup, or a version of it, is expected to make a lot of breast cancer patients with partial mastectomies quite happy. And there are few words to describe what such an approach, much more refined, could do for patients who have had full mastectomies. Stem cells may become a cure for one of current cancer therapy's most brutal side effects: disfigurement caused by surgery.

But, astonishingly, mutated versions of *normal* adult breast stem cells may also be what causes many patients' breast cancers in the first place, according to a steady stream of papers exploring something called the "cancer stem cell" theory, which started gaining steam a few years before Yoshimura's trial. Indeed, mutated versions of the adult blood and fat stem cells that many scientists use so successfully for a variety of therapeutic approaches have *also* all been fingered as possibly complicit in many cancers, from leukemias to liposarcomas (fat-based tumors). Furthermore, realization is dawning that the entire stem cell field may be owing a greater and greater debt to the cancer field, as oncologists increasingly find stem cells in their tumors; as stem cell scientists increasingly find tumors in dishes of stem cells that have been cultured too long; and as both look for markers of their stem cells, healthy or mutated, in each other's work.

That this strange and fascinating quid pro quo has been going on will become patently clear in June 2004, at the second annual International Society for Stem Cell Research (ISSCR) meeting. The largest stem cell meeting in the world will feature numerous talks and posters that heavily underscore the fact that the potency of adult stem cells may be both a cause for exhilaration and wariness; that, while adult stem cells and stem-cell-like progenitors are proving, in many clinical trials, to *cure* better than many therapies, they may also, by virtue of their age, be able to *kill* more efficiently. Yoshimura does not manipulate (force to proliferate exhaustively, or genetically modify) his cells. He just takes 'em out, cleans 'em up, and pops 'em back in. So it may not be necessary for doctors like him to conduct expensive tests looking for genetic and chromosomal abnormalities that are the hallmark of cancer. (Debate is ongoing as to which stem cell therapies will demand exhaustive pretesting, if the cancer stem cell theory pans out.)

But in general, it will become clear that the rapidly increasing number of doctors trying adult stem cells on patients may need to examine those cells thoroughly in many cases and may want to switch altogether to more youthful, less exhausted hES cells in others, when these are ready for primetime. (ES cells, in their "youngest" and/or most undifferentiated state, are randy and form benign tumors—in mice without immune systems—when implanted in that undifferentiated state. But when they are differentiated a bit to a more mature state, they settle down and their youth then becomes a profound advantage, for they are stronger and can exhibit more self-control than old, worn-out adult stem cells.)

For there is a growing body of evidence indicating that many malignant cancer cells look and act like normal stem cells and utilize the same genes. Both normal stem cells and malignant cancer cells are highly potent: they both survive longer than other cells, replicate more than other cells, and can morph into more different kinds of cells than other cells. More and more scientists are coming to believe that the most malignant cancer cells *are* normal stem cells, or are normal stem-cell-like progenitors, that have gone awry. As former University of Colorado pathologist Barry Pierce once famously put it, malignant cancer cells may be "caricatures" of normal stem cells: cancer stem cells. Says Rockefeller University immunologist Ralph Steinman, "Stem cells are controlled cancer."

Tissues can operate in slightly different ways, but the following is a general outline of how things work in the normal healthy body. Adult stem cells, which in almost all cases are the only cells in the body that are both permanent residents and can also replicate extensively, sit in protected niches in their specific tissues. These rare cells sit quietly, not replicating, until their tissue is injured or otherwise calls out for help. Then the elephantine, potent stem cell goes to work, slowly replicating itself, while at the same time turning out slightly more differentiated, rapidly proliferating, progenitor cells. Those junior "stemlike" progenitors move upward out of the niche even as they replicate, sensing as they go just how many new cells the tissue needs, then they differentiate into even more limited progenitors that can barely replicate at all. These progenitors differentiate into the final product: the mature cell that the tissue needs to replace. This mature cell, generally, does not replicate. A grown-up and sober member of the body's silent majority— its vast cellular workforce—it cannot waste the energy or the time: it must settle down to do the day-to-day business of the tissue.

This is how things generally work when all is well. However, proliferation wears down a cell. Every time a stem or progenitor cell replicates, it recopies the 3 billion letters of its genome. It often makes mistakes, called genetic mutations. Four or five mutations in one cell is often called "cancer."

Thus, logic dictates, it is the immature adult stem cell, of all cells, that is most likely to accumulate enough mutations to spur cancers. For the final product of the normal differentiation process, the mature cell, barely replicates if at all, so it acquires few mutations. Furthermore, the mature cell often dies off very fast in certain tissue systems (such as the skin, blood, eye surface, and gastrointestinal system, where mature cells are sloughed off every day) so it simply doesn't hang around long enough to form cancer. (In other tissue systems, the mature cell *does* hang around a long time [brain neurons], but it doesn't replicate, so it can't acquire more mutations, either.)

But when the slowly replicating adult stem cell acquires a mutation after dividing, the mutation can hang in there, since the stem cell is generally the only cell both permanent and replicative. Stem cells "are long-lived and are more likely to be the subject of mutations that are necessary for cancer initiation," wrote Baylor College's Jeffrey Rosen recently. The mutated stem cell lives to replicate another day—to make mistake number two, another day. And so on, until it collects enough mutations to go cancerous.

And those rapidly proliferating, in-between, adolescent stem-cell-like progenitor cells? They, too, can form cancers, says the theory, partly because they inherit those mutations collected by their stem cell mother, and partly because they *too* replicate, often even more rapidly than the stem cell does. Those progenitors that inherit one mutation from their mother stem cell when she is young and has only one mutation herself? They should do all right. For by nature, they tend to quickly stop replicating and differentiate into a sedentary mature cell that never replicates again and is often washed right out of the body. But those progenitors that inherit *many* mutations from an older, mutation-laden, stem cell mother? When those progenitors begin rapidly replicating, they can get into trouble. For when mutation-laden progenitors begin replicating, or recopying the 3 billion letters of their genome, all they have to do is make *one mistake,* to acquire that fourth or fifth mutation that can be the hallmark of cancer, to end up spinning out of control.

Other clues have been leading oncologists to believe that mutated stem/progenitor cells may be the main cause of their woes. But this is the seed of the reasoning. Stem cells and their immediate progeny may be potent cells for better *and* for worse. This notion is sobering. But it does not present an insurmountable hurdle for normal adult stem cell therapies aimed at replacing missing tissues and cells, many clinicians believe, as will be discussed later in this chapter.

Furthermore, the notion is absolutely electrifying the field of oncology. As late as the 1990s, according to NIH neural and embryonic stem cell expert Ron McKay, there was huge skepticism in the cancer community. For decades, the general belief had been that cancer cells were normal mature cells that had *de*differentiated back to an embryonic-like state. This made little sense as a general rule. Sure, it may happen sometimes. But it takes a lot more effort for a cell to lose control by reversing course than it does to lose control by simply going about its normal business at an abnormally fast rate. Still, it was the reigning dogma for a long time.

One reason was shockingly simple, say stem cell experts: oncologists didn't care about the cellular dogma. They didn't care about cells. They cared about the genetic dogma. They saw that normal cells had normal genes, that cancer cells had mutated genes, and assumed that every genetically normal cell was basically the same, and every genetically mutated cancer cell, as dangerous as the next. So the sport was to fish for mutated genes that all cancer cells had in common and go after them all. The quarry was every cancer cell. It was a strategy that was likely to keep oncologists fishing a long time.

The problem was that few oncologists were following the cellular revolution occurring in the 1990s with the discovery of the adult blood stem cell; they were still mired in the DNA revolution of the 1970s and 1980s. Few oncologists knew that developmental biologists were discovering there is a hierarchy of normal adult cells in every tissue, and a hierarchy in the level of potency cells possess. Thus, oncologists acted as though every normal cell was equally powerful, thus equally able to go powerfully awry. Indeed, oncologists can be thought of as believing that handing genetic mutations to normal cells is like handing out machine guns to an army of perfectly equivalent soldiers. But what stem cell biologists, with their understanding of the cellular hierarchy, were coming to see was that handing genetic mutations to normal cells is like handing out machine guns to members of a nursing home. Most of the members of the nursing home, just like the cells in the adult body, are old, set in their ways, and limited. They can't even pick up a machine gun, let alone figure out how to use one.

But, in every nursing home, there *is* a small minority of strong, open, vibrant people who keep it running—able young orderlies and janitors in their twenties who could certainly do a lot of damage with those machine guns. In the same way, comparatively young, able, potent, and open stem cells could certainly do a lot of damage with genetic mutations. Mature cells, the majority of the cells of the body? Not so much. It's all most of them can do to get their daily chores out of the way—and a whole lot of them die while they're at it.

What this meant, many stem cell biologists believed, was that oncologists should have been fishing for their mutated genetic prey within tiny pools of cancer stem cells, instead of within the entire universe of *all* cancer cells. But for a surprisingly long time, they did not.

"We've opened up the field of oncology, yet many of these cancer people don't know it yet," McKay laughed during a 1999 interview, speaking of stem cell biologists. "You can sit at a friendly dinner table and say, 'You need to know this,' but oncology people don't know it."

Still, by the time of the June 2004 ISSCR meeting, it will be clear that a lot of people finally do believe they know it, and are launching an exciting slew of preclinical and clinical trials in their attempt to prove it. What some oncologists are doing, essentially, is treating their cancer stem cells with some of the exact same growth—and growth-blocking—factors that stem cell researchers are using to control their normal cells.

And vice versa. The fields, in some ways, increasingly resemble two cheating kids taking a test in the back of a class, constantly peering over each other's shoulders for the answers to their questions. The cellular revolution that has been changing the way scientists look at health is changing the way they look at illness. And they are seeing that the stem cell's role may be equally critical in both.

*   *   *

ON JUNE 9, 2004, the day before the second annual ISSCR conference, the body of former U.S. president Ronald Wilson Reagan is flown to Andrews Air Force Base, driven to Constitution Avenue by hearse, and then rolled in a carriage up to the Capitol, where he will lie in state for two days.

Over and over in the media commentary, it is noted that this death is very different from the last death of a nondisgraced president (Lyndon Johnson in 1973). A huge panic ensued, for instance, when a light aircraft innocently wandered into a no-fly zone that was placed over DC after 9/11. Over and over it is noted how extraordinary the turnout in the streets is, and how elaborate the arrangements are. Some 100,000 people waited eight hours in California early in the day to say good-bye to the president, taking local police by surprise. Many more are expected to turn out at the Capitol.

Given all this, hES cell advocates have wasted little time. For days, stories have been noting that, seeing as Reagan died of Alzheimer's, perhaps the best way to honor him would be for President Bush to relax the restrictions on hES cells. A poll accompanying such a story on CNN on June 8 even found that 79 percent of respondents, or 90,928 people, voted yes and 21 percent or 24,651 people, voted no to the question, "Should President Bush remove restrictions on federal funding for stem cell research?" The pro-hES cell research numbers are the highest they have ever been.

Indeed, hES cell advocates are finding the timing of Reagan's death a little eerie, given that the day before Reagan died 58 senators, including 14 Republicans, had written President Bush a letter urging him to allow more of the 400,000 frozen IVF-clinic embryos slated for destruction to be used for research. Among the notable signatures on the Senate letter were those of abortion opponents Lamar Alexander (R-Tennessee), Ben Nighthorse Campbell (R-Colorado), Thad Cochran (R-Mississippi), and Trent Lott (R-Mississippi). Also signing was Senator Mary Landrieu (D-Louisiana), who had worked closely with Senator Sam Brownback (R-Kansas) to oppose embryo research, but distanced herself from him after winning a close reelection late in 2002, according to *The Washington Post*. The letter followed a similar pro-ES-cell plea from 206 members of the House of Representatives a few weeks earlier.

Around the country, it has been noted all week that the death of Reagan, whose wife Nancy has been in favor of hES cell research almost from the start, may spur the kind of momentum that might be needed to move Congress. Indeed, Senator Orrin Hatch (R-Utah), a key supporter of expanding stem cell research, said point-blank to *USA Today*: "Perhaps one of the smaller blessings of [Reagan's death] will be a greater opportunity for Nancy to work on this issue." Said Tony Mazzaschi of the Association of American Medical Colleges in the same article: "I expect the pressure on the president to adjust his policy to continue building."

Even the conservative *Washington Times* chimes in, citing a speech Reagan gave often—although the speech had nothing to do with stem cells. The writer, Robert Goldberg, director of the Center for Medical Progress at the Manhattan Institute for Policy Research, quoted Reagan as saying: "We could put a price tag on the value of these human benefits, but who would want to do that? Who can even imagine the wonders that lie ahead if we just have the faith and the courage to push on?" A CAMR member sends Reagan's words out to stem cell scientists worldwide.

Nancy Reagan's most recent words on the subject, when she spoke openly for the first time at a stem cell gathering the month before, are also quoted all week: "Ronnie has gone to a distant place . . . I just don't see how we can turn our backs on this. . . . We have lost so much time already."

But many watchers note that it is the silent figure of Nancy dressed in black, tiny, wearing out-of-date, too large glasses and grasping the arm of a huge soldier, which makes the most eloquent plea for stem cell research on this day. There is a sad half-smile on the face of the diminutive former first lady, who became a recluse for the 10 years of her husband's illness, as she watches her husband's coffin be removed from the carriage hitched to the riderless horse that, whimsically, sports Reagan's cowboy boots in its stirrups. The coffin being carried up the rotunda steps of a Capitol that is wreathed in the fog from a 21 gun salute, as a military band slowly plays "Glory Glory Halleluiah," has liberals and conservative commentators alike tearing up. Says PBS's Haynes Johnson: "This is Reagan's last great performance." Still, there is nothing smacking of performance in the look on Nancy Reagan's tired, affectionate face when House of Representatives chaplin Daniel Coughlin, speaking over the flag-draped coffin, says of the last years of Reagan's life: "The wheel turned all too slowly."

This particular surge of national interest in stem cells will not be resulting in a new congressional vote this year, 2004. Because it is an election year, Congress will end up tabling the issue. But the collective grief over the loss of a U.S. president, and what stem cell advocates repeatedly call the decade of unnecessary suffering he endured, adds an air of urgency and expectation to the second annual ISSCR conference when it begins on June 10.

.   .   .

MANY THINGS distinguish this conference, held in the World Trade Center on Boston Harbor, from the first ISSCR conference, held in 2003 in DC. It opens with a rousing speech by a Republican senator, for one thing—stem cell advocate Arlen Specter. There are some 1,500 scientists here, for another thing, twice last year's attendance of 700. Scientists hustling between sessions carrying their work bags normally look like students late for class. But the excitement level is pitched so high here, given the deluge of press and the doubled attendance size, that the participants more resemble happy passengers on a busy cruise ship—an impression admittedly enhanced by the frequent flashes of blue harbor out the windows and the pterodactyl sound of seagulls.

This year's conference, as well, comes at the tail end, and beginning, of three important coups. South Korea's Woo-Suk Hwang is here, a reminder that the first human cell has apparently just been cloned—the first truly flashy stem cell advance since the discovery of the hES cell in 1998. Also here are Israel's Benjamin Reubinoff and staffers from Lior Gepstein's lab, who will present historic work: the first two experiments, to be published in major Western journals, proving that hES cells are functional. Gepstein, who works at the Technion in Haifa, will announce that the pulsing hES cells described in this book's prologue took arrhythmic pigs off artificial pacemakers. His cells formed *natural* pacemakers. His presentation presages his upcoming *Nature Biotechnology* paper, the first major paper to prove hES cells work.

And Reubinoff, who as noted in an earlier chapter, has worked for years on hES cells and the brain, will announce that he has partially cured Parkinson's rats with neurons made from hES cells. His paper, to be published the month after Gepstein's, will be the second major one to prove hES cells work.

Still, perhaps the biggest difference between last year's conference and this lies in the poster room. It is huge, in order to accommodate the 460 posters, all on stem cells. Last year there were 200 posters, which snaked quietly around hallways and gathered in small corners. This year they demand the second-largest room in the conference and still have to be taken down and changed twice to accommodate all. Furthermore, last year only 24 of the posters were on hES cells, those posters coming from only four countries: South Korea, the United States, Australia, and Israel. This year there are 76 posters on hES cells, with the above four countries now joined by Switzerland, Taiwan, Singapore, Czechoslovakia, Sweden, Russia, and Canada.

Most stunningly: last year, there were a mere four posters on "cancer stem cells." This year, there are more than 35 "cancer stem cell" posters and formal presentations. Each day, as the old posters are replaced with the new, more such posters appear. The number of posters is surprising; the similarity of many of the conclusions is exciting.

Kim Bender of MIT, for example, created a mouse bearing a mutated rat oncogene (a cancer-causing gene). Within the rat's resulting pulmonary adenocarcinoma (lung cancer), he found that the stem cells that create most of the normal tissues of the lung possess some of the same markers as those found in embryonic lung precursors. Translation: simply by fishing out the most malignant cells in a lung carcinoma, he found, he believed, both the stem cell of the cancer and the normal healthy tissue, in one fell swoop.

Similarly, by fishing out malignant cells in the lethal brain cancer glioblastoma multiforme, and bringing them to the single cell level, Angelo Vescovi of the Milan, Italy, Stem Cell Research Institute found that his most highly malignant cells, differentiated, looked similar to differentiated normal neural stem cells. He found for the first time, he claimed, "true tumor neural stem cells."

Many of the posters are highly complex, reflective of the fact that this "new sub-field" has actually been creating and re-creating itself, quietly, for quite some time in the hands of a few believers. After reviewing literature that indicated that two different signaling pathways (chemical reactions within cells) have been implicated in certain brain cancers, and that one of those pathways has been found active in normal embryonic brain cell proliferation, Anna Kenney of the Dana Farber Cancer Institute in Boston set out to see if the two pathways actually work together to promote proliferation in both adult normal neural progenitors and tumorous neural cells. Not only did she find that this was the case—that the PI-3 kinase pathway and the Shh pathways work together—but she found a small molecule that blocks them: "a future effective mechanism for treating . . . medulloblas-tomas," she wrote.

Ian Mackenzie of the University of London studied 11 different head and neck cancer cell lines and found that "each develops stem cell colony patterns similar to those of normal keratinocytes [immature skin progenitors]." Furthermore, single stem cells from these cancers formed all the mature cells of the normal tissues upon differentiation.

Lubna Patrawala of MD Anderson Cancer Center in Smithville, Texas, found what she believed to be cancer stem cells in 15 different cultured human cancer cell lines, including cancers of the prostate, breast, ovary, colon, skin, and brain glia. She brought the cancer cells to the single cell level and watched to see which ones could form long-term cultures. Only 10 to 20 percent of clonal cells (exact copies of a single cell) in those cancers could continue growing in short-term cultures and a much *smaller* percentage could continue growing for a long time: the stem cells of the cancers. (Since stem cells are rare in the normal body, it would make sense they are rare in cancer.) She checked her results with literature to find normal tissue-specific stem cell markers. The results "have significant implication in designing tumor therapeutics specifically targeted to the small population of tumor initiating cancer stem cells," she wrote.

A University of Florida group led by Valery Kukekov plucked stemlike cells out of a brain cancer (glioblastoma multiforme) and a bone cancer (sarcoma) and found they were stem-cell-like progenitors that were similar to each other.

The posters just keep sprouting. The Brain Tumor Research Center of Toronto, Canada, announces it found a marker (CD 133) for a tumor stem cell shared by three different kinds of brain cancers (medulloblastomas, ependymoma, and glioma) and that cells from those cancers that do *not* express that marker can't spread the cancer to other mice of the same line. "Future therapies may be targeted at the tumor-initiating BTSC [brain tumor stem cell] rather than every cell in the tumor," wrote the center's Sheila Singh.

"Six out of seven of normal hematopoietic [blood] progenitors were discovered because of leukemia," Stuart Orkin of Johns Hopkins noted at one session, and several posters reflect this. David Taussig of the London Research Institute wrote about a clinical trial in which he found the progenitor at the heart of AML (acute myeloid leukemia), the most common form of adult leukemia, which he now believes is a mutated blood progenitor.

Harley Kornblum of the University of California, Los Angeles, reported that an enzyme, MELK, which regulates proliferation in normal neural stem cells is also expressed in brain tumors, including medulloblastomas. Blockage of this enzyme in a neural tumor line "dramatically reduced proliferation."

Paola Vecino of University Miguel Hernandez showed that Sertoli cells, located near sperm stem cells in the body, keep sperm stem cells and teratocarcinoma cells (sperm stem cell cancers) from proliferating. Sertoli cells, obviously, are a potential clinical therapy.

The above are only a small subset of the cancer stem cell work being touted here. Furthermore, several posters described what at first appears to be a bizarre phenomenon: some transplanted stem cells appear to be able to track down tumor cells. A Russian scientist points out that by accident she discovered her stem cells chased tumor cells. Karen Aboody of the City of Hope National Cancer Center found that a certain kind of human neural stem cell seems able to chase down both metastasized brain and breast cancers. Taiwan researchers made a similar observation using bone marrow MSCs and colon cancer. Xing Wu of Fudan University Hospital in Shanghai reported that MSCs don't just track and "quickly and extensively" infiltrate brain glioma cell beds, but that they fuse with them at a much higher rate than they do with normal cells (glioma, 46 percent; neurons, 17 percent; astrocytes, 21 percent). The Taiwan researchers came up with the best explanation for this, when it comes to MSCs: tumor angiogenesis. That is, it is by now an established fact that cancers must create blood vessels in order to survive. It makes sense, therefore, that tumors would send out chemical signals that attract blood-vessel-forming stem cells like MSCs to them. Work before and after this conference will support the whole phenomenon, including a paper by MD Anderson oncologist Michael Andreeff. "Maybe one of the reasons tumor cells are so elusive is they capitalize on stem cell properties. Sometimes the best way to get an enemy is to enlist a former friend who knows the enemy's routine. A spy, a turncoat," Burnham Institute neural stem cell expert Evan Snyder speculated after he made a similar finding.

Stem cells, in and of themselves, are not attackers the way mature T cells are. However, the above researchers have been taking advantage of what they see as their normal stem cells' fatal attraction to tumors by loading them with toxins. So when the healthy stem cells hook up with the tumors, the tumors have no time to direct them to form new blood vessels, because the stem cells release their loads and kill the tumors. The stem cells may be, in other words, natural suicide bombers.

All in all, the cancer stem cell field is moving shockingly fast. Many papers indicate that already new approaches accommodating this new/old vision are on their way to the clinic.

●   ●   ●

INDEED, A FEW steps away from this conference, at the world-famous Dana Farber Institute, a clinical trial targeting cancer stem cells has been ongoing for a year. A new "proteosome inhibitor" drug called Velcade is being given along with two standard chemos to AML patients who have slipped out of remission. Chemos, which go after rapidly dividing cells, have worked for decades on many patients with many different cancers; however, in many cases the cancers eventually return. The thought once was that this occurred simply because chemo is so toxic it must be stopped before all cancer cells can be killed. The thought was also that cancers then go chemo-resistant.

But the new thought is that many chemos don't work because the rapidly proliferating cells that chemos go after are not the permanent stem cells of the tumor. Most of the time normal stem cells are quiescent, after all. Furthermore, normal stem cells possess many survival mechanisms to retain their status as permanent residents of their tissues (since without them, mature cells would not be replenished, and the tissues would die).

"The leukemic stem cells are really different, with different properties from the bulk of the leukemia," says oncologist Philip Amrein, head of this new Dana Farber cancer stem cell clinical trial. "You see it, deal with it, all the time in patients. You can get rid of 98 percent of the cells fairly routinely, but what you find is the cancer keeps coming back. Why? Because the 2 percent of the cancer that is made of stem cells is not being affected." So Amrein has been giving his patients standard chemos to kill off the less dangerous, rapidly cycling cancer cells, and Velcade, which he hopes specifically targets and kills off the stem cells of the tumor.

"It's going fine," Amrein notes a few months after the conference, in December 2004. "We don't know what the right doses are for toxicity, so we're giving a low starting dose. Even so, we're seeing at least a 50 percent remission rate in patients with relapsed disease, and that's really quite good. We're having a nice complete remission rate and we're not seeing toxicity." What Amrein is really interested in is hiking those doses. This must be done carefully and slowly according to clinical trial rules. "A little tickler is this: as we've begun to finally increase the dose of the Velcade, the toxicity has actually gone *down* and remission rate has gone *up*. That's a dream come true if it holds up."

The above approach is the brainchild of Craig Jordan of the University of Rochester. Fresh out of his training in the late 1990s, he knew many leukemias were believed to come from stem cells (leukemia stem cell work began earlier than others simply because the blood stem cell was the first human stem cell characterized back in 1991). Particularly inspiring for him was a 1994 paper written by University of Toronto stem cell biologist John Dick, which indicated that the stem cell of AML had been definitively found. That paper essentially kick-started the cancer stem cell field. For when Dick irradiated the bone marrow of mice without immune systems, then laboriously placed tiny amounts of human AML into the bone marrow, he discovered that only 1 in 1 million of the cancerous cells was actually able to give the mice human cancer. The rest of the cancerous cells acted like normal differentiated cells and eventually died out.

Furthermore, when Dick then analyzed the truly cancerous cells, he found they expressed the same markers as the normal hematopoietic (blood) stem cell.

Jordan completely believed in the notion that "cells that are long-lived enough to undergo the many mutations required for cancer were the stem and progenitor cells," he says. "So we said, 'OK, let's look at the stem cells for AML [which the above-mentioned Dick had just found], and look at normal blood stem cells, to find out what is different about the leukemic stem cells.'" He wanted a therapy that would kill only the leukemic stem cell, not the healthy blood stem cells.

Jordan's group discovered that something called the NFkappa B signaling pathway (a chemical chain reaction inside cells) is turned on in normal stem cells only when they are activated, not when they are resting. The group also discovered this pathway is apparently stuck in the "on" position in AML cancer stem cells. This made sense because "NFkappa B is a survival mechanism in stem cells," Jordan notes. That is, normal stem cells turn the pathway on in special circumstances to block their death, and protect their status as permanent residents of the body. AML cancer stem cells, apparently, have learned how to keep this potent gene turned on *all the time.* He had found a perfect target. Jordan heard that a new drug called Velcade blocks NFkappa B, so he asked a doctor friend for some and discovered that, indeed, when Velcade is given to AML cells in rats, it gets rid of the stem cells of the tumor, and a standard chemo called idarubicin gets rid of the rest.

Not only that, but Jordan discovered that that standard chemo, idarubicin, actually causes NFkappa B to be upregulated (or turned on) in leukemic stem cells to an even greater degree than normal—which means that the chemo, in the past, may have actually caused leukemia stem cells to become *even more* resistant to death. He published a series of papers on his work beginning with a special plenary paper in *Blood* in 2001. "The combination of these two agents induces rapid and extensive apoptosis [death] of the LSC [leukemic stem cell] population while leaving normal HSCs [hematopoietic or blood stem cells] viable," he reported in *Proceedings of the National Academy of Sciences (PNAS)* in 2002.

"Very exciting for us," he concludes. Amrein at Dana Farber latched onto the work immediately, as did MD Anderson in Texas and the University of Kentucky. They are all trying the approach out in clinical trials for AML. And Jordan will announce in 2005 that the NIH has approved a special accelerated clinical trial for an even less toxic compound called feverfew, an herbal medicine that has been used for centuries, that does the same thing. This special category of aid is given only to trials the NIH considers extremely promising.

Then there is that famous, fairly new drug, Gleevec. Billed as the world's first drug specifically tailored to go after a specific molecule on the surface of cancer cells, its May 2001 approval by the FDA represented the fastest FDA approval ever. The November 2001 approval of the drug in Japan and the EU also represented record approval times. Precisely because of the potency that the drug immediately began showing against a leukemia called CML (chronic myeloid leukemia), the stem cell field began to postulate that the drug must be going after a molecule, not just on any old cancer cells, but on cancer stem cells. Indeed, some studies indicated that some immature cell was the target cell. And it was eventually discovered that the drug binds to a receptor that is found, not just on CML cells, but also on many normal stem cells, Bcr-Abl, which causes proliferation normally.

In recent years, however, it had become clear that when some patients went off Gleevec, the cancer rushed back. Was this because the drug failed to go after the cancer stem cell, or because the drug was simply a bit weak, failing to bind tightly enough to all the leukemia stem cells?

Novartis, which developed the drug, went back to the drawing board and came up with a similar drug that bound to the receptor on CML cells more tightly. In the petri dish, it was proven to be a whopping 30–100 times more effective at killing tumors. Francis Giles of MD Anderson will announce at a later cancer conference that this drug, named AMN107, caused the blood counts of over 90 percent of chronic-phase CML patients and over 70 percent of more advanced patients to return to normal. He, like Amrein, is increasing doses in the hope that he can make the remission permanent.

But if that fails, another stem cell theory investigator, Harvard University's Gary Gilliland, will already have *another* option available. He will publish evidence in December 2004 that the ultimate reason Gleevec fails is that it doesn't target the true stem cell of the tumor, which he believes is a normal blood progenitor. He has a molecule in the mutated version of that all ready to target: MOZ-TIF2.

There are a huge number of such mutant-stem-cell attack dogs being tried out. Almost monthly, there is a new announcement. A particularly dramatic announcement will come in April 2005, when Sidney Kimmel Cancer Research Center oncologists in San Diego will tell the American Association for Cancer Research that after examining 1,156 cancer patients with 10 different cancers, from prostate to lung to breast, they found that the same 11 stem cell genes (all of which were connected to a gene called Bmi-1) appeared in the most malignant forms of those cancers. Eleven Bmi-1 linked genes are "engaged in both normal stem cells and a highly malignant subset of human cancers diagnosed in a wide range of organs," the researchers will exult. Kimmel researcher Gennadi Glinsky will note that these "death from cancer signature" genes should allow doctors to better target their therapies. That is, patients with that 11-gene signature should not waste time on conventional therapies and should go straight to experimental therapies (presumably, those targeting that genetic pathway).

As new as the cancer stem cell idea seems, it has long haunted the oncology field. The notion has long, ghostly legs. In fact, it dates to 1877, when German experimental pathologist Julius Friedrich Cohnheim proposed a radical new theory of cancer origin. Shortly before 1877, the creation of greatly improved microscopes had been leading to a greater understanding of cancer. That is, oncologists were coming to the agreement that cancer was all about normal cells run amok (in contrast to early theories that ran the gamut from the notion that it was an excess of black bile, to an excess of mineral salts, to an infectious disease). This observation was permanently codifed in 1858 by Cohnheim's teacher, Rudolph Virchow. Indeed, Virchow was the one to establish that every cell, healthy or cancerous, "comes from a cell."

Cohnheim picked up the ball from there, quickly becoming inseparable from his fascinating microscope, familiarizing himself voraciously with the look of cells from all kinds of tissues. He was soon struck by the similarity between embryonal cells and cancer cells: the fact that both are unusually invasive, unusually proliferative, and can transform themselves into a highly unusual number of different mature cell types. He subsequently formulated the theory of "embryonal rests." Because embryonic cells are produced in excess in the developing fetus, cancer cells must be those leftover cells, he decided, trapped in the adult body. However, as fascinating as it was, his embryonal theory of cancer was largely ignored as oncologists continued to pursue a variety of other notions, including the idea that cancers were normal mature cells that dedifferentiated back toward an embryonal state—an idea that persisted in the mainstream until the 1990s.

But the cancer stem cell idea did begin to at least resurface among handfuls of researchers in the 1950s, when Leroy Stevens of the Jackson Lab in Mount Desert Island, Maine, began playing with mice who had developed extremely weird tumors called teratomas. The stem cell field—and it would appear now, perhaps the cancer field—would be forever indebted to his fascination with his weird mice. For it was thanks to the cancer of Stevens's weird mice that the very first mammalian embryonic stem cell and the very first cancer stem cell were discovered.

*   *   *

"'TERATOMA' IS the Greek word for monster," booms University of Kansas pathologist Ivan Damjanov, standing by a giant photo of a disgusting tumor—it has a tooth sticking out of it—lodged on a human pelvis. "None of us would be here today if two men hadn't started studying teratomas: Barry Pierce and Leroy Stevens. They are the grandfathers of the embryonic stem cell field. It's good to know who your grandparents are—although, when it comes to fatherhood, that's a matter of faith."

He beams at the room. It is two months after Reagan's death, and Damjanov is standing before an August 2004 wetlab comprised of some of the world's top hES cell scientists, and some of the world's newest scientists, at the above-mentioned Jackson Laboratory. Mount Desert is a typical northern Maine summer village. It is rife with lobster shacks manned by college kids patiently instructing tourists how to extract sand from their crustaceans; blueberry syrup and blueberry beer; wooden fishing boats; rough-hewn rock formations and low-slung firs that herald its closeness to the arctic. However, take a left off the main road that winds past the ocean and your nose will acknowledge that Mount Desert Island is also the home of the lab mouse. (Indeed, to stand in the Jackson Lab parking lot is to be transported back to childhood, to the days of hamsters and gerbils and being yelled at to clean the cage.) Since the early 1900s, some of the most important "disease mice" in the world have been made here in the Jackson Lab, including a mouse that would lead eventually to the discovery of the most important human cell: the hES cell.

So it is fitting that, for a week each year since 2003, Jackson has been offering intensive instruction in hES cell handling and manipulation to interested young students and postdocs from around the world. Funded by the NIH, the idea is to give young people a chance to work with the cells—which the government is still slow in releasing—by giving them hands-on experience in how to expand, differentiate, freeze, and understand them. It is, in a sense, the stem cell wetlab of wetlabs, known for its rigor, its great instructors, and its history as the place where the "cell of cells" began its slow journey from a spot buried at the heart of a mouse's tumor to the forefront of scientific understanding.

In a few hours, the 20 or so young scientists in this room will break out into the lab, and begin playing with their new, multipotent, human embryonic stem cells. They will curse when their cells spontaneously form rosetta-like neurons at the edges of their colonies. (The trick is to know when to switch the growing cells to another plate, so they won't pile up and differentiate.) Then, when no one is looking, they will return to their microscopes to stare in awe at the brain cells they made without doing a thing. And so on. But first, they will learn about the killer cancer that led to the discovery of their life-giving normal stem cells: the monstrous teratoma (or its more malignant counterpart, the teratocarcinoma).

And they will learn about it again and again. University of Sheffield hES cell expert Peter Andrews will begin his talk with a similar comment ("A lot of the history of ES cell work goes back to work on [cancerous] teratomas," he will say, flashing a similar gross tumor on the screen. "ES cells are degenerate embryonal carcinoma cells, or embryonal carcinoma cells are degenerate ES cells, it depends on how you look at it. In fact, the hES cell is more like the embryonal carcinoma cell than it is like the mouse ES cell"). HES cell discoverer Jamie Thomson will also make several references to the monster cancer when he talks about his hES cells.

Damjanov flashes more hideous-looking tumors onto the screen. They contain hair, bone, neurons, skin, eyes. For centuries, teratomas have fascinated doctors, many of whom believed they were a sign of the devil. Still, they were very rare tumors and rarer still in the mammal that was easiest to study because of its size and short lifespan: the mouse. Few scientists ever got the chance to study them. But in the late 1950s, Damjanov says, experimental embryologist LeRoy Stevens of the Jackson Lab was determined to change that. He would mate together as many rare mice with teratomas as he could to see if he could possibly create a strain that would reliably produce the tumors. He and others could then do endless experiments on the tumors. Because these tumors started small, yet even in that small size possessed cells from many tissues of the body, it seemed this cancer hadn't *de*differentiated from limited mature cells that normally can't morph into anything else, as the reigning theory of cancer dictated. It seemed this cancer had differentiated, if bizarrely, from some kind of wildly potent embryonic-like cell that can form all the cells of the body. This cancer apparently started from a very, very multipotent cell—an embryonic cell somehow trapped in the adult body of the mouse—whose forward-moving differentiation powers were simply partially blocked.

And if that was the case for this cancer, it should be true for all cancers.

By 1961, Stevens had succeeded in breeding many such mice and had begun shipping them around to interested embryologists and oncologists. Shortly thereafter, he discovered that this weird cancer indeed came from embryonic-like germ cells (the cells that form the sperm and ovaries in mammals). In 1964, Stevens's friend Barry Pierce of the University of Colorado, who had immediately acquired many of Stevens's mice, would publish a paper finding that the teratoma arose from a single cell. Pierce found that most of the cells from the tumor simply differentiated, when you handled them properly, and died. Only one cell malignantly continued on—and not only that, he said, it was clearly a "caricature" of a normal embryonic cell, and probably an embryonic stem cell, at that.

That is, by 1964, Pierce had indirectly proven the existence of the first cancer stem cell: the teratocarcinoma stem cell. "I couldn't get my first papers published," Pierce noted cheerily in an interview decades later. "Most people were not willing to buy into the idea that, while most cancer cells divide, most of those mature like a normal body, or a normal developing child. It was not a widely held topic at the time that only a small handful of cancer cells could be the truly malignant cells, or stem-cell-like cells."

But then came Ivan Damjanov and Davor Solter, a pathologist and embryologist, respectively, who were working in a lab at the University of Zagreb. The two confirmed in 1970 that the above cells were indeed caricatures of normal embryonic cells, when they placed normal mouse embryos under the kidneys of a normal adult mouse and found that these normal embryonic cells spun out of control, and looked very much like, the teratocarcinomas of Stevens's mice. And when Beatrice Mintz of the Institute for Cancer Research in Philadelphia reported in a 1975 paper that she had popped an embryonic *cancer* stem cell into a mouse womb and a *normal mouse* was born, it seemed clear that embryonic cancer cells and normal embryonic cells were related. If an embryonic cancer stem cell existed, it seemed clear enough, a normal, healthy, therapeutically significant embryonic stem cell must exist.

Understanding the potential import, Solter and Barbara Knowles, today director of research at the Jackson Lab, set out to characterize these so-called mouse embryonal carcinoma stem cells, hoping they would lead to normal mouse embryonic stem cell lines. They did this cleverly by popping a teratocarcinoma stem cell from one mouse into a very different mouse. The second mouse rejected it by sending antibodies to it. Solter and Knowles captured the antibodies that could recognize their teratocarcinoma stem cells and created armies of them (called monoclonal antibodies). In this way, following their antibodies, they were able to zero in on a unique antigen expressed on the surface of their teratocarcinoma stem cell lines: SSEA-1.

Using this antigen as a guide in 1981, two different groups, one in the U.K. and one in California, found the first normal mammalian ES cell, the mouse ES cell: the groups of Martin Evans and Gail Martin. Solter and Knowles would shortly after find two more antigens on the embryonic cancer stem cell lines: SSEA-3 and SSEA-4. Andrews would find antigen as well.

And James Thomson, in turn, would successfully use SSEA-3 and SSEA-4, among others, as his guide to isolate the stem cell of stem cells: the human embryonic stem cell, in 1998.

While all this points to the similarity between stem and cancer cells, it does not necessarily point to therapy. This is what Pierce was interested in. Throughout the 1970s and 1980s, he would write numerous articles pointing out that, since teratocarcinoma stem cells could be made normal in the developing embryo, this must mean that a perfect cancer therapy might involve taking growth factors found in the body and elsewhere that differentiate *normal* stem cells and using them to differentiate *cancer* stem cells into a normal, harmless, mature nonreplicating state. After that, they should naturally die, like most normal mature cells. The cancerous state may be "reversible" he noted, over and over. It is simply the result of the partially "blocked differentiation" of normal stem cells. "There is a quote of Pierce's, from a 1971 book he wrote, that I use in my seminars to show how old this idea is," says breast cancer specialist Jeffrey Rosen of Baylor University. "It is: 'The analysis of the cellular origin of carcinomas of different organs indicates that in each instance, a determined stem cell is required for tissue renewal, and is the cell of origin for carcinoma.' So that pretty much said it all, everything we know today—back in 1971."

But in part because a chemo—cisplatin—was found for the malignant form of Pierce's model cancer, teratocarcinoma, which had a near-perfect success rate, none of the natural differentiation factors in the body were ever seriously tried on it in a big clinical trial. (It is unethical to experiment on patients when a near-perfect cure is available.) Various scientists, starting with Andrews, in the 1980s found that adding a vitamin A derivative called retinoic acid to teratocarcinoma stem cells could turn them into neurons that did not seem to replicate, but this was never tried on brain cancer patients in a big way. Still Pierce remained convinced that differentiation therapy should work—if, in many cases, in conjunction with other therapies—on what he believed were the stem cell "caricatures" that lay at the heart of every cancer.

There was a brief flurry of interest in the idea of solid cancer stem cells and differentiation therapy in the late 1970s and early 1980s, following papers published by University of Arizona oncologist Sid Salmon finding that only 1 in 1,000 to 1 in 5,000 lung cancer, ovarian cancer, or neuroblastoma cells were able to form colonies in soft agar (normal cells can't grow separated out in soft agar this way, only wildly malignant cells can). These papers seemed to point to the notion that only a small handful of solid cancer cells could be malignant—*not* all cancer cells—which implied the presence of singularly potent cancer stem cells in solid tumors beyond the weird teratocarcinoma.

But scientists say, partly because no one had definitively found a normal solid-tissue stem cell outside mouse embryonic stem cells, most didn't believe other ones existed. Therefore, most believed a mutated version of those cells (cancer stem cells) could not exist. The mouse ES cell, and its cancerous cousin the teratocarcinoma stem cell, were probably the only ones, was the thought through the 1980s. Furthermore, "science is social. A new generation came along that was solely interested in molecular biology," Pierce laughed. "Few of those ended up understanding cellular nuances."

Solid-tumor stem cell cancer work languished.

Leukemia stem cell work proceeded more vibrantly thanks to the 1961 indirect discovery of the mouse hematopoietic (blood) stem cell (HSC). Its presence was only strongly implied by some famous, clever experiments done by a Canadian group in the early 1960s. (The cell's existence wouldn't be proven conclusively until Irv Weissman isolated it with a then-new machine called the fluorescence activated cell sorter (FACS), which enabled the cytometric analysis and isolation of single blood cells, in 1988.) However, the tantalizing possibility that a normal blood stem cell probably existed kept many hematologists pursuing the blood stem cell idea. (The fact that blood is so accessible also played a solid role, many believe. It is difficult to get ahold of tissue from cancerous human organs.) Since 1981, the efforts of hematologists produced a whopping 1,796 articles on leukemia stem cells, according to PubMed, out of the 3,692 "tumor stem cell" papers listed. (PubMed is the search service for the vast majority of the world's science and medical journals.)

Their efforts also produced one of the biggest success stories in cancer history to this day.

Beginning in 1970, Leo Sachs of the prestigious Weizmann Institute of Science in Rehovot, Israel, had been writing papers using language quite similar to Pierce's. By placing normal blood cells onto a bed of spleen cells, he had devised in the 1960s a system that persuaded those blood cells to differentiate. By the end of the 1960s, he was tossing certain leukemic cells onto the same bed and seeing something fascinating: *they* differentiated into harmless, normal cells, too, under those circumstances. "A block in cell differentiation in vivo thus appeared to be overcome in vitro," he wrote in 1970. He also, like some of the teratoma groups, popped a few leukemic cells into mouse embryos in 1982 and came up with normal embryos. Sachs urged hematologists to find "inducers" in the body that naturally differentiated immature cells to see if they worked to differentiate, and thus convert back to a normal state, immature leukemic cells.

In the mid 1980s, Laurent Degos of the Institute of Hematology in Paris heeded the call, trying out 60 different agents, natural and synthetic, on leukemia. His group hit the jackpot in the mid 1980s with retinoic acid, the natural derivative of vitamin A that plays a major role in embryonic development. Like a mother with a baby, retinoic acid appeared to successfully coax raging seas of immature acute promyelocytic leukemia (APL) cells into growing up and becoming calm rivers of normal mature blood cells—in the petri dish.

This was amazing given the following fact: The very first description of APL, written in 1957 by Leif K. Hillstad in *Acta Medica Scandinavia*, proceeded as follows: APL is characterized by "a very rapid fatal course of only a few weeks' duration" and "seems to be the most malignant form of acute leukemia."

However, the best form of the vitamin A derivative, called all-trans-retinoic acid (ATRA) was available only in China, so Degos struck up a partnership with Chinese hematologist Zhen Yi Wang. ATRA was given to its first APL patients at Rui-Jin Hospital in Shanghai in 1987. The results were "remarkable" Degos wrote in 2003. "ATRA could induce differentiation of malignant cells in APL patients until they reached the stage of complete clinical remission . . . the complete remission rates reach 95%." And the lack of side effects was "unusual for a treatment that induced complete remission." When the Tiananmen Square revolt occurred in 1989, France stopped allowing travel to China. But word of the success was such that an American company (Roche Nutley) stepped in, and the drug was made available worldwide. The upshot: by the end of 1992, "the European Cooperative Group considered any regimen for the treatment of APL that did not include ATRA to be unethical," according to Degos.

After a while, a problem emerged: patients could get resistant to ATRA. But the Chinese group had in stock another old Chinese medicine that seemed to differentiate immature leukemic cells: arsenic. When doctors tried the two together on mice in 1999, "the combination of ATRA and arsenic eradicated the disease. The mice died without any clinical, hematological, or molecular signs of APL," reported Degos. The cells appeared to not only differentiate immature leukemic cells into what appeared to be normal mature cells but also to then swiftly and quietly die like any good obedient, short-lived, mature adult blood cell.

The combined approach began working on patients as well, so much so that Wang was the recipient of awards worldwide by 1998, and by 2004, his young colleague, Zhu Chen, had been made vice president of the Chinese Medical Academy.

In 2004, Chen reported in *PNAS* that the combined treatment had prompted a complete remission rate in an impressive 90 percent of patients. This "synergistic targeting therapy model represents a molecular triumph over an otherwise lethal disease," Chen wrote. "It has been over three decades since Leo Sachs first postulated the possibility of inducing leukemic blast cells to more mature or terminal blood cells through a normal differentiation process. Now we know that differentiation induction is, after all, a mechanism by which key factors in differentiation arrest, growth advantage, and even apoptosis [death] re-regulation could be targeted."

Remarkably, it has even been found that cisplatin, the chemo that worked such wonders on Pierce's teratocarcinoma that it was thought unethical to try anything else, may work in part by differentiating the cancer. So the therapy used so successfully on bicyclist Lance Armstrong may have been a differentiation therapy that targets stem-cell-like cancer cells all along.

Still, it would not be until the late 1980s and early 1990s that some prominent oncologists and stem cell biologists in the rest of the solid tumor field would start considering the whole idea. First of all, says Jeffrey Rosen of Baylor University, it's very difficult to isolate single cells in solid tumors—which you would have to do to isolate and characterize a single cancer stem cell. It is generally done via the machine used by Weissman to isolate the blood stem cell: the FACS, which identifies and separates cells based on glowing antibodies attached to markers on their surface. Individual blood cells can be isolated easily this way. But "solid tissue cannot be broken down easily at all the way blood can. Solid tissue has tight junctions. In fact, a lot of surface markers are destroyed during the process of breaking them apart with enzymes for the machine."

Eventually scientists would find ways around this by doing things like gathering up metastasized, single tumor cells that break away from their tumors and land individually in lung fluid. They are also using monoclonal antibodies, which are getting better and better, that can zero in on surface markers on specific cells and stain them. And the ability to create better and better disease model mice has helped as well.

Still, the slower pace of the *solid* cancer stem cell world was also partly a matter of the mind. Since most people didn't think human adult stem cells existed, why would they think solid tumor stem cells existed? The affirmation in 1988 of the existence of the first normal adult stem cell—the blood stem cell—did get solid tumor people thinking, however. A small handful of neurologists thereafter, for example, began predicting the existence of a neural stem cell based on patterns in brain cancers. The NIH's Ron McKay, for one, came out with a paper in 1991 predicting the presence of a neural stem cell based on the multipotential characteristics of medulloblastoma, a cancer that seemed to so clearly mimic normal brain development.

But a more critical turning point occurred the next year, in 1992, when Sam Weiss and Brent Reynolds of the University of Calgary came out with a report that was momentous in its ramifications: they had discovered a version of the second adult stem cell, and it was, indeed, a neural stem cell (found in the adult mouse brain). Their cell could replicate and form different kinds of mature brain cells.

While many scientists had believed that stem cells in the adult body—outside the blood—were not possible, the *vast majority* believed the adult brain could not contain stem cells. If the brain were constantly replacing neurons, where would memory go? (There is still no answer to that one.) Still the Canadian's report was followed by others from the labs of Perry Bartlett in Australia, McKay, Harvard's Evan Snyder, the NIH's Mahendra Rao, and the Scripps Institute's Fred Gage.

If the adult neural stem cell existed—and it did—then *anything* was, presumably, possible. Scientists of all stripes started seeking out stem cells of all kinds, from normal to cancerous.

Indeed, few things argue for the intertwining of the two fields so much as the fact that *non*-oncologists began regularly looking for their normal stem cells in *cancers*. After the Weiss discovery of the normal neural stem cell, many neuroscientists began plowing through brain cancer literature to look for "niches" or places stem cells might reside, and they began finding them. "They went to some of the earliest studies of brain tumors," said Weiss in a 1999 interview. By studying the patterns of brain cancers, neuroscientists rightly predicted that there were not just stem cells in the brain making new neurons but stemlike cells called "progenitors," just as there were in the blood. More limited than stem cells, they could still form more than one of the different mature cells of the brain. This could be seen just by looking at brain cancers: some possessed many of the cell types of the brain, while others possessed only two. Brain cancers, Weiss said, "led us to believe that more than one type of cell in the brain had some of the incredible powers of the stem cell." Therefore, it seemed, some normal stemlike progenitors might also be the cause of cancers.

Still, many agree, the sheer force with which the "cancer stem cell" has been reborn is partly due, fittingly, to the ubiquitous Irv Weissman. The man who first definitively (prospectively) isolated functioning blood stem cells in 1988 (mouse) and 1992 (human), was so close to those cells that he kept seeing stem cell patterns in all kinds of cancers, especially the best known leukemias. At length, in a 2001 *Nature,* he dusted off the old idea with a paper called "Stem Cells, Cancer, and Cancer Stem Cells." In it he noted the numerous similarities between normal stem cells and malignant cancer cells—this time, down to the "signaling pathway" (chemical reactions within cells) level. He also carefully reshaped the theory by noting, as had some neurologists and hematologists before him, that stemlike progenitor cells, maybe even some that are stemlike but barely replicate, may be able to form cancers, too. The latter may be because, while such progenitors are not pure stem cells, they have only recently left the stem cell state, so their stem cell genes are not fully turned off. "What is actually happening there, is that genes involved in replication [a stem cell specialty] are actually still being used in progenitors for other purposes," says the NIH's Ron McKay. "So the difference between cells that are progenitors and the cells that are stem cells are not like night and day."

It was, however, Weissman's suggestion in that article that "cancer stem cells" might explain the scourge of every oncologist's life (chemoresistant cells) that hit the biggest nerve. "Irv's suggestion that resistant cancer cells may be cancer stem cells was, I think, a main reason this whole thing became of interest again," Rosen says. "Pierce had proposed long ago that the stem cell was the origin of cancer. But Irv proposed in that article that these cells are not only involved in the origin of cancer, but are important in treatment. They might be the cells that are more resistant to conventional chemotherapies. That is really what's gained the most attention in the last few years. The question has become, if you treat these cells, will you remove residual disease?"

Weissman's landmark article was rapidly followed up by an April 1, 2003, *PNAS* paper by one of his former students, Michael Clarke. Clarke, like Dick before him with AML in 1994, also used an irradiated mouse with a deficient immune system to discover that only a tiny fraction of breast cancer cells could pass on the cancer. Because a definitive normal breast cancer stem cell had not yet been found, he couldn't know if his cancer stem cell was a mutated version of a normal stem cell; however, it was found in an area of the breast that was believed to contain stems and progenitors. In a September 15, 2003, *Cancer Research* paper, Peter Dirks of the Toronto Hospital for Sick Children found an apparent stem cell for a number of brain cancers, all of which expressed markers of normal neural stem cells: CD133. Remarkably, a paper published almost simultaneously, in a December 2003 *PNAS* by the University of California's Harley Kornblum, also found a brain cancer stem cell that expressed CD133 among other markers.

Similarly, two papers came out in August and October of 2004—in *Nature* and *PNAS*—finding that the same genes, Hedgehog (Shh) genes, seem to mark what appears to be prostate cancer's stem cell when the gene is mutated or overexpressed. Both noted that the molecule cyclopamine could target that gene and wipe prostate cancer out in models. One result of the work: At the 2004 American Association for Cancer Research meeting, there was an entire symposium on the cancer stem cell. "Tantalizing," Dick calls this approach. "About to explode," Jordan says of it. "There's a lot of excitement out there," concludes Rosen, who himself came out with a breast cancer stem cell paper only months after Clarke came out with his 2003 breast cancer stem cell paper and who also came to similar conclusions. Rosen predicts this deluge of papers will continue. "If you just plot out the number of papers [on tumor stem cells] in the last three years, you'll see them going up exponentially."

Later in 2004, Weissman himself came out with a paper arguing that CML comes from a nonreplicating blood progenitor. The paper was slightly controversial. Oncologists say it will take a lot of persuading for them to accept that a nonreplicating progenitor can commonly form cancers. Many believe that different kinds of brain and breast cancers come from replicating—not nonreplicating—progenitors. Still, it too fits the model, since, as noted earlier, even the nonreplicating progenitor is close to the stem cell state given that it has retained the stem-cell-like ability to differentiate into more than one kind of mature cell. Genes can be turned on or off to different degrees, which means the stem cell replication gene may only be partly turned off in Weissman's progenitors. Furthermore, says Memorial Sloan Kettering Cancer Center researcher Eric Holland, "differentiation is a destabilizing event, too"— just as replication is. So a progenitor laden with its stem cell mother's mutations could experience its final destabilizing event when it differentiates into its progenitor state.

It appears that all a stem cell's fanciest magic tricks may occasionally be destabilizing under the right circumstances.

Weissman is, furthermore, quite sure that nonreplicating blood progenitors can initiate cancers. "We worked out all the blood forming lineages in mice," he said in an earlier interview. "And we discovered that we could induce myeloid leukemia by turning on an anti-death gene, not in the stem cell, but downstream, in a non-replicating progenitor. A progenitor picked up a property that only stem cells have in the blood-forming system. We also did a study where we looked at CML and AML in blood samples from Hiroshima survivors. Some had a characteristic translocation in their AML or CML cancer cells [a break in the same place in a chromosome]. These patients, even years after they were cured, possessed a small number of healthy stem cells that had the translocation characteristic of leukemia. They behaved like normal stem cells. So we can say every tumor has a stem cell, but it doesn't mean it always comes from the stem cell of that tissue."

Still, Weissman added, "it is true, some of the cancerous events had to happen in the stem cell, since they are the self-renewing cells in the blood system." Thus, even in cancers that spring up in nonreplicating stemlike progenitors, the initiating cancerous events may occur in the stem cells. Either way, oncologists in the future may head, in their patients, straight for the niche, not just where normal stem cells reside, but where progenitor cells reside, to make sure they isolate *the* critical seed of a patient's cancer. Happily, in most systems, stem cells and progenitors originate near or in the same niche.

Weissman has, since, also found two genes (Bmi-1 and Wnt) complicit in many leukemias that are *also* turned on in normal blood stem cells. He is likely to find many more, for he has established a $12 million cancer stem cell center at Stanford, in which stem cell biologists and oncologists are working together with a single goal: to find cancer stem cells.

Still, those oncologists and stem cell biologists who have taken so eagerly to this new/old theory tend to think that it is unlikely that most cancers can be cured via simple differentiation alone. Those early papers of Pierce and Mintz finding that the teratocarcinoma stem cell can be differentiated into normal cells when placed in the potent normal environment of the embryo? This has not worked in all hands: some mice born this way are cancer-ridden early on. And MIT's Rudy Jaenisch has found that while he can make some normal mice by cloning cancer cells, they or their progeny often end up with cancer. Collections of genetic mutations are not easily fully reversed. But the differentiation approach can clearly work beautifully in some cases, and it can work even better when combined with molecules that kill outright cancer stem cells, or with additional cancer stem cell differentiating agents (the arsenic/ATRA combo), or with chemos. The bottom line is that the cancer stem cell theory, if true, finally gives oncologists a target to aim for—the cancer stem cell bull's-eye—instead of shooting blindly at all the tumor cells they see, which doesn't kill the cancer off, and inevitably harms normal tissues.

"People are looking at their own data with new eyes," says Dick. "In order to cure cancer, it is necessary and sufficient to kill cancer stem cells," Weissman said in his 2001 *Nature* paper. It's been a long time since any top scientist has used the words "cure" and "cancer" in the same sentence.

.   .   .

As exhilarating as all this is for the oncology field, it is of course less so for the stem cell transplantation field. In April 2005, indeed, two separate reports in two different journals will note that some adult stem cells that scientists have been studying for transplantation purposes (for making new islet cells, neurons, blood cells, etc.) go cancerous when kept in culture for too long. It is something that has been known for a long time: all cells in culture can eventually spin out of control.

In a 1999 issue of *Developmental Neuroscience,* for example, noted neuroscientist Steven Goldman warned that after repeated passaging (population doublings), "The lineage potential, transformation state and karyotype of [neural stem cells] are all uncertain; after prolonged passage at high-split ratios, such lines may become unrepresentative of their founders, and at worst, transformed neuroectodreral blasts with perturbed growth control." (Translation: they can go cancerous.)

At that time, Harvard neuroscientist Jeffrey Macklis also noted that, when it came to adult cells, "Because we're trying to remove some of the cells' . . . proliferative restrictions, we're setting up the potential for uncontrolled proliferation." Many researchers have taken this into consideration from the start. "We have significant experience with transplanted NSCs that exhibit unrestricted growth," said Albert Einstein University neuroscientist Mark Mehler. "We are exploring the specific environmental conditions and intrinsic signals that differentially mediate these cellular events."

Furthermore, the potential tumorigenicity of *all* stem cells passaged too long, or mishandled, will end up having some surprising political ramifications. For in November 2004, five months after the second ISSCR conference, which focused so heavily on cancer stem cells, California stem cell advocates will win their fight for a gigantic $3 billion hES cell initiative. Voters will pass Proposition 71, a state constitutional amendment that calls for California to spend $300 million a year for 10 years on hES cells (or $1 million a *day*) on therapeutic cloning of human cells, and on any other stem cell work the NIH is neglecting.

The next installment in the national political stem cell drama will be an unusual one, indeed. After making an unprecedented public plea, in April 2005, for new hES cell lines to be approved for federal funding, several NIH staffers will begin interviewing in California, or with the new California Institute for Regenerative Medicine (CIRM). Arlene Chiu, a member of the NIH stem cell task force, will accept a job offer with the newly formed institute mere weeks after the vote. James Battey, head of the NIH stem cell task force, will announce that he might be happy to take a job as the head of the California institute.

In an April 2005 interview, another NIH task force member, neuroscientist Mahendra Rao, will admit that he, too, has been interviewing in California. (He will take a job with California's Invitrogen months later.) Not only that, he will note, but he knows of three other NIH stem cell specialists who have been interviewing in California.

And one reason for this will be: tumorigenicity. That is, in part because so many people have been forced to work with so few presidential hES cell lines, even some of the few *good* lines will have gone bad by 2005. "ES cells are much more stable than adult stem cells, ten times more stable, in fact," Rao will note. But passaging more than 100 times, among other things, is starting to cause some Bush lines to spin out of control. "This is not in the public domain, but I can tell you what is affecting some of these statements by NIH people: the actual science is suggesting that even ES cells are not perfect all the time. The number of lines available to federally funded researchers? That small number is going to get smaller, and the number of private lines is going to be much larger. From Doug Melton's private lab alone, there are more lines available than from all of NIH. The NIH lines number 22. Melton has made 26. And of the Bush 22, it looks like it is going to become less, very soon."

This notion—that even hES cells are not infallible—was introduced at some conferences in 2003 and in some papers in 2004. Some said they saw occasional problems after 50 passages (or 150 population doublings.)

This seemed a mystery until ESI CEO Alan Colman published a paper in 2004 noting that of his company's six hES cell lines, which were passaged up to 140 times (or a whopping 420 population doublings), only *one* has shown a chromosomal abnormality: at 38 passages (or about 114 population doublings). The reason for the difference, Colman speculated: culturing. ESI separates its cell colonies via a laborious, by-hand, cut-and-paste method. All those who reported more hES cell chromosomal abnormalities at repeated passaging used a *chemical method* to separate cells (trypsin). Furthermore, some people culture their cells at high density (which gives hES cells too many confusing signals), or at a density that is too low, which gives hES cells too few signals.

Notes Singapore National University hES cell pioneer Ariff Bongso, "It is becoming very clear that if you push ES cells hard with enzymes and try to expand them fast, they can go bizarre and change their karyotype [chromosomal arrangement]. But if they are allowed to grow according to their normal pace, via the cutting and pasting method, they don't show this."

Colman will agree in an interview in May 2005 that chromosomal abnormalities after repeat passaging "are not an intrinsic quality of these cells. And certainly, even in lines where problems occur over 20 passages? A single cell line, passaged 20 times, will create more cells than there are people in the world. . . . People often go wrong, thinking you need to go on forever making these cells, and the cells need to be stable over a huge number of population doublings. . . . They don't. . . . Eventually, 20 passages of a cell line will be enough to do anything you're aiming to do with a certain disease."

Jamie Thomson, the hES cell pioneer whose lab has solved the problem with regular chromosome checks, concurs: "These cells are remarkably stable under long periods of culture." Adds Geron CEO Thomas Okarma: "These are the most stable cells I have ever worked with."

Still, even given the vast superiority of hES cells to adult stem cells in this area, there are many good ways, many clinicians believe, that adult stem cells can be used with safety. Taking weary old adult neural stem cells (NSCs) out of the head of an aged patient with Parkinson's, Alzheimer's, spinal cord injury, brain trauma, or stroke, then forcing them to proliferate before returning them, may not be an optimal approach, given the above. Furthermore, over and over neuroscientists note that, when they do replicate their adult NSCs in a petri dish, they cannot get certain neurons when they transplant them back in the brain. Many old adult neural stem cells apparently do not like to be pushed around. Human ES cells, far and away, may well be best for brain or spinal cord stem cell transplantation, many scientists say. (Or human fetal neural cells, which can replicate more, although they are generally more controversial than any other approach, given that their source is abortions.)

But removing adult stem cells from younger patients' bodies, and returning them without replicating them too much—all of which Tokyo University's Kotaro Yoshimura is doing—may well work out, many scientists firmly believe. Indeed, Cytori Therapeutics will in April 2006 announce the first adult stem cell trial for mastectomy patients—also to be held in Japan. Again, logic would dictate that younger, more robust cells may always be preferable for a slew of reasons, from efficacy to safety. (And in the future, studies are hinting that a controversy-free way to derive hES cells may ultimately be devised, as will be outlined in depth in an upcoming chapter.)

But studies in cancer patients who have had adult bone marrow transplants, over decades, show that bone marrow stem cells, when taken from a patient and returned without manipulation, come with only a minimal rise in leukemias. And because most of those studies were done in cancer patients to begin with, it is clear the approach can be not only sound but also can deliver. Adult stem cells, as upcoming chapters will also show, may not be an optimal choice, but they can be a spectacular life-saving one.

Yoshimura's particular application of adult stem cells for breast enhancement and other fat-cell problems is brand new in a number of ways. Blood stem cells, a subset of which can form tiny blood capillaries, have never been used for breast enhancement before in a reputable clinic, and fat stem cells have not been used at all. It is early days. But the approach already has at least one vocal fan.

"I am very happy," a 37-year-old New York Yoshimura patient will note in April 2005. The patient has Romberg syndrome, a progressive facial degeneration of the skin and underlying fat, which can move on to the muscles and bone. He has had it since he was 14. "People used to ask me, 'What happened? What happened?' They thought I'd been hit in the face." Indeed, a "before" picture shows that the patient's right cheek was once severely indented, making the right and left halves of his face look as though they belonged to two different people. "I always wanted to fix it. I didn't have a problem going out with the girls. But it bothered me, I didn't have enough self-confidence sometimes."

He never jumped eagerly at the options open to him, because there were none that were remotely appealing. Open to him were simple mature fat injections, which often don't work and can require repeat injections as soon as six months later, he was told. There is surgery, where fat is taken from the back, along with blood vessels, but that involves much risk and scarring. "The thing about stem cells is that they are natural, self-sustaining, involve less risk than conventional surgery because they are just injected—and it is my own cells."

Yoshimura's stem cell approach was not cheap: $13,000. Pure fat injections would have run the New York patient from $3,000 to $5,000, and NYU's $20,000 surgery would have cost him almost nothing, because it would be covered by insurance here. And the patient knew he'd be flying blind, if he ran with stem cells. Yoshimura had not—and still has not—published his work. But the patient canceled his plans for conventional surgery at the last minute, borrowed money, and went to Tokyo for Yoshimura's injections in August 2004. The idea of stem cells was just too alluring.

After one month the patient went back to work. As of early 2006, the right side of his face will look perfect, to the untrained eye, although the patient will swear it is still slightly asymmetrical. "I know, most people can't see it. And I say it's perfect, compared to before." It is early days, and the patient knows it. "But what can I say? I am looking at myself right now and I am very happy."

He does not want his name printed. The girls now have no idea he ever looked any other way, and that's how he wants it. He wants to remain just another guy for as long as possible. He could have waited for embryonic stem cells to hit the clinical pipeline—whenever that will be—or even for this pioneering phase I adult stem cell clinical trial to move to phase II. But, while this patient's disease seemed to stop progressing when he was in his early twenties, it can slowly progress, until it affects the brain. The patient waited 23 years to be just another guy. It was enough.

8

The Medical Revolution of the Human Chimera

> The dream of the ancients from time
> immemorial has been the junction of portions
> of different individuals, not only to counteract
> disease, but also to combine the potentials of
> different species. This desire inspired the
> birth of many mythical creatures which were
> purported to have capabilities normally beyond
> the power of a single species. The modern world
> has inherited these dreams in the form of the
> sphinx, the mermaid and the chimerical forms
> of many heraldic beasts.
>
> —Heart transplant surgeon
> Christiaan Barnard, 1967

THE HALLS OF MASSACHUSETTS General Hospital's third-floor surgical wing have the suffocating feel of a labyrinthine bathroom, with their relentless paucity of windows and their relentless beige-and-white tiled walls. It is vaguely familiar-looking to anyone who has seen the movie *Coma,* in which organs were stolen from patients and then sold on the black market—and rightly so, for the film hospital was partly based on the world-famous MGH.

Outside OR 40 hangs a plastic strip, to which is attached remote sensors with the words "Surgeon," "Nurse," etc., on them. The sensors are important because the hospital needs to know where staffers are, especially during the type of operation that is about to occur in here. For some of the staffers working today's operation will be bolting back and forth between OR 40 and another OR down the hall throughout the procedure. The two operations must be carefully timed, because in OR 40 lies Derek Besenfelder, who is about to get a new kidney because his old ones failed due to the kidney disease Alport's syndrome. In the other OR lies Derek's mother, who is about to give up a kidney for him. Derek is only 25 years old, but he has been facing death for the last several years: 50 to 70 percent of kidney failure patients die, even those on dialysis machines. Mrs. Besenfelder, being 54 and thus old for a donor, is also comparatively frail, and her operation will be even more invasive than her son's, since he will not have to have his old kidney removed.

It's a dramatic scenario, just for this reason. But the fact is that this is not your average kidney transplant. While stem cells are being heralded as responsible for revolutions in understanding in many areas of medicine—which should lead to revolutions in many hospitals someday—stem cells started a quiet revolution in this hospital six years ago.

For if Derek Besenfelder were the average transplant recipient, he would walk away from this operation on about 30 different drugs that he would have to take every day for the rest of his life. Those drugs would severely compromise his immune system—to chemically force it to not reject his mother's foreign kidney—and leave him vulnerable to the slightest cold. His chances of getting premature cancers, heart disease, or diabetes would be hiked to near-certainty. And if the drugs didn't do him in in 10 to 15 years, this new kidney would: odds say that rejection of one kind or another would finally occur in that time, and he would have to get another kidney transplant.

This is what every transplant patient goes through right now. Every transplant patient, that is, *except* Derek, and eight of the nine MGH patients who came before him. For if Derek is like those pioneering patients, he will walk away a few weeks after this transplant on no drugs. This will happen because he will walk away from this transplant not just with a new kidney but with a new immune system, created out of his mother's adult blood stem cells.

Derek is living the revolution. Not only that, but he is living an adult stem cell revolution that may, in turn, help launch the long-awaited human embryonic stem cell revolution. For what immunologists will be doing with stem cells today is exactly what the human embryonic stem cell field needs to get its amazing cells into the clinic, many hES cell researchers believe.

Instead of making banks of millions of different hES cells to match everyone on Earth, or waiting for therapeutic cloning to be perfected, human ES cells may most naturally, and quickly, become clinic-ready, says ESI CEO Alan Colman, by "simultaneously making, out of a single line of hES cells, a preparation of blood stem cells and whatever tissue you want to transplant, then pre-treating the patient with those blood stem cells to induce tolerance to the other transplanted cells." Agrees Hadassah University's Shimon Slavin: "I am sure this is the answer for organ transplants and hES cells."

Indeed, Stanford University stem cell pioneer Irv Weissman says that this was the "vision" he and Salk Institute neuroscientist Fred Gage had when they formed the fetal stem cell company Stem Cells Inc. in the 1990s. In autoimmune diseases like diabetes, rheumatoid arthritis, and multiple sclerosis, the immune system kills off cells. "We knew that if we could ally hematopoietic [blood] stem cell transplantation and other organ-specific stem cell transplantation," thus replacing both the lost cells and the immune system that killed them, "then you have a different kind of medicine," Weissman says. "This is going to be, by far, the most important thing that stem cell biology can bring us for a while."

So this day in February 2005 is a big one. In the movie *Coma,* this (or a similar) labyrinth of ORs provided replacements for a handful of organs lost to a handful of different diseases. In real life, it may become the source of an unimaginable number of cures for an unimaginable number of diseases, a matter of science trouncing science fiction.

On a wall on the floor above, near a stairway, some 50 white lab coats are hanging. They bulge out from the wall messily, many with stethoscopes coiled inside the pockets: they are the coats of the surgeons and residents doing operations downstairs right now. It is eerie to see so many coats there, with their owners' names stitched on them: the number of surgeries going on at any one time in this city of a hospital rivals that of all hospitals together in other cities.

Also upstairs is a huge area rimmed with computers surgeons use to look up articles between procedures, while dipping into one of the many peanut butter jars and cracker boxes scattered on each table: hospital "protein on the go." Startling here is the fact that, while one or two groups of docs stand quietly chatting, the majority sit alone: this hospital is so big that many of the surgeons operating at any given time don't know each other.

Downstairs, in OR 40, two men are standing over Derek, who is lying on the OR table. The giant eyes of OR lights watch over them; the beeping sound of Derek's EKG serenades them. Surgical resident Tatsuo Kawai is tall and wearing magnifying "loupes" glasses, an antiseptic jeweler. Benedict Cosimi, chief of MGH organ transplantation, wears a headband with a headlight on it, an antiseptic miner.

Using a scalpel, Kawai slices a 12-inch swath through Derek's right abdomen, in the vicinity of the pelvis, into what is known as his rectus sheath. The line he has created is shockingly red in this muted blue-and-white room. He picks up a yellow Bovie knife, which spits alternating electrical currents through Derek's tissue, allowing him to simultaneously cut and cauterize vessels as a nurse pads the sites with gauze to soak up blood. Kawai slices through many centimeters of stringy white and red connective tissue like this, wending his way through various muscles, until he has exposed the bulging white bladder. Then he directs his attention to the area above the bladder, a cavity tucked inside the pelvis called the iliac fossa, where if all goes to plan, Derek's mother's kidney will be placed.

The names of everything Kawai sees, slashes through, and ligates in here are in Latin. Cosimi identifies each area as Kawai digs, as if pointing out ancient artifacts unearthed on an archeological dig. Second by second the surgeons' yellow gloves get redder, as does the growing pile of cotton swabs on a nearby table. The inevitable, cumulative impression is that all in this room are gradually leaving the modern, efficient, blue-and-white world and entering some other, darker, thousand-year-old world, one that belongs as much to centuries of surgeons in wooden ampitheaters as it does to Derek himself.

Scores of ringed forceps, many of which look like blunt-ended scissors, needle holders, clamps, and retractors are splayed in a neat fanlike shape at Derek's feet, along with a pan of water. Derek's face cannot be seen beyond the blue tent that rises above his chest, but his left arm is extended to one side on an extension of the OR table, palm up and vulnerable, loaded with tubes. The anesthesiologist's assistant is hunched over the outstretched arm, as if she expects Derek may at any moment try to fly away with it. But while this arm may not let Derek fly out of the room, it may let him fly through the rest of his life far more easily than the average kidney transplant patient. For if this operation and the one down the hall go well, into this arm Derek's mother's stem cells will be poured, hopefully creating a second immune system for Derek, one that will allow him to accept her kidney without the lifelong use of crippling, life-shortening immunosuppressive drugs.

It is approximately 11:20 A.M. Two hours earlier, immunologist Thomas Spitzer had plunged, and twisted, an instrument resembling a corkscrew into Mrs. Besenfelder's hip approximately 160 times, drawing out 800 cc of bone marrow, which contains stem cells. Then, a few minutes after 10:00 A.M., in a room identical to this, a surgeon had made a 12-inch incision in the area of Mrs. Besenfelder's eleventh rib. Her muscles were also divided via electrocauterization, opening up a portal to a similar ancient world, and a five-centimeter length of rib was removed. A swath of tissue called Gerota's fascia was sliced through to reveal her pink kidney.

The kidney was poked and prodded, examined for any evidence of overwhelming cystic disease. There was one cyst, but the kidney was found to be otherwise fine between 10:30 and 11:00 A.M. This news had been brought to the crew down the hall, signaling them to begin. All was well. The tenth patient in the world's only continually successful program to create a dual immune system—and thus tolerance induction—for a solid organ/bone marrow transplant had been cleared for takeoff. The operation to remove Mrs. Besenfelder's kidney could continue; the operation that would allow Derek to accept the kidney could begin.

Back in Derek's room, it at times looks as though Cosimi and Kawai are dipping their hands into Madge's Palmolive bowl—until they lift their hands out of the hole they have made in Derek's abdominal area and their hands are dripping with blood. Kawai begins tying off various epigastric vessels, performing a deft ballet with his hands as he dips the eyelash-shaped suture needle, which is balanced at the end of his forceps, into tiny vessels; he loops the thread, pushing the tie down until it is taut on the vessel surface; then he pulls the ends of the threads to the sides of Derek's chest, in the motion of someone opening an invisible book. As Kawai slices ever deeper, Cosimi is handed retractors that he uses to yank the hole wider and wider, pushing Derek's skin, which is covered by a transparent plastic, into frozen yellow waves. "The clamp is on the Foley, right?" Cosimi asks. "Yes," comes the reply from the anesthesiologist, who pokes his head over the blue tent that hides Derek's face. It is 11:40. "Is it in yet?" Cosimi asks, referring to Mrs. Besenfelder's kidney. "Not yet," comes the reply.

It is at this moment that the first deviation from a standard kidney transplant occurs. Cosimi asks for forceps, then picks out of Derek's open cavity a tiny bloody object that looks like a few smashed raspberries. "Iliac lymph nodes," he announces with satisfaction. "I need a saline cup." As the nurse hands him a cup, he tells her it should not go to pathology, but to immunology. Immunology will cut up the lymph nodes and then, eventually, test Derek's blood on them, he explains.

Hopefully in a few weeks Derek's blood cells, which will be at that point a mixture of his mother's and his own cells, will not react against his lymph nodes. This will help signal the fact that Mrs. Besenfelder's blood cells have "made peace" with her son's blood cells and vice versa, which will mean her son's immune system has accepted her blood and thus, her kidney. The reason for this will be as strange as it is (relatively) simple. Standard immunology textbooks teach that mature killer T cells and other mature killer blood cells are taught what is foreign—or dangerous—and must be rejected, inside a schoolhouse organ called the thymus when they are youngsters, that is, when they are young blood stem cells fresh out of the bone marrow. The small schoolhouse of the thymus will not accept and educate fully mature cells this way—and certainly not foreign mature cells. Once a cell is mature, it is an old dog, set in its ways.

But the thymus (and various other small schoolhouses scattered throughout the body) *will* accept and teach *immature blood stem cells* from another foreign person, in the way of a mother bear who is wary of all strange adults but adopts strange orphan cubs as her own. And once it has done that, the thymus will in turn (again in the way of an adoptive mother bear) eliminate those of her *own grown, adult children* who are programmed to kill those particular foreign cells. The hope is that, in a few weeks, Mrs. Besenfelder's infused blood stem cells will have been educated not to fight her son's body and blood cells, and Derek's blood stem cells will have been taught not to fight his mother's kidney and blood cells. The mature cells they grow into will remember this and behave in the same peaceful way toward each other.

There is a theory that all baby animals are born cute to maximize the odds that their parents will be drawn to them and nurture them. In this same way, researchers have found in recent years that all immature blood stem cells appear to look "cute" to the thymus and various other immune system educators. So cute, indeed, that the thymus can be willing, under certain circumstances, to kill off those of its own older children that are programmed to kill those foreign immature stem cells and their foreign mature relatives. (The thymus, and other parts of the immune system, can distinguish Mrs. Besenfelder's cells from Derek's by a precise series of identifying markers on the surface of her cells called human leukocyte antigens [HLA]. All humans possess distinguishing HLA on their cells that mark each as unique.) Indeed, the entire family of Mrs. Besenfelder's immature foreign blood stem cells, along with her kidney, will get the pass from Derek's family, thanks to his stem-cell-biased thymus. This is all, in more ways than one, very much a family affair.

That is the hope. And it is one that is likely to bear fruit. One reason: this procedure may well have natural roots. In 1996, Diana Bianchi of Tufts University discovered that many women retain the fetal cells of their children for decades after pregnancy. Indeed, in an article, Bianchi will bluntly state, "Fetal cells circulate in pregnant women and persist in blood and tissue for decades post-partum. The mother thus becomes chimeric." So these MGH doctors may simply be recapitulating what goes on in any mother, during and after any pregnancy. Ann Reed, chairwoman of rheumatology research at the Mayo Clinic, will tell *The New York Times* in May 2005 she has found that about 50 to 70 percent of healthy people are chimeras, if in an extremely minor way. "Some believe that if you look hard enough you can find chimerism in anybody," Reed will note.

Beyond that, however, the reason there is hope for Derek: this procedure has worked for eight of nine patients before him, a historic first.

The scrub nurse pops Derek's iliac lymph nodes into a plastic cup and labels it. Cosimi begins sucking blood out of the cavity with a vacuum hose. Maureen Bedle, a good-natured surgical nurse in dark blue scrubs, rolls a table with spare instruments around to another corner of the room. "Feng shui," she says.

Cosimi squirts saline into the hole in Derek's abdomen, which now literally looks like a well, with the shafts of four steel retractors interspersed with layers of what looks like red sedimentary earth. "Is it in yet?" Cosimi asks, referring again to Mrs. Besenfelder's kidney. Says Maureen, "He thinks if he keeps asking, it will come."

"They said it would be here by 11 and it's almost 12," he says. Responds Maureen, "And you believe them?" Cosimi laughs. Kawai spreads his legs wide to get down lower to Derek's side. They are scraping away the last of the tissue to expose the portion of the aorta (main artery taking blood away from the heart) and the inferior vena cava (main vein that brings blood back to the heart) connected to the iliac vein and artery that will be connected up to corresponding arteries and veins on the new kidney. Someone pops a shower-capped head in the door. "They say it should be two minutes." "If it's two minutes I'll eat the green stent," Maureen says, pointing to a thin, catheter-like hose. "You promise?" says Kuwai. Not looking up, he says to Cosimi, "Should we anastomose now?"

He gestures into the hole and points out the tiny iliac vein, and the tiny iliac artery, that he has cleared off and tied to the side of the bloody cavity, ready to be connected up to the new kidney. The two vessels look like fat, white-red worms that have been lassoed and are now dangling off the side of a dock, awaiting some big fish. The iliac artery even pulses, rhythmically.

With a gloved hand, Cosimi points at an area far below the two dangling vessels at a thin white worm: the lateral femoral cutaneous nerve. "We have to watch not to touch that," he says. "If it is cut, it can cause sensorimotor deficits [loss of sensation] in this kid's leg."

Then: "It's three minutes, Mo," sighs Cosimi, his eyes wrinkling above his mask. "Too bad, I would have liked to have seen you eat that stent."

At last, at 12:07, a resident enters and places what looks like a tin lasagna pan on Derek's draped legs. His mother's kidney is inside, awash in bloody crushed ice. The pink, veined kidney indeed looks like the fish the wormlike vessels predicted, if a fish that is belly up in a red arctic sea.

Above the tin, on the table, is the cup of Derek's iliac lymph nodes, swimming in saline. For a few seconds, as the surgeons tidy up Derek's cavity, the two containers, both filled with tissue, sit in an odd still life.

One is waiting to give life, the other is waiting for its chance to test what the quality—and length—of that life will be. The knowledge that Mrs. Besenfelder's stem cells are on the way adds to the general impression that a futuresque medicine is indeed being practiced in this famous hospital, and one that far surpasses the organ-market that was so fearfully predicted in *Coma,* which was made in the 1970s, when primitive drugs were just starting to make organ transplants possible. The medicine being practiced here in 2005, by contrast, is leaving synthetic chemicals behind. In a way, it is leaving the entire Big Pharma world behind. It is working with the body, not against it; listening to the body, not forcing it to comply. The effort is to create not an artificial, temporary, drug-ravaged home for an organ that will always fight its new environs, but a natural chimera, a legendary creature easily made up of two beings—and wielding the powers of two beings—out of a young 25-year-old patient.

◆　◆　◆

THE WORLD'S first successful organ transplant occurred in 1954 at what is now Brigham and Women's Hospital, a partner of MGH: the Peter Bent Brigham Hospital. It was a kidney transplant. But while it led to a 1990 Nobel Prize for its surgeon, Joseph Murray, it was a transplant between identical twins. Perfect tissue matches, where all the above-mentioned HLA antigens and their genetic alleles are completely the same, can only be obtained between identical twins. There aren't a whole lot of those around. It was understood that, with all other transplants, the recipient's immune system would have to be tinkered with, pummeled with drugs into tolerating the foreign organ.

That understanding actually first came to the world gratis Nobel Prize–winning immunologist Peter Medawar in the 1940s and early 1950s. He was the first to discover, as noted earlier, that the immune systems of newly born infants may be somewhat magical: that newly born infants could be persuaded to develop two immune systems, their own and one grown from the cells of another, if they received cells from the other individual at a very young age.

In 1951, Medawar's group also showed, however, that the hormones of the adrenal gland (or man-made chemicals with the same capabilities called corticosteroids or steroids) somehow curbed the immune system a bit, delaying skin graft rejection in animals. The world did not yet understand that it was probably magical immature cells called stem cells, on both sides of the equation, that were responsible for much of the tolerance Medawar generated in his newborn experiment. So doctors interested in transplanting organs focused on Medawar's steroid discovery; that is, that by tinkering with steroids and other drugs, they might be able to combat, if never defeat, immune rejection.

Throughout the mid-to-late 1950s, researchers toyed with steroids and total body irradiation (which was found, in WWII chemical warfare victims, to obliterate the bone marrow where the immune system's T-cells and other blood soldiers are born as stem cells). They also toyed with the drug 6-mercaptopurine and its derivative Imuran, which was found to block T cell proliferation. By the early 1960s, some doctors began to score success with kidney transplants between nonidentical people using these drugs. Most of these early trials were performed by a group at Colorado General Hospital that included Thomas Starzl, whose name today is synonymous with the words "organ transplant." By 1963, as a result of his limited success, there were only three organ transplant centers in the United States: the Brigham, Colorado, and the Medical College of Virginia. Indeed, *Time* highlighted these three centers in a famous cover story in May of 1963. However, when Starzl's group next tried the drugs on liver transplant recipients, they did not work (largely because of technical problems associated with the more complicated liver transplant).

So Starzl focused on a series of 30 kidney transplant patients. He began noticing something strange, and, in May 1963, submitted a manuscript to the *Journal of the American Medical Association* (*JAMA*) that proposed a radical observation: in some of these patients, the drugs seemed to pave the way for a *natural* tolerance to the kidney. He did not know why. The idea was so radical that *JAMA* sent his manuscript back for revision. Impatiently, Starzl published a similar article in *Surgery, Gynecology and Obstetrics* in October 1963. "The subsequent behavior of patients who have been brought through a successfully treated rejection crisis suggests the early development of some degree of host-graft adaptation, since the phenomenon of vigorous secondary rejection has been encountered only once," Starzl wrote.

Of the response to that article, and the one that eventually appeared in *JAMA,* Starzl would later write, "there was naked incredulity about our results." The notion that a natural, lasting tolerance could be induced in the adult would prove to be prescient by MGH decades later, but it was swiftly rejected at the time. Still, the field had to acknowledge that Starzl was onto something with the drugs. Because his team had administered, in tandem, lower doses of those three drugs (each of which could be lethal in full doses), it had produced more surviving kidney transplant cases than all other teams combined by 1963. Indeed, a registry compiled much later, in 1989, of the 342 nontwin kidney transplants done before March 1964 found that only 24 had made it past the 25-year mark, and 15 of those came from Starzl's group. The other nine came from six other centers.

Still, half of Starzl's kidney patients until 1963 had died. And looking back in 1989, it would be found that not a single patient who had received a nonrelated donor transplant—or a cadaver transplant—before 1964 lived 25 years. This meant that Starzl's approach needed much improvement when it came to cadaver kidney transplants and no one had any idea how to deal with most other transplants, since most other transplants must come from cadavers or brain-dead donors.

Indeed, shortly, it became apparent that even Starzl's successes weren't quite what they seemed. In 1964, Starzl began noticing that his surviving patients were not the picture of health. The steroids caused a rearrangement of body fat so dramatic that Starzl would refer to it as "nature's revenge for defeating his purpose of rejection." Patients developed "moon face," a swelling of the face, and a "buffalo hump" on the shoulder blades. Children's growth was arrested. Steroids also caused stretch marks on the abdomen, cataracts, weakened bones, degenerated muscles and nerves.

And the immune system damage caused patients to become highly vulnerable to all kinds of bacteria and viruses, including those that live naturally in the body. There were infections of the lungs, liver, and brain, among other organs. A virus caused 5,000 transplant patients to develop a cancer, B cell lymphoma, by 1990. Viruses accumulated in the wards of transplant patients, eventually attacking the doctors themselves—including Starzl, who almost died from the hepatitis he contracted on the ward in 1964. Knowing he might die, Starzl hastily finished his autobiography and then flew off to bring a bound copy to his father.

Finding himself alive after a few weeks, Starzl that year went on to develop another drug that would become a mainstay of antirejection therapy. This one came closer than all the ones preceding to be a child of the immune system—as stem cells would turn out to be—rather than an artificial chemical that blocked or killed it off. A Russian scientist in the 1800s had injected cells from a guinea pig's blood into a rat; noticed the rat rejected the guinea pig's cells; then labored to identify the cells in the rat that had rejected the guinea pig's cells. The Russian had discovered very specific antibodies in the rat that countered the attacking T cells of the guinea pig. Inspired when he read this, Starzl injected human T cells into a horse, then winnowed out of the horse the very precise antibodies that attacked the foreign human T cells. He (and others simultaneously working on this approach) called them anti-lymphocyte globulin (ALG). It was the first highly targeted co-optation of the body's own specialized immune armies to do the job of protection from rejection, and Starzl scored more successes with it. It dramatically increased the survival rate of kidney patients in the early period after the transplant.

One of Starzl's students at that time was Benedict Cosimi. Inspired by Starzl's work, Cosimi moved to MGH in 1964 and developed there a safer form of the antibody, a monoclonal antibody called OKT3. He tried it on his first patient at MGH in 1980. For years it would be considered the final third of the most widely used cocktail of immune rejection drugs: OKT3, prednisone (a steroid), and Imuran.

It is unsurprising that the developer of this therapy, which represents one of the first successful uses of one immune system to fight another, should be Ben Cosimi, the surgical head of the team now attempting to completely, naturally, solve Derek Besenfelder's rejection problem with stem cells.

.   .   .

IT IS 12:15. "We still have to deal with the cyst," Cosimi says. Timing matters. Every second that Mrs. Besenfelder's kidney sits in the tin, it is degenerating. He turns the kidney this way and that with his gloved hands, pulling out from the top of the kidney a longish cut artery. From the bottom he then pulls out a short cuff of vein that looks like a clam's neck, or a headless turtleneck, and Mrs. Besenfeler's long thin yellow ureter, which connects the kidney to the bladder. Derek's own dying kidney is invisible, tucked much higher up in his abdomen. It will not be removed, because it doesn't need to be, and there is no need to risk the infection that could come with its removal. So instead of replacing the old ureter tunnel into the bladder, Cosimi and Kawai will be cutting a small hole into Derek's bladder, creating a new tunnel for the new ureter.

"The renal vein is too short, they were not kind to us," Cosimi murmurs, turning back to the headless turtleneck on the kidney.

As a resident ties off an unnecessary vein on the kidney, another turns the kidney over, revealing a yellow cyst on one end. He gently squeezes it with tweezerlike forceps and the cyst pops. Yellow liquid oozes out. It is 12:17. Cosimi and Kawai return their attention to the cavity in Derek's abdomen. Kawai carefully makes a slit in Derek's iliac vein. It is flushed with heparin to ensure no blood clots result when it is hooked up to the kidney vein. Then Cosimi picks up Mrs. Besenfelder's kidney, places it in Derek's gaping cavity, and holds it against one wall with a pair of forceps, while Kawai begins suturing, delicately slipping the kidney vein into the slit in Derek's iliac vein, and sewing it into place with an eyelash-shaped needle that is again gripped tightly at the end of a long needle holder. "This one will only need immune suppression for three months at the outside, if all goes to plan," Cosimi says.

"That's all?" a resident asks incredulously.

"We should check the time out of ice," Cosimi says, looking up at the clock. "It's 12:23." Kawai is done connecting the veins. Cosimi leans over to inspect his job. There is a white ropey line along the intersection of the two veins. The blue thread can't even be seen. "I used to do a lot of hemming," says Kawai, smiling. Cosimi jokes about Kawai's past job in a dry cleaner.

Another resident enters and informs the crew that Mrs. Besenfelder's team, which is still closing her wound, is sorry that it couldn't provide more vein. Cosimi laughs. "You told them we were complaining about the vein?"

It is 12:28. The kidney is looking very white now. It has been out of Mrs. Besenfelder's body less than an hour, and out of the cold water only a few minutes, yet it is already decomposing. Kawai is now holding Derek's thick red artery, which was throbbing earlier, to the other side of the cavity. Cosimi checks various clamps to make sure the newly sewn vein is not leaking blood, then turns his attention, with Kawai, to the two arteries to be linked up. A syringe of heparin is injected systematically into Derek's body to keep blood from clotting throughout. The very white new kidney, leaning against the wall of the cavity, is seen with its hole where the cyst was facing upward, like a headless chicken.

The crew moves on to the arteries, repeating the procedure until mother and son arteries are joined. "Now you want to see a miracle?" Cosimi asks. "Note the white color of this kidney. The minute we remove the clamps, it will get all pink."

And indeed, as Kawai and Cosimi remove the silver clamps, Mrs. Besenfelder's white kidney suddenly lights up like a rose-colored Christmas bulb, flush with Derek's blood. It is 12:45. The kidney was only out of the cold solution for 30 minutes. Kawai reclamps the vessels and makes a slit through the renal or iliac artery. Warm, then cold, solution is flushed through the kidney to make sure the sutures are holding. Then Kawai sews up the hole; the clamps come off again; the kidney blushes ruddily again. "We're back on the air," Cosimi says.

•   •   •

THE YEAR 1967 was a big one in organ transplantation. That year, as a result of confidence in the new drug cocktails and better organ preservation techniques, Starzl performed some of the first successful transplants of another organ: the cadaver liver. His patients died in a matter of months, but they passed the acute rejection phase. This prompted a swarm of liver transplants worldwide, just as in 1963 Starzl's initial kidney transplants had prompted the building of kidney transplant programs worldwide. It was also in 1967 that Christiaan Barnard performed the world's first human heart transplantation in Cape Town, South Africa, having spent a year in the United States watching kidney transplants and studying the new immunosuppressives.

Those two developments together inspired the first successful lung transplant in 1968, and the formation of a pancreas and intestinal transplantation program at the University of Minnesota. However, another setback occurred. Most liver, intestine, heart and lung, and pancreas transplants failed, and would continue to fail well into the 1970s. From

1969 to 1971, Starzl would test a drug called cytoxan (or cyclophosphamide) that would turn out to be an acceptable, slightly gentler replacement for Imuran for those patients intolerant to the latter. But it did not represent a major advance.

And it would not be until 1976 that he would discover that the liver problem was a technical one: in liver transplantation, the route from the pancreas to the liver should not be disturbed because the liver needs insulin from the pancreas, which improved things slightly in that field.

As a rule, then, the 1970s were bleak. Kidney transplants represented the only successes in the field, largely because they came from living donors (people have one extra they can spare). Organs from cadavers failed to go the distance, over and over.

Then, in 1978, British transplantation expert Roy Calne tried the drug cyclosporine on kidney transplant patients. There was some excitement about this drug, which essentially simply *blocked* the ability of one kind of T cell, the CD4 helper T cell, to emit proteins called cytokines that caused other T cells to proliferate. If this worked, it would be a far more targeted, and gentler, way to achieve a more potent immune effect, since OKT3 and Imuran both killed off all or most T cells, including those that may actually be good to have around in a transplant patient.

Cyclosporine did not work for Calne as a single agent. But when Starzl tried smaller doses of it along with correspondingly smaller doses of steroids, it did work. It was the same general approach he had used with steroids and Imuran. Alone, neither could be tolerated in high doses; together, in lower doses, the side effects were different enough that the combo was not too toxic. Cyclosporine use on the first 22 recipients would be a "stunning success" Starzl would report later, with only one patient dying from an unrelated heart attack. His liver transplant trial with cyclosporine and prednisone was even more impressive, given the fact that liver transplants so often failed: 11 of 12 lived over a year.

But every step of the way, each new drug had generated a wellspring of opposition from established doctors fearful of change, and the wave of protest that came with Starzl's typically swift, aggressive use of cyclosporine finally wore him down. On January 1, 1981, he moved on to the University of Pittsburgh, which promised to encourage his experimentation. There, he would score some of his greatest coups.

Cyclosporine, with its more targeted activity—followed by better preservation methods and the establishment in the mid 1980s of the national network for cadaver organ distribution—would turn the corner for other transplants. It is still the most widely used immunosuppressive drug.

Still, and of course, as time went on it was clear that cyclosporine came with massive side effects, too. It damaged the kidneys, caused high blood pressure, excessive hair growth, and tremor. In the late 1980s and early 1990s, a number of other drugs were found. But side effects still remained prohibitive enough that recipients of living donor kidneys that weren't exact matches, even as late as 2005, on average could hope only for 10 more years, and recipients of kidneys from exactly matched cadavers, only 12 years—with severe quality of life problems persisting throughout. As late as 2005, in other words, the human body continued to fight new organs, necessitating that its immune system be hammered into submission, year after year and day after day, with drugs.

•   •   •

BENEDICT COSIMI's show is hardly over. Crawling out of the cavity in Derek's abdomen is a long thin reddish-yellow worm, the ureter, which must be attached to the bladder so Mrs. Besenfelder's kidney will be able to sort out waste products coming into the kidney via Derek's artery, send them into the ureter and the bladder, and recycle his remaining blood back into his body via the vein. Liquid urea is already spilling out the new ureter: a good sign. Cosimi says, "Let me see that Bovie please," and, handed the yellow cauterizer, begins slicing a tiny swath into the white bladder. Then he steps back and allows a German resident to nick a hole into the cauterized tissue. A horde of men enter in baby-blue gowns. One is holding a large bag of magenta blood: Mrs. Besenfelder's stem cells.

It is 12:54. Kawai opens his gown with a flourish, dumps it into a can, and leaps out of the room, late for another appointment. Cosimi, with tweezerlike forceps in one hand and scissorlike forceps in the other, is tying off more tissue, once again pulling the threads to the side as if opening an invisible book. The German resident snips off the end of Mrs. Besenfelder's ureter.

"Shod please?" Cosimi asks. He is handed what looks like scissors with blunted yellow tips. "Not many lymph nodes in here," he says. "The cytoxan must be doing a good job." (A few days ago Derek began taking the immunosuppressive cytoxan, or cyclophosphamide—a very specific T cell killer, as noted earlier. He will take this and some other suppressives for one to three months before he hopefully goes off them all altogether.) It is 1:08. The tube carrying blood out of Derek's cavity is coiled casually around the German resident's legs, as he continues to tunnel the ureter into the bladder and sew it in. The kidney is now looking very healthy, Spam-colored and pulsing with tiny red arteries. By 1:14 the ureter is securely sewn inside the bladder. At 1:18, Cosimi takes a very long thin white catheter (a stent) and slips it into a tiny hole the German resident has made in the ureter.

This is the second major thing that makes this no average transplant. Indeed, the crowd of residents and nurses begin talking, and asking Cosimi about the stent. "This is unusual," Cosimi confirms, "But in the normal transplant, there are a lot of immunosuppressives that would protect against swelling in the ureter. Because he will be on few immunosuppressives, and then none, he needs protection. It is a disadvantage to do this, but it's no big deal. We'll be taking it out in a few days."

At 1:25 Cosimi says, "Now we should be seeing more urine." The anesthesiologist checks a plastic box on the floor. "Ah yes," he says, "10 cc." Cosimi asks for more fluid. "Saline?" the anesthesiologist asks. "Sure," says Cosimi. The kidney is working.

Immunologist Thomas Spitzer is standing by Derek's head, where the bag of magenta bone marrow has been placed. This bag, of course, represents the third and most important difference between this and normal kidney transplantations. There are 800 cc of Derek's mother's unmanipulated, whole bone marrow in the bag. It contains the stem cells that will hopefully become educated by Derek's thymus not to attack his tissue and will in turn educate the thymus to instruct Derek's blood cells not to attack her kidney or blood.

As Mrs. Besenfelder's blood is unmanipulated, it also contains Mrs. Besenfelder's mature killer T cells. The drugs Derek is taking—and will continue to take for a while—will selectively kill off both her T cells and Derek's own to allow their stem cells, like dating celebrities, to slip quietly into that thymus, avoiding the clamoring of obnoxious T cells, to get to know each other.

As he stands with his arms at his sides, Spitzer explains that for about two weeks, Derek's blood should remain in a state of "mixed chimerism," that is, a mixture of his and his mother's stem cells and few remaining T cells. Between two weeks and one month, this state may subside; Derek's blood may well remain chimeric, or it may revert back to only his own cells. But even if his blood becomes his own again, it will hopefully have become permanently changed. His cells will now forever have been educated in his thymus to accept his mother's kidney, thanks to the infusion of her stem cells. And many of his cells that are even potentially tempted to attack his mother's kidney will be wiped out in the thymus. Thanks to stem cells, Derek will be the proud possessor of two immune systems: his mother's and his own.

• • •

It was in 1991 that a handful of transplantation surgeons began to catch on to the idea that maybe drugs weren't the answer to their problems, but the immune system itself. That year, Nicholas Tilney from MGH's partner hospital, Brigham and Women's, reported that two cancer patients who had received perfectly matched, allogeneic (foreign) bone marrow transplants for cancer, years later needed kidney transplants. When given kidneys from the same marrow donors, both went without immunosuppressives. There were similar reports scattered through the 1990s. But they were rare, and generally when surgeons tried to repeat what they had done, they failed.

But MGH's Megan Sykes (then at the National Institutes of Health) had been watching for a long time—hard. For years, she and a few other basic researchers had been making organ tolerance work with some regularity in mice with bone marrow stem cells. But mice are not men so it was assumed by most that this could not work in humans. Still, Sykes had already begun in the 1980s to play with the various new immunosuppressives discovered by Starzl et al., trying to find a way to consistently wean mouse organ transplant patients off the drugs—having induced tolerance with foreign stem cells from the same mouse organ donors—and trying to find out why it worked, when it did.

Before that, in the late 1970s and early 1980s, the main two people in the world who were consistently inducing organ transplant tolerance in animals were Shimon Slavin in Israel and the man who would become Sykes's boss, David Sachs, also then at the NIH. Playing with ionizing radiation and the chemical immunosuppressives of the traditional organ transplanters, both had figured out that a state of "mixed chimerism" was needed to get drug-free tolerance, "mixed chimerism" being blood containing a good percentage of stem cells from both the donor and recipient. Slavin had achieved it in animals by using high-dose irradiation of the recipient's bone marrow, followed by injection of whole bone marrow from the donor. Sachs too had achieved it in animals by using lower-dose irradiation of the entire body, then injection of bone marrow with the killer T cells deleted out.

Both approaches were too lethal to be used in the adult human. Then, in February 1989, Sachs published an article indicating he had discovered a far kinder and gentler way to ablate the immune system via a nonmyeloablative bone marrow approach. He discovered that if you shot a small amount of radiation into the all-important school-house thymus before transplantation, you could achieve a state of "peace" between the two immune systems for a while. (Presumably this was because there were potent mature killer T cells hanging out in there that had gone unnoticed before and that had been getting in the way of the peace, or distracting the thymus from its job educating new stem cells.)

By lightly irradiating that thymus before the bone marrow transplant, Sachs found, he could lower the dose of whole-body irradiation. This, added to T cell antibodies, achieved a state of peace, or tolerance, after allogeneic bone marrow transplants, and much more gently than it ever had before. "We had a very reliable regimen for chimerism and tolerance that was not very toxic," Sykes says. "We had nearly completely overcome the barriers in both directions."

Throughout the 1990s, Sykes (who came from the University of Toronto in 1985 to work for Sachs at the NIH, then moved to MGH with him in 1990) tinkered with the approach. The regimen she and Sachs came up with allowed them to get rid of whole-body irradiation altogether, making the approach even gentler. Their regimen was as follows: Issue a few days' worth of cyclophosphamide, a potent T cell killer, and ATG (antithymocyte globulin), an antibody that also goes after T cells. Partially irradiate the thymus. Transplant kidney and bone marrow from the same donor. In the host, all those anti-T-cell regimens would continue to go after both the T cells of the donor *and* the host for a while, allowing the stem cells to negotiate their peace. A bit of the gentler T cell blocker cyclosporine was added for a few weeks thereafter. It allowed T cells to form but not proliferate, keeping them reined in for an extra chunk of time.

The approach, essentially the same one used on Derek in February 2005, created mixed chimerism and total graft acceptance with no immunosuppressives in 100 percent of mice, consistently. While Slavin in Israel was pulling off similar stunts, and MD Anderson's Richard Champlin in 1997 would be the first to publish a first-rate paper on the viability of the nonmyeloablative approach (if a different one) in humans, few, if anyone, were achieving such a pure state of mixed chimerism and tolerance so militantly, and so reliably, in animals.

In September of 1998, the MGH team was ready to try the approach on a patient. Because it was radical, Sykes, Sachs, and the clinical bone marrow transplant chief, Thomas Spitzer, chose a patient who was not eligible for a normal kidney transplant because she had the cancer multiple myeloma (which had caused her kidneys to die). Shortly after, they tried it on another myeloma/kidney failure patient.

In 1999, the MGH team came out with a paper making a claim that seemed near miraculous: not only were both those patients accepting their new kidneys without drugs, sustained only by stem cells from the same donor as their kidneys, but their cancers had gone into remission. The stem cells of both the donors and the hosts, freed for a while from attack by pesky T cells, had first established a peace for both "sides." Then the new dual immune systems had joined forces, apparently, to attack what they seemed to both see as their common enemy: the patients' cancerous cells. Together, the mature cells that grew out of the collaborating stem cell armies wiped the cancers out.

"Now that was a 'voilà' moment," says Sykes.

MGH proceeded to try the same recipe on four more kidney failure/myeloma patients. When these worked as well, as the team reported to conferences throughout the early 2000s, it was clear the approach could now ethically be tried on kidney failure patients *without* cancer. Derek

Besenfelder is the fourth kidney-failure-only patient to receive a new kidney this way—with a slight modification of the protocol—and the tenth patient in MGH's double bone marrow/organ transplant effort, period. Of the nine before him, eight are completely without immuno-suppressives. Only one patient saw his kidney rejected.

The approach is working. And while a handful of other top transplant programs (at Stanford University, the University of Miami, and the University of Pittsburgh) are also investigating natural bone marrow tolerance induction, only MGH's approach is scoring one success after another. In a recent paper, Samuel Strober of Stanford, who once worked with Slavin, noted that his unit has finally begun trying something close to the MGH approach on its patients.

.   .   .

A FEW months before Derek's operation, Thomas Spitzer, who represents the clinical immunology arm of the MGH effort, was relaxing in his office overlooking the Charles River. With his tweed jacket, his stethoscope curled around his neck like a sleeping animal, and the sailboats gliding silently by, he appeared every inch the cautious, slow-moving average doctor—hardly the revolutionary dancing on tables to celebrate his overthrow of the Old Guard. As the interview proceeded, it became clear why: the regimen is working, but not even the MGH crew is exactly clear why.

"The Sykes model is different from most others in that she found, from the beginning, that mixed chimerism is more powerful than full chimerism," Spitzer said, folding his hands in his lap. (Full chimerism is a state whereby the donor's immune cells *completely* eradicate the host's immune cells, instead of sharing bone marrow space.) "Sykes's mixed-chimerism creates a bi-directional effect that is critical in animal models, and easily obtained. But trying to mimic that in humans is more difficult."

He explained that some of the eight initial patients lost their mixed chimerism a few weeks after their transplants, going back to their own cells. Each time, the crew had held its breath. In animal models, this should mean that tolerance should disappear. But it has not happened. Eight of nine patients remained tolerant, and the kidney failure/cancer patients have remained in remission. It's a triumph, "but it's also a mystery," Spitzer said. "Our first patient has gone six years without myeloma, and without any immunosuppressive drugs for her transplanted kidney. But she had mixed chimerism for only 100 days." This matters because the belief is that the donor stem cells make all the difference because they are permanent residents, not transitory residents like short-lived mature cells. Their "forever" status is supposed to allow them, for the lifetime of the patient, to continually remind the host immune system that they and their foreign organ are there and must be tolerated. But the MGH team recently did find a tiny amount of chimerism in a few isolated tissues in these patients: it is possible that chimerism is indeed the mechanism in all patients and current detection methods are just not good enough yet to see it.

More likely, though, the consensus is that the MGH approach has resulted in something amazing: those foreign stem cells educated the thymus, and thus the host stem cells, so well in some patients that their own cells can now do the work of two immune systems at once. Some papers put out by the lab of Stanford University's Irv Weissman, among others, have pointed to this.

One way or another, MGH is clearly succeeding where others have failed. The ramifications of it have yet to fully sink in, Spitzer said. "Every time these double transplants occur, we do get excited," he said, smiling. "But we didn't celebrate right away. In fact, it's all only slowly becoming real, just a little more real as time passes and things are still OK. I'm not sure we've ever truly celebrated. Every year we get together and talk about our excitement. But we haven't had a party yet."

•   •   •

"DRY AS a bone," Cosimi says, pulling the retractors out of the hole in Derek's abdominal area. It is 1:45 P.M. Cosimi pokes a small hole in Derek's side easily and pushes a thick catheter through to drain excess blood over the next few days.

During this period, the anesthesiologist has done the third and most important thing that differentiates this transplant from others: he has attached that bag full of magenta blood, chock full of stem cells, to Derek's outstretched arm and then hung it over his head. His mother's stem cells are already beginning to slip into his bone marrow. Soon, they will begin their journey to the schoolhouse of Derek's thymus.

As the doctors close the wound, and the nurses count the surgical instruments to make sure none were left inside Derek, the bag remains over Derek's head, a plump, living flag. Spitzer explains that Derek's regimen is slightly different from that of the first patients. He began receiving cyclosporine yesterday; he will be on it for two months. He will be on a steroid, prednisone, for 10 days. Before that, he had received cyclophosphamide and a more potent anti-T-cell antibody called Medi507, along with an anti-B-cell antibody called Rituxan for a few days. His thymus was partially irradiated yesterday.

By two months, Derek should be off all drugs. He will have a tough time in the first month, because while his new dual immune system is taking hold, he will possess few killer T cells, so he will be at high risk for infection. His hospital stay is expected to be, therefore, one month, a longer initial hospital stay than for patients getting a normal organ transplant.

But the approach may soon get much milder and much gentler. For the protocol that has been approved by the FDA will soon allow MGH to use stem cells only, not the entire bone marrow mix. This should eliminate the mature cells in the donor bone marrow, which may let patients go on far fewer drugs up front, because the drugs will only have to tamp down the recipient's immune system, they will not have to go after donor T cells (because there won't be any). This is the approach that Stanford's Irv Weissman favors, having tried it successfully on many animals.

"There is more risk involved in general with this up front [compared to the standard lifelong drug-taking regimen]," says Spitzer, still watching the scene with folded arms. "Blood counts, initially, are lower." But given that MGH's first double transplant patient, Janet McCourt, is still drug- and rejection-free six years out, it is possible that Derek's kidney will last as long as he does—which, since he will be drug-free, should be decades longer than the vast majority of organ transplant recipients.

At 2:02 the surgeons are almost done closing the wound. Someone turns on a radio in the corner. Many staffers from both ORs have gathered in here now, chatting about this new procedure, asking questions. As Derek is transferred to a gurney, he looks very young, his face clean of wrinkles, his thick eyelashes dark on his eyelids. The blood is transferred to a pole on his gurney. He will roll through the halls like this, the bag of blood waving slightly above him, a passive soldier in the war for his immune system that has now officially begun.

In the hallway, on the way to the changing room, Spitzer smiles for the first time since accompanying the cells into the OR. "This approach is especially good for the younger ones," he says. "There is a reasonable chance their transplants will go the distance."

◦   •   ◦

MILTON, MASSACHUSETTS, is a beautiful suburban Boston town of pert white nineteenth-century colonials and sprawling, hundred-year-old brick prep schools. Even today, a few hours after Derek Besen-felder's transplant on a cold February day, Milton exudes New England beauty. Rivers are sluggish with giant, snow-topped ice floes; the trees are bare but huge, limbs raised as if in a kind of boxer's triumph: spring will come, more beautifully here than most anywhere else. It is a town that reeks of polite, perennial privilege.

Janet McCourt, 63, at first exudes the air of the town she lives in. With her long white hair, black turtleneck, black jean jacket, and fashionable preppie half-work-boots, she looks 20 years younger than her age, reeking of health and quiet confidence. She could easily be any one of the scores of preppy mothers on their way to their alma maters— Milton or Fontbonne academies—to pick up their children. But as soon as she settles back and begins to talk at a local Starbucks, it is clear that her complacent attitude has been hard-earned, not handed down.

Back in 1997 McCourt was on dialysis. "I was dying, and I didn't want to live. Dialysis was a living hell." She had been a model when she was young. In the mid 1990s she ran a construction company, worked out every day, and "was on every board you could think of. I was busy every minute of every day."

Every once in a while, though, she would break a bone. Then she began feeling sick a lot. She went to the Jordan Hospital in Plymouth, Massachusetts, three times. Each time they told her she had the flu and sent her home. "But after a while, I knew I was dying," she says. "My eyes were gray. My skin was gray." She went to a local doctor who gave her a blood test, then called her a few days later to tell her: "You have no kidney. Go to the hospital—not Jordan—right away." She went to South Shore Hospital where they put her on dialysis, and then did a battery of tests to find out what was wrong. On January 3, 1997, at age 55, she was just feeling better when her doctor came into her room to give her the results of one test in particular: her bone marrow biopsy. "He had a long face. Normally he's smiling happily. He said, 'Do you want to know the outcome?' I said, 'I think you just told me.'"

She was in the last stage of multiple myeloma, a blood cancer that often goes after the kidney. It was, in other words, two death sentences at once, because if the myeloma didn't kill her, the failed kidney would—and for a cruel reason. Patients with cancer are not allowed onto kidney transplant lists, because there aren't enough kidneys to go around for otherwise healthy patients. And patients with failed kidneys are considered too near-death to receive the optimal, if still-punishing, advanced myeloma treatment: a standard bone marrow/chemo regimen.

So they put her on steroids, to at least slow the cancer, and hemodialysis. Hemodialysis involves a tube, running into a permanent shunt implanted in the arm, that pulls waste and fluid out of the blood, and returns nonwaste to the patient. But because she had low blood pressure, the normal four-hour-a-day, three-days-a-week regimen was quickly upped to four hours a day, every day, for Janet. "I went from a size 6 to a size 26," she says, a result of fluid buildup from the kidney failure and the steroids.

This was unacceptable, so she went on peritoneal dialysis, where she gave dialysis to herself. This still took up hours every day and required intense dedication: her bedroom had to be sterile and she had to change the bags leading from the hole in her stomach several times a day. "I started just wearing sweatpants, and never going out," she says. "I looked like hell, and felt like hell. My life was spent in that little bedroom. Someone asked me once, 'If you could chose one or the other, cancer, or kidney failure, which would you chose?' I answered right away. 'I hate dialysis. I choose cancer.'"

But McCourt didn't have a choice, she had both. Indeed, at one point she had a relapse of her cancer that was so bad her hospital suggested she have hospice take care of her. She said no. But she'd had to give up her company. Because she was between insurance plans when she fell ill, she had no insurance and had to sell everything and spend it all before the state would pick up the medical tab. "I lost everything, from my company to my looks." Too dispirited to do dialysis in her home anymore, she went back to getting dialysis in the local clinic. But that was increasingly depressing. "Everyone is depressed in those places," she says. She was ready to give up.

Then, in October of 1997, her daughter called, pregnant. "My kids were all in their thirties, and I had been saying for a long time, 'When are these kids going to give me grandchildren?' It was a turning point for me. I went to my doctor and said, 'You have to keep me alive.'" She wanted to try some experimental therapy—any.

He told her it was up to her. He couldn't research all the trials out there. So she began researching alternative therapies. "I tried everything . . . there wasn't anything I didn't try. If there was a power tape out there, I did it. If there was an anti-cancer food, I ate it. Acupuncture . . . I did it all. I went to my doctor two months later, and he said, 'Your cancer is in remission.' I thought, 'Wow. Can I do it myself?' From then on, I realized I had to open all my own doors."

In that spirit, that December she went to her nephrologist and asked him to find out if, since her cancer was in remission, she would be eligible for a kidney transplant at MGH, considered the best transplantation hospital in the state. "He said, 'No clinic in their right mind would do that.' So I thought, 'Well, if I were still in business what would I do?' I realized that if a client told me he didn't want my business, I'd show him why he needed me." She tried cold-calling MGH: no go. "They wouldn't even look at me," she says.

So one day, when her doctor was on vacation, she spied her opportunity. A replacement nurse in the dialysis clinic was starting her first day, and she didn't know about McCourt's myeloma. "So I went up to her and said, 'I missed my appointment with the kidney transplant coordinator at Massachusetts General, Dr. Nina Rubin. Could you make an appointment for me?'"

The nurse said she would, and she did. McCourt was finally in. A week later, she sat across from the MGH coordinator. The coordinator, once faced with McCourt, said she would do tests to see if her remission was solid enough to make her a candidate for transplant. But a week later she called to say it was not: the cancer was still there. There would be no kidney transplant.

But by now, McCourt was determined. "I said, 'Okay,' but a week later I went back to her office, sat down, and said, 'I'm taking myself off dialysis, so you simply have to find someone who might want to try something on me—either a bone marrow transplant for the cancer, or a kidney transplant for the kidney failure.'" Taken aback, Rubin promised she would write a letter to the bone marrow transplant division at the hospital, but told McCourt not to get her hopes up. A few days later, McCourt received a call from Spitzer, MGH's bone marrow chief. "I'd like to see you, why don't you come in?" he said.

She did. "He told me all about this new procedure; that it had never been done on a human before; that he couldn't give me 50/50 odds of surviving; that he couldn't even give me 10/100 odds." It was the double bone marrow/kidney transplant, which had only been tried on mice at that point, but which Spitzer hoped would both cure her of the cancer and allow her to tolerate her kidney without immunosuppressive drugs.

She said she would do it.

"He said, 'Then this is what you have to do. Find a living related matching donor for the kidney and bone marrow. If you can do that, then we'll talk.' It was all I needed."

McCourt had seven brothers and sisters. Three were willing to go in and get tested. But of those three, two told her almost immediately that they would not do it, even if they tested positive. One had small children, and the other's spouse wouldn't allow it.

"When I went in there, and he told me there was good news, there was a match, I knew it may not be good news," McCourt says. "Then he named the sister who I knew would do it. I started crying."

It was June of 1998. MGH organ transplant surgeon Benedict Cosimi was ready to go at any time. McCourt's sister, however, wanted to wait for September because she was a professional golfer with a tournament that month. While that could possibly allow the cancer to spread further, McCourt immediately said, simply, "Yes." "Her whole family was against it. Her three kids were afraid they'd lose her. People told her, 'Your sister is going to die anyway, and this way you'll lose both a sister and a kidney.' I could wait." Over the next weeks, McCourt tried to tell her sister, who was 45, time and again how much it meant to her, but her sister said, each time, "I don't want to talk about it."

McCourt was raring to go. She understood the risks, which couldn't be higher, as she was warned by Spitzer over and over. Yet, McCourt says, "My grandson had been born while I was on dialysis. I didn't want him to see me the way I was. I wanted him to remember a grandmother active and busy, not one who was very sickly and overweight. I'd prefer, if it came to that, he remember who I was, not who I had turned into." She laughs. "You know, to this day, I'm not sure that wasn't a bluff, that statement that I would go off dialysis. I'm not sure I had the energy to fight."

She had plenty of fight—and plenty *to* fight—after that, however. The day of the operation, on September 22, 1998, her sister was in a room right next to hers, separated by a pane of glass. The two middle-aged sisters made faces at each other, spoke in sign language.

But for McCourt, the dramas never seemed to let up. "That morning, Spitzer came in, and it was scary. He said they were thinking of canceling the operation. Perhaps it was just too dangerous after all, he said. I said, look, don't do that on my account. I think he just wanted to make sure I had no second thoughts." She laughs again. "I did not."

But after the operation—for over two years after the operation—the drama continued. For she woke up hours later in terrible pain, and it took doctors weeks to realize why: their morphine wasn't working because McCourt was hooked on oxycontin. She'd begun taking it for the cancer pain "like it was going out of style. I was used to being on such high doses that morphine couldn't touch the pain."

McCourt's sister was out golfing two weeks after the transplant, but McCourt was in for another 2½ years of hell. Miraculously, the cancer went into remission immediately after the transplant of her sister's kidney and the elixir of her sister's bone marrow stem cells. But McCourt, like so many cancer patients, had become fully addicted to the painkiller. She went to live in her mother's house. Her mother had recently died and the family wanted to sell her house, but they agreed McCourt could live there while she recuperated and as she tried to withdraw from the drug. At first, during that time, she simply withdrew from the world again, from her family, friends, and a relationship of 25 years.

True to form, eventually, she began fighting her way back into it. Toward the middle of 2000, she decided to go off oxycontin slowly, a little bit each day, over the course of one year. For she had gotten into college (the University of Massachusetts) and had decided, even in the midst of her addiction, to attend. By May 2001, after a year of school, it was all over. Janet McCourt had beaten her addiction. It was only then that she began to focus on something: for 2½ years she had been walking around with no cancer, and no sign of rejection of her new kidney despite the fact that she was on no drugs whatsoever.

But like Spitzer, even in February of 2005, more than six years out, about to get her BA at age 63, she doesn't let herself celebrate. "I just live day by day," she says calmly. "They say you should live like you're dying—that's how I live. I spent a few months in Italy last year on a fellowship. I wouldn't have done that before this. And I do thank God that I was able to do something good in my life. But it's interesting, I wish I could do more. Once you've done something like that, you don't necessarily feel lucky. You think, well, what else should I do?"

She thinks a bit, then says slowly, "But I guess the main thing that I think is: everybody should take their lives into their own hands."

The sentiment echoes that of stem cell scientists worldwide, so many of whom are working on stem cells against stiff odds.

Meanwhile, this technology that may put an end to life-shortening immunosuppressive drugs and that could make hES cells both clinically—and financially—feasible, is being put to use in what is, in the short run, perhaps an even more unbelievable fashion. The power of allogeneic blood stem cells to negotiate a state of truce in some bodies has also been harnessed to actually launch a state of war in others: the war against cancer. For it did not come as a shock to Sykes and Sachs back in 1999 that their allogeneic stem cells had not only induced tolerance in myeloma/kidney failure patients like McCourt but had also been responsible for the death of the cancer cells. They expected it, having started to try their approach on cancer patients in the mid 1990s.

"The allogeneic (foreign) bone marrow response is the strongest immune response there is," says Sykes, simply.

The NIH's Richard Childs puts it the same way. "If somebody were to ask, 'What is the most powerful immune effect that can be induced against a malignancy that is curative?' If you know anything about immunology, you must answer, 'Allogeneic bone marrow transplantation.' Thousands of patients with hematologic [blood] malignancies have been cured through that immune effect."

And if Childs has his way, millions more will join them.

·   ·   ·

THE TORTUROUSLY long escalator ride up from the depths of the Washington, DC, Metro's buried "Medical Center" subway tunnel stop is science fiction–esque. It has the aura of a postnuclear underground world, from which one must bravely surface to test the air. Certainly the expectation is that, at the end of the ride, some fabulous super-modern medical complex awaits, seeing as the richest force for medical science support in the world is up there. But the view instead is that of a simple bus stop with a small sign reading, National Institutes of Health.

Beyond, too, the NIH grounds can look deceptively like an ordinary college campus, its old and new brick buildings cradled in green hills that are interlaced with picturesque, meandering brooks. But anyone who has visited often in the past decade has witnessed something extraordinary: through good times and bad, there is always construction going on here, always a new building or wing going up. There is always money here.

Indeed, to enter the NIH's National Cancer Institute (NCI), or Building 10, has recently become an extraordinary experience. It was once a hulking, old-fashioned brick building with a drab, if breezily open, lounge area, topped by hundreds of old labs. But the NCI was in recent years given $750 million to renovate—which it has put to use.

X-ray machines, and squads of security men, now greet patients and scientists and doctors. The vast lounge indeed looks futuresque, with sweeping, comma-shaped aluminum tables, padded aluminum stools for chairs, gigantic palm trees, and a bright, sun-filled atrium that reaches up several stories and sings with the occasional, happily trapped bird.

In the old building (now tucked in back of the new one) it was possible as of only a few years ago to find one's way through the warrens of labs above by making out certain landmark ancient freezers consigned to the halls, or certain hallway chalkboards where scientists gathered to talk because it was too crowded in the rooms. Now it is virtually impossible to navigate by visual landmarks. The hallways are tall and yellow, windowless, numerous, indistinguishable. This is partly because, as

Richard Childs's secretary mutters, "We're not allowed to put anything on the new walls without permission from several committees. . . . For this reason, I'm sorry, but there is nowhere to hang your coat. We're awaiting permission to put up a hook."

In his early forties, Childs is young to have his own lab and clinical quarters in these prestigious, if strangely pristine, digs. But in the September 2000 *New England Journal of Medicine (NEJM),* shortly after completing his training, he published the results of the world's first clinical trial showing that allogeneic (foreign) bone marrow transplanted cells like those of McCourt's sister or Derek's mother, when killer T cell infusions from the same donor are added later, can cause solid tumors to regress. Childs quickly vaulted to the head of the immunology class, invited to write reviews on the history of immunology in prestigious venues, and receiving impressive funds with which to launch more such clinical trials.

Throughout the 1990s, while Sykes and Sachs were busy finding a kinder, gentler, "nonmyeloablative" way to direct allogeneic stem cells to induce tolerance to organs, they—along with many others—were also busy trying to direct allogeneic stem cells, starting with the same approach, to essentially *fight* blood cancers. They were doing it by first inducing tolerance between the two sets of stem cells—as between McCourt and Derek and their donors—so that when they added T cells from the same donor later the battle wouldn't get too vicious, spreading to other organs; the battle would proceed in a just-so, careful fashion. Childs likes to put it thusly: "Essentially what we've all been doing is giving a transfusion of stem cells and immune cells and just letting them go. If anything, we're just controlling them, putting a governor on them for the first couple of months, to keep them from going bananas. It's all a testament to the power of the immune system."

As noted earlier, for a couple of decades before 1990, bone marrow transplants were being used on blood cancer patients suffering from leukemia and lymphoma. But for a long time oncologists preferred transferring the stem cells from one identical twin to another, because it was thought that imperfectly matched allogeneic transplants like McCourt's and Derek's required the host immune system to be completely ablated before the transplant occurred—and that complete ablation killed as many as 40 percent of patients. Thus, the bone marrow stem cell approach was actually used on blood cancer before 1990 as mainly a stop-gap, a way to give the most desperately ill of younger patients a temporary new immune system after cancer-fighting chemo had wiped it out, until the patient's own bone marrow stem cells regrew and kicked back in.

But in the early 1990s, as noted above, Sachs, Sykes, Slavin, and Richard Champlin of MD Anderson were finding ways to "nonmyeloablatively" wipe out only a *portion* of the immune system, which would allow stem cells on both sides of transplants to negotiate some peace while they were fighting—and keep patients from being killed by the treatment.

And in 1990, Hans Jochem Kolb of Munchen University in Germany was a coauthor on two papers that further rocked the cancer world. First, a multicenter group he worked with found that, while imperfectly matched allogeneic transplants were more toxic to leukemia patients than perfectly matched transplants from identical twins, if imperfectly matched patients could survive the treatment, they were cancer-free far longer. The study looked at 2,254 patients with three different kinds of leukemia. Instead of using allogeneic stem cells as a palliative, the group suggested, maybe they could actually be used to fight cancer on their own. The paper implied that in many, many instances, patients actually weren't being cured by chemo, but by the stem cells themselves. The power of one person's immune system could be harnessed to blast away at another person's blood cancer. "These results," the multicenter group reported in an understated fashion in 1990, "explain the efficacy of allogeneic bone marrow transplantation in eradicating leukemia, provide evidence for a role of the immune system in controlling human cancers, and suggest future directions to improve leukemia therapy."

Furthermore, that same year Kolb published a paper in which he found that he could make allogeneic bone marrow transplants, accompanied by irradiation and T cell depleting regimens, even more powerful if he tossed in some of the donor's T cells a few weeks later. Called donor leukocyte (or lymphocyte) infusion, or DLI, it would later become fairly clear why this worked. If you could first get stem cells from donor and host to accept each other, go into that state of tolerance, the two sides wouldn't pitch a titanic, messy, disorganized battle against each other, and that later donor T cell army, which had not been educated to accept the host, would be free to come in and wipe out the cancer cells—since the host's cells had by then been educated not to attack any of that particular donor's cells.

"You need that initial bone marrow transplant, and you need to get rid of the T cells of both host and donor, to get to a tolerant state so that, by the time you give your donor lymphocyte infusion, the recipient won't reject the donor lymphocytes," Sykes says.

This way, many of the host's healthy blood cells were wiped out along with the host's cancerous blood cells. But that was okay, because the weeks-long delay before the donor lymphocyte infusion allowed the donor stem cells to take root in the bone marrow. Unlike the situation with organ transplantation, where the ultimate goal has been, in theory a permanent mixed chimeric immune system (since host and donor tissues live side by side forever), an ultimate goal for some in blood cancer seemed to be to slowly move from mixed chimerism to full chimerism, to eventually replace the dual immune systems with a fully donor immune system. That way, by completely wiping out all host blood cells, one could be as sure as possible that every bit of blood cancer was gone. (Others say mixed chimerism is the ultimate goal in cancer treatments as well. Studies are ongoing.)

Why, regardless, didn't those uneducated donor T cells go after all the tissues of the body, since all tissues of the host's body were foreign to them? (As noted earlier, this phenomenon is called graft-versus-host disease and can be deadly.) The answer seems to be that the weeks-long wait serves another purpose. In addition to allowing the donor stem cells to take root, the waiting period allows the inflammation caused by all the preconditioning (irradiation and drugs) to die down in the rest of the body. Inflammation attracts T cells. Without that distraction, the small army of uneducated T cells injected into the bloodstream focuses simply on the lesser, if ongoing, inflammation in the blood, as the two armies formed by the competing sets of stem cells continue to duke it out.

"We're studying very actively why you don't get graft-versus-host disease" by injecting donor T cells weeks after the donor bone marrow transplant in blood cancer, Sykes says. "The overwhelming reason we're finding is that there is no more inflammation" in tissues outside the blood system weeks after the bone marrow transplant. "It's quiet, nothing out there to distract the T cells to go to those tissues and attack them. So the attack stays inside the lymphoid system, the bone marrow, the blood—where the tumors are."

The result is that increasing numbers of blood cancers are being cured by allogeneic transplants. The cure rate for chronic myelogenous leukemia is as high as 85 percent in some centers because of the approach. The cure rate for others is less—for example, 15 to 20 percent for acute leukemia patients who have failed chemotherapy. Sykes believes many more blood cancer patients can be cured this way as soon as ethical guidelines allow doctors to try the approach on patients who just got their cancer, not those who have relapsed and are about to die with no other options, as it is often being used now. (Approaches dubbed "experimental" can generally only be tried on patients who are in the last stages of illness, not the first.) "We have the worst possible patients," Sykes says. "Given that, we've been very surprised at how good our results have been." Because it can take over a month for the new immune system to kick in, "It makes sense to do this in people who don't have a disease that's going to progress rapidly in that period." Still, 30 percent of MGH's refractory lymphoma patients—all of whom should have died—have seen complete remissions. Sykes expects to see that number rise considerably, as do other blood cancer doctors and researchers.

But oncologists still fear trying this approach on solid tumor patients. For solid tumor patients, one might have to deliberately keep the inflammation going in non-blood-system areas, to deliberately attract donor T cells to tumors spread all over the body. In that case, graft-versus-health-disease might continue apace all over, killing the host.

But starting with Slavin in the 1980s, a handful of researchers here and there began showing it could work to a small degree in mice. And the NIH's Richard Childs showed for the first time it could work in 30 percent of advanced kidney cancer patients, though it was often accompanied by a graft-versus-host disease effect that frequently sickens them. Now he's tweaking his approach, trying to figure out how to make those donor T cells go after tumors only, not other tissues, and trying to understand why it's working. He's accomplishing this by studying, in minute detail, what happens to old immune systems when they are suddenly joined by new immune systems formed by stem cell transplants.

*   *   *

WITH HIS purple shirt, purple tie, and thick short hair that sweeps off the front of his face, à la Opie, Richard Childs is unusually fresh-faced and happy for a scientist so young. Often, young scientists can wear a haunted, pursued, overworked look. But Childs was lucky to get into the blood stem cell field when he did, he says, leaning back comfortably in his new office at the NIH.

"There are paradigm shifts [revolutionary changes] every three to five years in this field now," he says. For example, he notes, it was once thought that the allogeneic bone marrow–versus–cancer effect was all about the donor cells attacking the patient's cancer cells. But recent studies are finding that a certain kind of mature *host* immune cell called the dendritic cell (DC) is needed to get that new foreign immune system revved up to attack the host blood cells—and the host cancerous blood cells. When, indeed, in all that preconditioning done with drugs, scientists make mistakes and wipe out all the host DCs too soon, the donor T cells don't attack the host. "We once thought all that was dictated by the donor, but it's the patient's dendritic cells that dictate it. The patient's DCs are a critical determinant. The patient's DCs prime donor immune cells. To put it another way, the donor *really* sees the host DCs as foreign."

This is backed up by a Sykes study in 2002 that found that those delayed, uneducated donor lymphocyte infusions, administered weeks after bone marrow transplant, work against blood cancer much better in hosts who retain mixed chimeric blood for a few weeks, as opposed to hosts who prematurely slip into full chimerism (or fully donor blood). In mice, antitumor activity is "markedly superior in mixed chimeras as opposed to full chimeras, demonstrating the importance of host-type [dendritic cells]," she wrote in 2002. One reason for that, she says, is that mature DCs are the orchestra conductors of the immune system. It is the job of dendritic cells to hold up little bits of tumor to its own T cells, for example, to show its T cells what the cancer looks like so they can hunt it down and attack it (much in the way you would give a T-shirt of a prisoner to a hound dog, who can pick up the scent of the prisoner off it).

Foreign T cells coming in later—in that delayed uneducated donor lymphocyte infusion—apparently swim up to host DCs as if they were their own orchestra leaders, see the tumor bits being held up, and proceed to wipe out everything in the blood that looks like the tumor to them. This can end up being many or most of the host's blood cells, since the tumor and the host blood cells seem equally foreign to uneducated donor T cells.

Furthermore, Childs, Sykes, and others believe, host DCs may be important for another, even more fascinating reason. While host DCs raise the alarm for donor cells to attack the host, the host DCs and some other host cells may be woken up, aroused by the attacking cells, and become able to see their own tumors much better themselves. Cancer is smart. It often renders its own immune system "anergic," or unable to see and attack it. But Sykes has found that about 30 percent of her successful bone marrow transplant cancer patients went from mixed chimerism to full chimerism and then lost chimerism altogether, yet remained cancer-free.

What this may well mean, a handful of oncology transplanters are starting to believe, is that allogeneic stem cell transplants may also "wake up" cancer-weakened host immune systems. In some cases this works so well, that's all some cancer patients need the allogeneic stem cells for: to deliver a jolt to their own systems. "It seems that the process of rejecting the donor cells has in some way woken up other cells that recognize the tumor," Sykes says. "The immune response against the donor helps to bring out an immune response against the tumor."

Indeed, Johns Hopkins University oncologist Ephraim Fuchs has found that allogeneic stem cells, tried on cancer-ridden mice born with *no* immune systems, didn't work. The donor cells didn't even begin to gear up. "This shows that you have to have a host immune system to cure the cancer," he contends. Clearly, in initial phases, allogeneic stem cells work at least in part to inspire and rejuvenate the host immune system. Fuchs thinks that host helper T cells are the key. He believes tumor cells render host helper T cells the most impotent and that the allogeneic transplant replaces broken helper Ts with new, unbroken ones, revitalizing the host immune system—including host DCs.

Either way, it is clear: allogeneic stem cells don't just form new immune systems in the body. They can shake awake old immune systems.

Stem cells continually surprise scientists.

Certainly, though, all this can make the battles the two sets of cells wage against each other quite titanic. And with the enthusiasm of a general, the NIH's Childs has begun breaking down and analyzing the tactics and strategies of both sides. He's found the same tactics are used by the immune system, when prompted by similar drugs, whether the fight is against cancer or other stem cell diseases.

He pitches a picture across his desk. "I call this picture, 'Four Cured Guys,'" he says. The picture shows a group of four young men, of differing nationalities, casually dressed, grinning.

"They all had paroxysmal nocturnal hemoglobin (PNH) urea," he says. "It's a disease of abnormal stem cells, where some of them become red blood cells that are super-fragile, and your blood in general has a predisposition to clot. One of these guys had a lung clot, the other a clot that bled into the brain. You feel weak; you have fever and chills. The quality of life is horrible. The average survival is 10 years from the time of diagnosis. So I gave 'em hematopoietic cells (including stem cells) from a matched donor. They were all cured."

He tosses the paper across the table. Published in the February 15, 2004, *Blood,* it reveals that the procedure Childs used is similar in intent—if not in specifics—to the approach used by MGH. For the intent is to induce mixed chimerism and tolerance up-front since, once again, tolerance is required to some degree up-front to keep ensuing battles from becoming too vicious. He gave "the four cured guys" some chemo (fludarabine) and two standard immunosuppressives (cyclophosphamide and ATG) for a while to wipe out as many of their native T cells as possible. Then he gave the bone marrow transplants, deliberately complete with T cells, and issued cyclosporine to keep those donor T cells in check and allow the new donor bone marrow stem cells to take root and get educated. (His approach differs from Sykes's in that he did not try to wipe out all donor T cells up-front; he just kept them in check. Essentially, instead of giving T cells later in the above-mentioned donor lymphocyte infusion, he gave them up front, but kept them from doing anything for several weeks, during the inflammation period, with cyclosporine.)

Childs explains that, over the next weeks, as the donor Ts were kept in check, those new stem cells destined to become new T cells slipped into the thymus to become educated to become tolerant.

What happened then, he says excitedly, literally jumping about in his seat as he refers to charts, pictures, and graphs on his laptop to his right and his computer to his left, "is the most impressive thing you've ever seen." He continues: "Because what happens next is a complete mye-loablation of the host immune system, but it's done *immunologically,* not via cytotoxic chemotherapy. Ultimately the T cells you've been holding in check ablate the old stem cells in the marrow. But you do have to wait. About 100 days." He flips his laptop around, showing a graph. "Right after the chemo has partially ablated the sick PNH immune sys-tem, and after we insert the new stem cells, the patients' blood counts crash. Then they come back up again, for about three months, and patients can get discouraged, because most of the cells are the sick PNH cells."

But eventually the host cells begin to decline: the donor stem cells begin taking root. Seeing that, he takes away the cyclosporine, and within days the uneducated donor T cells have roared into action, going after every foreign thing in sight—although, as with Sykes's approach, because he gave stem cells first and waited for some tolerance to set in, he has been able to keep inflammation, thus graft-versus-host disease, down, largely confined to the blood system.

So the donor T cells that were dumped in with the stem cells couldn't expand that first 100 days, because of the cyclosporine—but they could operate to a degree, slowly killing off recipient cells even as the donor stem cells kept growing into mature healthy cells that slowly replaced the diseased cells. "Those are done growing at about 80–100 days." In some ways, he notes, it is more a ballet than a war.

He calls up another patient chart. It shows sick host PNH neu-trophils, sick mature cells that for many reasons mirror the slow decline of sick host stem cells, which are being attacked by the slow growing donor T cells. "At about day 25 post-transplant you see the PNH popu-lation is shrinking, by day 43 there are even less. . . . After about day 60 there are enough donor cells in there that it's clear some tolerance has been built up, its fairly clear that graft-versus-host disease won't occur. So we start taking down the cyclosporine and Bam! the donor T cells, which we've been holding back, now go and wipe out all the rest of the [sick] PNH stem cells. In seven days, a big battle is pitched, and the immune system rapidly becomes normal, all donor. The donor T cells attack the PNH stem cells, and the donor stem cell pool begins expanding like crazy."

This, Childs says, is what happened in his historic 1999–2000 allogeneic stem cell/kidney cancer trial, the first to show that allogeneic stem cells, followed by allogeneic T cells briefly kept in check by cyclosporine, can go after solid tumors. One major difference: with solid cancer, it can take longer.

He calls up CT scans of a kidney cancer patient's lungs. On day 1 of the stem cell transplant, there is a small metastasized tumor nodule in the lungs. On day 100, it is bigger. "We told this person at day 100 that he needed to get his affairs in order, basically," Childs says. "This was at the beginning, and in a standard chemo protocol, this guy would be called a disease progressor and would be taken off the study. But we were smart enough to know that we didn't have it all figured out. The immune system could work differently. So we watched him and sure enough, by day 180, as you can see, the tumor is shrinking. By day 270, it's gone. Everything was gone by day 270."

Some other kidney cancer patients responded much more quickly. "Here's a guy . . ."—Childs calls up another series of CT scans—"who had hundreds of tiny tumor nodules in his lungs. Five months later, they were all gone. This guy is alive now seven years."

Overall, he's seen a 40 percent response rate in his kidney cancer patients over the last six years. This is extraordinary, given that patients in experimental trials are only eligible for them after they've failed conventional therapies. As with Sykes's protocol, they are in the last stages, and their own immune systems are nearly kaput, worn out from chemo, generally unable to help.

Why is it working? In a review Childs noted that, first of all, new immune systems haven't had the chance to be "tolerized," or rendered "anergic" by tumors. They are fresh, new, raring to go. And cells from a foreign immune system are much more likely to recognize one's own tumors as foreign than one's own T cells, simply because those tumors are more foreign to the foreign immune cells.

Despite this, Childs does see a lot of graft-versus-host disease. And

60 percent of his patients do not respond at all. He agrees with critics that his next step should be to train his allogeneic T cells to go after specific markers on tumors so they won't go after healthy tissue. In fact, scientists worldwide are exploring ways to do just that.

One of the most promising approaches was dreamed up by Johns Hopkins University's Ephraim Fuchs. After transplanting allogeneic blood stem cells into mice with metastatic breast cancer (they received chemo first to induce tolerance to keep initial battles from getting too vicious) he gave them small irradiated bits of the mice's tumors and granulocyte-macrophage colony-stimulating factor (GM-CSF), which spurs the growth of dendritic cells (those T cell ringleaders). Healthy dendritic cells call the attention of T cells to tumors by snarfing up tumor bits and showing them to T cells. Again, like bloodhounds who have sniffed a criminal's shirt, the T cells then know to go after the tumor. He followed that with a donor lymphocyte infusion.

By rousing dendritic cells of host and donor, lymphocytes including T cells of host and donor, and tumor bits, he hoped to rouse both immune systems to recognize and kill the breast tumors.

It worked. Tests revealed he had created and aroused breast-tumor-specific T cells of both host and donor origin, although most came from the host whose immune system was apparently rejuvenated by the allogeneic cells. With this approach, he cured 80 percent of the metastatic breast cancers peacefully, that is, without inducing graft-versus-host disease. This was exciting, especially given the fact that there is virtually nothing effective out there for metastatic breast cancer right now. Fuchs hopes to try the approach on humans soon.

Another scientist, Drew Pardoll of Johns Hopkins University, recently genetically engineered bits of tumor right into the DNA of bone marrow stem cells, then shot those into mice. The stem cells grew up into T cells and dendritic cells that were exquisitely aware of just what the solid tumors looked like. Essentially what he did was analogous to pinning bits of a criminal's T-shirt to thousands of bloodhound puppies, who live with that smell throughout their entire development, coming to intimately know the smell of their prey. The results were similarly spectacular. Pardoll, too, hopes to move to clinical trial. "We're very excited about it," he says.

Childs will try another approach on patients. In his (now three) allogenic stem cell trials for kidney cancer, he has found that a certain kind of immune cell called the NK cell seems to go after his kidney cancers most vigorously. He will add concentrated armies of this subset of allogeneic NK cells to his next stem cell armies in a coming trial.

"One thing is certain," Childs says, "The most powerful immune system is the allogeneic immune system." And the most powerful way of transferring the allogeneic immune system from one person to another is via the potent seed of the stem cell.

The reason for this is again relatively simple, say stem cell fans. Metastatic cancer reaches into the most remote corners of the human body. A sustained, long-term attack is critical. T cells are normal mature cells: they can only replicate up to 40 times, and then they die. Even so-called "memory T cells," which remember past diseases and are roused with those diseases' return, die after 40 replications—they can only survive in the body if not roused to replicate. (Which is why there has been concern of late about childhood small pox vaccines. Their effects may not last through adulthood.) Of all cells, then, stem cells may be the most likely to provide the kind of sustained immune response—without rousing deadly graft-versus-host disease—that metastatic cancer demands.

And once the cancer is gone after an allogeneic stem cell transplant, it tends to stay gone, in Childs's experience. "When a complete response in an allogeneic stem cell setting is mediated via the immune system, the response tends to be durable, unlike a chemo response," says Childs. "You give chemo to someone and their tumor shrinks? You hold your breath, because you know it's generally a matter of time before it starts growing again. But with a stem cell transplant, if you see something shrink, it's through the immune system, and you have a lot more confidence that it will be sustained. It's different."

All told, says Childs, "We know that we can mount a donor immune response specifically against a tumor. We know that there are populations of immune cells that go only after the tumor, and other populations that only cause graft-versus-host disease. So putting it all together and getting it all to work with high specificity . . . people in this field really feel strongly that it's just a matter of time."

And Childs is prepared to wait as long as it takes. He's been prepared since the first time he witnessed a chimeric immune system, which he had created, treat a dying kidney cancer patient some six years ago. "It was about 87 days after the [allogenetic stem cell] transplant. I got a CT scan on him and sat there thinking, 'OK, that nodule looks a little smaller, but maybe it's just a cut on the scan.' Still, I sat there for 45 minutes until I thought, 'Well, I'll rescan him in two weeks.' That was incredibly naïve of me, seeing as how the vast majority of research fails to benefit cancer patients. Usually you do a CT scan every six weeks to two months. Still, two weeks later I put the CT scan up . . . and everything was gone. He was in remission. I almost fell over. It was just off-the-scale unbelievable. Whether another patient had responded or not, it was a defining moment. Seeing something that powerful would still have led me to do this kind of research for the rest of my life."

* * *

HUMAN CHIMERA Derek Besenfelder has headphones on, and is watching a DVD on his silver laptop, sitting up in bed. He has a five o'clock shadow, and he is drowsy with morphine. But the day after his kidney transplant on February 15, 2005, he is already feeling better. "Oh, yeah," he says. "For a long time I felt lethargic, really tired, my kidneys were only functioning at 10 to 15 percent, so all the toxins weren't being cleared out, even by dialysis. For two or three years, it's been hard. This is the best I've felt in a year." His mother's kidney is already working.

Still, like fellow human chimera Janet McCourt, he is dancing no jigs. He is the tenth patient to receive the dual bone marrow/kidney transplant at MGH; unfortunately, the only 1 of the 10 to reject the kidney was the patient just before him. The patient lost both his chimerism and the use of his donor kidney. Doctors don't know why. This has unnerved Derek.

He raises his right arm, with an IV in it, to rest it on the top of his head. Several times over the course of the next 20 minutes he will do this, absentmindedly, half-confused boy, half-macho man not wanting to sit with his hands folded on his lap in a girly way. Three telemetry units that dispense fluids and record his vitals stand at his side, like mute robots, guarding him. Everyone enters in mask and gloves, as will be the case for weeks, especially in the sterile room he will move into upstairs, tomorrow. Even his laptop will have to be sterilized up there. A morphine control unit lies in his lap.

Derek looks far too young to be in here. Yet almost all Alport's syndrome males end up in similar situations by their twenties. Still, Derek may never have to take those lifelong immunosuppressive drugs that shorten lives after transplant.

He has some weeks to go, however, and he is aware of it. It's been a long haul. "I was OK until college," he says, "then my function dropped. I started to get tired. I was on lots of meds, Epogen shots and Procrit to boost the red blood cell level." Finally, his kidneys were functioning so poorly that he was put on dialysis. "It was the worst, I would not wish that on my worst enemy," he says, emphatically. "Four hours, three times a week. They stick you with these huge needles. I was feeling really crappy all the time. I couldn't sleep. After college, all my friends got jobs, started to travel. I had to move home. It was stressful." When he got out of college, he could only take part-time jobs so he could accommodate the rigorous demands of dialysis.

Over a year ago, his parents heard about MGH's bone marrow/kidney transplant trials and decided it was the best option for Derek. He agreed. "Normal kidney transplant patients have to take 20 pills a day the rest of their lives. Their bones get brittle, and I like working out. I'm 25. I want to live a full and complete life." Still it did take courage to do this, he admits. "Obviously I guess you grow up faster when something like this happens to you. And someone once told me, 'You can have anything you want, just not everything you want.' I always remember that."

His mother would be his donor, "there was no question from the start." She came in to visit Derek at 5:00 A.M. this morning, unable to wait to see her son. "They wheeled her in. She was in pain, and in a wheelchair because they had to saw off part of her rib to get the kidney. She's a little bit battered. But she was OK. She was happy, and I was happy. I hadn't been able to sleep for two nights in here."

The reason for his insomnia is heartbreaking, yet a routine part of most kidney transplants. "My parents are the type of people who would do anything for their children. My mom is an amazing woman. If anything had happened to her because of this transplant, I'd never forgive myself."

But while Derek couldn't keep his mother from risking her life for him, his decision to enroll in this trial brings the entire organ transplant field closer to the day when family members won't have to risk their lives for each other this way, ever. For all around Derek, other kinds of stem cell researchers are working to make the "used" organ transplants of today a thing of the past.

The famous four Vacanti brothers, some of whom are associated with MGH, are trying to create entire kidneys from hES cells. In 2004, they reported implanting into rats artificial livers made of polymers and cells including stem cells. If and when they succeed with a kidney, they will need to figure out how to make those organs tolerable to recipients, and MGH's pioneering approach may well be it. That is, they may well generate tolerance to their new hES-cell-made kidneys using blood stem cells created from the same hES cells that made the kidneys.

Likewise, Douglas Melton across the river at Harvard University is working feverishly to create islet cells from hES cells to cure his two children of diabetes, a condition that often leads to organ transplants. If he succeeds, he too may well embrace the MGH approach to induce tolerance.

And George Daley, also at Harvard, is busily creating the blood stem cells from hES cells that would make it all possible, that is, that would make possible human chimeras out of an even more potent mix than Derek's: transplants of both organ and blood made entirely from embryonic cells. "In the era of tissue therapy," he wrote in a 2003 *Experimental Hematology*, "an important role for ES-derived HSC may be the ability to induce hematopoietic chimerism in individuals undergoing transplants with other types of ES-derived tissues. . . . This effect would enable a large patient population to access cellular therapies from a limited bank of approved ES cell lines."

Indeed, it might be, in the long run, both far safer and far more effective to create both new organs and matching new blood stem cells from hES cells, to create embryonic/adult chimeras, instead of adult/adult chimeras. As noted earlier, adult blood stem cells are old, and can contribute to problems like athero. Derek's mother, indeed, has a lot of athero, some of which was trapped in both the kidney and the blood she gave her son, the MGH surgeons said during the surgery. And as noted earlier, old adult stem cells can also be the source of cancer.

Furthermore, this month, an overlooked *Nature* paper with explosive ramifications will come out forcefully demonstrating the above point. For his paper, "Rejuvenation of Aged Progenitor Cells by Exposure to a

Young Systemic Environment," the ubiquitous Irving Weissman of Stanford University hooked up the circulatory (blood) system of an old mouse with a much younger one. That is, he created a form of chimera out of them similar to the human chimeras being created today. "Tissue regenerative capacity . . . declines with age, and in tissues such as muscle, blood, liver, and brain this decline has been attributed to a diminished responsiveness of tissue-specific stem and progenitor cells," he wrote. But when young blood surged through his older mice, both the tissues he studied—muscles and liver—healed far faster, and far better after injury, than those of old mice receiving old blood in the same way. "Regeneration" was "significantly enhanced. Our studies . . . demonstrate that the decline of tissue regenerative potential with age can be reversed through the modulation of systemic factors," that is young blood, or blood stem cells from a young source.

But the above are only a few of the hES cell researchers who could well end up embracing the MGH approach to move the hES cell field into the clinic. ESI CEO Alan Colman and Stanford University's Irv Weissman, as noted earlier, believe the MGH approach and others like it may bring hES cells to the clinic first. And at this time in particular, MGH's approach couldn't seem more welcome. For, this month, as Derek is undergoing his tolerance therapy, Massachusetts' governor has been decrying efforts to push through the state legislature a bill like California's that would let state funds be used to create "designer," or nonrejectable, hES cell transplants for adults via therapeutic cloning. Indeed, in a move Melton called "disastrous," Governor Mitt Romney announced plans to submit a bill that would *criminalize* all embryo creation for hES cell research. This would criminalize therapeutic cloning, the only other way scientists can think of to safely, and efficiently, create ES cells that won't be rejected.

Happily, there will prove to be enough support for the bill in the state legislature to block Romney's veto. This will be big. Only a few weeks earlier a Harvard academic wrote an article in *The Boston Globe* describing a "climate of fear" in the halls of the nation's most vaunted institutions. "Consider the recent experience of an academic colleague who submitted a routine reimbursement request for a book about stem cells," David A. Shaywitz wrote. "'Stem cell research at this time is quite controversial,' an administrator wrote back. 'At the moment, we can't pay for anything related to stem cell research because we are completely federally funded and could be at risk to lose our funds if we do.'"

Shaywitz continued: "I have seen the consequences of this mindset personally, when I was thinking about applying for a government grant to study how (federally approved) stem cells might be converted into insulin-producing cells to treat diabetes. I asked an experienced NIH administrator whether he thought I should include supporting data generated by colleagues working with nonfederally approved cell lines in the funding application. He said no because these data might give the perception that non–federally fundable cell lines were being used as a means to obtain federal funds. Hearing this, it occurred to me that in a scientific environment ruled by perception and fear, is it any wonder that scientists are starting to worry about who they talk to and what books they order? I decided to seek funding elsewhere, and I now receive only private support."

The new bill should help.

But in the U.S. realm of therapeutic cloning, there's always something. In a few weeks, on March 8, 2005, the United Nations will finally vote to condemn "human cloning" as contrary to "human dignity." The declaration will be far vaguer than the right wing, led by the United States, wanted. And it will not be binding. The vast majority of countries will either vote against it, be absent, or abstain. It will be seen as a victory for "national rights." Since the United States led the charge against all kinds of cloning, however, it will leave U.S. scientists insecure. Will Congress criminalize the practice soon, bolstered by the UN decision?

Still, even if cloning should be criminalized on the federal level, it by now seems unlikely to many that Congress will ban outright further hES cell research, that is, work on spare IVF-clinic embryos. Some 20 states have stem cell bills in the works. Even Romney, along with many other state governors in the wake of California's success in scoring $3 billion from state voters as of November 2004, supports work on hES cells from spare IVF-clinic embryos. More and more state legislatures, this month alone (February 2005), are exhibiting defiance. Congress, several members of which are up for reelection in 2006, is not likely to tell state after state they may not fund their own hES cell work. There are many small but significant victories ahead for "straight" hES cell researchers.

But once those hES cells are ready for the clinic, an antirejection remedy must be in place for the majority of therapies. Therapeutic cloning should not be outlawed, many scientists believe. Among other things, therapeutic cloning should offer scientists the chance to study the genetic basis of disease by dedifferentiating genetically mutated diseased cells back to the embryonic state, then watching them grow back up; watching, that is, for what originally went wrong.

Still, if therapeutic cloning *is* outlawed on the federal level, ironically, it may well be thanks to adult stem cell scientists like those at MGH that embryonic stem cell scientists may be able to launch their clinical trials anyway.

And that time is approaching in the United States, hurdles be damned. Geron plans a clinical trial for hES cells and spinal cord injury patients in 2006. If it is blocked, rumors are swirling the company has a plan B: it will take its clinical trial to another country. South America has been mentioned. Stem Cells Inc. plans a human fetal neural stem cell clinical trial in 2006. Indeed, scores of the United States's top hES cell scientists have established highly unusual partnerships with scientists in other countries, particularly China. That way, should their clinical trials be blocked or should the Senate take the unlikely step of banning hES cell work altogether, U.S. scientists will have options. Even the head of the new $3 billion California initiative has demonstrated "great interest" in a spinal cord injury clinical trial network being set up in China, says a top neurologist involved in that effort.

Fascinatingly, scientists have been predicting for years that all branches of biology must become more interdisciplinary and collaborative for true understanding of complex biological systems to occur. Legal restrictions and politics are forcing collaboration on the stem cell field. While most scientists decry the intrusion of politics into their work, there is a rising excitement and curiosity: what will come of it?

Peter Medawar, the first man to deliberately create a mammalian chimera half a century ago, may have worded it best: "One of the distinguishing marks of modern science is the disappearance of sectarian loyalties. Isolationism is over; we all depend upon and sustain each other." In other words, the more chimeric the stem cell field becomes, the more successful it may become.

9

STEM CELLS AND THE HEART: SETTING CARDIOLOGISTS' PULSES RACING

Unbelievable.

—University of Pittsburgh cardiologist
Amit Patel on the gains of his patients after
stem cells were injected into their hearts

HEART FAILURE PATIENT RUTH PAVELKO, 55, whose former cardiologist said two years ago that she had two years to live, lies in what looks more like NASA ground control than a hospital OR.

Standing over her is a familiar baby-blue battalion of doctors and nurses, one of whom is driving a catheter up her leg and into her beating heart. But unlike most surgical types, these Texas Heart Institute personnel are staring up at four monitors above Ruth, not down at her. It is as if they are charting a ship's path through outer space, not a sensor's path toward Ruth's left ventricle (main pumping chamber). And, unlike most ORs, which are as bright as tanning salons to accommodate the most important things in them—the doctors' eyes—this lab is dark to accommodate the most important things here: those eerily glowing monitors.

In a fascinating alliance between cutting-edge military technology and cutting-edge biology, the monitors are letting these doctors see the uneven electromechanical terrain of Ruth's heart—areas pulsing with rivers of blood, areas that four heart attacks have turned into deserts of dead tissue. This unique view, which cannot be seen during a therapeutic procedure any other way, will let the doctors release, with unprecedented military precision, payloads that may transform that devastated biological landscape with unprecedented medical potency: stem cells.

"Here we go," says presiding cardiologist Emerson Perin. He is watching a standard fluoroscopy (moving x-ray) screen on which Ruth's bloated ventricle pumps in a vague, film-noir shadow behind the staples of a past open-heart surgery, which resemble the rungs of a ladder. Ruth's ventricle grew huge trying to pump blood out of narrowing blood vessels. She has diabetes, which gave her accelerated atherosclerosis (fat clog buildup in vessels) and four heart attacks in five years. Half of a wall of her ventricle barely beats, the group found earlier, after a sensor on the catheter tip explored it, measuring changes in voltage and movement millimeter by millimeter, creating a detailed function map in 37 minutes. The ventricle is 20 percent starved for blood and very scarred; the rest lacks blood flow to varied degrees.

The goal: find areas that aren't moving well (determined by a computer that reads motion off the catheter's sensor and offers LS, or loop stabilizer, numbers) but possess adequate electricity (which the computer reads off electrodes in the catheter tip and transforms into "voltage" numbers). Then the team will rejuvenate those damaged, but still viable, areas with stem cell "surgical bombing."

Earlier, Perin had shot his catheter into about 70 regions inside Ruth's ventricle, collecting the above-mentioned electromechanical data for each area. Now, guided by the virtual map he made, he will return to those sites that need stem cells and his assistant, research fellow Silva Guilherme, will release the bombs.

"STs," reports Guilherme, whose thumb is poised above the catheter's stem cell injector, as he eyeballs four monitors that offer the virtual map, EKG waves, vital signs, and the fluoroscopy shots. (STs are sinus tachycardia elevations: slight, steep EKG changes that let the crew know they are positioned against the inside of the ventricle wall.) The catheter is positioned well. The crew awaits the computer's verdict.

"Voltage 12.32; LS 2.09," says a technician manning the computer. "Needle out"—that is, out of the catheter and into the heart—says Silva. He gently starts pushing the stem cell injector. "Wait, no PVCs!" says the technician. PVCs are premature ventricular contractions, marked by wide EKG swings, another indicator that a wall area is still alive.

"Pull out," says Perin. "It's probably scar." A silence falls. According to a growing number of studies, when certain stem cells are placed in scar they create more scar: a potential danger. Stem cells placed in good tissue are wasted. They must be shot into living-yet-damaged tissue that responds to touch with just-so levels of voltage and motion.

Perin twists the catheter. Its bent head is seen tapping at the inner walls of Ruth's ventricle on the vague, film-noir fluoroscopy screen, causing numbers and waves to flutter on four others. "STs," Guilherme says, trying again. The catheter is well positioned.

Again, the crew waits for the computer to crunch numbers.

"Voltage 12.85; LS 1.03," says the technician. "Needle out," says Guilherme, pushing the stem cell injector. "PVCs!" say Guilherme and the technician in relief, simultaneously translating a key EKG wave. "Needle out. . . . Injecting . . . needle back in. . . . Injection complete," Guilherme says.

There is, all around, that progression of sounds heard at NASA when contact is made, or in church when the boring sermon is over: the sound of a crowd exhaling in unison, then laughing to realize how many had been holding their breath.

The first of some 1 million stem cells have landed in the heart of a patient enrolled in the first U.S. bone marrow (BM) stem cell trial for congestive heart failure. It is unknown whether they will change the patient from a terrified woman who gets heart attacks just driving a car to a beach-going jogger, as they have other patients in as little as four months. But if scores of somewhat similar trials abroad in other kinds of heart patients are a clue, there is a good chance the cardiologist who said Ruth will die this year will be wrong. There is a good chance that, outside BM transplants for cancer, procedures like Ruth's—which are seeing such success that one cardiologist calls them "unbelievable"— will go down as the first broadly successful clinical use of stem cells in history. The trials are proliferating around the world at an astonishingly rapid rate.

Furthermore, as is the case with cancer, trials like this are revealing that stem cells may maintain a love/hate relationship with heart disease, the United States's number one killer: that they may both help heal it and help cause it. Indeed, trials like this seem to be on the verge of busting paradigms on both sides of the health/disease equation.

To paraphrase the 1992 Clinton campaign mantra, cardiologists seem to be learning, like doctors in many other fields, that "It's about stem cells, stupid."

•   •   •

SINCE 2000, stem cells have set cardiologists' pulses racing. That year French cardiologists, in a startling move, infused the first adult stem cells into heart attack patients to try to rejuvenate their cardiac muscle. The approach seemed to work, but some patients suffered arrhythmias, presumably because the stem cells came from thigh muscle, which contracts differently.

Still, there was no time for disappointment. In April 2001, National Institutes of Health (NIH) researcher Don Orlic and others issued a report in *Nature* that really got cardiologists panting. After Orlic injected into damaged rat hearts adult BM stem cell mixtures somewhat similar to Perin's (and somewhat similar to the cells, mentioned in the last chapter, conquering cancer and inducing tolerance), the hearts were significantly rejuvenated sans arrhythmias. Orlic's report said this occurred because his BM stem cells had differentiated directly into blood vessels and *trans*differentiated (or morphed into an unexpected kind of cell) into heart muscle. He claimed that 68 percent of the dead areas of the mouse hearts were made up of new muscle and blood vessels, many of which evolved directly out of his BM stem cells.

The same day, Silviu Itescu of Columbia University published a report in *Nature Medicine* making a similar find—that injured mouse hearts were significantly rejuvenated by another subset of BM stem cells. But in this case, he found the cells simply differentiated into blood vessels. While most believed then and now that BM stem cells most likely turn into new blood and blood vessels, not heart muscle, there was excitement. It had been thought that adult hearts simply could not regenerate at all.

The result: From 2001 through 2003, there was an explosion of phase 1 BM stem cell trials comprised of small patient groups, mostly unblinded and unrandomized, in over a dozen countries abroad, most in Europe. Most of the trials involved recent heart attack patients, and most reported success. In patients with coronary artery disease, heart health numbers relentlessly went up when BM stem cells were infused into the circulation via the coronary arteries.

It is always exciting when one phase I trial after another scores. But these were particularly exciting because of the reputation of the "drug." There seemed no limit to the heartaches cardiologists could cure with adult BM stem cells. Still, this is stem cells: there was controversy. It raged inside the field so fiercely that when Perin treated his first patient, he "was sweating bullets."

Perin is based at Texas Heart Institute. But in 2000 he was also doing work on the new Cordis NOGA electromechanical mapping system (which would be used later on Ruth Pavelko) in Rio de Janeiro, where he once lived and had a partnership with a hospital. Borrowing from Israeli technology that uses the earth's magnetic field to track missiles, NOGA's metallic sensor—tracked via magnets above and below patients—was giving doctors a stunningly clear view of the heart during therapeutic procedures. Using standard, two-dimensional fluoroscopy x-ray for placing stents in arteries (stents are tiny meshed objects that temporarily keep arteries open), "we'd never had any idea where we were," Perin says. Fluoroscopy can't give detailed motion or voltage readings. Other machines can, but they can't get both at once, and they can't get either while patients are in the midst of therapeutic procedures. Agrees cardiologist Amit Patel of the University of Pennsylvania: "NOGA is an incredible mapping catheter."

Via NOGA, Perin was finding that 50 percent of the time, stents placed blindly via fluoroscopy hurt the contractibility of the heart because atherosclerotic plaque collected on them.

He had planned next to use NOGA to inject healing growth factors into hearts. But when he heard at a 2000 European meeting that some scientists were shooting BM stem cells through animal coronary arteries, he wondered. Stem cells are growth-factor factories: perfect for him. Yet when slipped into blood, would some of the cells further clog arteries? Some papers had found that certain BM stem cells may add to atherosclerosis this way, in the same way they can simply create more scar when added to scar. (Atherosclerosis—fatty clogs—is the main cause of heart disease.) "It can depend on how you are looking at it, whether stem cells are good or bad," agrees NIH cardiologist and stem cell researcher Richard Cannon. It might be best to inject them right into heart muscle and only hard-to-see, scar-free muscle, Perin thought. That way, you avoid the prospect of the sensitive cells being confused, either by too many signals coming from inflamed atherosclerotic tissue or the total lack of signals coming from dead tissue.

He believed he could do it with NOGA. "I thought, 'I need to try this.'" In December 2001, before most European groups had published their results, he stood poised to do so.

But he chose Rio for his first tests, not the United States. Orlic's paper on rats was causing a scientific storm. No one had been able to duplicate his finding, in an animal, that BM stem cells can morph into both new vessels *and* new heart muscle. So some basic scientists believed the cells created only blood and blood vessels; some believed they only emitted growth factors; many, including some scientists at the U.S. FDA, believed they shouldn't be used on patients until they were completely understood.

Yet, says Perin, basic scientists "don't have to facing dying patients. I do. Every day I have to look heart failure patients in the eye and tell them I can't help." So he chose to do his first trial outside the U.S. maelstrom—in Rio, where his NOGA was ready—confident that for end-stage patients sans options he was doing the right thing.

Still, he was aware of the risks. Encouraging reports on stem cells and the human heart were all largely hearsay at this juncture. Furthermore, one of the first patients the Rio hospital offered him almost gave Perin himself a heart attack. "He was ashen, unable to breath, and cachetic [starving] because you can't eat if you can't breath. His ejection fraction [left heart pumping ability] was 10—any less than that and you're dead. [Normal is 55.] I thought he was going to die. I called the doctor in charge and said, 'This guy isn't even able to lie flat. I'll go ahead and do a map on him, and if I find an area of viability *maybe* we'll think about it,'" Perin says. "Well, I found a little area in the back that showed viability. So I called the doctor again and said, 'Should we do this?' I was really worried." Given the controversy surrounding the cells, a single death could end the whole trial.

Persuaded eventually by the notion that the sickest patients stand to gain the most, "I took a deep breath, injected the guy, and went back to Texas. When I came back three months later to inject some more patients, the one who was so gray? He was tan and smiling. One month after that, he was jogging on the beach. That's when I called the improvement we started to see, 'the tanning factor.' And that's when I thought, 'This is no placebo effect.'"

Within a year, 13 of 14 patients saw substantial improvements (most occurring by four months) with 13 out of 14 crossing over the "MVO2 (myocardial oxygen consumption) of 14" line. MVO2 is a key oxygenation measurement taken during exercise that helps determine whether a patient should go on a heart transplant list. "If you have an MVO2 of 14 or less you're in trouble, you're going to die pretty soon," said Perin. Not only did 13 of 14 cross over that line, but another key heart function measure—ejection fraction, measuring the left ventricle's pumping ability—went up 6 percent. Four out of the five patients who were already on the heart transplant list, dropped off. "It was un-frigging-believable, groundbreaking stuff."

Perin's paper was published in May 2003 by the top U.S. heart journal, *Circulation,* with the gray patient's heart on the cover. By then, several European papers had been published documenting the success of BM stem cells infused—during standard heart procedures—into the blood of patients who just had a heart attack. "Virtually everyone has reported some benefit," notes Cannon.

But Perin's was the first published paper to show that BM stem cells may help much worse-off human hearts, ones in chronic congestive heart failure. And his approach, unlike that of the Europeans, set few cells free into the circulation, perhaps dodging the risk of adding to atherosclerosis. Indeed, Perin would seem prescient in a matter of months, when a South Korean trial that released cells into circulation, instead of injecting them into heart muscle, was suspended because of rapidly reclogged arteries.

True, the FDA did wake Perin up one morning, alarmed: staffers had seen his paper and assumed his Brazilian trial had been done in the United States. Once reassured that it had not been, staffers began asking questions. Shortly after, the agency OK'd Perin's request to launch the first advanced randomized and blinded BM clinical stem cell trial in the United States—the kind of rigorous trial that tests for placebo effect—for congestive heart failure patients. The trial was launched in the spring of 2004.

But Perin's problems were hardly over. Now he had to prove that the cells worked—and how.

*　◆　•

THE DAY before Ruth Pavelko will attempt to extend her life with stem cells, FDA Pig 911 (his name has been changed because this study has not yet been published) is in the basement of the Cooley building of the Texas Heart Institute, being prepared to sacrifice his life *for* stem cells. Pig 911's cath lab is identical to that of Ruth's many stories above, except that in rooms here come the sounds of shifting hooves of farm animals, along with the occasional horror-movie shriek of Pig 911's fellow pig-patients demanding to be fed.

At the top of his OR table, 911's huge pink head and hairy chin are still, his eyes are closed, and his large pink tongue is hanging out and pushed to one side to make room for the endotracheal tube attached to a ventilator that is breathing for him. Pigs undergo general anesthesia, unlike the patients, who receive only Versed to minimize complications. An IV drip of lydocaine is attached to his ear to prevent arrhythmias.

Toward the middle of the table, cardiology fellows and observing doctors from Brazil and Spain are standing over the large, blue-sheeted rest of Pig 911, staring up, as they will later in Ruth's operating room, at a shadowy fluoroscopy shot of the pig's moving ventricle on one monitor, comparing it to a film of his ventricle 30 days ago, just before he received his stem cells. Thirty days before *that,* the pig had had a stainless-steel constrictor filled with colloidal gel placed around his circumflex, a branch of the left main coronary artery. The gel in the constrictor has slowly filled with fluid in the days following, slowly cutting off his blood supply until it is occluded more than 90 percent in this artery. This condition mirrors that which occurs before congestive heart failure: ischemia.

The constrictor, a dark ring rising up and down rhythmically, as though bumping against an invisible beach on invisible waves, is clearly visible now on the fluoroscopy screen.

"Look at the ventricle walls, a clear improvement in movement," says one of the cardiologists, nodding at the two fluoroscopy monitors. Then he nods at a third. On it is a very different kind of picture, a computerized version of 911's heart 30 days ago as it was mapped by NOGA. While it looks cartoonlike—indeed, like a cartoon version of a multicolored football sliced in half—it is clearly the star of the show. "But the NOGA map we make, when compared to that old one, will tell us what we really need to know about each tiny area of the heart. Often parts of the heart don't move, but they are still alive."

Still, this "old-fashioned" fluoroscope gives extra dimension to the movement calculations, so a cardiologist hooks a dye injector machine to the catheter running into Pig 911's heart and presses a button. Dark fluid fills the aorta and the branching coronary artery system. It is as if lightning had struck a leafless tree for a moment on a dark night, making it fully visible just long enough for a picture to be taken. The picture disappears. This coronary angiogram lets the doctors know the ring constricted the artery properly. Now dye is shot into the full ventricle, and another piece of film is shot—this one, to check movement. It is replayed over and over next to the 30-day-old film. "Heart rate," says Franca Angeli, a Brazilian doctor here to study all aspects of Perin's operation before she starts a stem cell center back in Brazil. Pig 911's heart is beating too fast. "Pigs' hearts are very, very sensitive. They get very stressed," she explains, blinking behind her mask. "Most labs don't like to work with them. . . . Now look, you can compare more clearly, see? One wall wasn't working."

She points excitedly at the "before" film—the pumping is lumpen, lopsided. "Thirty days after we injected the cells—it is." And indeed, there is less of a lumpen thump to the movement. "We see our best improvement in the second month, but there is a lot of improvement at this point, exactly 30 days."

Microscopic examination after autopsy, enabled by the attachment of glowing antibodies, she says, should show many new blood vessels in Pig 911's heart, once it has been dissected out. Pig 911's new stem cells were transduced with a gene that glows, called DIL. The crew will first cast light on the tissue to see where the glowing stem cells have congregated. Then they will send in antibodies that glow a different color, antibodies that look for mature blood vessels. If all works as it has in the past, the antibodies will head straight for the areas glowing green, the areas where the stem cells will have congregated. The infarcted area of 911's heart should be rife with new blood vessels and/or tissue.

Perin enters the room in cap and gown. It's NOGA time. He takes over the catheter, placing his legs far apart on the floor in that instinctive surgeon way—like a gunslinger just before the showdown—and begins twiddling. "It's like a video game," he says cheerfully, then: "OK." "OK," the technologist replies. "OK," he says again. "OK," she responds again. Slowly, a new multicolored, three-dimensional virtual ventricle starts growing on the screen next to the old virtual ventricle. Whenever Perin pokes the heart too hard, the beeping sound of Pig 911's heart rate accelerates, and he pulls the catheter back, then forward again. Over and over—66 times, getting 66 multiple readings in 66 different areas of the heart—they play this scene, Perin choosing the spot for his sensor to map and the tech locking the electrical and mechanical coordinates into the computer.

Midway through, a vet surgeon wearing loupes on his glasses wanders in and watches. "This is it, huh?" he says. "Yeah," says Perin, eyeing both the real catheter in the fluoroscopy monitor on one side and its virtual NOGA twin icon, which looks like a cartoon bullet, moving in perfect tandem with the real thing. "But next week we get the XP version of this. We'll be able to pick up eight points at once. Mapping will take 10 minutes instead of 30."

"Cool," says the vet surgeon. "You guys get the best stuff." This is a reference to the fact that James Willerson, head of the institute, has taken such an intense interest in Perin's stem cell work that money has been flowing toward his program. Wherever one goes in this hospital, staffers have heard about Perin's stem cell work, stopping even reporters to ask for medical updates.

After half an hour, Perin and the tech have rebuilt the three-dimensional structure of Pig 911's heart from the inside. Perin ambles over to the tech's monitor to behold the NOGA before-and-after. They are looking at two cartoon ventricles. Red and yellow depict dead and dying areas; purple and blue depict healthy areas. The sensor, with its .7-millimeter resolution, had picked up and thrown onto the monitor some huge red and yellow areas before the pig received its cells. Now, those areas are diminished, the ventricle bathed in a large sheet of soothing violet. "Before the stem cells, median voltage on the antero-lateral wall was 7.6," says the technician, leaning forward, looking at both the cartoon hearts and the numerous graphs that have popped up to better explain them. "Now it is above normal [which is 16]:18.4. A drastic improvement." Wall movement improved similarly.

"Beautiful," Perin beams.

Pig 911 is part of Perin's large and relatively unique new bench-to-beside attempt to figure out how stem cells work on the heart, ASAP. Ruth, like the Brazilian patients, received 30 million of her own BM mononuclear cells, about 1 million of which were stem cells. Pig 911, along with three other pigs, is getting 100 million BM mononuclear cells, or more than 3 million stem cells, three times Ruth's amount. Four are getting 50 million; four controls are getting saline; and four more are getting 200 million cells, seven times Ruth's dose. When this dosing study is over, Perin will have nailed the optimum dose of BM mononuclear cells for rejuvenation in congestive heart failure.

Yet that's only the beginning. Four other studies have begun. The cells Perin is using now on patients and in his pig dosing preclinical trial are the same cells being used by most of the Europeans, that is, the mononuclear cells of the marrow, containing all its stem cells. It is generally believed that the "star" stem cell in that mix, the one that is responsible for most of the improvement seen, is the CD34 endothelial progenitor cell, a BM stem cell that creates small blood vessels.

But Perin is trying four other stem cell types on animals. They are: placental, cord blood, fat derived, and mesenchymal stem cells (MSCs). Of greatest interest to him right now are the latter, which represent a different, purified subset of his BM cells. This subset is largely uncharacterized and is generally believed to create bone, skeletal muscle, and cartilage cells—not heart muscle. But a few scientists believe MSCs may become even larger blood vessels when purified and perhaps—perhaps—functioning new cardiac muscle (although this is controversial).

Perin is intrigued by that MSC subset of bone marrow. Last month, he released a study finding that in dogs with ischemia (heart areas starving for blood flow) purified doses of allogeneic MSCs ("foreign" MSCs taken from another dog) indeed cause new blood vessel formation, just as CD34s do. But soon he will begin similar trials with autologous MSCs (taken from and returned to the same animal) to see if MSCs possess a *super*special quality—beyond the possible formation of large blood vessels. Osiris, the company that gave Perin the cells, has published work suggesting that MSCs, unlike all other stem cells, avoid immune detection when transferred from animal to animal. Perin will compare his autologous to his allogeneic results.

"Beautiful thing, how these cells work, one way or another," Perin says smiling broadly, his mask dropped onto his chest.

But as interested as he is in the verdict on the exact mechanics of subsets of BM stem cells, the restless Perin isn't interested in making patients wait for those verdicts. So another of these new stem cell batches is moving into patients very fast. Perin is starting a new stem cell clinical trial in Spain this year, keeping the U.S. FDA informed as he goes to facilitate a swift move back to the United States when the time comes. He's also planning a stem cell program in the United States for patients who are on artificial heart machines called left ventricular assist devices (LVADs) as they wait for heart transplants to see if he can get them off the waiting list. Things are moving so fast here in Texas— where things are generally unusually large, not fast—that this seems less like a bench-to-bedside stem cell facility than a bench-to-bedside stem cell factory.

Pig 911 is wheeled out of the room. He was quietly put to death via IV potassium while the group was talking.

While on the way to Pig 911's autopsy, the team stops before a room where the next pig is being prepped for NOGA. Docs have pulled out of his chest an event recorder that, when activated, gave a full readout of all the pig's abnormal heart rhythms after 100 million cells were shot in his heart 30 days ago. "To see if the cells provoke arrhythmias," says a technician. "His last loop recording says there were none that were unusual—for a pig." Pigs are testy; they can get sinus tachycardia "if someone comes into the cage and goes 'Boo,'" the Brazilian doctor explains. "But there was no ventricular tachycardia." "Good," Perin says.

In the autopsy room, good-natured redhead Judy Ober, a senior research scientist, has cut through the now-deceased Pig 911's ribs with a huge set of pruning shears and has scraped patiently away at scores of blood vessels with a razor to clear the heart away from its cavity, which is awash in blood. Then she lifts out the heart, says, "Heart in hand," places it in a tub of cold water to slow its deterioration, and begins scraping away at its yellow pericardium using forceps and scissors. After she has sliced off its yellow aorta, which has the consistency of lobster meat, she carefully slices the heart into about 17 sections. Somewhere beyond, pigs have begun screaming again. Ober is unmoved. "They're pigs. They scream if you look at 'em."

When Ober is done with 911's heart, the pieces that remain look like sausage rings with holes in the middle (the middle of the heart is hollow). Cloudy areas inside the rim are the ischemic areas, she explains— the areas that remain damaged by ischemia. She begins slicing the rings now into ever-smaller pieces. Franca Angeli begins talking excitedly about Perin's aforementioned January 2005 dog study with MSCs. She is besotted with the notion that not only were the MSCs not rejected (a big deal, since they came from another dog) but that the crew was able to get pictures of new blood vessels that apparently formed out of the stem cells themselves.

The first claim—that MSCs can avoid immune detection—is not quite earth shattering, yet. The group did not try injecting the dogs with various cytokines, or natural proteins found in the body, which have been shown in the past to cause transplanted allogeneic embryonic stem cells to mature and, at that point, get rejected. That is a critical test. That supposed immune-protected quality of MSCs is useless if they come into contact with a cytokine that matures them in a way that gets them rejected.

Still, the fact that in a few days the FDA will fast-track a product from adult MSCs (Prochymal), which Osiris says mediates immune rejection, will strengthen Perin's case. Sylviu Itescu of Columbia University has built a company around a mesenchymal stem cell he calls a mesechymal progenitor, he is so confident that MSCs can avoid immune detection. And across the street here, Michael Andreeff of MD Anderson has just published a paper that also finds that MSCs avoid immune detection. MSCs could be unusual cells, able to avoid immune rejection without drugs or an accompanying complex BM transplant, like no other cell in existence. It remains unproven.

The second claim of Perin's January dog paper is more immediately of interest, if true. It would strengthen the more believable claim of some that large new blood vessels can form in the hearts of animals directly out of the MSC subset of bone marrow cells. Again, thus far, the belief has been that another bone marrow subset, CD34 endothelial progenitor cells, are the starring stem cells responsible for the success cardiologists have been seeing. And CD34's, it is believed, create only small capillaries, not huge blood vessels. Possibly, purified subsets of MSCs could heal hearts faster and better because they form bigger and better blood vessels.

Regardless, Perin's group, says Angeli, has figured out a clever way to come close to proving that MSCs form new blood vessels in patients' hearts.

As Pig 911's head unceremoniously disappears into a trash-bag-lined, red hazard box, Angeli leads the way past the NOGA room—where doctors are now looking for the new pig's femoral artery for their catheter—into a windowless office. On her computer screen she pulls up pictures. The first shows glowing green cells, the MSCs that have been given the green DIL gene. They have formed circles. The next shows the same tissue stained for blood vessel cells, that is, that have been attacked by antibodies that go after certain molecules on the surface of blood vessel cells: Factor 8, actin, CD34, and CD45. They also form circles. The third shows the first two photos superimposed. Nearly all the stem cells have congregated within blood vessel walls. All the dogs improved. This appears to be a picture of the MSC bone marrow subset, which has morphed perfectly into new blood vessels in dogs' ischemic hearts.

"If you did not believe in stem cells before, you do after you see things like this," says Franca.

*   *   *

FOR ALL the excitement generated by Orlic's 2001 bone marrow stem cell paper—the first one to show that BM stem cells may rejuvenate hearts—it was a young Japanese postdoc in Boston, Takayuki Asahara, who actually got the ball rolling. It was Asahara who first discovered, back in 1994, that BM stem cells could be used for something other than making blood. Until then, it was believed that the heart could not naturally rejuvenate and that the only way to *force* it to was to somehow persuade existing *mature* blood vessel cells to proliferate via various inserted genes, or administration of cytokines, causing better blood flow.

Asahara was the first to discover that, actually, it was the aforementioned subset of BM stem cells called CD34 endothelial progenitors that most robustly generated (small) new blood vessels. Asahara busted yet another major paradigm with his work. Before him, cardiologists were quite simply looking at the wrong cells to make new vessels—in the same way that so many other specialists, in so many fields, have been looking at the wrong cells when trying to find, in mature cells, the source of both disease and healing.

He did not set out to bust paradigms. In 1994, he was just a bored postdoc who had started his schooling in Japan and was training in the lab of Jeffrey Isner, a renowned Tufts University genetics researcher. He was in charge of injecting the gene VEGF into injured mouse hindlimb arteries, waiting, then opening up the arteries to see how this gene caused repair. The belief was that he would find that mature cells from an outer layer of the vessel had migrated toward the VEGF in the center of the vessel.

"I made more mice than I needed one day, though," he said in a phone interview in the summer of 2004. "I was curious, so I sacrificed some in the acute phase, only three days after gene transfer." (Usually the group waited longer to sacrifice the animals.) "I found some accumulating cells on the surface of a VEGF gene-transferred artery. Usually, if you transfect this gene [then sacrifice the animal weeks later] you don't have any accumulation on the surface of injured arteries." Also, *colonies* of cells were covering the denuded arteries. Mature, differentiated endothelial cells don't form colonies the way stem cells do. And the colonies accumulated in the middle of the artery, not the edge.

"The concept then was what 100 percent of people believed: that re-endothelialization [inner blood vessel repair] starts from the edge of injured arteries, from mature smooth muscle cells in the artery itself. But if that were true, then I should have seen mature differentiated endothelial cells at the edge of injured arteries, on their way to migrating into the denuded area of the arteries. These little colonies were far from the edge of injured arteries. So I thought maybe it might be some *immature* endothelial cells derived from blood cells, which were recruited by the VEGF. Nobody had thought this before."

Before going to Isner, his supervisor, he did some research. "I researched the origin of immature blood vessel cells. There was no description in the papers of vascular [vessel-forming] progenitor cells in the adult, of course. [Because people didn't believe they existed.] But I did find that if you open some embryology textbooks, we can find stem or progenitor vascular cells in the embryo. I found that hematopoietic [blood-forming] and vascular stem cells had a common origin between them: the hemangioblast." Perhaps what he had found, Asahara surmised, was a colony of hemangioblasts trapped in the bone marrow since the embryonic stage.

It was an audacious thought, for this was 1994. The existence of the first human stem cell had only been definitively proven in 1992, when Irving Weissman characterized the blood-forming hematopoietic stem cell. As noted, this had ignited the imagination of a few savvy scientists in a variety of fields, such as neurology, who immediately began searching for stem cells in other tissues. But most thought that the blood stem cell would be the exception not the rule: the existence of no other human stem cell had been proven.

So Asahara approached his supervisor cautiously. "I told Jeff we had a really interesting picture in the electron microscope and that maybe we should try to find the origin of the cells. And Jeff looked and said, 'Oh what a sexy idea.' But he also said that the existence of a vascular stem cell in the blood had been discussed for years. Some top people had claimed to have found it, and tried to prove it, but failed. So he told me it was a very interesting idea but dangerous—so I should keep doing my other work at the same time."

Asahara did that, working by day on the VEGF project, then knocking off at 5:00 P.M. to do experiments looking for a stem cell that most people said did not exist. "A half a year later I got some really interesting preliminary results and brought them to him. He was really excited. He told me to stop my old project, and just do new studies for this one."

The experiments that Asahara did over the next two years would turn out to be some of the classic experiments people conduct today to find a stem cell. First, he looked for cells that bore the same markers that blood stem cells possessed, thinking many stem cells may share some. Simple culturing did not work. He mused and had an idea: he hadn't seen those proliferating colonies until he'd injured an artery. Maybe these stem cells were only roused in response to injury. So he transplanted cells that bore the blood stem cell marker CD34 into ischemic mouse models—mice with blocked arteries in their hindlimbs. "And I found this way that these CD34 positive cells can differentiate into blood vessels in the ischemic [blood-starved] condition."

He was lucky, he says. Irv Weissman had only just discovered the series of markers for hematopoietic stem cells that Asahara could pick and choose from. Established scientists seeking the vascular stem cell had been looking in vain in the 1980s, when such markers were still unknown. And many of the markers, Asahara found, matched those that embryology textbooks said existed on fetal vascular stem cells. In that period he also looked up "cocktail" recipes—various growth factors like VEGF—for stimulating blood vessel growth in embryology cookbooks and tried those. They worked. "Such in vitro and in vivo data suggested the possibility that we had the vascular stem cell." This cell would be found to be able to both repair injured vessels by relining them with new mature endothelial cells and prompt the growth of new small vessels.

But Asahara would suddenly find himself with competition from an older, more experienced scientist. "In 1996, at an upcoming international vascular biology meeting in Seattle, I read from an abstract that this scientist was planning to make a presentation at a satellite symposium that mentioned some possibility of a vascular stem cell derived from bone marrow. So I went to that meeting and I had a discussion with him. His study was very similar. He also used the CD34 marker. So I told him I was doing the same thing. But at the time he paid no attention to me, I guess because I was so young, and because he was a hematologist. They were more advanced in stem cell biology areas at that time."

Asahara was discouraged, but not for long. "When I was ready to go back to Boston from Seattle, I had a special lecture. From Judah Folkman." A distinguished yet maverick scientist who had been insisting for years that tumors, like any other tissue, need blood vessels to grow, Harvard University's Folkman had yet to make the international headlines that he would in 1998, when *The New York Times* reported that James Watson, founder of DNA, believed that Folkman would cure cancer "in two years" with substances that block the growth of blood vessels into tumors, thus starving them.

But Folkman was already a legend as the kind of scientist every postdoc secretly hopes he'll be strong enough to become: one who risks the scorn of his peers by fearlessly busting paradigms. Folkman had famously encountered, and weathered, much derision for his tumor theory through the years. "I went to hear him talk, then I had to run to the airport. I called a taxi and was waiting, when there Judah Folkman was, running. He found I was taking a taxi to the airport, so he asked me to take a ride together. I had almost two hours talking with him. I was just a young scientist, so I felt a big pressure. But he was a really kind person. He asked me what I was doing, and when I told him, he was really excited."

But like Isner, Folkman—who was in a position to know—warned Asahara that "the issue is very delicate. He said I had to do very, very careful studies. So he suggested a few of my weak points. I did what he suggested, then a half a year later, I submitted the paper to *Science*. And I guess I believe . . . he reviewed it" for *Science*, Asahara says tentatively. Then he laughs. So for more reasons than one, it would turn out, "his initial suggestions were really helpful to me."

What were the ideas Folkman had proposed? "General, conceptual. After I was speaking to him he suddenly realized, I could see, the significance of the concept. I guess because he was seeing that kind of phenomenon in his studies. He told me tumors are very active angiogenesis [blood-vessel-forming] sites, so the stem cell plasticity of blood may be reasonable" since it would make sense that tumors collect the cells they need to create blood vessels from circulating blood. "He asked, 'Have you checked tumor angiogenesis for your endothelial stem cell?' So after that, I tried to see if endothelial progenitors are related to tumor angiogenesis. The answer was yes. It was encouraging."

The young Asahara's historic paper was published in *Science* in 1997. The paper described the experiments and announced the existence of the adult endothelial (small-blood-vessel-forming) CD34 progenitor cell. In 1999, he published a second paper, in *Circulation Research*, suggesting the source of the cell: the bone marrow. The same year he published a paper in *Nature Medicine* announcing he'd figured out how these progenitors are mobilized out of the bone marrow and into circulation.

"The initial response was really striking," Asahara says happily. "People were surprised. Everybody thought this would be an expanding area, if it was true. However, the experiments following were not easy, they took me two years. Now everybody can do it because I established the techniques. But at first people claimed they could not get the same results in their laboratories. . . . The first problem was that it was hard to culture CD34 cells to get them to turn into endothelial cells. There were some technical tricks we found, though. For example, the proper cell density [in the petri dishes] mattered. It should be low. We weren't able to publish all the tricks in the paper because *Science* papers were very short. Everyone who came to me, I gave information, but some people could not follow. So it took a long time for everyone to be able to follow."

Other tricks: Asahara had found, by trial and error, that CD34 positive cells should first be cocultured with CD34 negative cells to grow well, presumably because this best mimics the environment in the bone marrow in which they are born. "We now know that cell-to-cell contact is very important for differentiation of stem cells. But at the time, nobody had such a concept. Purified, it can be done, but it is very hard."

After he published those first papers, other scientists got to work in earnest, and about six months later, the first paper confirming Asahara's results came out. Since, he has discovered that CD34 endothelial progenitor cells can alleviate hindlimb ischemia. His work has become the basis for many of the heart clinical trials seen today, since most believe that the most important cell in the recent heart/bone marrow clinical trials is his CD34 endothelial progenitor. Because of his research, he has been given two labs in Japan, one with 15 staffers and the other with 20. He is doing basic and clinical research at the now-famous bench-to-bedside Riken stem cell center and at the University of Tokei. He will soon be starting several new CD34 clinical trials for the circulatory system.

*   *   *

A FEW HOURS after scoring his success with Pig 911, toward the end of the day, Perin relaxes in his office, which is on the thirteenth floor of a building with an entrance flanked by lush oak trees screaming with thousands of invisible Texas starlings. All day the birds sweep in and out and around the buildings of the Texas Medical Center as silently, and as unified, as a massive, mindless school of minnows. But when the sun starts to set, they all disappear into the oaks and launch into a deafening cacophony, as if near exploding with the need to make some good old

Texas hay after a day spent being so unnaturally hushed and disciplined, making endless rounds of endless hospitals.

Apart from his horseshoe bookends, which are as ubiquitous in this medical center as horseshoe paperweights, the most striking thing about Perin's office is his huge spare NOGA system, and the view, which appears to encompass all the suburbs of Houston. When cabbing out from the airport, the first thing one notices is that there appear to be two cities on the frank flat horizon. But the two cities are just the two giant halves of Houston—downtown and the Texas Medical Center. Indeed, the Texas Medical Center (TMC) is the largest in the world, serving 5.5 million patients a year, as most staffers will tell you within a half-hour of introduction (following that with the mandatory, "We do everything bigger in Texas"). Perin's office is in the tallest building of the largest medical center in the world.

But while the TMC may be biggest, and while its heart institute has remained near the cutting edge since the golden years of the 1960s when Denton Cooley was doing the United States's first heart transplants here, the heart hospital hasn't been on the *very* cutting edge for a while. Clearly, Perin aims to change that.

As he sits comfortably plumped in his rolling chair, looking vaguely Buddha-esque with his wide face and dark eyebrows, a diet Dr. Pepper by his side, he notes earnestly that he is not in competition with the other landmark U.S. BM stem cell cardiac clinical trial going on at this moment: that of Douglas Losordo, a Tufts University and Caritas St. Elizabeth's Hospital cardiologist in Boston who has worked with both Asahara and Isner on the CD34 endothelial progenitor cell. On heart patients, Losordo is using *only* purified CD34s, not the entire mononuclear cell component of bone marrow like Perin and the Europeans. Perin's mix of mononuclear cells are less than 2 percent CD34. "Sure," he says, shrugging, when asked if Losordo's mix, so concentrated with blood-vessel-forming CD34, may be more potent than his.

But then he notes other differences between the two trials: Losordo is getting his cells by releasing them from the bone marrow of the patients using a cytokine (protein), GCSF. (Some of the cells released by the GCSF *could* be contributing to athero, so his patients wear defibrillator vests.) He then handles them even more in order to get the CD34s out. Perin's cells, by comparison, are taken directly out of the bone marrow, so very little manipulation occurs before injection into the heart. Furthermore, he points out, Losordo's mix does not include that other interesting BM subset, MSCs, as Perin's does. Again, Perin and others are finding that the MSCs seem to form blood vessels, too—perhaps larger ones than CD34s form—and may add to new heart muscle (although, again, this is unproven). And some wonder if different stem cell subsets aid and abet *each other*. If so, Perin's more diverse stem cell mixture may win.

Finally, Perin notes, while Losordo's patients have chronic angina and are thus sicker than the European BM stem cell patients, Perin's are the sickest of all: they have left ventricular dysfunction. They are deep into heart failure. So his current randomized and blinded U.S. study should show, as his earlier and smaller Brazilian study did, that stem cells can help hearts that have long been ill, not just recently ill. All told, he says, he is happy with his choice.

Then he turns, excitedly, to his computer. "I've got to show you this," he says. He calls up pictures of the heart of a patient who died of a stroke unrelated to the cells in his Brazil study. Perin cannot yet mark, with a glowing green gene, stem cells he uses on people in the way he can on animals. The FDA is wary of genetically engineered cells. Still, he has discovered something in this dead patient's heart that no one else has found (since no other human hearts from formal BM stem cell clinical heart trials have been dissected out, directly viewed, and reported on).

On Perin's monitor is a microscopic photo of the dead Brazilian patient's heart, 11 months after receiving his BM stem cells. In the anterolateral (half of the front) portion of the ventricle where Perin had injected 30 million cells, it is lit up with antibodies that mark blood vessels. But on the back portion of the heart, where Perin injected no stem cells, there are one or two spots.

Furthermore, Perin had sent antibodies in to mark all the cells expressing on their surface the vimentin and desmin proteins, which are only found on developing, not mature, cardiac muscle cells. The markers were found surrounding the blood vessels, including the heart muscle, of the anterolateral section where he'd transplanted the new cells. "You don't have to be a rocket scientist to know what this is," he says. "Vimentin and desmin shouldn't be there. But they are. When I show this to people who know cardiology, they say, 'This is not normal.' We can't know, yet, if the new tissue and blood vessels are old cells that have been rejuvenated by the added stem cells, or are cells that transformed directly out of the added stem cells, themselves. But it's undeniable stuff. It's the first powerful indication, in the heart of a human heart failure patient, that transplanted BM stem cells may be responsible for the creation of both new blood vessels, and new cardiac muscle."

While some small studies do indicate the ischemic heart can express these proteins of development when trying to repair itself after a heart attack, such expression doesn't occur to this extraordinary degree.

"Something new went on in this guy's heart."

He pulls up next a film of two tanned older gentlemen jogging in their T-shirts on a wide smile of a beach in Copacabana. The one on the right, he says, is the gray man who's heart made the cover of *Circulation*. Then he pulls up a NOGA map of the man's left ventricle before the cells. It is half-red (dead) and half-green (filled with voltage, but no wall movement). He pulls up the "after" map. The red area is still dead, but half of the green area is now purple: it is totally healthy tissue now. "That's where we targeted his cells," Perin says, pointing to the "before" green area. "We woke up the wall motion. It was hibernating myocardial tissue that wasn't moving but was, clearly, still viable."

Perin excitedly pulls up a few more maps of his Brazilian patients and after a while, something else becomes clear. "There was some movement of the cells away from the transplanted areas in a good half-dozen of the patients. We don't know why," he says.

The implication is, however, that the cells are on their way to attempt to rejuvenate other injured heart areas that Perin didn't even try to target. It would appear to be yet another instance where stem cells may have worked a magic scientists didn't expect.

*   *   *

"SHE SLEPT GREAT. But I went to bed at 9:00 P.M. and was up by 2:00 A.M.," says Andrew Pavelko. A 56-year-old long-distance truck driver with gray hair and mustache, steel-rimmed glasses, and kind eyes, he is backed into a corner of a patient unit at the Cooley building, hands in the pockets of his khakis. It is 7:00 A.M. in the morning of January 15, 2005, a few hours before his wife, Ruth, will receive stem cells for her heart. Ruth was the third patient in this study, getting her "cells" on May 13. But she found out a few weeks ago, when her six months were up, that she had been on placebo. She was the third patient in the trial, but the first control. Up until the third month she had thought she had received the cells. But soon after, she began to feel tired, and by the end of her six months, she was not surprised to find out that she had not received stem cells.

A woman whose family history of both heart disease and diabetes resulted in four heart attacks, a failed double-bypass operation, and 13 heart stents all by age 55, it didn't take Ruth long to decide to go through the stem cell procedure for real this time, an option open to all patients in the Texas trial. She is the first crossover patient, thus, the first congestive heart failure patient in the United States to get bone marrow stem cells injected into her heart muscle—knowing they actually are stem cells.

Despite all the historic firsts that she is at the heart of, Ruth sits calmly in bed with her hands folded on her stomach, patiently answering questions that she has clearly answered before. "Nothing to eat since midnight?" a nurse asks. "No." "Can I take your oxygen?" "Yes." Ruth sticks out her finger before a nurse can even rustle up the oxygen clip. She laughs. "I'm used to this. It will be 99." It is.

A nurse has her sign a form, then retreats to her laptop, and another moves in to take her blood pressure. Ruth, wearing rimless glasses, squints at the telemetry unit. "185/87," she says. This is high. Normally it is 130/70 "because of all the drugs I take." Is she nervous? "A little," she says, smiling.

The second nurse asks if she smokes (no) or is allergic to any drugs (no). The second nurse presents another form.

In moves her anesthesiologist. Did she eat anything today? Just a little water. Does she smoke? No, she quit. When? Nine years ago. Any allergies? No. Diabetes? I have diabetes. Any stents? 13.

Another nurse moves in. From 0 to 10, what level of pain could she endure?

Even this question Ruth has heard before. "Four," she says.

A bag of Ruth's clothes arrives. Ruth's husband signs for it.

While waiting for orderlies, Ruth describes her first heart attack at

50. "It was Christmas, I had been shopping, and was sitting in my car in Sam's parking lot. I opened the door and suddenly started throwing up. It felt as if my lungs were filling up. I couldn't breath. A nurse was walking by. She said I was as white as a ghost. She called an ambulance." Was she shocked to find out she'd had a heart attack? "Yeah. I didn't know that was what it was like." The orderlies arrive. It's time for Ruth to be wheeled into her first OR of the day, where stem cells will be taken from the bone marrow in her pelvis.

Several minutes later Ruth is wheeled up to OR 19 in St. Luke's Hospital. Another nurse with a clipboard asks her questions in the hall, having parked her near several carts filled with OR supplies. In comes hematologist George Carrum. "Ready?" he asks. Says Ruth: "Yup. I was ready the last time, too."

In the OR, Ruth lies on her stomach and a square of her lower back is exposed. She is made comfortable with artfully arranged pillows. Then she is issued some fentanyl, a sedative.

There are three huge syringes lying on a table. Heparin in each will keep Ruth's BM cells from coagulating. Her back is washed with yellow antiseptic betadine, and she is strapped down. Carrum had entered the room the way all surgeons do: with his wet hands in the air, as if getting ready to catch someone falling from above. Then his hands had made swan dives into two out-held rubber gloves. Now he nudges, with his head, an overhead light into position, takes a long needle filled with lydocaine, and plunges it several times into Ruth's exposed pelvis.

Next he picks up an instrument that is the general size and shape of a corkscrew and plunges that into Ruth's pelvis, twisting it back and forth. There is the slightest cracking sound. Leaving the corkscrew wedged into Ruth's bone, he then takes one of the heparin-filled syringes and shoots it down a tunnel in the middle of the corkscrew. As he begins drawing dark maroon blood out of her marrow, Ruth's legs begin kicking and she moans. "Yes, you'll feel a pinch," Carrum says. The first syringe filled, he withdraws it; pulls out the corkscrew; drills it into another section of Ruth's pelvis. He fills that syringe, then a third. Each time, there is the sound of crunching bone. "We're done," he says. As grueling as the scene was for onlookers, it apparently was not for Ruth. She is snoring.

"Where's my friend?" Carrum is saying, just as chief lab technician Jeff Wilson walks in. "There you are." Wilson takes Ruth's BM cells and scurries off. He will be taking the cells to a "clean lab" in the Baylor Center for Cell and Gene Therapy. It is run by Adrian Gee, one of the United States's top geneticists.

In this clean lab, called a Class 10,000 lab because it allows in no more than 10,000 5-micron particles per cubic foot of air, sensors are spaced so that essentially only two people in white Tyvek jumpsuits, which resemble astronaut uniforms (and were indeed developed for space), may enter any given subarea at a time. More, and contamination alarms will sound. There are no windows, so the sun can't alter the temperature; the air is pumped out as steadily as the humans inside contaminate it with their breath; every box of test tubes, gloves, and reagents is sprayed with ethyl alcohol; and the whole enclosed complex is hosed down from floor to ceiling every week.

It is a place, Wilson will note later, where "you can't tell whether it is 3:00 A.M. or 3:00 P.M." He likes this, since cells are processed 24/7 here, "and when you are working at 3:00 A.M., you don't want to know it." (For those nights when such knowledge is unavoidable, the front office offers three prominently placed, huge dispensers of Aleve, Tylenol, and Advil.) The clean room is also a place lined with huge liquid nitrogen tanks, which spew −180 degree clouds into the air when opened, like witches' cauldrons. They hold billions of stem cells, now mostly CD34s, in plastic bags the size of decks of cards.

In this room, Wilson and another tech, working in a class-100 laminar flow hood (which allows only 100 5-micron particles in a cubic foot of air), will dilute the 63 milliliters of Ruth's bone marrow aspirate to 750 milliliters with a ficol gradient. In each of 30–50-milliliter conical bottom tubes they will dribble carefully, using a pipette, 25 milliliters of Ruth's bone marrow onto 17 milliliters of ficol. They will then pop these mixtures into a centrifuge, which will spin them at 1740 rpm (or 400 G, 400 times the force of gravity) for 30 minutes.

When done spinning, the tubes will contain Ruth's red blood cells on the bottom, a clear layer of ficol on top of that, a cloudy layer of Ruth's mononuclear cells on top of that, and yellow plasmid topping it all off: a strange, watery, living sundae. Wilson will then siphon off the cloudy middle layer, called the buffy coat, which, when all the tubes are tallied, will contain together some 30 million of Ruth's mononuclear cells, the most important among those being over 1 million of her BM stem cells. These extracted cells will be placed into a large 250-milliliter conical tube, where they will be diluted—and end up looking like a jar of pink lemonade. They will be cleaned several times, tested for various contaminants, and delivered a few hours later to the catheter lab, where Dr. Perin will inject them into about 70 different areas of Ruth's heart.

        •   •   •

AS RUTH'S cells are being processed, veterinary surgeon Judy Ober, who earlier conducted the autopsy on Pig 911, is scrubbing up in the basement of the MD Anderson animal facility across the street. Beyond, in an animal OR, a hound named Phineas is asleep on a table, all four paws wearing human plastic gloves, an endotracheal tube in his mouth. Ober will be fitting Phineas with a metal constrictor to give him the same condition mirroring ischemia, a heart failure precondition, as Pig 911 earlier. Unlike the pig, Phineas will in 30 days receive stem cells from fat. He is the ninth dog to receive varying doses of the fat stem cells. All the dogs in this series are hounds, because they have a lot of fat from which to draw the cells. The reason to try fat stem cells is obvious: if they are as effective as bone marrow cells, patients will not have to undergo that painful bone marrow procedure and they will have many millions more for use (depending on the patient's heftiness).

Phineas is a beautiful creature, his colors like those of a calico cat. He and his fellow dog patients all came from a prison, where their job was to play with prisoners who exhibited good behavior. These dogs are now being used for certain parts of these stem cell studies for good reasons. Pigs have no exposed vessels, so a central line must be placed into their bodies for blood tests. Pigs can get so stressed that they die during treatment. There are other good reasons to occasionally use dogs.

Despite all this, it is heartbreaking to see a dog on this table, and Ober admits she does not get to know the dogs before she performs her surgeries. "It's easier that way."

Franca Angeli, who is doing the ultrasound on Phineas, waves an observer over. "Isn't this beautiful?" she says. She is taking shadowy films of the inside of Phineas's heart, with its three opening and closing valves, so she can compare and ensure that, in one month, when the metal constrictor has done its job, one of the left ventricle walls will have slowed down. The valves appear to be as delicate as papier mâché. At any moment it seems as if one of their wafer-thin arms could snap off and drift away.

Having banked her beautiful film, Angeli leaves to do an echo for another animal surgery, and Ober begins the thoracotomy (minimally invasive heart surgery) on Phineas. With a yellow cauterizing knife, Ober burns a hole through several layers of tissue in the dog's side. She fits her hand into the hole, then wedges a steel spreader in and cranks the dog's ribs apart. She pulls out what looks like two pink mushroom caps (the lungs) and says, "Lungs look good." Then she picks netting off a tray, gathers the lungs inside the netting, pulls them to the side, and stitches them lightly to the skin, so they will not get in her way.

There, beneath her hands, causing blood to slosh rhythmically in its cavity, is Phineas's pulsing heart, which is rife with distended red arteries and purple veins, the bulging eyeball of a hidden giant shocked to find itself exposed.

Ober delicately pokes a needle through the edge of the heart and stitches it to a tough layer of tissue rimming the cavity to steady the organ as much as possible. She will be operating on it as it is beating, but, she says, never taking her eyes off the dog, "You just have to get into the rhythm of it." Because the dog's left anterior descending (LAD) artery, around which she will place a constrictor, runs between the left and right ventricle then breaks off to the right, "You have to be careful to place the ameroid constrictor somewhere beneath that juncture, so you won't affect the right ventricle," she says. (The human scenario must be mimicked as closely as possible. It is the left ventricle—the main pumping chamber that sends blood from the lungs out into the body—that is most damaged in human heart failure after the LAD artery has clogged.) With a suture needle, she quickly slips a thread underneath what looks like a white worm—the LAD artery—and then pulls the thread off to the side and clamps it to some skin, which elevates the tiny LAD artery off the surface of Phineas's heart.

The dog's arteries during this procedure "have swelled up quite a bit because of the lydocaine we dropped on them. We may need larger constrictors," Ober tells her assistant, who runs off to find some. Wielding forceps with impossible deftness, Ober then picks up a tiny metal ring with a slit in it—the constrictor—and slides it onto the dog's elevated LAD artery. "Never mind, perfect fit," she says. She places another constrictor in, for safekeeping, twists it, cuts a thread.

The artery settles onto the heart, constrictors in place: done. The gel within the constrictors is already filling with fluid. As a result, in 30 days, Phineas's most important coronary artery will be so compromised he will be in the initial stage of congestive heart failure, letting him undergo the procedure that Pig 911, Ruth, and countless other patients and animals underwent before him, and will undergo after: the precise, NOGA computer-aided injection of millions of stem cells into his ventricle. Thirty days after that, like Pig 911, he will be sacrificed and the impact of the stem cells will be determined.

Ober releases Phineas's lungs. They literally stick out of his rib cage, their color a pure white having been deprived of oxygen for several minutes. As the air is pumped back into them, they rise and fall, rise and fall, and begin to gain color. They end up looking like the wings of a massive pink butterfly struggling to break off and fly away from the dog's exposed, roiling, suddenly dying heart.

•   •   •

THE DAY after her heart was surgically bombed with stem cells, Ruth is sitting in a chair in a Texas Heart Institute ward, dressed in a black suit and red shirt, her cheeks flushed red with excitement. Still, she is clutching a wad of Kleenex. The NOGA stem cell catheterization procedure itself didn't much faze her, as her husband notes, because she has had 20 like it in the last five years (a similar catheter was used in more traditional ways prior to the stem cell procedure: that is, to place 13 stents in her rapidly clogging arteries, and to take pressure measurements of her heart).

What fazes Ruth tends to be moments like this, the 20 mornings after the catheterizations, just before she's about to go home. All 20 of those mornings she had hoped that things would change for her. All 20 of those mornings, she had been wrong. A lot of hope has been permanently taken from her, as seems clear when she bursts into tears when asked about the worst part of her life. She clearly can't decide, there is so much. She hates being unable to make the walk to the mailbox in the front yard of her La Porte, Texas, home, she says. She hates packing "handfuls of stuff" in the morning for the journey downstairs, making sure she's bringing all she'll need all day because she gets dizzy climbing back up. She hates taking over 40 pills a day, which she keeps in a small blue body pack that she wears everywhere she goes. She hates relying on others—her son most days, because her husband is a truck driver away for two weeks at a time.

Ruth also hates, she notes, waiting for the next heart attack. And she hates knowing that, if these stem cells don't work for her, the only option the conventional cardiologist has remaining in his medical bag for her is a list: the heart transplant waiting list.

"I'm sorry, I'm crying at the drop of a hat today," she says.

Ruth's story is like that of millions of diabetics. It all started back in 1987 when her thyroid was taken out because of a nodule. She was 38. "I had high blood pressure ever since. But everything was normal until I turned 50, in 1999." It was then that she had her first heart attack. She had a bypass operation at that time and felt better. But "then six months later, I needed to have four stents put in." In the years since, three heart attacks followed in such rapid succession that about two years ago, "her doctors just gave up on her," says husband Andrew. With each heart attack, she lost more heart function. So when she was brought into the ER the last time, her husband had her transferred to Dr. Perin.

Perin had told them there were plenty of new things that could be tried, Andrew Pavelko notes. He first sent her to a doctor who specializes in diabetes, who gave her some pills that made her feel better. Then, on May 13, 2004, Ruth received what she hoped were stem cells. It was this December that she discovered they were placebos and was offered the chance to get the cells for real.

"I tell everyone, if you end up at other hospitals after repeated heart attacks, get stabilized, and come down here, fast," says Andrew Pavelko intensely. "Don't let anybody blow smoke at ya any other way. This is the place to be."

"When do you want to come back, Ruth?" a nurse asks then.

"Never!" she calls out, laughing.

Everything about this scene is ebullient, unhospital-like. The ward itself looks like the hallway of a European villa, with black-framed, frosted-glass sliding doors and black art deco lamps. Paperwork here is computerized: nurses tote laptops on wheels and ID patients via bar-codes on their wristbands, to cut down on mistakes. It was all designed after consultation with patients. Even the ceilings were built in an attractive style, after patients pointed out an obvious universal fact: they spend a lot of time looking at them.

All this, coupled with the fact that Ruth's husband is already telling people that standard heart therapy is about "blowing smoke," makes this morning feel like a bright new day, not just for heart patients but for the stem cell field in general.

Is it? Will stem cells work on the heart? Many believe the answer is yes. This includes another young Japanese postdoc, one who made the major, if still controversial, discovery that BM stem cells may carry on both a love *and* hate relationship with the heart; that the stem cells that may save Ruth may have contributed to her woes in the first place; that Perin's targeted approach may thus be right-on: Masataka Sata.

. . .

ON MARCH 24, 2004, *The Japan Times* reported on what at first seemed a most un-Japanese occurrence: a Molotov cocktail had been thrown into the U.S. Embassy in Minato Ward, Tokyo, a few days ear-lier. Because it occurred on the anniversary of the Iraqi invasion, police speculated it was war protesters.

Otherwise it seemed just another wildly amiable, mildly rainy Tokyo day. Outside the University of Tokyo, workers emerging from manholes and wearing immaculate white jumpsuits bowed, as is the custom, to passersby. Lovely brick university buildings, when they weren't flanked by plum and cherry blossom trees, were surrounded by hundreds of neatly parked bicycles and motorbikes that were left politely unchained. Businessmen in suits deftly maneuvered their bicycles in and out of traffic with one hand, holding umbrellas over their heads with the other, their paths seemingly random, relaxed. In the streets, the most com-monly spoken phrase was "Hi, hi," (yes, yes)—because it is the height of rudeness to say no. Lunching women covered their smiles when they laughed.

Yet while Molotov cocktails are hardly the Japanese way—the

nation's crime and violence rate is astoundingly low—Japan is still, famously, a place where extremes of both graciousness and determination can coexist at all levels of life. And this is true of its stem cell field. Unlike in many other countries, the government here has done little-to-no trumpeting of its extensive work on stem cells, yet it has built the largest bench-to-bedside stem cell complex (Riken) in the world, in part to fast-track hES cells to the clinic. Likewise, it is often overlooked that Japanese researchers have quietly made some of the most disciplined, detailed, and ultimately seminal observations in adult stem cell research in the past few decades. Riken's Takayushi Asahara is one. The University's of Tokyo's Masataka Sata is another.

The huge main waiting area of the University of Tokyo hospital is a marvel of polite efficiency. At its center is a militant array of comfortably padded seats like the first-class section of an airplane. Every counter is topped by a neon sign announcing its purpose in both English and Japanese ("Pay," "General Admissions," etc.) The tiny stalls in its ladies' room thoughtfully provide both old-fashioned bidets and computerized toilets with some astonishing talents (including the ability to heat up and spray water around the room).

When the tall and gangly Dr. Sata ran into this lobby with his white lab coat flying behind him and holding a silver laptop open in his hands, he seemed out of place, a blast from science's past. He looked up from his laptop absentmindedly when he came to a halt in the lobby, as if unsure where he was, or ready to plunk his computer down right there on the floor and conduct his interview, apparently sure that all that mattered was his mind and his data. Then, glancing about, he bowed slightly and backed up, the cue to follow him, and led the way to a place more suited to his absentminded demeanor: a basement area of the hospital, with halls that boasted peeling paint and were as freezing as a crypt.

An open door by his lab looked out onto a portico where scores of ubiquitous unlocked bicycles were parked, in fan shape, under ubiquitous cherry blossom trees. But the lab itself would have looked like a boxed off corner of a warehouse, if it weren't for the fact that the boxes were laid on, and stacked around, expensive equipment: immaculate white refrigerators, centrifuges, a new FACS machine (which sorts cells), confocal microscopes, and laminar flow (sterile) hoods.

Sata was young—40—which explained the basement lab. But he was also famous, thanks to his work on BM stem cells and the heart, which explained the top-flight equipment and the good-sized staff (10, with 2 techs to man the FACS machine alone). It was unusual for a young investigator to have all this in Japan, which was still steeped in hierarchy. But in 2001 and 2002, barely done with his training, Sata published two papers in a top journal, *Nature Medicine*. The papers generated much buzz, with one becoming a "hot paper" that fired up many citations, much controversy. For Sata's papers found that the BM stem cells being used in stem cell/heart trials around the world may also play a role in the Western world's number one killer: atherosclerosis, the main cause of heart disease.

Sata cleared papers and a large plastic heart off a table with apparently the same youthful energy that had literally propelled him up Mount Fuji four times and counting, he would note later, grinning. He plugged in his silver laptop and pulled up a picture of a coronary artery on a shadowy coronary angiogram. There was a break in its shadow, a stenotic (clogged) section. "We dilated this clogged artery with [balloon] angioplasty, but six months later, restenosis," he said. "This is common." (Balloon angioplasty involves shooting a catheter like Ruth's into a coronary artery, then opening a balloon there, pushing atherosclerotic clogs to the artery's sides.)

That is, Sata explained, it is generally agreed that vessel walls begin to stenose, or clog up, because they have been injured in some way: "by hypertension or mechanical injury or . . . Western . . . food." He paused, clearly wondering if he had caused offense to the Westerner speaking to him. When the Westerner said, "You mean McDonald's?" he laughed. "Hi, hi, the McDonald's," he said, and thereafter he obligingly referred to all killer foods as "the McDonald's."

The difference between the prevailing wisdom and Sata's about athero ended there. "The prevalent idea has been that atherosclerotic lesions are caused by mature smooth muscle cells in the middle wall of the arteries—right next to the inner wall with the stuck atherosclerosis —that migrate toward the inner wall and dedifferentiate to an immature state," he said. "I found this is not true. I think atherosclerosis is largely stem cells, many of which come from the bone marrow to repair—and fail."

He called up what looked like shots of swirling multicolored lava. They were stained tissue biopsies of the atherosclerotic layers of mouse coronary arteries. Athero is known to build up fast after organ transplants. So Sata created mice that all possessed a gene, Lac-Z, that stains when a certain dye is added, then took out their hearts and placed them in normal mice. Four weeks later, he found that 90 percent of the athero clogs were comprised of cells that came from the circulation of the normal mice, not from the outer walls of the Lac-Z arteries of the Lac-Z transplanted hearts. When he did it the other way around, the same thing happened: the atherosclerotic young smooth muscle cells did *not* come from the artery walls themselves, but from blood circulation. "As expected," he said, grinning.

But Sata almost lost the race. For a year his paper circulated, landing nowhere. No one believed him. "A reviewer at *Circulation Research* called the Lac-Z staining approach artificial—so we did it again, this time with a female-to-male model. The same thing happened."

Still, no one would take the paper. Then word went out that Brigham and Women's Hospital cardiologist Peter Libby, the dean of atherosclerosis, was about to publish a similar paper. A colleague of his, indeed, had presented similar results at an American Heart Association conference. "I was very upset. But then *Nature Medicine,* which had a draft of my paper, called in January 2001, asked me to change a few sentences, and published the paper in April." (Ironically this paper was published the same month that Orlic's impactful paper was published finding that BM stem cells also *heal* the heart.) *Nature Medicine* published Libby's paper a few months later. "I won the race."

And a few months after that, a Canadian group published similar results in the *NEJM,* if this time, in human organ transplant patients.

Still, as is so often the case in the stem cell field, there was no time to gloat. With so many people now on board, the race was on to find the source of the immature cells. Did they come from the bone marrow, the circulating blood, or other tissues outside the heart? "We thought the ultimate source must be the bone marrow, so we transplanted bone marrow from a wild type [normal] mouse into a Lac-Z mouse," Sata said. They did this after they damaged the femoral artery of a mouse, similar to the way angioplasty damages human arteries. "We proved definitely that the cells were bone marrow. Sixty percent of the neo-intimal [atherosclerotic] cells were Lac-Z positive," that is, from bone marrow, and had differentiated into immature smooth muscle cells that helped clog the vessels. When Sata's team repeated this heart transplantation study, that number jumped to 80 percent.

Finally, to find out if BM cells contribute to ordinary atherosclerotic plaque caused by bad eating, the crew injected marked BM stem cells into atherosclerosis-prone mice, then fed them "the McDonald's"—a diet of 15 percent butter, among other things. The team found that 40 percent of the atherosclerotic plaques came from bone marrow.

And to take a whack at narrowing down the cultprit even further, Sata repeated the first experiment (using a wire to injure the femoral artery, mimicking angioplasty), but this time he used what is believed to be just the hematopoietic (blood-forming) stem cell subset of bone marrow: the same as Orlic's cells. These did not contribute as robustly as whole bone marrow (only 43 percent), but they still contributed, and most were immature smooth muscle cells.

A few papers followed that backed up Sata's work. Then: trouble. A researcher named Yanhua Hu of St. George's Hospital in London published two papers contradicting Sata. In the first, Hu did find that, like Sata, when he transplanted an aorta from one mouse to another, indeed, the classic paradigm did not hold; that is, the athero did *not* come from mature smooth muscle cells in the same blood vessel, it came from circulating cells. But it did not come from bone marrow cells. In the second paper, when he transplanted vein tissue and grafted it onto the aorta of another mouse, Hu found that half the cells came from circulation—again busting the paradigm—but again, not from bone marrow.

At first stumped, Sata hit on a plan, and in an October 2003 *Circulation Research* paper, he noted that he had found that, with severe and sudden vessel wall damage as occurs in angioplasty, bone marrow cells contributed to atherosclerosis by traveling to vessels and converting into clotting, immature smooth muscle cells. But with more gradual damage, as in tying off a coronary artery or placing a cuff around an artery, bone marrow cells did *not* contribute as substantially to athero. It is possible, then, that Hu got the results he did because replacing the aorta creates less of an inflammatory crisis than deliberately mauling an artery or transplanting a large organ, Sata suggested.

As a result, Hu went back to the drawing board. He did an even more detailed study of aorta transplantation. He found as before that the athero clot in this milder form of damage came from circulating cells, but this time he *indeed* found, after all, that some were BM progenitors. Most importantly, the micro-vessels that structure the clot, keep it together, and feed it all came from BM progenitors—*100 percent*. This was published in December 2003. In May of 2004, Hu would publish in the *Journal of Clinical Investigation* his discovery of what those other circulating cells may be that contribute to athero in the transplanted aorta: stem cells similar to Orlic's that can make both blood and blood vessels. He didn't find them in the bone marrow but on the outer wall of the root of the aorta, the bottom of the huge hose leading from the heart to the body. Those stem cells, he said, can migrate through the aortic wall and into its intima medial space—the atherosclerotic space—of blood vessels and form both healthy mature smooth muscle cell there *and* atherosclerotic clots. Presumably, then, stem cells from this spot in the aorta and perhaps the outer wall of all vessels contribute heavily to atherosclerotic lesions, especially when the lesions weren't caused by a severe injury.

In general, by 2005 Hu would come to believe that 5 percent of all CD34 endothelial progenitor cells that make new small blood vessels come from bone marrow, and 95 percent come from nonmarrow sources: the outer wall of the blood vessel. Regardless, his papers make clear that blood-vessel-forming stem cells, no matter what their source, can contribute to atherosclerosis.

The elegant back and forth of papers continues. Most begin the way so many stem cell papers do: "The paradigm was once . . . , but now it has changed. . . ." It was all energizing to Sata. "I received e-mails from all over the world about that work," he said proudly.

He was bent on studying even further the paradigm introduced by Hu. That is, he believed that MSCs from the bone marrow contribute to the mass of atherosclerotic clots, and endothelial stem cells (CD34s) form the vasculature that feeds them. His team was finding osteoblasts from bone marrow in the atherosclerotic plaque and vessel walls of old mice and none in those regions at all in young mice. (Bone-forming osteoblasts come from MSCs.) His team was also seeing endothelial progenitor cells there. Sata believed that certain "bone marrow cells eventually go inside the atherosclerotic region and feed them, as occurs with cancer."

"Plaque is more alive than we thought," Sata said excitedly. "We are very confident this is the case. Atherosclerosis is like cancer." That is, like cancer, athero may be made up, in part, of stem cells gone awry and fed by blood-vessel-forming stem cells from the blood that are lured to them. Indeed, some researchers continue to find that growth factors like VEGF turn stem cells into blood vessels and can exacerbate cancer, atherosclerosis, and retinopathy (caused by too many blood vessels in the eye). So some are studying whether targeting and blocking these growth factors will cause athero to decrease. One, rapamycin, is already working, said Sata.

Excitedly, Sata added that he, like the young Japanese postdoc who discovered the CD34 cell, also received some bolstering from that great renegade Judah Folkman, the Boston doctor famous for discovering that tumors need their own blood vessels. For Folkman reported in *Proceedings of the National Academy of Sciences* that his anticancer drug angiostatin, which kills tumor blood vessels, also kills vessels feeding plaque. "We will establish that plaque needs blood vessels," Sata concluded.

It may still turn out that, while stem cells commonly found in the bone marrow may contribute to athero and cancer, only smaller subsets of those are actually responsible. It is also possible that the same cells are responsible for *both* healing and clogging, and the timing of their arrival on the scene dictates their behavior. (As in: young stem cells are good; old stem cells are bad.) "It's a very delicate issue," CD34 endothelial cell progenitor pioneer Takayushi Asahara will note a few months later. "It is looking as if now some fraction of endothelial progenitor cells may be able to differentiate into smooth muscle cells [which make up the bulk of the atherosclerotic clot]. And they can, it may be true, enhance atherosclerosis with neo-vascularization [that is, they may feed atherosclerotic clots with new blood vessels]. One paper out there has found that endothelial progenitor cells did good and bad in one experiment, in *Circulation*. . . . We are looking into it now." Until an answer is found, he notes, it is best to be careful in clinical trials. Injecting cells directly into the heart (as Perin does it) may be the best way, he says.

• • •

BUT OF COURSE, Sata and Asahara are hardly alone in their belief that stem cell research may prove key to cardiology, both leading to better therapies for athero and revascularizing blood-starved hearts. Cardiologists worldwide are jumping on board. Douglas Losordo, whose team worked with Asahara at Tufts University in the 1990s, is conducting, as noted earlier, a trial like Perin's, though using Asahara's purified subset of Perin's cells: CD34s. He injects them, via NOGA, into the hearts of patients with chronic ischemia at Boston's St. Elizabeth's hospital. "If you ask me," Losordo said in December of 2004, "There is no debate in one area": transplanted BM stem cells are rejuvenating hearts. "There has been very little evidence otherwise."

Losordo adds that while his trial is blinded, its structure has given him a hint as to how the cells are doing. That is, for every six patients who get cells, two don't, and "the majority of patients have had a symptomatic improvement." Furthermore, in a bench-to-bedside stem cell facility not unlike Perin's, he's doing preclinical work showing that even old, debilitated BM blood-vessel-forming stem cells, when transduced with new genes, can get blood vessels blooming.

Indeed, indirect support for the notion that the CD34 subset, when healthy, *naturally* forms blood vessels in hearts has come from work by the NIH's Richard Cannon. Cannon has found that patients with coronary artery disease naturally possess fewer blood-vessel-forming endothelial BM stem cells—and less effective ones—than healthy patients. This could mean it's all a vicious cycle. When we are young and athero-free, our CD34s are our guardians, rushing to injured areas to rebuild blood vessels and create new ones. But as we age, the number and quality of our guardian CD34s drops, allowing inflammation to rage unchallenged in vessels. At that point, when our waning CD34s finally do arrive, they can just get caught in the traffic jam, adding to the problem.

Still, Cannon, too, is jumping aboard. "We're thinking of doing a BM stem cell transplant trial at the NIH." And Elizabeth McNally of the University of Chicago, who recently came out with a paper claiming that BM stem cells can't form new cardiac muscle in a meaningful way, says she *did* see new blood vessels. "A subset of bone marrow cells could very well be forming new blood vessels in these patients' hearts," she says.

Perhaps even more excitingly, since 2002, the University of Pittsburgh's Amit Patel has conducted *three* randomized, blinded trials in South America and Asia for a total of 100 heart failure patients, bypassing the FDA for a while, as Perin initially did. (Patel admits he did this in part because he expected the FDA to drag its heels, although he adds that he, like Perin, had relationships with foreign hospitals.) Patel gave 50 million CD34s to each patient in the first two trials, far more purified than Perin's doses but less pure than Losordo's. Patel believes a mixture of cells can be key. "The Texas Heart guys have figured out something that is important; that is, it's not how pure your cells are that matters to a certain degree. Sometimes when people have gotten greater than 90 percent of one kind of cell, other cell types were washed out that help augment the function of the existing cells."

While Patel, as of spring 2006, will have yet to announce the results of the third trial (in which he used largely CD34s from peripheral blood), he has announced results from the other two (papers for both have been accepted for publication). In the first trial, he treated 10 ischemic cardiomyopathy heart failure patients with CD34 and some CD45 bone marrow stem cells (which may turn into heart muscle, some studies suggest) along with bypass surgery. (He treated 10 patients with surgery alone.) After the bypass, in a process taking about 10 minutes, the surgeons then injected the cells into 25 to 30 sites of apparent muscle damage. (NOGA was not used.) Before surgery, the average ejection fraction (measuring left ventricle pumping) was about 30 percent. (Normal is 55 percent.) At six months, average ejection fractions were 46.1 percent for the stem cell group and a mere 37.2 percent for the surgery-only group.

In the second trial, Patel gave 15 patients with idiopathic cardiomyopathy (no known cause) 50 million CD34s each (and 15 patients, serum) through three tiny incisions in the chest. Among those getting cells, the average ejection fraction rose from 26–27 percent to 45–46 percent, and the BNP (B-type naturitic peptide) level dropped from the 800s to the 300s (60 is normal). All cell patients improved by two heart failure classification levels. "Frankly unbelievable," Patel says.

Patel will launch his first U.S. trial in 2005, receiving permission from the FDA in May—even before his first papers are out. Overall, results have been so positive that infighting is going on among nations. "I don't tell people the names of the countries involved in particular order," Patel says, laughing, because they have been so impressed with these trials that "one country gets offended if another comes before it when I'm listing them. So what ends up happening is that I just say, 'South America and Asia.'" (The countries are: Thailand, India, Italy, Netherlands, Ecuador, Uruguay, and Argentina.)

The list of U.S. and foreign institutions with Texas-sized ambitions for stem cell/heart trials grows constantly. As noted earlier, a trial for intractable angina is scheduled to begin by early 2006 at Columbia University using mesenchymal precursors (MPCs), a very immature form of the MSC that Perin tried on dogs. Like so many other cardiologists, Columbia's Silviu Itescu did his first heart/stem cell clinical trial in Australia. He says his MPCs can form large arterioles, not just smaller capillaries, also heart muscle, perhaps, and seem to be able to avoid immune rejection. This is unproven: "There is no evidence that MSCs avoid immune rejection," Harvard's Doug Melton will note in October 2005. Still, Itescu hopes to change that. He also hopes to administer MPCs to patients on the heart transplant list who are on LVADs. The goal: get them off both the transplant list and the LVADs.

Itescu has formed an MPC company, confident his cells also proliferate faster than other BM stem cells (also unproven). If true, this would indeed mean MPCs may be patented and sold, a potential market in addition to a potential therapy; mesenchymal stem cells in general could become truly commercial stem cells. Johns Hopkins's Joshua Hare believes it. Hare's team began slipping allogeneic (foreign) MSCs into the bloodstream of recent heart attack patients in the spring of 2005. In animals they found that allogeneic MSCs, administered after heart attack, reduce tissue scarring by 75 percent. They also believe allogeneic MSCs avoid immune rejection and may create both cardiac muscle and blood vessels. The team is not necessarily avoiding athero problems, however, since they are slipping the MSCs into the blood. Time will tell.

In the meantime, more randomized trials of BM cells like Perin's are going on in Europe. Some are huge, enrolling hundreds. In Brazil, the government announced in February 2005 it was launching the largest BM stem cell heart trial yet, enrolling a whopping 1,200 patients.

Still, just as the jury is out over whether BM stem cells can be bad for hearts—contributing to atherosclerosis—it is out over whether, and how, the cells are *good* for hearts. Basic science research by Stanford's Irv Weissman, the University of Chicago's Elizabeth McNally, and the University of Indiana's Loren Field in particular indicates that, probably, the very small subset of BM stem cells that Don Orlic's group originally used cannot transdifferentiate into heart muscle, if it may well have formed new blood vessels. Indeed, in general the belief is that BM stem cells of all kinds are either leaking healing growth factors into patients' hearts or forming new blood vessels—they are not transdifferentiating into functional new heart tissue. (They may sometimes fuse with heart tissue, or occasionally express heart tissue genes, or prompt new heart tissue to form out of *other* cells, but they are not yet becoming funtional heart tissue.) For that, cardiologists may have to turn to hES cells.

But intensive, worldwide research is ongoing. And regardless of the mechanism, many patients appear quite happy so far. "The thing we're going to continue to find is that stem cells are both part of our everyday disease processes and our everyday healing," says Losordo. "It's amazing this didn't appear obvious sooner."

*   *   *

AFTER PAVELKO has gone home, Perin relaxes in an office. "I don't really care," he says, to a question about a study finding that certain subsets of bone marrow cells may not become heart. "Why get so uptight about it? Our objective it to help people get better. Those guys who work only with rats, they get so lost in the world of 'Is, Isn't.'"

Things are moving fast for Perin. Today he was courted by a company hoping he'll try another kind of stem cell (his sixth) on animals. In a few weeks he will make a presentation at the American College of Cardiology conference of a finding that, in dogs with acute ischemia, injection of MSCs into heart muscle is more effective than releasing MSCs into circulation, as he predicted. At that conference, there will be some 50 abstracts or presentations on stem cell therapy for the heart, most of which are positive; 14 of which focus on MSCs. ("We're kicking a——," he will note.)

Most of Perin's life, as son of a Houston petroleum company executive, has been spent in either Brazil or Texas. Born in Brazil, he came to Texas to live at age 2, then went back to Brazil at age 13, and came back to Texas when he was about 27, in 1985. While in school in Brazil, he had thought he'd like to specialize in reconstructive surgery, but after finding that most reconstructive surgery in Brazil is cosmetic, "I was outta there." So he switched to interventional cardiology, because he did like working with his hands and "couldn't sit in an office, prescribing drugs all day."

"Losordo and I," he says, speaking of Doug Losordo, head of the other major 2004–2005 heart/BM-derived stem cell trial in the United States, "are kind of like the oldest sister who dad doesn't let go on dates. The kids who come along later will have it easier." He is speaking of the months it took the FDA to approve both of their stem cell trials, which might not even have happened when they did if Perin had not first taken his cells to Brazil. "But we have a fabulous relationship now. Life is getting easier."

Perin's wide-set eyes are always slightly closed, as if he is always on the verge of smiling, but he is outright grinning as he leads a tour of the new stem cell facility being built for his crew. Set on the tenth floor of the new Cooley Building, his new center will boast another hugely expensive "clean room" (three in one complex is rare). The clean room resembles a giant glass mouse maze. Huge beakers glow with a golden liquid in the sunlight. There are more impossibly vast views of Texas all around. Perin stalks about excitedly—then bolts out the door.

Emerson Perin is unable to sit still. It led to his decision to become an interventional cardiologist who moves about all day, instead of a GP who sits behind a desk. It led to his decision to impatiently bypass the United States to get his controversial work done. At 45, he's not slowing down anytime soon. He's starting a new clinical trial—using yet another stem cell that the U.S. FDA isn't ready for—in Spain in a few months.

And Ruth Pavelko isn't slowing down either. The housebound housewife who once couldn't make it to her own mailbox will be walking laps around a nearby park three months after receiving her stem cells.

10

THE PROLIFERATING
UNDERGROUND CLINIC AND
THE REGULATION OF CHINA

*Almost every place I visited in China had
isolated its own hES cell lines and I visited
more than 20 hospitals. Over 1,000 Chinese
patients have received stem or stemlike cells in
the last five years.*

—U.S. neurologist Wise Young

*After the injections, it's like a miracle. We must
spend $300 calling home.*

*—Sherry Ashmore, on her chronically ill
son's shocking if temporary improvements
when receiving "embryonic cells" in Ukraine*

SHAO RONG GAO IS FIERCELY driving, gently cajoling, and occasionally just plain jiggling a $50,000 machine: the piezo drill cloning
micromanipulator. It is March 2005. The piezo, which was first utilized
in Teru Wakayama's pioneering mouse experiments, looks like a home-
made amateur musical instrument, especially as it is "played" by Gao.
Gao alternately, and rhythmically, grabs at and twiddles two black
spheres—each of which looks not unlike the rubber end of Harpo
Marx's horn—dangling off two sides of the machine. Simultaneously, he
eases back the plunger of a plastic syringe snaking out from the bottom
of the machine, as his feet tap at two pedals far below and his eyes stare

into a microscope. It looks as if he is honking horns, crashing cymbals, and playing an organ all at once, a one-man band.

Gao is not making music, but he *is* a creator. He is making life—a tiny teeming universe of cloned life. The researcher, who went to Jerry Yang's university in China, trained in Ian Wilmut's lab, and works now for Yang in Connecticut, makes 150 to 200 cloned embryos a day, armies of embryos that are all copies of one mouse. It is a task so delicate and difficult that the machine occupies a special $4,000 "antivibration" table. Gao has been known as the "fastest cloning hands in the U.S." since the title's former holder, Wakayama, went back to Japan in 2002. Here in the Center for Regenerative Biology, the gifted Gao is trying to decipher cloning's mysteries, following Western procedures and rules. That is, he will continue cloning mouse cells until the state lets him switch to human cells.

A few months later, a man who could be called Gao's doppelgänger, neurosurgeon Hong Yun Huang, stands in an operating room in Beijing's Chaoyang Hospital. Huang went to school in Beijing, trained with Rutgers University neuroscientist Wise Young, and then, unlike Gao, returned to China. Asleep beneath Huang is a patient whose neck was injured 18 months ago, leaving him a quadraplegic. Huang proceeds to do something he could never do with such ease— if at all—in the United States. He makes cuts on either side of the man's injury, exposing his muscles, his spine, the membrane beneath the spine and the spinal cord midline beneath that. Nearby sit cultures of human cells (human fetal olfactory ensheathing glial cells and human fetal cells with brain stem cell markers). Seen under a microscope earlier, the mixtures looked like waterlogged, pearl-studded wedding veils, their perfect round nuclei embedded in graceful, lacelike networks of long, playful tentacles. Some 500,000 cells are injected on either side of the neck injury, then the patient is closed up. Four days later, he will regain all movement in his right wrist. He will later be declared to have moved up one notch on an international paralysis scale.

Huang has given human fetal cell injections to over 400 spinal cord injury (SCI) patients. His Chinese waiting list is 3,000 strong; his international waiting list is 1,000 strong. He is following Chinese rules for this procedure, but he is not strictly following Western or international rules. If he were, he would in all likelihood still be trying his fetal cells on mice, not men.

Gao and Huang, in a way, are bookends representing China's past and potential future. Once, talented students like Gao went West to train, then stayed. China's science was not something to hurry home for. Now, many China-born researchers—like Huang—are going back, pulled by big bucks and promises, and pushed, in regenerative medicine, by the United States's unsatisfying stem cell laws.

But Huang represents China's potential future in another way. While many developing nations, China among them, have been setting up clinics that are poorly regulated by Western standards and administer uncharacterized, mysterious so-called embryonic and fetal stem cells to patients, China has been doing some changing. In Russia, Ukraine, and the Caribbean increasing numbers of questionable clinics are catering to desperate patients tired of waiting, among other things, for the United States to revise its strict hES cell stance. But China has been making some concerted efforts to look to the West for guidance. Indeed, Huang himself recently let scientists from a top U.S. spinal cord clinic judge his work and publish their views. (The views: Huang's approach seems to offer an impressive "rapid" if "limited" recovery, but much more work needs to be done to identify and systematize the cells, track patients, include "placebo" patients, and arrange for independent assessment, and this should have been done all along. Others have noted his patients seem to eventually reject the cells; that he should look into tissue matching.) Huang sought those views even though he had no financial need to: he makes $20,000 for every foreigner's operation.

China has not yet pulled off the kind of published human cell cloning coups that South Korea appears to have pulled off at this time. It has not matched the published hES cell coups of Israel and the United States. But it has been spending bigger-than-Singapore-sized bucks on hES cell and cloning work. It has a lot of scientific talent overseas—unlike tiny Singapore—and is demonstrating great success luring it back. And it is establishing key alliances with U.S. stem cell scientists, who range from fed up with, to curious about, to mesmerized by the China buzz. "There's a strong U.S. interest in China collaborations," says Young. In the recent past, China's stem cell regulation enforcement has been so lax it could make the most liberal Western researchers blanch. The state still seems to avert its eyes when its scientists create hES cells from fresh embryos instead of spare IVF-clinic embryos, perform limitless monkey experimentation, dodge global clinical trial guidelines. There has been sketchy fetal work. But Westernization is occurring, says Young, who is setting up an unprecedented U.S./China clinical trial network to round up, legitimize, and study Chinese data— including elusive stem cell data. "It is exciting, almost unbelievable that things are working out as they are."

China is what Young and others call "a hidden power" in the area of hES cells, cloning and even adult stem cells. Nowhere is this perhaps more apparent in the United States than in Yang's lab in March 2005.

.   .   .

GAO IS UNHAPPY. Earlier in the day he had to stop work for a nationally syndicated TV crew that moved into the lab. His boss, Jerry Yang, recently pulled off two more cloning coups. Now Gao has to show a visiting writer how to clone mice. It is a tough day.

But Gao, with a sigh, waves the writer into a seat beside him and points at a computer-sized screen to his right. "Watch there," he says. A small digital camera is turned on, and magnifies the petri dish in his expensive micromanipulator (of which this lab boasts three). The screen reveals a whitish-gray background—the bottom of the dish. Scattered across its surface are dark lumpen cumulus cells (mature adult cells that protect eggs) which look not unlike fields of rocky comets. They are suspended in a large droplet of a PVP (polyvinyl pyrrolidone) solution to make them slippery, manipulable. (Ian Wilmut's team established the idea that adult cells should probably be in the resting and semiresting G0 or G1 stages before being cloned: most adult cumulus cells are always in this stage, so Gao has not had to pretreat them.)

All told, the scene in the petri dish does not look unlike a film-negative of a solar system, with outer space white-gray, and comets, planets, and stars all glowering darkly. A pipette enters slowly from the right, controlled by Gao via the spherical dangling object on the right. Once positioned near a "comet" (an adult cumulus cell), Gao turns to the left a knob at the tip of the left-hand spherical object, which generates a negative pressure in the pipette. The "comet" is sucked into the pipette. Gao now jiggles the right spherical object. A few bits of dark "rock" float out of the pipette.

"Cytoplasm, see it floating out? That's bits of cytoplasm from the cumulus cell broken up. That tells me the nucleus of the adult cell has been shaken free of cytoplasm and is left behind in the pipette. See it?" Left behind is a translucent, tiny blob, the nucleus. It is as if Gao has sucked the energy out of the comet and trapped it in his pipette. There is some excitement in his voice.

Gao swivels the petri dish away from this scene and a different small "universe" twirls into view. On the left side of the screen now appears a large stationary, docking pipette. Twiddling again, Gao uses the pipette that he had filled with the adult cumulus nucleus to, literally, swat a large, mottled moonlike cell into view: a metaphase II oocyte (unfertilized egg) that was previously enucleated. (Another thing that Wilmut helped establish was that the egg should also probably be in a certain phase—metaphase II—before cloning can work.) Gao had earlier added hormones to this egg, hormones not unlike those used in human IVF clinics, to get it to that phase.

Pumping a bit on the plastic syringe with his left hand, causing air to move out of the petri dish toward the left, he gently draws the large moonlike oocyte to the stablizing pipette on the left, and docks it there. Then he takes the pipette on the right, which still holds the "energy" of the adult cometlike cumulus cell—its translucent nucleus—and gently pokes it at the zona pellucida (outer shell) of the oocyte. At the same time, he presses the foot pedal on the left. A beep sounds. The pipette vibrates. He then presses the foot pedal on the right to increase the vibration. He moves the pipette toward the thick shell of the zona, and the vibrating pipette zaps a hole in it. He pushes the pipette through the hole. As the pipette nears the second shell of the moonlike oocyte (the oolema shell) his right foot pushes the right pedal to make the pipette vibrate less vigorously: this membrane is softer. The gently vibrating pipette now pokes a hole in *that* membrane, through which Gao sends the pipette.

The pipette moves slowly straight through the inner sanctum of the moonlike oocyte. At the end, far enough from the furthest edge of the shell to make sure he does not perforate it, but far enough from the site of entry so nothing important can dribble out, Gao injects the transparent adult nucleus in the pipette into the egg by twisting to the right the button on the spherical object hanging on the left (the injector). The adult nucleus immediately disappears into the pockmarked face of the oocyte.

Done. Gao has transferred the translucent, intelligent "energy" from the "comet" into the hollow center of the "moon." He has created a cloned mouse embryo. After placing the new embryo into an incubator the temperature of a mouse's womb (37 degrees C.), he will do nothing for the next hour. During that incredible hour, the two entities will introduce themselves to each other on their own, rearrange themselves on their own, hook up on their own. Prompted by mysterious ingredients in the egg—ingredients every stem cell scientist in the world would kill to identify—the chromosomes from the adult cumulus cell nucleus will begin to condense, as they do in normal metaphase II, one-celled embryos that are preparing to divide for the first time.

At the end of that hour, Gao will add the chemical strontium. This will cause calcium to flood the egg as it does after sperm penetration, and prompt "activation" of the newly "fertilized" egg. That is, old adult genes will begin to turn on or off in new ways in their new embryonic home. Set in motion will be a rapid series of chemical events that will turn that lifeless moon into a complex, living, energized "planet," an entity far greater than the sum of its two parts: an embryo that is a copy of an old mouse. And all this will be done without that massive jolt of electricity that was used for Dolly and is still used for the larger mammalian eggs of cattle and rabbits. The mouse egg is more delicate, so requires the lighter touch that Japan's Wakayama brought to cloning.

"The oocyte cytoplasm is very powerful," notes Gao. Yet mysterious as they are, the cytoplasm's powers do make sense to Gao: it is not necessarily "magic" that egg cytoplasm can turn the clock back on this older nucleus, he says. When egg meets sperm, at least one job of the cytoplasm is to bring the mature sperm back to a "young" state, to "reprogram" it. This is what it does when used to clone the nucleus of adult, nonsperm cells, although obviously by the end many different things have occurred.

The cloned embryo will sit in the activation medium for six hours. Then it will be placed in a culture medium for four days. At the 72-to-90-cell blastocyst stage, the embryo will either be placed in a womb to become a cloned mouse or have its ES cells removed, depending on the experiment.

The whole business starts off as a simple business proposition, one any CEO could understand: Gao is in some ways simply a business-man who has co-opted a factory and is feeding it new raw materials to create a new product. The difference is that this is a biological factory, and these are biological raw materials. Everything here is alive. So the raw materials begin interacting with the factory, talking to and direct-ing it; the factory interacts with the raw materials, talking to and directing them; until eventually the factory and the raw materials incorporate themselves into *each other* and become the final product: a cloned embryo.

Gao pulls the pipette out of the newly cloned embryo, unceremoni-ously swats it out of view with the active pipette and swats a new, unfer-tilized oocyte into view.

"There," says Creator Gao.

"That is the hardest thing in the world," says a student from Taiwan.

The student is standing near research associate Fu Liang Du, who came here from China's Jiansu Agricultural University and is hunched over a dissecting microscope. He is sorting viable cloned rabbit oocytes. In each of half a dozen dime-sized wells, filled with pink liquid, are about 100 cloned rabbit embryos in the two-to-four cell stage. With a rubber tube, he has been sucking out the embryos that died. ("I swallow a lot of them," he jokes, smiling.) More than 90 percent of those remaining will die once implanted in the rabbit womb, a testament to the fact that cloning techniques require vast improvement. The high death rate is the main reason many top scientists agree that human reproductive cloning—which would result, theoretically, in another human—as opposed to therapeutic cloning, should be banned. The tiny rabbit embryo clones look like eyes-only faces with no mouths—classi-cally depicted alien heads.

"So did you get all that?" Gao says, looking at his watch. As noted, it has been a tough day for the obsessed cloner. Beyond the new coups, which have had TV cameras scooting about, the Connecticut legislature votes soon on whether to defy the Bush edict and supply labs with $100 million for more expansive hES cell work—and allow human therapeu-tic cloning work with the money. And Yang has for the first time publicly threatened to leave if the state does not come through. He has been issued another invitation to work in China. He is considering the offer.

A few hours earlier, dolled up in a mismatched jacket and tie, Yang had headed amiably across the courtyard of his new regenerative medicine complex and into his cloning lab. It had been briefly transformed into a slapdash TV studio, with bright lights, a camera on a tripod, and a techie all in black. The crew was here to discuss the second of Yang's recent coups, the first to actually be published next month, in *Proceedings of the National Academy of Sciences (PNAS)*. After analyzing the beef from two of Yang's cloned bulls and the milk from Yang's four cloned cows, using USDA standards, Yang's crew found that milk and beef from cloned cattle are identical to normal milk and beef. There seems to be nothing odd about them. It was the first time this had been shown in a carefully controlled experiment, and it was likely to cause a stir.

Yang donned a white lab coat, sat in front of a piezo, and within minutes was swearing up and down to the interviewer, laughing, that he would consume cloned milk and beef and that—yes, that too—he would feed it to his wife and his son. He then swore it up and down a second time, after headquarters called to say the interviewer had been talking over Yang and that the scene needed to be reshot.

"With the meat I would make hamburgers for myself; I would drink the milk . . . No safety concerns," he said. Indeed, he added, the beef scored 8, where only a score of 6.5 is required. Ten percent of the milk scored higher than requirements, as was true of the donor animal, Aspen. Furthermore, Aspen was a champion milker, putting out 35,000 pounds a year, twice the average. The clones are champion milkers, too, Yang noted.

A few hours later, having accompanied the TV crew to the barn to visit the Holy Cows, Yang rushes into his office in a windbreaker and a blue-and-white striped shirt. He stares, says, "You had no lunch? Wait, I have an apple," and rushes off. He returns 15 minutes later with an apple that has been peeled and sliced, along with a cup of Taiwan tea with leaves filling it to the brim: Taiwanese scientists with whom he is collaborating keep pressing tea upon him. He has hordes of it. "I'm sorry," he says, out of breath. "The apple was in my car."

Yang is in an exceptionally good mood today. His salivary cancer returned since his remission at the time of his grand opening in September 2003. But some stereotactic, targeted radiation treatments have since eliminated most of his tumors, except for a few in one lung. And the tumors are low-grade, moving slowly. He has bought himself more time.

Furthermore, Yang is optimistic about the state legislature's debate. "I think they may pass the bill," he says happily. "Senator Dodd has been very supportive." This is key, he says, because things have not changed in the United States over the past year: animal cloning continued to fall through the funding cracks. He wants to start in right away turning human skin cells into embryonic cells via cloning, as he has been doing "for so many years, in animals."

Also keeping Yang cheery: His most recent job offer to work in the lab of Lin Song Li, one of two stem cell scientists in China receiving the most support from the state government. Li's lab has 50 people in it, with 20 scientists, much larger than Yang's 7-scientist lab. It is working on human therapeutic cloning, hES cells, and human fetal cells, a cornucopia of work forbidden, by law or political pressure, in federally funded United States academia. For now, Yang is just collaborating.

"At this stage, I'm not saying I am leaving," he says. "U.S. research is still strong. If the United States bans therapeutic cloning research, that will be another story. I am collaborating with many labs, talking to many Chinese agencies." One of those agencies appointed him to the advisory board of a new interdisciplinary biology institute. The 1 billion yuan National Institute of Biological Scientists is to be made up of 35 foreign scientists, all receiving higher-than-average Chinese salaries. The hope is that this army of Western investigators, including stem cell experts, will infuse a jolt of Western standards into China's life sciences. Ten members of the advisory board are Nobel Prize winners.

Then Yang explodes, as he often does nowadays, with alacrity. His wife likes to joke to people that he is not a romantic in his personal life: he forgets birthdays and such. But he is certainly one of the great romantics of the regenerative medicine field. "Therapeutic cloning—it will work in human. Think of it, when Dolly was born just nine years ago, poor success rate. They tried 277 eggs: one Dolly. Now the success rate in cattle is 30 to 40 percent. When Louise Brown [the first test tube baby] was born in 1978, the IVF success rate was 10 to 15 percent. Now it can be 50 percent to 80 percent. . . ."

He beams and needlessly straightens a cow on his desk, like someone so happy he is embarrassed and must point out that not everything is perfect: look, a wayward cow. Yang is surrounded by cows. People love to give him cows, be they wooden, porcelain, ceramic, glass. Few top scientists are so approachable that people feel comfortable offering whimsical gifts like this, in veritable herds like this. "All kinds of good news," he says, smiling lopsidedly.

All kinds, indeed. For Yang's other cloning coup—creating cattle hES-like cells from both cloned and normal cattle embryos—was done by Chinese postdoc Li Wang. Because she came from the lab of Guang Xiu Lu in China, he figured she could handle the challenge. Lu's giant staff of 120 has been deriving their own fresh hES cells, in addition to trying to get hES cells from cloned human cells, for years. While only the South Koreans would appear, at this time, to have gotten hES cells from cloned human blastocysts, according to internationally accepted journals, Lu's staff still has much experience with hES cells and simple cloning techniques. So despite the fact that U.S. scientists had been trying for decades to derive cattle ES cells, Yang thought Wang was up to it. He was right. Wang pulled off the challenge in under two years. The paper comes out in *Biology of Reproduction* next month.

"Two different sources," Yang says, laughing, holding up two fingers in a victory sign. "ACT, Texas A and M, researchers in Chicago, they've all tried it. But Li Wang did it."

China's power is apparently starting to come out of hiding.

*   *   *

IN MUCH the same way that the view in Gao's microscope seems a parallel universe, the history of Chinese stem cell research has at times silently paralleled that of the West. At rare moments, China's stem cell work has even preceded that of the West. The 1990s work of Zhu Chen of the Shanghai Institute of Hematology is an example. Chen "transformed leukemia therapy," as *PNAS* noted in April 2004, by continuing the work of Zhen Yi Wang. Chen scored one of the most dramatic successes in oncology when, in the late 1990s and early 2000s, he realized the potential of retinoic acid—and later, arsenic—for differentiating certain leukemia stemlike cells into a more harmless state. His Shanghai lab now has 12 principal investigators, 26 clinicians, 34 technicians, and a whopping 76 MD/PhD students.

In the hES cell world, until recently, the Chinese record has been far more spotty by Western standards. One reason, it is believed: the nation, due in part to its political upheavals, has lacked a long, unbroken tradition of standardizing techniques and methods.

The West traditionally credits Robert Briggs and Thomas King of Philadelphia's Institute of Cancer Research with cloning the first vertebrate, the American frog, *Rana pipiens,* from embryonic frog cells in 1951. *Science* in 2004 named Shandong University embryologist Di Zhou Tong the man who produced the world's first cloned vertebrate, an Asian carp, in 1963. Strictly speaking, Briggs and King did clone the first vertebrate, it is generally agreed. But Tong could be said to have been the first to clone a vertebrate from an adult cell without causing controversy, since a 1962 report of Oxford University's John Gurdon, who claimed to have cloned the first vertebrate from an adult tadpole cell, was disclaimed by some because the cell could have been an embryonic cell trapped in the tadpole's intestine.

Regardless, it is certainly agreed that Tong produced the first cloned fish, and the first interspecies clone in 1973, by inserting Asian carp DNA into a European crucian carp egg.

What makes the above all the more impressive is that it occurred in the middle of the Cultural Revolution of 1966–1976, which, with its purging of intellectuals, seriously damaged Chinese science. "China's scientific capacities quickly fell behind those of developed nations and vital personnel were lost," wrote an international panel in *Nature Biotechnology* in December 2004. Reform began in the late 1970s and early 1980s, but continuing political unheaval, a history of dictatorship and conformism, and "the Confucian tradition of respecting customs and hierarchy has cast a long shadow over modern China," as it was put in a March 2004 *Nature* article by Mu Ming Poo, director of the Shanghai Institute of Neuroscience. Furthermore, "Traditionally, the Chinese government has taken care of the entire life of a scientist, from college graduation to retirement, regardless of their performance. The result has been the absence of pressure and a lack of incentive to excel."

Thus, the analysts acknowledged, while Chinese biotech is on "solid footing . . . R&D organizations and activities" and "legal and political forces" are "in flux." Still, there has been a massive desire to play catch-up, and China's recent role in the Human Genome Project showed this was possible. Late to join, the nation nevertheless in a few years built two major genomics facilities filled with state-of-the-art equipment. It sequenced 1 percent of the human genome with 99 percent accuracy. In 2002, China sequenced the rice genome. All this indicates "how rapidly China has reached world standards in sequencing, and this in turn reflects its general advanced state of development in health biotechnology," the international team wrote. After joining the World Trade Organization in 2001, China has shored up its patent laws, boosting Big Pharma confidence (if not gelling it: patent problems linger). As of the end of 2004, there were 500 biotech firms in China employing 50,000 people. Half of those had been established within five years.

The central government is pushing hard. In past years, some 80 percent of all science patents were owned by universities, with only 6 percent the result of joint enterprises between academia and industry. To counter this, the government began offering incentives for collaboration that paid off. In 1991, 13.6 percent of China's papers in international journals came from more than one institute. That number was 30 percent in 2002. And the government has been loosening its tight control over science funding, resulting in fewer bureaucrats and more scientists making decisions—a key move.

Still, because biotech is a high-risk, long-term endeavor, and because patent irregularities still exist, venture capitalists have remained reluctant to invest, preferring more mature technology. In a March 28, 2005, article on offshoring in the life sciences, *The Scientist* quoted Glenn Rice, CEO of Bridge Pharmaceuticals, as noting that biotech is a "zero-defects game." "Zero defects meant that a faulty drug can have deadly consequences, whereas a faulty computer part is often just annoying." In the past this discouraged Big Pharma from offshoring. But "the barriers to offshoring in life sciences are starting to crumble," *The Scientist* continued, as Asian countries in particular begin building EU- and U.S.-approved facilities, and start better educating students. Indeed, this is happening across the life sciences to such a degree that Cambridge Healthtech Advisors in 2004 predicted that, by 2008, the proportion of Western biotech companies that will outsource at least 80 percent of their manufacturing will double.

Given that China's gross domestic product grew 7 percent from 1998 to 2002, China stands to gain.

China also has a massive pool of educated expertise. There were 300,000 Chinese students overseas in 2004, one-third of whom were in biotech. After China began offering incentives to return, 18,000 students in all disciplines did so in 2002, compared to half that in 2000.

Finally, say analysts, the obvious: China has lots of patients. To find scores of patients with particular profiles for clinical trials will not be difficult. "Singapore, Taiwan, and Japan will (not) be successful if they don't take advantage of the patient population in China," Shanghai Genomics executive Jun Wu told *Nature Biotechnology* in December 2004. And because there is such poverty in China, there are vast numbers of "clean" patients untainted by past therapies. China has hordes of "true placebo" patients, noted yet another March 2004 *Nature* article on the subject.

Still, the money and the ambition are fairly recent phenomena. Most of the "shadow firsts" that China has been scoring in the area of stem cells have not been earth-shattering. In January 1998, for example, 11 months before Jamie Thomson's hES cell report would come out, Shu Nong Li of Sun Yat Sen University reported in his university's journal that he had isolated the hES cell the year before.

"I saw their data. It looked pretty good to me," says Bruce Lahn, a University of Chicago geneticist. "He got the cell, its morphology [visual appearance] looked good, and it formed teratomas. He passaged [multiplied] it for more than two or three generations. But he didn't use the same rigorous characterizations as Thomson's group." Sun Yat Sen University researcher Andy Peng, who worked in Li's lab, noted the reason for Li's characterization woes was a lack of proper reagents—historically a common problem in China. "At the time, he couldn't get all the right antibodies [to identify ES cells], so he just marked out AKT, not SSEA 1,3,4 [ES-cell-specific markers]. So it wasn't a complete work."

Equally unfortunately, later there was an accident, says Lahn. "During a vacation, everyone left the lab and the liquid nitrogen tank dried up, so they lost the line they published along with a bunch of other cells. Everything in that freezer was destroyed. This is sort of the half-assed Chinese style, you know, nobody cares and nobody even knows and then when it happens it's like, 'Oops.'"

Then there is Hui Zhen Sheng of Shanghai Second Medical University, who shocked people when she reported in a 2003 Chinese journal that she made hES cells from cloned human/rabbit oocytes. The goal was to solve the human egg shortage. But top Western journals rejected the paper. "If she had published while she was at NIH [her former employer], it would have made a big journal," Yang contends. "In some cases, it is a matter of trust." That is, the West still doesn't trust the science of the East.

The issue may be more complicated. In October 2003, Sir George Radda, then head of the U.K.'s stem cell funding body, expressed alarm over the fact that the experiment had been done. "If ethical issues are not controlled over there, it might derail the efforts of other countries," he noted. And Douglas Melton of Harvard University noted that Sheng's cells did not last as long as hES cells, and may not form all tissues. Melton welcomed the advance, however, as did many other Western researchers who retain great respect for the former NIH researcher.

Yang admitted, in a March 2004 *Nature* article, that science in China overall had to change, if the nation wanted to lead in biotech. "China has probably the most liberal environment for embryo research in the world," he said. "In addition, the relatively easy access to human material, including embryonic and fetal tissues, in China is a huge advantage for researchers." But, Yang noted, while China developed the technology to produce transgenic animals and pharmaceutical proteins years back, it had not as of early 2004 commercialized the work, despite the fact that drug development takes only 5 years in China, compared to the United States's 15. As of 2001, it spent only $12.5 billion or 1.1 percent of its GDP on R&D, compared to 3 percent, 2.8 percent, and 2.7 percent for Japan, the United States, and South Korea. Rules and regulations were often not followed, and not enforced. There were not enough institutional review boards (IRBs) in hospitals, nor enough central regulatory agencies.

Most importantly, Yang noted, China had not taken proper advantage of the fact that "collaborations with China are becoming very attractive to researchers based in the West." This is critical, Yang noted, for fostering trust at Western journals and educating Chinese scientists in the ways of Western patents, collaborations, protocols, "principles and attitudes."

Still, throughout 2004, the government continued to pour money into stem cells, and continued to listen and change. One result: a *Pipeline China* report published in October 2004 noted that China's IP (intellectual property) reform, hiked government spending, and the shorter and cheaper drug approval process—as low as $120 million compared to the United States's $800 million—had spurred the creation of 139 new drugs in China, an impressive 60 of which were biologics. (Although this does drag behind South Korea, given China's size: as of December 2004, South Korea's 40 pharmaceutical companies had 130 new drugs in clinical trial phases I and II, says *Nature Biotechnology.*)

The number of *hai gui* (Chinese returnees) began to increase, and there was a 60 percent hike in senior scientists, said Yang. In September 2004 the same U.K. delegation that drooled over South Korea was awed by China. "The amount of money going into Chinese stem cell science is amazing," said Global Watch mission delegate and King's College London hES cell researcher Stephen Minger in March 2005. "The level of infrastructure they have is mind-boggling. Most places had a large number of enthusiastic and motivated PhD students. Even Singapore didn't seem impressive after we saw China."

The center that perhaps left the deepest impression on the U.K. crew—which included scientists from GlaxoSmithKline and Pfizer— was Beijing's Institute of Zoology. It is run by Qi Zhou, who has also been courting Yang. Zhou helped clone the first rat when he worked in the lab of French researcher Jean-Paul Renard (who also cloned the first rabbit). Zhou's animal ES and cloning institute claims "a 70 percent blastocyst rate in mice [the ES cell phase], ten times more than anyone in the world. He's superb," Minger said.

At first, the U.K. crew was unimpressed. "We went to the most run-down ratty labs I've ever seen. This was our first lab and I was thinking, 'Yeah, this is exactly what I expected.' Then the French professor who was showing us around said, 'Let me show you the other side of the hall.' There were glass-enclosed labs within crappy rooms. She said, 'We were going to completely refurbish our labs but there were problems with asbestos, and we're moving into brand new labs in six months anyway. So we just built a glass box inside.' It was a world-class cloning lab in there."

Zhou's lab had 20 staffers and collaborations with about 100 more scientists including Yang. It had seven micromanipulators (compared to

Yang's three). Zhou's people are working on interspecies and panda cloning, hES cells, and human fetal cells. It published a cover story in the respected *Human Reproduction* in 2004 on human fetal epidermal cells. The state has made a substantial investment.

"That group is absolutely way out there in terms of efficiency of cloning," says Minger. Indeed, a student from this lab went on to Peter Mombaert's lab at Rockefeller University and helped him clone a mouse from terminally differentiated neurons, which became a seminal *Nature* article in 2004.

Another lab that favorably impressed the U.K. delegation was that of Lin Song Li, who, with Hui Zhen Sheng, is receiving the largest state grant. Li came back to China in 2002, having trained at Stanford and Washington universities. His Stem Cell Research Center at Beijing University has over 50 people, and an associated multi-million-dollar company, SinoCells. He has derived four hES cell lines and hired a U.K. human fetal neural stem cell expert. In January, the government began offering $1 and $4 million grants to European Union scientists to collaborate with Li's group on their hES cells.

The lab "has everything you would ever want," says Minger. "I expected it to be 1950s science and it was anything but. Three guys from Big Pharma were with us, including Glaxo and Novartis. These guys don't get wowed by anything. You sort of look at what their basic kit is and it makes my lab look pitiful. These guys are used to seeing big massive displays of equipment—and *they* were wowed. It was as good a lab as any I've seen in the world. And that was repeated everywhere we went."

This included the lab of Hui Zhen Sheng, the rabbit/human cloner. Besides giving her the other top state grant, the government also began offering, in January 2005, $1 to $4 million grants to European Union researchers to work with her on cloning human cells. Her lab has 28 scientists and 32 students. She is working on human fetal muscle stem cells and hES cells, of which she has derived six of her own lines. A focus is Duchenne muscular dystrophy. She has a small associated company.

Then there is Guang Xiu Lu, the scientist with the best story. Lu is the scientist whose postdoc scored the cattle coup for Yang. She is the daughter of Hui Lin Lu, who was a leading Chinese geneticist until the 1960s when the Cultural Revolution put a halt to his work, genetics having been declared bourgeois. Gang Xiu Lu had trained to be a surgeon, but was unable to find work when she graduated. After the Cultural Revolution ended, in 1981, she started one of China's first IVF clinics. In the late 1990s, she began trying to derive her own hES cells from fresh eggs and to clone human cells, before most any lab in the world. She succeeded in at least cloning cells to the blastocyst stage in 2000, although she only published the results in an obscure Chinese journal. She has created at least four hES cell lines.

Her huge lab is located in the remote town of Changsha, where she stayed after graduating med school to take care of her ailing father. By the end of the 1990s, she was able to expand her clinic. Her revenue rose to about $2 million a year from just $50,000 in 1995, according to *The Wall Street Journal,* which also reported that a company, Citic, bought a $3 million majority stake in her clinic/lab.

After she became the first Chinese cloner to make international headlines, in March 2002, funding from regional and industrial sources flowed. She has not published in any top Western journals, but her institute, with over 100 people, qualifies as perhaps the largest hES/cloning lab in China. "No one [else] has a staff of over 100," Yang notes.

The irony is clear. Lu's family once suffered under the Chinese clampdown on genetics work. Now she benefits from the U.S.'s clampdown on hES cell science.

Then there is Alex Zhang, who did his training at Northwestern and Stanford universities and returned to China in 2001 to head up the Cell Therapy Center at Xuan Wu Hospital in Beijing. That someone so young should be given his own 12-person lab is unusual enough. But Zhang has also struck up collaborations with top scientists at Johns Hopkins, Sloan Kettering, Stanford, and the University of Wisconsin. This is partly due to his access to fetal human cells. But it is also due to yet another advantage that China has over the United States: cheap access to primates for testing. Zhang has a relationship with a center that has 30,000 monkeys. His lab is an expensive "clean lab" built according to good lab practice (GLP) standards. This means his cells can be used in pilot clinical trials. He has been making human fetal neural stem cells for Parkinson's disease and human fetal islet stem cells for diabetes.

"In the next one or two years, I think my U.S. collaborators will be coming to do work here," says Zhang. At first they will come because "our monkeys are a big plus for our collaborators." But later, they will come for clinical trials. "I think most of my U.S. collaborators would like to see pilot or parallel clinical trials in China." This is not just because of the more permissive environment. "China is obviously a big market for industry. If stem cells become an industry, China will be big in stem cells."

Zhang understands that he is attractive to such high-profile researchers because of his Western education. "I think overall the standard of research is not quite as high as the United States here yet," he says. "I am more accepted by my colleagues, I think. This is partly because, while most Chinese scientists received part of their training in the West, I received all of it there. I am blessed."

Among other things, Zhang is part of a $3 million state grant headed by Hui Zhen Sheng. Zhang says that's big money: "We don't pay for our postdocs. Their salaries are paid by the institutions. A big plus for research." Yang agrees: "Some 80 to 90 percent of NIH grant money goes toward salaries and overhead. In China, it goes to the scientist and the science. So $4 million is more like $40 million."

Still, Zhang says, there are problems money—and monkeys—can't solve, only time. "It's a different environment. Here the students just listen to the supervisor, and the supervisor is supposed to tell the student to do step one, step two, step three. In the United States, students are self-motivated and rigorous in terms of thinking. 'This control or that?' 'Should I consult with this person?' That is lacking here. It takes a while to cultivate such an environment. I've traveled a lot in China and it is still not there. I now tell my students: in order to graduate you have to point out my mistakes. Then I will let you go." Poo agreed in his March 2004 *Nature* article. "A habit of raising questions needs to be fostered."

Yet China has a head start in a more notorious area that frightens, intrigues, and frustrates Western scientists and clinicians. Partly because hospital IRBs and ethics committees have been long in coming to the nation, clinicians of all kinds have been trying all kinds of so-called stem cells on their patients without using globally accepted clinical trial guidelines.

But some have been following the rules. The U.K. delegation met with what it considered one of the best of the stem cell clinicians: Jian Hong Zhu, who once did work on angiogenesis with the famous Harvard oncologist Judah Folkman and did some postdoc training with former Harvard stem cell neurologist Evan Snyder. Zhu, the delegation reported, "is performing groundbreaking research into the autologous application of adult neural stem cells in traumatic brain injury." Zhu works at the largest brain injury hospital in China, Huashan Hospital, which is associated with the prestigious Fudan University in Shanghai. He was presented a few years ago with a patient who got into a fight in a restaurant and ended up with chopsticks stuck in his brain. ("It transpires that chopsticks are sometimes utilized in Shanghai restaurants to culminate otherwise unresolvable differences of opinion," the report noted.) Seeing he had a rare opportunity—since neural stem cells are in short supply—he took the brain matter oozing out of the patient's brain, expanded its neural stem cells in a petri dish, and returned them. The patient recovered his ability to walk after a year. When the U.K. group visited in September 2004, Zhu had treated 11 patients with adult neural stem cells and compared them to 11 controls. The treated patients supposedly fared far better than the controls (although as of the fall of 2005 no paper was yet published in a Western journal). In the West, it is much harder for doctors to just up and try new procedures.

Zhu pulled off another coup when he magnetically labeled, with superparamagnetic nanoparticles, a patient's cells. He was able to show where the stem cells went in the brain of the patient, said the U.K. delegation.

Zhu had yet another ace up his sleeve: easy access to monkeys. "These clinical studies build on a series of early studies performed with adult monkeys with NSCs derived from various brain tissue sources," the U.K. report said. "Professor Zhu has conducted extensive preclinical studies in mouse and monkey brains with expanded adult NSCs and has measured proliferation and neuronal differentiation, including patch-clamped records of GFP-labeled cells in monkey slices. Like others, he finds that the frequency of neuronal differentiation is low and that most cells (over 95%) differentiate into [generally undesired] astrocytes. He finds, however, that some neurons show mature synaptic properties, and EM has revealed synaptic connectivity, so functional integration is clearly possible. To our knowledge, Professor Zhu is the first clinician to transplant autologous adult human NSCs as a therapy in traumatic brain injury patients."

Minger, a neurologist, was transported, noting that Zhu has carefully followed his patients for 2½ years, using Western standards, PET scans, and MRIs, and has submitted a paper to Western journals. "He is doing world-class clinical research. He is the only guy in the world doing adult neural stem cell transplants in humans worldwide and nobody knew anything about it. The work is ongoing but very, very cautiously. You can only expand those cells in about one-third of the cases. It's amazing."

Zhu has been able to cash in on many of the advantages offered by China, Minger notes, including its size. Injuries with brain matter outside the skull, so allowing NSC extraction, are rare. But "he is in a huge neurosurgery clinic treating 7,000 patients a year. So he sees more than others."

Many other clinicians have been doing some similar things in China—indeed, around the world—bypassing standard clinical trial rules. Unlike Zhu, however, far too many of those have ignored a huge slew of established global procedures that would make their work repeatable, analyzable, acceptable. Minger has not been the only one watching. Wise Young, who was once dubbed one of the best scientists in the world, has.

And in the fall of 2004 he decided to do something about it, launching the China SCI Network, which is teaching top neurologists how to conduct proper clinical trials on spinal cord injury patients in exchange for data on scores of hitherto secretive trials. It is an unprecedented project.

One of the go-getters he has signed up is Andy Peng. Peng worked in the lab of the first Chinese to isolate hES cells. He moved on to the Chicago genetics lab of Bruce Lahn, then came back to Sun Yat Sen University to head up his own lab—Stem Cell Biology and Tissue Engineering—which has its *own* huge monkey facility. There are 20 scientists associated with the center, and 10 PhD students. Peng has isolated two hES cell lines and is working on human fetal neural stem cells and MSCs. Shu Nong Li, Peng's former boss who isolated China's first hES cells, is working for him. The lab has $2 million.

Other labs at Sun Yat Sen have done numerous clinical trials with stem cells. Shao Ling Huang, who was a student in the lab of Shu Nong Li when he isolated China's first hES cells in January 1998, has tried

MSCs coupled with cord blood stem cells on leukemic children. "He got pretty good results. I have his data," Peng said. "Usually recovery from cord blood stem cell transplantation is three months. With MSCs, it was one month. Of the 13 kids treated, 9 were good, 4 died—and only 2 of those from the procedure, rejection. He is doing more clinical trials for this." Huang used both human adult and fetal MSCs. The results were similar, Peng said. "This isn't necessarily surprising. Fetal NSCs are better than adult because they proliferate better. But adult MSCs proliferate pretty well."

On six to nine SCI patients, Shao Ling Huang has also tried human fetal OEG (olfactory ensheathing cells, which wrap around axons and may help them grow; axons transmit impulses between nerves and muscles). And he will try fetal OEG *plus* fetal neural stem cells—a cocktail Wise Young is dying to try—on humans when he is done with his primate trials, Peng said.

Another scientist at Sun Yat Sen, Cheng Zhang, gave a cord blood stem cell transplant to a patient with Duchenne muscular dystrophy in the fall of 2004. "He got good results, I think he will do more, as well." (This was published in a Chinese journal.)

Many labs that have already conducted unusual stem cell clinical trials have been drawn into Wise Young's vast network. Young is the first U.S. scientist to do something about ever-growing reports of Western patients giving up on the United States and heading for untried, unregulated "stem cell" clinics abroad.

Those clinics are everywhere—China, Russia, the Dominican Republic, Barbados, Ukraine—and have been worrying clinicians everywhere. They have also been worrying basic scientists who fear that unregulated clinics will do so much failing they will turn the public against legitimate stem cells. The phenomenon seems to have started in Ukraine and Russia in the early 1990s, exploding after the 1998 isolation of the legitimate hES cell. And it has not let up, for some bad reasons and for some utterly heartbreaking, utterly understandable reasons.

        •   •   •

"HIS BREATH smells like a baby being born after he gets them," says Keansburg, New Jersey, resident Sherry Ashmore serenely, referring to the $15,000 "embryonic stem cells" that her son Ricky receives every six months.

It is June of 2001. Sherry takes a drag of her cigarette. With her pink nails, pink T-shirt, jeans, and spiked high-heels, she looks out of place on the deserted boardwalk near the Asbury Park convention hall. The empty beach is garbage-strewn, lined with boarded-up stores, and towered over by a grim array of Soviet-style high-rise brick tenements with "For Rent" signs in many windows. A parade of "No Swimming" signs lines the beach.

Still, Ashmore does belong here, and not just because a loud, rowdy, good-hearted New Jersey fund-raiser is being thrown in the hall for her 10-year-old son, who may die of Duchenne muscular dystrophy in his late teens. For this run-down place strikingly resembles one thousands of miles away to which she has traveled five times, over two years, for her son's treatment. While the U.S. government has debated the ethics of hES cells, Ashmore's son, like thousands worldwide, has been quietly receiving some uncharacterized version of them in what is, by Western standards, an unregulated clinic in poverty-stricken Ukraine.

Behind Sherry in the dark hall, a local singer in spangles is singing "crazy for feeling so blue." The microphone is too loud, making the voice swell over the band and echo tinnily down the deserted beach.

Each time she takes her son to Ukraine, Sherry says, so-called hES cells are pulsed into his body via an IV for 1½ hours. Then what she has been told are human fetal myoblasts—fetal stem cells that form muscle—are injected into his buttocks, abdomen, back muscles.

"Before the procedures," she says, "he's not able to rise from a sitting position. He falls three times a day. He has trouble with stairs. After, he can get up by himself and only falls a few times a month. He's much better at climbing stairs." She explains that the Ukrainian doctors say he needs to return every few months because he's still growing, and the cells "don't compensate for growing. . . . But after his injections, it's like a miracle. We must spend $300 calling home after each one. In the beginning he'd cry a little because at first, it hurt. But now he doesn't get upset no more."

The band has left the stage, and has been replaced by a squad of karate students in shiny red robes. They begin doing somersaults to a disco tune.

"He has better concentration after, too. His physical therapist, who works with him two times a week, says she's seen a major difference. This time the treatment wore off after three months. . . ." Sherry expels her cigarette smoke. "But usually, it's five or six. They treat a lot of different diseases, AIDS, diabetes. They've been curing diabetics to the point where they don't even need insulin, I was told."

When the Ashmores first saw Emcell, the Ukranian clinic, its appearance frightened them. "It looked like that abandoned hotel over there," she says, pointing to a tall, worn brick building behind her with two elaborate crumbling concrete turrets, like the bride and groom on a huge wedding cake. "I didn't know what I was going to do. Even the elevator was old." But on the seventh floor, all of which is occupied by the clinic, she was reassured: it looked like a hospital floor. There were computers. There were IVs.

The fact that the two clinic chiefs met her at the airport and took her to dinner helped, as well. "Professor Smikodoub saved the life of Professor Karpenko's mother-in-law with the cells, it brought them close together," she says. "They walk their dogs together in the park." The treatments, she says, cost $10,000 each; the flights, $7,000 for the family each trip. They are paid for by bashes like the one behind her, which kind townspeople throw every few months for Ricky.

That Ricky received cells both via IV and injections directly into muscle makes some sense to U.S. researchers. Why not optimize the chance for the cells to reach as many damaged muscles as possible? But other parts of the story raise red flags. First, if they are indeed hES cells, they can go cancerous if not first differentiated. It is likely Sherry is wrong and the cells were differentiated. But it is impossible to know, because the clinic has published no articles in international journals.

There is another problem. That Ricky's embryonic cells are only briefly working probably has nothing to do with his growth and everything to do with the fact that they are being rejected, Western doctors say. As noted often in this book, only 1 in 40,000 people are tissue matches, which is why most organ transplant patients must take crippling, lifelong regimens of immunosuppressives. (The exceptions include kidney transplant patients at the MGH program described in an earlier chapter—and perhaps, people with brain or spinal cord disorders. The latter *may* accept unmatched cells without immunosuppressives because the brain and spinal cord are protected to some degree by the blood/brain barrier. But no one knows to what degree, and many clinicians are starting to believe the protection is minimal.) Regardless, Ricky's problem is outside the barrier. Immunosuppressives would likely be critical to keep Emcell's cells from being rejected. Growth factors emitted by the cells could help for a short time before rejection. But it is unlikely they can survive without immunosuppressives, and he isn't on any.

A voice booms, "Ricky Ashmore!" Sherry rushes into the darkened auditorium. She stands next to a man and boy who look exactly alike. Each is big (in one case, for his age) and wears baggy pants, sneakers, and a T-shirt; each has a long slender nose, an open face and restless, dirty-blond hair. Each sits with his hands in his lap, waiting. The two Rickys. The main difference between them seems to lie, at first, in the T-shirts with which they express themselves. Dad's is an impassive black; the son's is green with frogs on it.

But when they are called again, Ricky Sr. bolts to his feet, while Ricky Jr. grabs the seat in front of him and slowly pulls himself to a hunched standing position. Ricky Sr. takes a giant step; Ricky Jr. takes several tiny shuffling steps to catch up. The second difference between the two is brutally apparent. The 10-year-old has a body that is far older than his father's. His leg muscles began wasting years ago. Next up will be his shoulders and neck. He will probably die by age 25, the muscles in his lungs having stopped being able to pump.

It is easy to understand why the Ashmores travel to Ukraine for their "magical" treatment, no matter whether it's ready for scientific primetime. They have no other options. The Ukrainian clinic has published no legitimate articles on the cells, despite having done a claimed 1,500 transplants in 10 years. (That figure will rise to over 2,000 by 2005.) Doctors from coast to U.S. coast warn that without internationally respected papers, there is no way to tell if the therapy works or, indeed, is dangerous. An ALS association will contact Emcell and report there is no way to tell what kinds of cells are being used, why, or how they are being used. But the medical establishment has no options for the Ashmores. Drugs can't help, and hES cell research is moving too slowly in the United States. So the Ashmores will continue to throw town parties, continue to bring Ricky to Ukraine.

Emcell has been giving uncharacterized, unknown human "embry-onic" cells to desperate patients worldwide since the mid 1990s. But it was not the first clinic to do so. In the early 1900s, embryonic animal cells were big in Europe. The practice died out when it became appar-ent that untreated animal cells can bring viruses and anaphylactic shock.

Then, in the early 1990s, a group of Russian doctors got into the act. A discredited U.S. cosmetologist named Michael Molnar, along with Russ-ian researcher Gennady Sukhikh, started a fetal tissue clinic called the International Institute of Biological Medicine in Moscow. This clinic tried uncharacterized (by Western standards) human fetal brain, pan-creas, ovary, and blood cells on children with Down syndrome and on adults with diabetes, Alzheimer's, and sterility, among other disorders.

Fetal tissue therapy in the formal clinical trial setting was being done on a limited basis in some Western nations at the time. In 1993 President Bill Clinton lifted a U.S. federal-funding ban on fetal tissue therapy, and Sweden was trying the therapy on a few patients with Parkinson's disease. (Fetal tissue is generally described as tissue com-ing from aborted fetuses, between 4 and 12 weeks. By contrast, again, hES cells are five days old.) Well over 200 Parkinson's patients were treated worldwide, in the formal Western clinical trial setting, with fetal neural tissue. While Harvard University neurologist Ole Isaacson, in 1999, called the Western work "the most dramatic conceptual advance in neurology since the 1950s," he conceded that the approach "may be superceded by the use of stem cells" for several reasons. Indeed, the field has concluded that placing fetal tissue into a brain is not enough. Fetal tissue is in many ways mature already, and some-times cannot respond properly to chemical cries for help in the adult brain. It is not uniform, thus not reliable. And many find using older tissue like that objectionable.

Then in 1998, with Jamie Thomson's discovery of the five-day-old hES cell, many Western scientists decided *this* was the best way to go. First, U.S. hES cells come from spare IVF-clinic embryos destined to be thrown away and that stats say would never have become a fetus. (Some 70 percent of human embryos naturally never make it to birth, as noted.) Second, five-day-old hES cells can proliferate endlessly, unlike older fetal cells, so scientists don't have to keep seeking more cells and can generate standardized doses. Third, hES cells are more potent, pliable, and responsive. Fourth, hES cells do not come from abortions. And finally, an hES cell cannot form a baby if placed in a womb.

Ethically, politically, and scientifically, hES cells seemed superior, although some highly limited, careful, regulated study continued in the West with fetal tissue.

Regardless, the Russian clinic's cells and methods were very different from those of the West when it came to both fetal tissue and hES cells. As reported by several news outlets in the mid 1990s, Molnar's license to practice plastic surgery was suspended in several U.S. states. The cells came from older fetuses, four to five months old: viable. The group did not or was unable to publish in Western journals, so zero methods were held up to international peer review—especially meaningful since various Russian groups apparently tried human fetal cells on a few thousand patients even before the 1990s. The institute's controversial fetal activities supposedly halted by the late 1990s, presumably due to the negative Western press.

But bizarrely, the institute has resurfaced as part of Russia's recent, supposedly more "legitimate," fetal and embryonic stem cell effort. In June 2000, doctors respected inside Russia launched a national stem cell effort with government backing, largely focused on human fetal neural stem cells. The official Web site proclaiming the launch of this new national effort named the once-discredited Institute of Biological Medicine as a collaborator whose human fetal work it also cited as a basis for the new effort. Andrei Bruhovetsky, a Russian Academy of Sciences doctor in charge of the effort, transplanted human fetal brain cells into spinal cord patients for a while, then switched to adult bone marrow stem cell transplants at $5,000 to $50,000 a pop. He has yet to publish in a respected international journal. And the former head of the institute during its controversial early days, Gennady Sukhikh, works with more established Russian institutes today even as he continues peddling human fetal cells to patients for thousands of dollars.

Samuil S. Rabinovich, with the Institute of Clinical Immunology in Novosibirsk (Siberia), is implanting human fetal OEG stem cells into spinal cord patients. He too cites the controversial Institute of Biological Medicine in his work. Yet the Siberian clinic *has* demonstrated some promise. It has published at least one paper in a nonlocal journal—

reporting on 15 patients in its fetal neural stem cell trial for spinal cord injury patients. And it is open to questions. Victor Seledtsov, a doctor with the clinic, noted in correspondence in November 2004 that the clinic had treated 100 patients with neurological disorders, and 90 spinal cord injury patients, with human neural stem cells, seeing "very encouraging results." He said the cells are "effective in treating SCI, cerebral palsy, consequences of traumatic brain injuries and consequences of neuroinfections."

But Seledtsov admitted the group does not characterize the cells fully. When asked for a precise description of the cells—since this did not appear in the group's 2003 *Biomedicine and Pharmacotherapy* paper, an unusual thing—he said "where reasonable, we characterize the cells by flow cytometry (CD34, CD38 and others). . . . To my opinion, the transplantation of complex of different cell type should be more clinically effective than the transplantation of single cell type." That may be so, but if the implanted cells are not characterized, it is nearly impossible to repeat the procedure and all but eliminates the possibility of anyone verifying your claims.

Western scientists say the paper doesn't give them an idea whether, or how, the cells are working. Rutgers's Wise Young has read it, and even had Russian papers cited in it translated for him. "I don't know what they're transplanting There's no question they are fetal cells, but whether fetal precursor or embryonic or astrocytes, I don't know. The markers they use are not the best, and they haven't fully characterized the cells. The only proof of pluripotency is demonstrating it in tissue culture and transplanting to animals. I don't think they've done that."

Still, the above is at least legal in Russia—and Bruhovetsky will be meeting with Young for a serious talk. New, *unregulated* clinics have sprouted up all over Moscow. In March 2005 the Associated Press reported that hundreds of wealthy patients were rushing to unlicensed fetal cell (which called themselves "stem cell") salons for "beauty" treatments because the salons interpreted Moscow laws as allowing extraction and storage of human cells, but not yet explicitly outlawing commercial treatment with them. Andrei Yuriyev, deputy head of the Federal Health Care and Social Development Inspection Service, has been investigating some 20 cosmetic clinics offering the cells. One, like "Beauty Plaza," a private clinic run by a Dr. Alexander Teplyashin, was

advertising fat and bone marrow "stem cell" injections for many diseases. "We are taking advantage of the loopholes in the law," Teplyashin
said. "What is not forbidden is allowed." A clinic called Cellulait, run by
a Dr. Roman Knyazev in central Moscow, was advertising $2,850 injec-
tions of aborted fetus stem cells into thighs, buttocks, and the stomach
to fight cellulite. Other clinics were charging as much as $20,000. Fur-
thermore, there have been charges that some of these salons were pay-
ing poor women to have abortions in order to gather up their fetal cells.
In April, the Ministry of Health announced that 37 out of 41 clinics
offering "stem cell" treatments in Moscow were acting illegally. (Most
stayed in business anyway.)

But it is the Ukrainian clinics that may ultimately end up wielding
the biggest stick in the so-called underground stem cell world. For the
launch of Emcell and the Institute for Problems of Cryobiology and
Cryomedicine (which has been freezing human tissue since 1972 and is
run by Valentin Grischenko) eventually led to the start of perhaps the
largest underground "stem cell" clinic in the Southern hemisphere.
Called Medra, the clinic is located in the Dominican Republic. It was
started by an American, William Rader. The clinic claims to offer fetal
and embryonic stem cells and has historically required patients to wire
$25,000 to a Swiss Bank account before the clinic will even discuss
their cases by phone. Rader was known in the early 1990s as a therapist
running a chain of anorexia clinics. Formerly married to Sally Struthers,
he once was a cowriter on an *All in the Family* script. In the mid 1990s
he was a reporter for the Christian TV show *Lifestyle*.

In 1997, after observing work at the Ukraine's Emcell clinic, Rader
started his first fetal clinic in the Bahamas with help from a U.S.
celebrity or two, using imported cells.

More than a year before Thomson would publish his discovery of the
"true" hES cell, Rader hit up Kristina Kiehl Friedman, wife of Levi jeans
family member Robert Friedman, to help establish the clinic, according
to a former researcher for Friedman. Friedman was cofounder of Voters
for Choice with Gloria Steinem. (Steinem had nothing to do with Rader,
Friedman would later testify.) Using the line that pro-lifers would never
allow such cells for therapy in the United States, Rader persuaded
Friedman to come up with some funding and patients. For the next
three years, Rader delivered some form of human fetal cells into the

bloodstream of 400 patients with a variety of disorders, he claimed. He was closed down in 2000 by the Bahamas Ministry of Health, but reopened shortly after in the Dominican Republic, and claims to have opened more clinics in Prague and Mexico.

Rader is soldiering on. He has published nothing on stem cells and plans to publish nothing, he said in 2005, because a "conspiracy" would build against him if he did. He has claimed to have cured Alzheimer's. As of early 2005, he was charging $30,000 a procedure. His ads appeared on legitimate stem cell Web sites as late as the fall of 2006.

Year after year clinics that are unregulated, by Western standards, multiply. The Ukranian Institute of Cryobiology mentioned above, which was launched in 1972, teamed up in 2005 with U.S. investors and Caribbean tourism officials to set up an Institute of Regenerative Medicine in Barbados. It is run by Yuliy Baltaytis, a Ukrainian physician who had collaborated with Rader in Europe and the Bahamas, and also charges $25,000 per injection, according to *The Los Angeles Times*.

It offers therapy using Ukranian "stem cells" from 6- to 12-week old aborted fetuses on patients with autoimmune diseases, diabetes, and neurological disorders for $25,000. It plans a more formal clinical trial of stem cell therapy for congestive heart therapy in association with Queen Elizabeth Hospital, the main hospital on the island. Ukraine Cryobiology Institute head Grishchenko told U.K. reporters in early 2005 that 20 years of clinical experience indicated the therapy was safe and effective. He claimed 1,740 patients had been treated in Ukraine for diseases ranging from diabetes to Parkinson's, and that "significant improvements" were seen in 68 percent; "partial improvements" in 28 percent; no improvement in 4 percent. Once again, they were not formal clinical trials, and no papers were published in globally respected journals.

It is shocking that so many doctors, in so many different countries, bypass accepted standards in patient care, "advancing nothing," as British neurologist Geoffrey Raisman told MIT *Technology Review* with reference to one such trial.

Yet also shocking is the plight of the Ricky Ashmores. With the help of Keansburg, New Jersey, Ricky Ashmore went to the Emcell clinic three more times, for eight visits total. But by the last visit, he was only looking and feeling better for three to four weeks after treatment. And none of the visits spared him: he ended up in a wheelchair shortly after the Asbury Park fund-raiser in 2001. This was right on schedule, given that most Duchenne boys end up in wheelchairs by age 12. "That's when I gave up," Sherry Ashmore will note in October 2005. Emcell people "still kept talking about miracles. They didn't acknowledge what wasn't happening." Emcell never told Sherry that her son was probably rejecting the cells even though, in the Ukraine press, one of the doctors would state this generally seemed to be happening. "If I had thought it was doing anything lasting I would have kept doing whatever it took. But it wasn't." Still, Sherry recommends it for others. "I'd do it again for him, if I could afford it. If you have the money, three to four weeks of something is way better than nothing."

Point being: nothing is what Ricky has now. By October 2005 he will have been in a wheelchair for years. There will still be no drugs for him. Some respected researchers are finding that bone marrow stem cells can do tantalizingly good things for mice with muscular dystrophy. But cautious Western doctors aren't trying them on Duchenne patients yet, and Ricky is running out of time. He is 14.

Cue Wise Young.

●　　●　　●

"IN 2004, THE U.S. spent less than $230 million on all human stem cell research, embryonic and adult. That same year, Singapore spent $1 billion. . . . We do not have a clinical trial network to support our 250,000 spinal cord injury patients nor our 8 million Alzheimer's patients. We do not have a cell source that is feasible in the U.S. We are facing a moral catastrophe."

When the diminutive powerhouse Wise Young speaks, he does so with his entire body. His diction is pointed and perfect. He makes chopping motions on the podium one minute with both hands; the next, he pinwheels his hands in the air.

But unlike the average human, he also utilizes something close to his entire mind, both its calculating and emotional halves. On March 28, 2005, when he climbs onstage at the New York Biotechnology Association meeting, he cuts to the chase right away, unhampered by the average scientist's inability to communicate from a podium without addendums, footnotes, qualifications. With Stanford University's Irv Weissman and the NIH's Ron McKay, Young would appear to be one the field's most gifted speakers.

"There are only 40,000 to 50,000 units of bone marrow in the registries here," he continues. "And we certainly don't have the ability to scale up bone marrow cells. Yet embryonic stem cells indefinitely expand. . . ." And the United States has severely curtailed their study. "It is absolutely crazy."

Young begins chopping at the air, as if he is being attacked by demons or very large invisible mosquitoes. Furthermore, he says, as a result of the 2001 Bush policy, huge numbers of spare IVF-clinic embryos have been and are being thrown away. "Wasted." He fairly hurls the word off the podium.

Then Young goes on to note that some babies are resistant to AIDS. They have provided only 50 to 60 units of blood. But if their cord blood stem cells could be expanded 100-fold—which no one can do right now—then huge numbers of scientists would have access to the cells to see if they could figure out what makes them resistant. "Yet $30 million in umbilical cord blood research funding was cut out of the recent budget by Bush," Young says.

Much study of both embryonic and adult stem cells is needed, he continues. "There is an orchestral ballet of signals that lead these cells. We need to learn them. We need to embrace stem cell therapy, invest in it, or we face a moral catastrophe of enormous proportions."

After the talk, Young is surrounded by people asking him to come speak to their schools and groups. Like his former patient Christopher Reeve, he is in need of a few Superman skills. Young is founder of New Jersey's $100 million stem cell research institute, head of a Rutgers University neuroscience stem cell lab, founder of the International Neurotrauma Society and the *Journal of Neural Trauma;* founder of a Web site for SCI patients that gets thousands of hits a day, and a leader in a movement to establish an Eastern North America stem cell initiative with Canada. And in the fall, Young started a clinical trial network for SCI scientists and doctors, including stem cell personnel, in China. He will train clinicians in China how to conduct clinical trials in the internationally accepted manner, and they will share their data on their trials, including stem cell trials. The idea is to at last standardize Chinese clinical trial work, at least in the spinal cord injury area, and collect data on the feasibility of approaches that aren't being studied in the United States, including fetal stem cell approaches.

Of all of his projects, this last is clearly his baby.

"I was absolutely shocked," Young said in a December 2004 interview. "I went to 10 institutes in China in the last three weeks to start an SCI clinical trial network, but every dean I met with also said, 'Would you like to see our stem cell institute?' It's hardly just clinical trials over there. They are investing heavily in research and buildings; they don't have enough manpower but are recruiting like crazy; they are hot on stem cells." Furthermore, he would note after an earlier visit, "almost every place I have visited has said it has isolated its own hES cell lines. And I have visited more than 20 hospitals."

Also to his surprise, Young said: "Many of the older professors talked to me knowingly about cloning. Ever since the first fish was cloned in China decades ago, there has been interest in cloning. I see so much effort, space and commitment in stem cell research that I'm beginning to think that in a year or two they will become a powerhouse. There is a hidden power there."

Basic-science investments aside, Young has found that "over one thousand patients in China have received stem or stemlike cells in the last five years. I am really excited about this. And I'm really excited that the first stem cell therapies [outside of cancer bone marrow transplants] to be done consistently in clinical trials will be done in China."

The idea of establishing a Western network for Chinese clinicians came to him years ago, after he kept hearing from SCI patients about unusual Chinese clinical trials. Young was born in Hong Kong in 1950, and educated at Stanford University. As a neuroscientist at New York University, he co-led a 1990 study showing that 20 percent of function can be saved in spinal cord injury patients if they are given, within eight hours of their injury, methylprednisone, which tamps destructive inflammation. "It can often be the difference between breathing unassisted or relying on a respirator, walking or spending one's life in a wheelchair," *Time* said in 2001 when it named him one of the best scientists in the world.

Still, the drug only returned 20 percent of function, and there is no network for SCI clinical trials in the United States. The reason: money.

Drugs must be proven and reproven in the United States before clinical trials are considered, Young says. And SCI has had a pall over it. Until stem cells made the news, many believed there would never be a way to get paralyzed patients moving. Even despite the stem cell hype, "only three hospitals in the U.S. have any kind of cell transplantation experience in spinal cord, and no moves whatsoever have been made to set up the infrastructure necessary to get a clinical trial going. It would cost $100 million to do a cell therapy clinical trial for 1,000 patients, $100,000 per patient."

But in China, Young noticed many things. The cost of clinical trials was minuscule. Many Chinese SCI patients couldn't get access to any good medical care at all. Also: "China is busting at the seams with money." And it seemed that stem cells of all kinds were being tried on patients with diseases of all kinds. The problem was the usual: few publications in Western journals. So Young decided to build a network and hunt down candidates for it himself.

Through 2004 into 2005 he met with doctors using stem cells. In Beijing, he met with his former protégé, who opened this chapter. Hong Yun Huang was putting human fetal OEG cells into spinal cords. The ideal therapy may be OEGs plus neural stem cells, Young believed. And Huang wasn't characterizing the cells nor following patients well. But his work was promising. Young signed him on.

Young also met with one Tian Sheng Sun at Beijing Army Hospital, who had also been placing human fetal OEG into SCI patients. He was getting some somatosensory improvements, if few motor improvements. Young signed him on, as well.

There was word the naval hospital in Beijing had given fetal NSCs to 30 patients—but no one there would confirm. Other clinics that had administered fetal NSCs to SCI patients were elusive. Still, Young was able to find many clinics that had been, or were, administering stem cells of varying kinds. In the Henan People's Provincial Hospital in Zhengzhou, a doctor named Qing Yong Zhang had injected adult BM stem cells into 180 SCI patients, and some 400 patients altogether. No papers have conclusively shown neurons from BM cells are functional. But it is possible they emit growth factors that could help damaged native neurons regrow. The results of the Zhengzhou work were not yet known, but Young was impressed. "A whole floor of the hospital is devoted to adult BM stem cells, and they're doing it for SCI, stroke and other diseases. Quite an active BM mesenchymal stem cell transplant program. A whole team was growing up the cells, and was approved by the government for clinical trials. It is all being done under good manufacturing protocols [GMPs]. So they're documenting, and it's being done well and right. . . . Worldwide, there have been four to five preclinical papers reporting that BM stems are beneficial in animals with SCI. People are trying to get funding in the United States. It's a matter of getting facilities together to grow the cells and transplant them properly. Yet at Zhengzhou they have already solved many problems. They grow the cells routinely; freeze them so they aren't tied into immediate transplantation; can schedule patients any time. And autografts avoid immune rejection and ethical problems." Young signed on the Zhengzhou hospital.

Young also signed on doctors giving fetal human Schwann cells (neuronal supporting cells) to SCI patients. But at Sun Yat Sen University, where many clinical trials have been ongoing, Young was impressed with Peng's basic-science lab; less impressed with some of the clinical work (unrelated to Peng). "Once when I asked for data on a clinical trial, they called in a resident who pulled out of his lab coat a sheet of paper. That was their data, a sheet: patient's name, cell phone number, some jotted notes. They are getting phoned reports on patients. These are the kinds of things we're seeking to change. We want to get them into the right habits. If they enroll a patient, they better track the patient all the way, or they won't get paid." Still, there were other promising things going on. He signed up the university.

Young encountered many problems, some good. "Leadership at universities were frustrated by the fact that these doctors were doing therapies with no attempt to establish their efficacy," Young said. "They really want to develop an evidence-based approach. The hospital leadership everywhere said, 'Just tell us what to do and we'll do it.' There has been a lot of competition between these hospitals. They don't trust each other. But when somebody from outside comes in, they say, 'We'll be equal partners in this network.' We signed up 10 of China's top hospitals in three weeks."

By February of 2005, that number had increased to 12. By March, 15. And non-SCI Tianjin and Zhengzhou neurosurgical teams have asked to join. "We are getting phone calls from people treating all kinds of disease. There is tremendous enthusiasm. It's almost unbelievable."

The protocol for this unprecedented network is almost a recipe for curing the problems mentioned in so many articles about Chinese bioscience. In the spring of 2005, doctors from all 15 institutes met with Young's U.S. team and were taught how to prepare SCI patients for clinical trial according to international standards. Some 600 SCI patients were eventually chosen. Sometime after, participating hospitals were to begin creating a general database about patients as they received different therapies, including stem cells. At various intervals, all would enter updates. A traveling team would visit each institute periodically to oversee.

If all goes according to plan, in 2006 Young will have something unique: accurate Western-style data on the progress of patients across China, many of those stem cell patients. After reviewing that data, the network will then launch giant Westernized clinical trials of the best therapies.

The network is receiving praise from within and outside China's borders. One fan is neuroscientist John McDonald, now at Johns Hopkins. McDonald was the first scientist to prove that brain cells created out of mouse ES cells can partially restore function in paralyzed rodents. As of the end of 2004, McDonald was collaborating with Woo Suk Hwang of South Korea. He had also begun an hES cell project with monkeys in Colombia. The short-term goal is to infuse mature cells made from hES cells into monkey spinal cord with a cannula, so major surgery isn't required. But the ultimate goal: clinical trials with hES cells and SCI patients, and McDonald may conduct the first trials in Colombia if need be. "Naturally we're building in a parallel path right? Our preference is to do our first clinical trial in the States, but if it's not doable here, we'll have a parallel path."

He is rooting for Young. "We should do everything as scientists that we can to take advantage and gain data from places like China. We shouldn't be sitting back and complaining. We should be out there saying, 'Let's gather as much data from this as we can, because this is not going to be done anywhere else.' It's interesting. Cell therapy is a bit of an unconventional approach so typical funding agencies don't want to fund this stuff here. But this is exactly what needs to be done. Money should be put into this because we will harness important information. We're crazy if we don't get that data."

●   ●   ●

BACK ON Jerry Yang's Connecticut farm, on March 30, 2005, all appears serene as usual. Out front, a sign featuring the silhouette of a giant chicken advertises the upcoming Southern New England 4-H Poultry Show. Across the street, a yellow banner reading He is Risen flutters gently in the wind, tethered between the white pillars of the Storrs Congregational Church. A cloned cow moos.

But in a few days, all will be less serene. Some 300 articles will be published from Wichita to Malaysia about Yang's finding that cloned milk and beef are normal. The headlines will range from "Rare, Medium-well or Cloned?" to "Nothing Fishy About Meat and Milk of Clones." Some articles will celebrate. Some will complain more tests should be done than those required by the USDA. Bob Schieffer, the CBS national news anchor, will offer the scientific assessment, "This gags me."

Also in a few days, news will circulate nationwide that lawsuits from groups with pro-life ties recently put a temporary halt to the $3 billion California hES cell and therapeutic cloning initiative, which was passed by California voters in November of 2004 and was supposed to start distributing funds in May of 2005. The money distribution is now not expected to begin until the fall of 2005.

While the Connecticut legislature will soon agree to pony up $100 million for hES cell and therapeutic human cloning work in the state, on March 17 U.S. Senator Sam Brownback reintroduced a federal bill that would criminalize all efforts to clone human cells with a $10 million fine and up to 10 years in jail. The bill twice passed the House in earlier years, but was stymied in the Senate. This time there is a Republican majority in the Senate. President Bush has said he supports it. If the bill passes, Yang will not be able to clone human cells, despite the ruling and cash outlay of his own state legislature.

Still, papers throughout March have been filled with the news that there is support in Congress for a bill that would at least increase the number of presidential hES cell lines—from spare IVF-clinic embryos, not cloning. However, optimists are reminded in these same articles that in his State of the Union address, Bush, who wields the veto pen, said he saw no reason to change his policies. And in a few weeks, Yang will receive some very bad news: his top cloner, Gao (he with the fastest hands in the West), will accept a job in China. Indeed, the job will be with an institute Yang helped establish.

The bottom line: things are not as serene as they seem down on any U.S. cloning farm.

Of all the cows in Yang's field, only two stand twisted together. They are so tightly entwined that they almost look like one cow with many legs and two heads. It is Aspen and her first identical clone Amy. Anyone involved in the hES cell/cloning debate at this time would be tempted to say the clones are leaning into each other in such a weird fashion, it is as if they are trying to disappear into each other, fed up with the endless noise their existence incites.

Aspen's clones all gestated in the womb of a surrogate cow. In their youth, they were all were kept from each other. Yet: "They're at it again," says a graduate student. "They always seek each other out like that. Aspen's natural children don't seek her out, and she doesn't seek them out. But the clones do this. Cool, huh?"

11

SMASHING BIOLOGICAL CLOCKS: THE STEM CELL'S FIRST SOCIAL REVOLUTION?

*This is like finding out that the world is
actually flat.*
                    *—Cornell's Kutluk Oktay, on hearing a
            mammalian egg stem cell may have been found*

It makes the hairs stand up on my arms.
                    —Harvard's Jonathan Tilly, same topic

THE PROSPECT OF MATERNAL BLISS, for millions of young/old women aged 35 to 43, comes to the door in a FedEx box the size of a riot-gear shield.

This is appropriate, for this box of bliss contains weapons. In addition to large stores of needles, the box offers instruments of chemical warfare, powerful drugs which, when unloaded into the woman's body, will fight not only her dying reproductive system but also her comparatively young endocrine, cardiovascular, and nervous systems, forcing them to rally around and get behind an activity that will be shocking to them all. For the job of those FedExed weapons is to provoke the woman's body into developing and briefly stockpiling scores of old eggs simultaneously in her ovaries—instead of the usual one a month—in the hope that bombarding sperm with a blitzkrieg of rapidly aging eggs will increase the odds of a successful bull's eye, and birth. The long-term effect of this on the woman is unknown. In vitro fertilization (IVF) is still a very modern form of modern warfare.

Each weapon in this battle for bliss is, furthermore, expensive, and each has been fought over with insurers if the woman is lucky enough to live in one of the few states that mandate at least partial IVF coverage (as of 2002, 3 states mandated complete coverage; as of 2004, 14 states mandated at least partial coverage). Once the drop has been made, therefore, the IVF warrior must check her arms shipment to make sure it includes something along the lines of the following (it varies): 30 3cc/ml syringes with 1-inch-long needles and 20 half-inch needles to be plunged a couple of times daily, over the course of one to several weeks, into the buttocks, thighs, stomach. Five smaller needles already locked and loaded with Antagon. Five vials of Menotropin, and five vials of sodium chloride, for later loading into syringes. A plastic box with the words "bio-hazard" and "contaminated" spelled out on it in ghostly raised plastic lettering. Two large vials of injectable Novarel (HCG). Alcohol pads and gauzes. Four 16 mg tablets of Medrol. Six 100 mg tablets of Doxycycline. Fourteen caplets of 200 mg progesterone. Four syringes with $1\frac{1}{2}$-inch needles. Vials of injectable Gonal-f, 1200 IU multidose. Fifteen more 27-guage syringes and one prefilled syringe of bacteriostactic diluent.

Every drug here has an equivalent; every clinic, its preferences; every insurance company, different preferences. Many recipients of this box have thus battled for each drug, drug by drug. And many do fight. One mistake could be an error they can't afford.

For this is a $5,000 to $10,000 weapons cache, depending on which part of the nation the IVF warrior purchases it in. It will service a single attempt to have a child, a single menstrual cycle. Stats say the vast majority of women aged 35 to 43 entering this first round of the battle will lose it and will have to order up another $5,000 to $10,000 box of maternal weaponry. And this represents only the gun-and-poison, syringe-and-drug part of the war. Each round of IVF, each menstrual cycle, costs an *additional* $5,000 to $10,000 for clinic services. The official U.S. average per round tops $12,400, says the Society for Assisted Reproductive Technology (SART), of which the vast majority of states mandate zero coverage. Most women aged 35 to 43 can't even think about entering this war. If you didn't have a child in time, tough.

The fact that this box is the size of a riot-gear shield is appropriate for another reason. Many women receiving it are at the end of another series of battles. For U.S. IVF, unlike IVF in much of the rest of the

Western world, is also largely federally unregulated, due in very large part to the Christian Right. Thus, clinics are free to service who they *want* to service and they tend to prefer women who are guaranteed successes—younger women who possess a simple fallopian tube problem, for example—since the main way they score clients is via published success rates. So, many IVF warriors aged 35 to 43 successfully negotiating their first arms drop have already been rejected by at least one clinic on the basis of age, forcing them to settle for a lesser clinic.

Indeed, many clinics deny single women treatment because many Americans believe it is immoral to have a child out of wedlock. And many clinics can and do deny, or make impossible, treatment to single women who want to bear a child with someone they know (citing complex, some say prejudicial, laws regarding unwed mothers). Only banked frozen sperm from strangers is allowed for singletons in many U.S. clinics.

The battle doesn't end there. Those IVF warriors possessing the predisposing mutated BRCA-1 or–2 cancer genes may want to avoid standard IVF, some experts believe. At Cornell University's top IVF clinic, a few patients with a mutated BRCA gene developed breast cancer after standard IVF treatment. Thus, IVF warriors with breast or ovarian cancer in the family can end up fighting insurers yet again; this time, for $3,000 genetic tests. These must be preapproved; take from one to three months to get approved; and often *aren't* approved for healthy women. The warrior can face yet another decision: gamble with her life, or spend a few thousands more and wait another few months for results. (She can also wait months for a simple mammogram—Cornell is shockingly egregious this way.) All as the clock ticks. Every month wasted can be the warrior's last.

Still, when the rookie warrior finally receives her arms cache, she can often react by hurling it into a closet. Indeed, many who have reached this point are advised by IVF clinics to rest and regroup. The reason is that she must now prepare for the down-and-dirtiest part of this war, which is to face the fact that she will probably lose every round. For this victory—scoring those arms—is, odds say, pyrrhic. Every woman 35 to 43 has to prepare herself for the wordless, marrow-deep sadness that so often comes with the knowledge that, stats say, she faces: she's too late.

She is the average age of 38.5 and has an average of 41 years of life to go. But her reproductive system is rapidly dying, and odds are that she will never have that child she seeks now no matter how many $10,000 to $20,000 rounds she fights. For while IVF works well on women under 35, helping nearly 50 percent to conceive, there is what gynecologists call a "precipitous decline" after. The average 40-year-old's chance of conceiving after a round of IVF is 15.2 percent, according to the CDC. The average 41- to 42-year-old's chance is 10.7 percent. The average 43-year-old's chance is not tabulated by the CDC, because the number is statistically insignificant, hovering at zero.

"There have been no significant advances in the field of women's fertility in 20 years, since the first years IVF was created," says NIH hES cell scientist Mahendra Rao. (The most significant advance: infertile women can buy a $30,000 donor egg to have someone else's child, not their own.) Rao and others say this is largely because the federal government not only does not regulate the field, but puts next to no money into it—again, largely due to Christian Right objections—leaving most fertility research in the hands of private clinics. "And privately funded clinics horde data; data is proprietary in private industry. So knowledge is not built." Still, once again, it is partly a matter of the mind, for women haven't exactly been staging marches to protest their government's lack of aid. Few have believed anything really could be done about it. IVF has become, in some ways, what many young/old women do to mark the death of their reproductive lives: the "last shot" they take to try to avoid regrets.

Regardless, as part of the regrouping stage in this strange and frustrating war, many a young/old warrior has dreamed of a future where 40 will not mean the end of one dream or another: to have more children (if she'd wanted more); to have both love and a child (if she didn't love fast enough); to have love, a child, and an engaging career (if she compromised the latter to make the motherhood deadline). For 100 years, the warrior muses, 40 has marked the death of a critical part of the female life. (Before 1900, the average age of death was between 47 and 49, when menopause occurs. Before 1900, in other words, nature had the sexes conceiving right up to death, in tandem.) For some 100 years, women have experienced reproductive death midway through life, while their partners stay reproductively young, comparatively, forever. For some 100 years, with each year that modern medicine has extended life, a physical and psychological chasm has widened between the sexes.

That this is unnatural is shocking, given society's total acceptance of it. But it is a fact, for humans and all other animals. Female baboons living in areas with few predators die naturally at 27. They reproduce easily through 21, then go through three years of births and miscarriages until 24, when they begin having irregular menstrual periods. Menopause strikes at 27, the year they die, as was once essentially the case for women. If women's fertility kept exact pace with this model, mathematically they would begin having fertility problems at 56, not 35; stop reproducing at 67, not 42. Sticking *strictly* to the model is perhaps unadvisable. But nature clearly never intended the sexes to undergo such divergent reproductive declines. Harvard University reproductive biologist Jonathan Tilly calls midlife menopause "a clinical side-effect. . . . We've extended human lifespan, but we've forgotten the ovary." Agrees Cornell reproductive biologist Roger Gosden, "Irreversible ovarian failure is almost unheard of in nature. Wild creatures breed for as long as they live." Midlife menopause is man-made.

It is a profound social issue. If that 40-year chasm (the number of extra years men can have children) disappeared, many aspects of modern life could conceivably change. Fewer than 5 percent of CEOs were women in 2004 (and only 1.8 percent of Fortune 500 CEOs were women in 2005) even though females have graduated from college in greater numbers than men since 1982, and from law school in equal numbers since the late 1980s. When push comes to shove, many women still choose children over careers (although not over jobs: financially, most must work). Would the glass ceiling be shattered if women weren't exchanging careers for less-demanding jobs to squeeze children in? Would cracking ceilings before creating families make many women less regretful, more emotionally stable, and their families the same? Would the 50 percent divorce rate go down if the sexes were on similar physical and emotional timetables, having families when both sexes want them, not when one must?

Twenty percent of families in the United States are headed by single mothers, and these represent 50 percent of U.S. families living in poverty. U.S. women earn 78 cents on men's dollar, and that has been dropping. Worldwide, 70 percent of the poor are women. Whatever the reason, it takes more time for women to become financially stable. Would postponing childbirth until financial stability make women more stable, their families more stable, society more stable? It seems likely.

The list continues. Women in menopause, which can last a decade, can go on an emotional roller coaster ride. To what degree is this due to hormones? To what degree is it due to grief over the premature death of fertility, a death that nature never intended? Hormone loss delivers a blow to women's health from which they never recover. Heart disease, osteoporosis, cancers, low libido, and other physical problems spike after menopause. Men never experience that sudden decay.

Of course, there is a chance that if this physical, psychological, and economic imbalance should ever right itself, society wouldn't change. Or it would deteriorate, with most couples procreating in their sixties and seventies, burdening society with orphans. (Men *are* cashing in on their option to have children somewhat later: between 1980 and 2002, the rate of births to men 40 to 44 went up 32 percent; to men 45 to 49, 21 percent; to men 50 to 54, 9 percent. But they are not using the option after 54.) It is possible that the much-caricatured emotional divide between young men and women on "the baby issue" (women want them; men must be "trapped" into them) is genetic.

There is no way to know. There are few serious papers calculating how different society would or would not be. Few have ever believed it was possible.

Until now. Harvard's Jonathan Tilly is not only making famous the remarkable No Menopause Mouse, which is created with ES cells, he also believes he has found an adult oocyte stem cell that generates new eggs into adulthood and can be tweaked to pump out many more. Other researchers are progressing in the development of alternative ways to use stem cells to fight infertility. All this, added to the fact that scientists may be closing in on a viable, uncontroversial method for producing hES cells, means that someday *most* women may visit IVF clinics. They will do it to have a child from their own personal egg stem cell bank after 40, and/or to postpone menopause, and/or to bank away their child's own personal ES cells, to protect that child the rest of her or his life, no destruction of embryos involved.

· · ·

THE EVENTS that led to Jonathan Tilly's March 11, 2004, *Nature* paper claiming the existence of the adult mammalian oocyte stem cell were, he noted with a laugh in August 2004, "Mind-boggling."

The rise of Tilly, an energetic man with an electrifying speaking manner, has been meteoric in the ranks of science. Upon finishing his post-doctoral work in molecular biology at Stanford University in 1993, he served for two years as an assistant professor of reproductive biology at Johns Hopkins University before being tapped as a professor at Harvard and assistant director of MGH's Vincent Center for Reproductive Biology in 1995. By 1998, he was director.

He is not the kind of renegade scientist of whom one would ever say, "He entered the fray reluctantly." An articulate and passionate 42-year-old, he possesses a munificent, Thomas Friedman–esque ability to think out loud with eloquence—and the ego to go with it. Even the testicular cancer he was diagnosed with in 2000 seemed to just spur him on. The cancer is one of the most curable, and they got all of his. But the import of another reproductive imbalance—men can freeze sperm before cancer treatment to protect fertility, but women can't freeze oocytes—hit him hard. Tilly had just had his first child when he got cancer in 2000. At the time of the August interview, he was about to have a second child, four years after. Tilly was impacted. "Investigating a crime against female fertility," he sternly headlined a paper on cancer therapy and female fertility, two years after his own experience.

(Admittedly, his papers' titles rarely incite the "kill-me-now" boredom that so many of his peers' can. "Realizing the promise of apoptosis [cell-suicide] based therapies: Separating the living from the clinically undead," reads another of his papers. Then there's: "Commuting the death sentence: How oocytes strive to survive." And: "Ovarian follicle counts—not as simple as 1, 2, 3.")

So Tilly lacked for nothing—confidence, credentials, and a sharp sense of empathy for the potential beneficiaries of his infertility work ("I lived it; I experienced it," he'd tell a reporter later)—when he entered his lab in 2002 to check an experiment. He had irradiated female mice to gauge the rate at which their fertility declined. Since this is measured by counting oocytes, the experiment was simple: at intervals, count eggs in irradiated mice and healthy mice; compare.

But Tilly shocked even himself when he finished his calculations. He also shocked the field's leading practitioner of the effort to extend women's reproductive lives, Cornell University's Kutluk Oktay, who repeated to the press just after the discovery was unveiled, "This is almost like finding out that the world is actually flat." Said the University of Edinburgh's Evelyn E. Telfer in *Reproductive Biology and Endocrinology* that May, Tilly's discovery "forces us to reassess long held beliefs." Allan Spradling, a leading Drosophilia (fly) oocyte stem cell researcher with the Carnegie Institute of Washington, said in a *Nature* article accompanying Tilly's paper: "Contrary to long-held views, female mice contain a population of germline stem cells that are required to maintain the overall number of follicle [sac-enclosed egg] numbers throughout adult life. This important finding seems destined to greatly enhance our understanding of mammalian oogenesis." Dr. Marian Damewood, president of the American Society for Reproductive Medicine, said in a statement that the discovery is potentially "the most significant advance in reproductive medicine since the advent of in-vitro fertilization more than 25 years ago."

The praise kept coming. "Textbook rewrite?" *Science* asked. "Astonishing," *Nature Reviews Molecular Cell Biology* said. In *Stem Cells and Development*, Soon Chye Ng of the National University of Singapore would claim Tilly's paper "shook the foundations of reproductive biology."

What Tilly discovered was—to hell with the irradiated mice—his healthy, control mice lost so many eggs that their ovaries should have been depleted at two weeks old. But they weren't. Oocytes kept popping up in mouse ovaries until a few months before their death. "The Dogma," as reproductive endocrinologists and gynecologists have referred to it for 50 years, dictated the opposite. The Dogma said that all female mammals are born with all the eggs they will ever have; that there is no replicating adult oocyte stem cell; that women simply keep losing their static supply of eggs until, in their forties and early fifties, they lose their last, causing them to plunge into ovarian failure (menopause) at 50, with all that implies. That is, after 50, in addition to experiencing hot flashes and depression, women's health declines sharply in some areas, no longer protected by the hormones oocytes once generated.

"It wasn't planned by any stretch of the imagination," Tilly said in August 2004, his voice still charged with excitement five months after the publication of the *Nature* paper. "My lab has focused 10 to 12 years on the other side of the coin: apoptosis, or cell death. All we did was cell death in the ovary. A few years ago the No Menopause Mouse was created, right? [The mouse lacks the *Bax* gene, thanks to ES cell technology. This deficiency keeps mice ovulating—and even giving birth with Mouse IVF aid—*right up to their deaths*. Developed initially by the Washington University oncology team of Stanley Korsmeyer for cancer studies, Tilly adopted the rodent for infertility work.] Because of The Dogma, in this lab we assumed that with No Menopause Mice we had just *slowed oocyte death* into old age, not that we had allowed the mice to create *new oocytes* into old age."

Tilly knew more about oocyte death than almost anyone. In 1999, he reported that the No Menopause Mouse made three times more oocytes than normal mice. In 2000 he edged closer to the clinic when he showed that he could just give mice a molecule called S1P (sphingosine-1-phosphate) to protect oocytes from death caused by radiation. In 2001, he revealed how cigarettes kill oocytes. (They turn on *Bax*. When *Bax* is blocked, cigarette smoke no longer kills oocytes.) In 2002 he proved that pups of S1P-protected mice were normal. And a few months after this interview, in February 2005, he would find oocytes naturally die partly because *Bax* builds throughout life, and partly because a chemical called ceramide builds up in the follicles enclosing oocytes, then works with *Bax* to kill off the oocytes for good. Tilly's S1P would halt this death.

Tilly filed for a patent on S1P. What he lacked was a perfect way to deliver it, because S1P is toxic when administered systemically. He was experimenting with a time-release capsule in monkeys that might release it in the ovary only, and make it a commercial reality.

So Tilly, in late 2002, thought he was fine. He had found a molecule that might slow the death of human oocytes into old age, perhaps someday prolonging fertility and postponing menopause. He had come up with a drug that might someday radically, if artificially, change women's lives. He was not looking for a cell that could *very* radically and *naturally* change women's lives. But Tilly got curious.

"We realized at one point that nobody had ever actually done a full comprehensive study of the *amount* of death in the ovary," he said. "People had made deductions about oocyte loss based on taking an ovary at a certain age, and later, and saying, 'Well the number has changed by 30 percent, so that means 30 percent of the pool has died.' They were assuming changes based on how the healthy pool changed. So if the number of healthy oocytes declined by 32 percent, then the assumption was that 32 percent of the existing oocyte pool died. In our lab, though, we scratched our heads and said 'Why didn't anyone measure actual oocyte death?' Being a death lab, we decided to do it."

Daily, Tilly's team removed ovaries from healthy mice from birth to adulthood. It monotonously did what no prominent lab had ever done before: religiously counted both the living *and* the dead. "At first we found exactly what everyone else did, that it goes down 30 to 32 percent from birth to puberty. We said, 'That's great, our data are consistent with everyone else's, we're happy, and we fully expect to next measure a corresponding number of dead oocytes.' But then we found that, at *any given time,* 30-plus percent of the oocyte pool was dying. Not cumulatively. At *any given time.*"

There were two possible explanations, Tilly continued. One: the ovary was unique in the way it ensured its cells were eliminated. Perhaps, when the ovary's cells die, the organ doesn't clear them out, but lets them "accumulate as corpses. What we could have been seeing in the ovary was that, as these animals moved into adulthood, they allowed an accumulation of dead oocyte corpses, sort of like a biological funeral parlor where no cells go away once they die."

However, Tilly said, "That isn't how death happens. When a cell dies, the body wants to get rid of it, get it buried, get back to normal. If cells sit around once they die, if they're not cleared, you get an inflammatory reaction, identical outcomes to primary necrosis [death]." Thus, Tilly's crew "had a conflict of two dogmas. Either apoptosis was failing and corpses were sitting around, or we had all completely misjudged how oocyte dynamics is regulated; we'd all been blinded to the fact that there's a lot of renewal offsetting a lot of death in the ovary."

So Tilly's crew did more measurements. "We concluded the ovary is pretty damn efficient. When an oocyte dies, it's cleared out at most within three days and probably within 24 hours. That was consistent with studies of every other tissue in the body. If any cell dies, it's removed within a day."

The way Tilly established this: he knew that when ovaries form, oocytes get enclosed in follicles. During follicle formation, half of all oocytes can die because they don't get enclosed. "So if you look at an ovary during the early stages of life, in mice, there are roughly 8,000 oocytes at birth in an ovary. Then three or four days later, that number drops to 3,500. So you would argue that within four days, 4,000 oocytes died. If the hearses were not effective, and all these corpses were piling up on slabs, you would imagine that we would find 3,000 oocyte corpses. But when we looked at the ovaries all we found were 200 corpses. So the ovary essentially was able to clear out 3,000-plus oocytes within a three-day period."

The team was careful. What they had counted was death of oocytes *not* enclosed within follicles. In adult life, oocytes are already enclosed in follicles. So the team decided it was possible that maybe oocytes enclosed in follicles weren't cleared out so quickly.

"So we did a second study with a chemical agent that we used in a study in *Nature Genetics* in 2001 which showed that a chemical accelerates apoptosis in the oocyte pool. It is DMBA, dimethylbenzathracene, which is found in cigarettes and factory emissions. In that earlier study we found that when we give DMBA to follicle-enclosed oocytes, within three days we can eradicate the ovarian oocyte pool. The agent hijacks the same genetic machinery [involving *Bax*] that oocytes normally use to kill themselves. You get very rapid ovarian failure. So we reasoned that if we can get a whole cohort of oocytes dying, we should either see a piling up of corpses, or they will be cleared out. So we gave DMBA, and took ovaries out every day, and counted healthy versus dying oocytes. Within that three-day period the resting pool went down to zero. We saw a transient peak in atresia of follicle-enclosed oocytes within 24 to 48 hours, but by the third day there were no more. They were cleared out. So using two different approaches, our conclusion was that, irrespective of whether an oocyte is follicle-enclosed, or not, if it is dead, it is dead and gone. The concept of apoptosis being the mechanism by which the body removes cells without causing an inflammatory reaction held true in the ovary."

Still, they were talking about smashing The Dogma. The team soldiered on. "We said, 'OK, if one-third of the body's oocytes are dying at any given time, and if they are being cleared out every three days, it doesn't take a math whiz to figure out the ovary will fail in less than two weeks. So we continued to measure atresia well into adult life. It stayed high. It wasn't like this was an aberrant peak. It stayed high well into adult life, and that got us thinking that maybe people had missed this, that they assumed the rate of death. That got us into the *Nature* study. We did four more experiments to prove there was mitosis [putative stem cell proliferation], meiosis [putative stem cells, or stem-like cells, shedding a set of chromosomes to become oocytes], and follicle formation. We got the same answer each time. We never intended this. This started as a study on death dynamics in the ovary. But it turned out to give birth to the idea of life in the ovary."

Tilly's "voilà moment" was so big, it was tinged with fear. "I've been so open, I tell people everything. I share at meetings. I'm free with data. But when this happened, I put a gag order on the lab. Still, right now, the hairs are standing on my arm just thinking about it. I sat back in my chair, took a breath, and thought about what this meant if it were true. It overwhelms you. If you think about how, for the last 50 years, we've counseled patients, interpreted infertility problems, managed infertility, viewed the menopause. . . . All that's wrong. It's *wrong*."

Tilly's voice pitched higher like a kid with a secret. "And it hit me like a ton of bricks that we have a lot of work in front of us. So I called my department chair and let him know what had happened. He almost dropped the phone. He [Isaac Schiff] is a world expert on menopause and he immediately got the ramifications. He said, 'I'll support however you want to go about this.' So we kept it pretty quiet. The first time I spoke about it was at a meeting in Texas in January. A fertility and cancer survivor's meeting. I was asked to talk about our work developing small molecules to protect mature oocyte pools in cancer patients. I talked about that, and then I gave them a glimpse of what we thought might be coming down the road to preserve or prolong fertility: adult germ stem cell technologies. There was a hush. A couple of hundred people were in there. And—I'll never forget this—a colleague came up to the microphone, a pioneer in cryobiology and grafting of ovarian tissue, Kutluk Oktay. He said, 'John, you know, I hear a lot of talks and I see a lot of things, and very rarely do they make the hair on my arms stand up on end. But I have to tell you . . . this is like hearing that the earth is actually flat.'"

Later Oktay would note that Tilly's work might explain something he had seen in his clinic at Cornell. Oktay performed the first human ovarian frozen tissue transplant to result in an embryo; the paper came out at about the same time as Tilly's. What surprised Oktay was that the transplant produced eggs for over a year, beyond what Oktay predicted based on the follicles he first counted. After Tilly's paper, Oktay wondered if the egg hike had something to do with those impossible adult female germline stem cells.

"It blew him away. How could we collectively as a field have missed this? I told everybody how we got into this; that I was taught the same thing they were taught; that I never questioned it in my life. Look at any textbook. The Dogma is the opening paragraph for the biology of the ovary."

Tilly's crew set out to discover how the field could have made what appeared to be such an error. "We went back into the literature of the 1950s when this controversy was still going on. The debate really ended in the mid 1950s. From that point, it was accepted that there was no renewal in the ovary. Frankly, we found that neither arguments for nor against renewal were convincing. But a very powerful person came out and said, 'There is no renewal,' and everybody listened."

That person was Lord Solomon Zuckerman, a University of Birmingham scientist with "a lot of pull with the British crown. I know some people who knew him and heard him speak. Essentially he told people, 'This is the way life is.' And from then on, this is what everybody believed life was." Zuckerman published a paper in 1951 claiming there was no germ cell renewal. "He shot down any paper before that that suggested there could be germ cell renewal, and a number of papers did suggest it." In hindsight, Zuckerman didn't offer data that contradicted the notion, Tilly said.

It all led to the obvious question: Why does fertility begin to fail so drammatically around 40 and why do women have menopause at 50? The answer is not that they have a static number of oocytes from birth on, he said. It's not that eggs are harder to make than sperm. Like men—like most if not all of the rest of the body—women have stem cells that create new oocytes for decades, he now believed. They are just born with fewer stem cells than men. "It's a numbers game!" Tilly said happily. "The organ system most well studied is blood. Hematopoietic stem cells (HSCs) support new blood cell production on the order of millions. There's so much new blood cell production in your life, it's crazy. Tons of HSCs in the bone marrow. You look at spermatogenesis, the making of millions upon millions of sperm. There are hundreds of thousands of stem cells in the testes. It's completely different. It's mass production."

But when it comes to oocytes "it is a numbers game, because you don't need mass production. The limiting factor is there are so few stem cells. Adult stem cells have a limited replicative lifespan. They can only divide a limited number of times and then they shut down. That's how all adult stem cells function. So to us, it's not an earthshattering concept. Men can lose half their stem cell pool and still make millions of sperm because they were born with so many. If women lose half their stem cell pool, that should have a huge impact on their pipelines, because they're already fighting a losing battle."

The reason this matters clinically: mature oocytes cannot replicate. But stem cells do. Furthermore, stem cells, like sperm cells, can be efficiently frozen—unlike oocytes.

All this matters for many reasons, noted Tilly, including the fact that it would appear to be in nature's best interest. "Humans have been around a long time, and if you look at life expectancy, we've done medical wonders to extend it. One hundred years ago life expectancy was in the 40 to 50 range. If you made it to 50 you were having a big celebration. When does menopause happen? At 50. Menopause happens at 50 because you were, until recently, dead at 50. We've been able to extend human life, but we've forgotten one very important variable. The ovary. We haven't extended its lifespan. It's still doing what it was told to do thousands of years ago. The ovary is living for its projected lifespan."

Why then can many men still procreate until their nineties? "In many animal species, the purpose of the male is to fertilize a female and then go fertilize another female. The female nurtures the young until she lets them go. Females are responsible for nurturing the young, breast feeding, raising the young to the point of self sustenance." So males are born with enormous supplies of stem cells because they are unlimited in the numbers of females they may impregnate. They have plenty of "spares."

Tilly noted his historical research did turn up clues as to the existence of the mammal oocyte stem cell. "In males there are germline stem cells galore." And adult female germline stem cells have been found in fish, fruitflies, and birds—including the chicken. (Indeed, "the rockfish lives up to 150 years and never experiences menopause," says

Cornell reproductive biologist Roger Gosden, who wrote a paper on this in 2004. The reason? Two words: "stem cells.")

Continued Tilly: "But the most interesting paper was from Gertrude Vermande–Van Eck in an obscure journal in 1956. She found germline stem cells in female rhesus monkeys after Lord Zuckerman made his proclamation. She said, 'I am going to do my own thing.' She had three cohorts of rhesus monkeys, one just before puberty and two that had already started cycling. They were a couple of years apart, yet, irrespectively, the oocyte pool was static, didn't appear to change: 58,000 oocytes per ovary. This was surprising to her. If oocytes are dying and going away then there has to be decline with age, but these animals spanned years and didn't change. So she said, 'Maybe atresia is happening at a low rate,' so she measured atresia the way we did. She looked for condensing, fragmenting, dying oocytes. Once dead, you can't miss 'em, they stand right out and look at 'cha. She found 4.5 percent of the oocyte pool dying at any time. That was considerably less than our mice, so when we first read her paper, we thought mice must be abnormal. Primates must have a very low rate of atresia.

"But when we reread, we realized: she took into account clearance rates, and said, 'You know, we better find out what is going on.' So she gave the monkeys a dose of radiation that she knew would kill oocytes. Then she took the ovaries out at intervals after radiation. She counted dead oocytes, and found oocyte death went up after the first few days, peaked within a week, and by two weeks all dead oocytes were gone. She concluded atresia is complete in two weeks. So she plugged all this into an equation: 58,000 oocytes at puberty, 4.5 percent dying at given time, clearance every 14 days. Take off 4.5 percent of the pool every 14 days, and you get a beautiful exponential decay curve, a projection of when ovaries will fail. Guess how long it would take a monkey ovary to fail? Two years!"

That's a big problem for The Dogma. "Big discordance. So way back in 1956, she concluded that, in contrast to Lord Zuckerman, oogenesis does persist in primates. Their whole lives. And you find other examples, beautiful studies done 50 years ago. But they went under the radar screen. There's a lot of fear of change in science. Scientists are taught that once you discover something, that's the way it is, you move on to the next variable. After so many years, people don't want to believe things like this because it would make them rethink their own careers. They don't want to acknowledge that if they are so brilliant, how the hell did they miss something so basic as oogenesis? It is like the neurogenesis field. Nobody wanted to believe the brain makes new neurons." But since the discovery of the human adult neural stem cell, it's been clear: "Hell, of course the brain makes new neurons."

So Tilly is now sure that cigarette smoke affects fertility by killing off stem cells? "There is no doubt. We've been working on S1P. We've been working for 9 to 10 years on it. If we could figure out how oocytes respond to insults like chemo and radiation, we could figure out how they initially sense that stress. Then if you put up a shield or blocker, you wouldn't permit stress to be sensed. The oocyte would not know it was exposed and would survive. That how S1P works. The oocytes are normal even though exposed. Animals protected by S1P can make normal babies, suffer no damage. Well, for nine years we've been believing that the way S1P does this—and we're now doing primate trials because we want a clinical trial—is by protecting the oocyte pool. But the contents of the ovary would be dead in three weeks, and it wasn't. We need to protect germline stem cells. That is what S1P apparently does.

"In the *Nature* paper we used busulfan, a chemo agent. We used it because *in the male, it targets specifically germline stem cells.* When we gave it to female mice we found they had ovarian failure; the oocyte pool went away in three weeks. People would say, 'OK, busulfan killed the oocyte pool.' We tested and found that actually, it *doesn't* kill the existing oocyte pool. [Mature] oocytes don't give a crap if busulfan is around. Busulfan kills germline stem cells. This tells us that insults to the ovary don't necessarily work by only killing oocytes, but germline stem cells. So we have to revisit all of our own work. We spent nine years on S1P and oocytes and it's probably been interpreted only partly right. S1P more than likely is also protective of the germline stem cell pool. We have yet to prove that, but we have to believe that's the case and now we have to go back and deal.

"So busulfan wipes out the germline stem cell pool. In humans it guarantees irreversible ovarian failure. We looked at clinical trials. When busulfan is used in chemo cocktails, 99 percent of the time there is irreversible ovarian failure. When busulfan is *not* there, just other bad chemicals, irreversible ovarian failure happens only 50 percent of the time. That's indirect proof that what we're targeting in the human is the same as mouse. It's stem cells. Busulfan does the same damn thing to mice and humans. So we said if we can find drugs that target germline stem cells, why not find drugs that *don't?* Drugs that would be tumoricidal, but wouldn't cause sterility. It is very likely we could begin to screen chemo drugs for ability to be tumoricidal, but not affect the stem cell pool. Fertility-friendly chemo cocktails."

So where are these adult germline stem cells? In the *Nature* paper, Tilly noted that large ovoid cells expressing the germ-cell-exclusive gene *Vasa* were on the surface of his mouse ovaries. He believed they are not the stem cells, but perhaps progenitors rising out of stem cells. As to where the actual stem cell is, "We spent a lot of time trying to figure out where the niche is. Every stem cell has its Shangri-la, its niche, and if those surface cells are the true germline stem cells, where is their niche? It is doubtful the surface is a niche. In other systems, the niche is hidden. Anyway, we've spent the past year trying to find the niche. We found it. We hope people will look at all this with an open mind. We're starting to put together a more convincing argument that, like other stem cell systems, the ovary has its own characteristic features that agree with many tenets of stem cell biology."

Thus, having found the niche in which the stem cells reside—and having isolated some 200 oocyte stem cells therein—"We are approaching things from two angles. One is to define the germline stem cell: what markers does it express, what is it, how can we purify the cells, What does it do, where is the niche? How does the ovary communicate with the niche and ask for new ocoytes? The basic biology. The other major thrust of the lab is the clinical biology, getting things to the clinic as fast as we can. Do human stem cells exist, and can we develop models to test for their ability to make new oocytes?"

Having dropped that bomb, Tilly paused. Clearly, he was hinting that his work since the *Nature* paper had moved fast. He wouldn't give details, because that could interfere with publication in a serious journal. But, he continued, "The beauty of this is that it is adult stem cells. If we were able to purify these cells, it would be in the clinic tomorrow, because it's the patient's own cells. They're already putting ovarian tissue back in people. Why not put their own stem cells back in?"

It would not be necessary to take out an entire ovary to retrieve germline stem cells for future use. "That's the big technical hurdle right now that we've been pushing through. How to get the stem cells out with as little intrusion as possible. That's the follow-up that I presented at the ovarian workshop just last month. This is what I was hoping people in our field would look at and say, 'Jesus, they did it, the progress is there. . . .' Anyway, do you have to take out the ovary? Absolutely not."

He paused again. "We have a paper nearing completion and I cannot tell you what is in it. It presents something that will so blow your mind . . . I think most of the field believes the follow-up to the March 2004 *Nature* paper will be either that we've isolated the cells, or figured out how to regulate them, or done something to further validate the concept that they exist. The field is going to be in for a shock. I almost am having difficulty accepting it myself."

He laughed. "The beauty is, this has nothing to do with George Bush's moratorium. This is adult stem cell technology. So we bypass all the ethical B.S. and all the philosophical B.S. This is like taking off a piece of skin and putting it right back in."

Would it affect women already in their forties? "It's intriguing. If you proceed according to the Old Dogma, then forty-somethings become a cadre of patients you really can't touch because you've passed the mark. But we're not in the Old Dogma anymore. This set of findings opens possibilities that six months ago I wouldn't have fathomed. Yet they're staring us right in the face."

His new finding did not involve cytoplasmic transfer, he said. In cytoplasmic transfer, oocyte cytoplasm (the magical soup of the unfertilized egg in which the nucleus sits, and which is responsible for cloning) from a young patient is injected into the egg of an older woman. The idea: rejuvenate it. Jacques Cohen of the St. Barnabas Medical Center in New Jersey injected cytoplasm into several patients' eggs, and 15 children were born by the early 2000s (30 patients have been born worldwide). Mitochondria, the tiny energy units of every cell, are scattered throughout the cytoplasm. Most of us inherit mitochondria from our mother only. Sometimes heteroplasmy—where cell cytoplasm contains mitochondria from two different beings—can happen naturally, as *Nature Genetics* pointed out in 1996 when the remains of Tsar Nicholas II were identified by his heteroplasmic cells. But that is rare.

Worried that the children were heteroplasmic—possessed of mito-

chondria from that young donor plus mitochondria from their own
mothers—the FDA demanded doctors apply for permission in the early
2000s, which has halted U.S. work. Indeed, in 2001, Cohen reported
two of the children were heteroplasmic at one year, adding that they
were healthy. He also reported the "pregnancy rate was higher than
expected (12 out of 28 cases), in a patient population with low fecun-
dity. In some clinical cases significant improvement in embryonic development was seen." But because some believed not enough animal work
was done first, the approach remained controversial.

Also controversial was an approach whereby the nuclei of an older
couple's sperm and egg are injected into the enucleated egg of a younger
woman. This way there appeared to be a better chance that mitochondria at least largely from one person would be passed on. Jamie Grifo of
New York University achieved a pregnancy this way advising an IVF
team at Sun Yat Sen University in China. No child was born due to
problems unrelated to the procedure, but the fetus had the mitochondria of a third party. It created a small brouhaha after results were
announced in the summer of 2003.

Still, in 2005, a U.K. team would receive government permission to
put nuclei from older couples' embryos into younger women's eggs. (The
team: Doug Turnbull and Mary Herbert at the University of Newcastle.)
Some mouse work indicates this approach is effective. The approach
remains attractive to reproductive biologists because the cytoplasm of
older women's eggs is clearly defective.

Regardless, Tilly believed transferring young cytoplasm to an old egg
without more animal work is a bad idea. "There are a few examples in
animal literature where, when you mix mitochondria in cells, they get
sick. Reproductive biology is crazy. It's the only field in which you're
allowed to outpace animal work with clinical application. . . . That's
why we spent 10 years on that S1P study. We didn't have to spend 10
years on two generations of 500 offspring looking for micronuclei and
cytogenetic damage. But we're talking about making children. There's
no in-between."

Wouldn't there be some experimentation going on in women using
Tilly's germline stem cells, which he clearly believed he found? "It's the
patients' own cells. We're basically going to take them out and put them
back" later, having frozen them at a young age. "They're going to be more

amenable to freezing because stem cells can freeze with no problem. Oocytes you freeze; they pop."

The ramifications went beyond "applications of other stem cells, because one-half of the world's population is going to hit menopause. It's like death and taxes: it's gonna come. You're talking about a technology that will potentially allow one-half of the world's population to make an informed decision about whether or not they want to hit menopause."

Whether women should is not yet clear, he said. Because of problems with the commercial, part horse-urine hormone replacement therapy (HRT) Prempro, many women dumped all HRT. (They dumped even more physiological—human—kinds that may work far better.) Out of fear, many are "back to square one when they hit menopause. They're risking now osteoporosis and heart failure and Alzheimer's and cardiovascular disease. And the list goes on. HRT [at least, Prempro] is not an option for many women as a way to get through menopause. Can we come up with a different way? We can come up with drug after drug, but that's not what nature would do. Nature would have the ovaries continue to work.

"Regarding postponing menopause, I think it's possible that *whether you want to do it is open to debate*." That's why he's been studying the No Menopause Mouse. "We have the follow-up. One of the things people keeping asking me is: could you postpone menopause? And I say, do we *want* to? Physical problems that come around menopause also come around the time women are aging. So how do you separate which health risks are due to ovarian failure?" When the No Menopause Mouse was born, "we realized that we finally have a female that is old, but its ovaries are still working. So we can finally answer the question, 'Is keeping the ovaries working a good thing? Do we see a better body mass, bone quality, quality and quantity of life?' After nine years, we finally finished that study. We did every end point you can imagine in these No Menopause Mice. Each cohort takes about 2½ years to age before we can do what we want, so this has taken a godawful amount of time. But we looked at everything: blindness, deafness, cognitive function, bone mass, the list goes on."

That study was concluding as well. "We have three big studies coming out. They are different but they all come together. If we know that postponing ovarian function is good for aging females, then the stem cell work has bigger ramifications. If it's not good, we talk about stem cells as fertility preservation in cancer patients and as the source of a couple more years of fertility for forty-somethings. We're putting everything in order to make conclusions that are sound, not based on the assumptions this field has rested on for 50 years."

Still, after the first paper, it was hardly all kudos for Tilly. Critics in the field contended, among other things, that oocyte death analysis is subjective; that oocyte death is hard to see. Oocytes put into culture can look unhealthy for 24 hours, then bounce back. (Tilly's response: dead is dead. In past papers he has shown the way to identify dead oocytes, being one of the rare oocyte death labs out there.) Maybe, critics said, but the experiment where Tilly's ovaries were wiped out with a chemo that kills sperm stem cells? It offered only indirect proof that oocyte stem cells exist. And the fact that Tilly found markers for meiosis near cells on the surface of the ovary may simply mean that a few fetal oocyte stem cells hang out after birth, critics said. The bottom line: many didn't believe the field could have missed all that new oocyte production in the mouse. The field wanted Tilly to show it the stem cell—and then make a mouse pregnant with its progeny, to prove function.

Tilly finally did show them the cell and ended up causing an even bigger uproar. For in a paper in the July 2005 Cell, he claimed that the oocyte stem cell had been missed all along because it was not located in the ovary, but in the bone marrow. He found that after administering bone marrow cells to sterile rodents, their ovaries filled up with oocytes—incredibly fast.

Many in the field were downright incredulous this time. The protests reached a crescendo in September 2005, when 16 researchers signed a letter of "concerns" to Cell. The researchers said they were surprised by the speed of the oocyte production Tilly found in adult mice. It was faster than oocyte production in the embryo. (In a written response, Tilly was fairly brief: these are adults, not embryos. They are different.) Another worry of the researchers was that the green fluorescent protein (GFP) marker he used to track his bone marrow cells could have bled onto already existing oocytes in the ovaries. (Tilly tidily turned that around by noting that when he gave GFP bone marrow cells from males to his sterile mice, they did *not* grow oocytes.) The group then said that many of the germ cell markers he used to identify putative oocyte stem cells are found elsewhere. But when Tilly asked for papers proving this, there was no response. Tilly neatly defended himself against most criticisms save one.

Prove it with a pregnancy.

*   *   *

ON OCTOBER 29, 2005, days after his rebuttal comes out, Tilly is busy. He is at home with his new baby, and getting ready for another one. It is (yet another) coup for the researcher who once had cancer treatment, and his 40-something wife: three children total.

"It's what it's all about," he says of the kids, then turns to the experiments leading to the new controversial paper. When his crew went to test the stemlike cells on the ovary surface described in his first oocyte stem cell paper (the 2004 *Nature* paper), they discovered they disappeared after 30 days, at mouse puberty. They could not be creating new oocytes in adult mouse life. The crew did see a few of the cells in adult mice, but not enough to cause lots of oocyte production. So he went on a wider search, looking for germ stem cells in the adult ovary via a marker common to embryonic cells—SSEA-1. His team found something surprising: they could only find it in the core of the ovary.

"The core, or medulla, was exactly where we didn't expect to find the stem cells. It's thought to be a dead zone," he says. "But we found to our amazement the SSEA-1 cells in there expressed *many* germline markers. It seemed like a slam dunk. We thought we had the cells. We didn't. The pool was small and the cells weren't actively proliferating or entering meiosis" as would be expected if they were constantly making oocytes.

At the same time, Tilly says, "we were conducting other experiments to further establish the fact that ovaries have regenerative power. We insulted ovaries with a chemo that kills oocytes, doxorubicin. We knew from mouse studies they are a very effective oocyte killer, yet in the clinic women treated with doxorubicin very rarely suffer reproductive harm. We reasoned that doxorubicin kills oocytes but never touches germ stem cells. We injected the mice and within 24 hours their oocyte pool was knocked down by 80 percent as we expected. This added more fuel to the concept that it is ridiculous to think that oocytes can't be reborn in adult life. It can and does happen quite nicely.

"The crazy part was that we could show that in mice, 1,000 new oocytes could be formed in follicles within a day or two, a staggering number. Apparently because this was an insult model, the ovaries responded quickly. They stayed normal even two months after. Then we asked, 'Instead of insulting the ovary, can we simply add to the existing oocyte pool?' We chose a histone deacetylase inhibitor used in the clinic on cancer [patients], trichostatin A (TSA). Hematopoietic stem cells (HSCs), the best characterized adult stem cells, expand in number when it's added. We gave it to mice in puberty, young adulthood, and in perimenopause nearing reproductive failure. We found that when you give a shot of TSA and wait 24 hours, the primordial finite resting pool of oocytes *doubles in perimenopausal mice,* without a measurable effect on growth or death rates. This was a staggering number of oocytes being produced and enclosed within follicles in a short period of time. It reconfirmed in our eyes the validity of the *Nature* paper even more. But it posed a problem. Here was evidence that oocytes could be made very quickly, and that small, quiet pool of SSEA-1–positive cells in the core didn't seem responsible."

The crew thought, and realized that the core is where blood vessels enter and exit the ovaries. "Perhaps the germ cells we saw in there were just *passing through*. Perhaps there were germ cells circulating in the body that support the ovary. It sounds wild, but everything we did for two years before this was wild. We had decided to have no preconceived notions." Some studies postulate that the bone marrow, via the blood, may sustain some body parts by delivering stem cells. "Why not the germline?

"So we got thinking about the factory in the bone marrow. We did gene expression profiling of adult mouse and human bone marrow, and found many germline and embryonic stem cell markers: vasa, nobox, dazyl, stella fragilis, Oct4. The markers were there, and this encouraged us, so we did the key experiments in the *Cell* paper. We decided that if we transplant bone marrow into sterilized animals it should regenerate ovarian tissue and it did. We did it two different ways. First we gave mice two of the worst chemos, cyclophosphamide and busulfan. Together this blows the ovaries out. But the bone marrow worked: new oocytes. Then we wanted a mouse that was already genetically sterile, the ATM gene knockout mouse. Five different labs have shown that if you knock out the ATM (ataxia telangiectasia mutated) gene, as soon as a germ cell enters meiosis it dies. So females can produce no oocytes at all. We took these animals and gave them cyclophosphamide and busulfan on top of *that*. We transplanted bone marrow with intact ATM genes into these females—and oocytes and follicles regrew for over a *year*, when the mice were in perimenopause. If you transplant at six weeks of age in a mouse, and you are still seeing oocytes and follicles 12 months later, that says something significant happened."

An interesting thing, Tilly adds: his postdoc was chatting with a neurogenesis postdoc when the other postdoc mentioned her lab had given mice busulfan and cyclophosphamide because it wanted to see if bone marrow could rescue neurons. "He said, 'Do you think we could look at the ovaries?' She said, 'You are more than welcome.' The three females were 11½ months after their transplant. They were already conditioned and transplanted. It was done completely independent of us. And it worked: beautiful new oocytes in follicles. Independent corroboration doesn't get any better."

Peripheral blood cells that already escaped the marrow also worked on sterilized mice.

"There is a paper from 1921 that indicates if you take an ovary out of a chicken, you get even *more* oocytes [33 to 68 percent more]. Flies, fish, birds, mice regenerate oocytes throughout life. Why would something track from invertebrates to vertebrates to vertebrate mammals but not primates?"

Furthermore, says Tilly, his group is digging up more and more stories of women regaining fertility after bone marrow transplants. "In the literature we found hundreds of examples where women undergoing sterilizing chemo and then bone marrow transplant have seen their ovaries regain function. And I called a couple of hospitals and asked, 'Ever see this?' 'Occasionally,' is the response. We found examples of women in their forties who got pregnant after bone marrow transplant, one at 43." And bone marrow transplants for cancer do not purify the subset of stem cells Tilly finds support fertility. The group has also found that certain germ cell markers in blood rise and fall with the menstrual cycle.

"We think menopause is a clear example of a tissue losing its stem cell support. Either menopause is about the factory in the bone marrow shutting down, or transport via blood going defective, or the ovary itself changing with age. Or a combination. The test is simple, you take old bone marrow and transplant it into a young sterilized recipent. We are doing it for our next paper."

The group has even gotten some of its sterilized mice ovulating after bone marrow transplant; that is, they formed corpora lutea (pregnancy-supporting, homone-secreting masses), proving return of some function. Tilly has not been able to get older perimenopausal mice ovulating with bone marrow. This may be because older mice make eggs but do not ovulate because some endocrine glands, involved in hormone signaling, are not in peak shape. But he *has* been able to double the egg output of older mice, via the aforementioned TSA. So improvement is clearly possible. He's seeking that—and pregnancies, the proof of complete function.

"Who says we don't have pregnant mice?" he asks coyly. "It is not published yet. Give us time. These animals are conditioned with a nasty chemo that affects the uterus, throws off the endocrine environment and the reproductive axis."

There still may be stem cells in the adult ovary, Tilly says. They didn't do an exhaustive search; resources are limited. "My lab for 10 years focused on cell death, and in 3 years I have switched my whole lab and all my funding to stem cells. We've had to put our manpower into home runs and right now the home run to me is extragonadal germ cells. If we can harvest blood from women, purify out and store egg-producing cells, it would revolutionize ovarian function, especially if the patient freezes them at 21 then uses them at 40. The egg cells won't know 20 years have passed. Harvesting from blood is a hell of a lot easier than harvesting from ovaries."

In the meantime, the No Menopause Mouse paper gauging the health of mice never undergoing menopause will soon be sent to a journal. Tilly is also now testing his S1P on primates, to see if it can essentially create No Menopause Monkeys (and become a drug for women). He is also working on TSA, which doubles oocyte pools even in older mice, trying to find a hormonal equivalent for women. All this, and he is trying to figure out where and how his stem cells turn into oocytes—and get that pregnancy. But the bottom line, he contends, is that he is there. "We have the cell."

In May 2005, a scientist claimed to have beaten Tilly. Antonin Bukovsky of the University of Tennessee asserted in the online journal *Reproductive Biology and Endocrinology* that he collected cells from the surface of ovaries of five women ages 39 to 52, added estrogen, and grew new eggs he said were viable. Others, however, said the work was not conclusive. It did not mean Bukovsky was not right; it meant, only, that he had not proven he was right.

Will the fertility situation mimic that of the brain? Until recently, most scientists were vehement in their belief the brain possessed no stem cell—and if it did, it was either not functional or was limited. Yet in the early 1990s the existence of an adult neural stem cell (NSC) in humans was proven, and research has soared. Many studies show they are functional. In a 2003 *Nature*, San Raffaele Scientific Institute neuroscientist Angelo Vescovi showed that NSCs, injected into spinal fluid or blood, can repair some of the damage done to myelin sheaths surrounding degenerating axons. Harvard's Jeffrey Macklis showed in 2001 that cutting the brain in certain areas not known for neurogenesis (neuron formation) can attract NSCs to the areas, where they form neurons. In 2005, a University of Auckland group revealed in *Neuroscience* that there is a threefold increase of neural progenitors (stemlike cells) in the brains of autopsied Huntington's disease patients over controls, indicating that progenitors *try* to cure that disease. In a 2005 April *Journal of Neuroscience Research* Hideo Okano found that transplanted human NSCs ameliorated spinal cord injuries in monkeys. In one area of the brain alone, 500 to 1000 new neurons are made in 72-year-old humans each day.

For 100 years, it was believed there were no adult brain stem cells. The reason: few looked. When they looked, they found. The situation for neurons and oocytes may be similar, with a key difference: A huge drawback of NSCs is the fact that millions upon millions are needed for transplantation, and it is very difficult to get adult NSCs to proliferate that much.

By contrast, in theory, all you need is one oocyte stem cell to create a child.

Some pro-lifers are unhappy. In an article that calls women using their own stem cells to make eggs "artificial ova creation," pro-life campaigner Matthew O'Gorman, president of Sussex University Pro-Life

Society told Lifesitenews.com in May 2005: "The artificial harvesting of eggs is synonymous with the intention to manufacture human beings for research. This is unethical, unnecessary and unacceptable."

One way or another, women may one day soon be able to postpone infertility and/or menopause for a spell, thanks to stem cell technology. If Tilly's adult stem cell work does not pan out, perhaps his S1P work (done with the help of No Menopause Mice made from ES cells) will. Or his work finding that TSA, which works on HSCs, also seems to expand germ cells. Then there are the oocyte-like cells created from ES cells by the Max Planck Institute's Hans Scholer, and the oocyte-like cells made from skin cells by the University of Guelph's Li Julang. Or perhaps using cloning technology to transfer older women's oocyte nuclei into younger women's enucleated eggs will prove safe. The growing realization that our cells are more pliable than we knew could forever alter the arc of women's lives.

Meanwhile, women may soon visit IVF clinics to ensure not just their own futures but also their *babies'* futures, by banking away some personalized, noncontroversial hES cells for them.

*　*　*

THERE IS IRONY in the tale of Yuri Verlinsky's beginnings. He is now a high-profile Illinois IVF specialist who has apparently created more hES cell lines than any single scientist in the world: over 120. He has also whipped up an approach that, refined in the future, may compensate every IVF warrior for her woes by letting her save some personalized hES cells for her children, thus avoiding immune rejection and possibly all controversy.

Yet Verlinsky's journey has hardly been typical. He was born in Siberia and lived in Ukraine until 1978. A good student, he was Jewish, so he was not accepted as a day student to the university of his choice in Ukraine: only 2 percent of all Jewish students were admitted, according to his bio. So the U.S. geneticist whose clinic would gross $9 million a year by 2003 worked as a taxidermist while attending the University of Kharkov at night, earning a master's degree and a doctorate in genetics. When his training was over, he sought the assistance of the Hebrew Immigrant Aid Society. Within seven months he and his family arrived in the city where he would become one of the world's most daring and controversial reproductive biologists and stem cell researchers: Chicago.

For a decade Verlinsky worked as a standard IVF specialist. Then, in 1990, Alan Handyside of the University of Leeds published a paper announcing that the first baby had been born with the aid of preimplantation genetic diagnosis (PGD). Most IVF babies are born in the following way: good sperm and good eggs are selected. They are mated in a petri dish and placed in a womb. But just eyeballing the embryo is not always good enough. Many embryos possess chromosome and gene abnormalities IVF doctors know nothing about, since a cell must be destroyed for in-depth analysis.

But Handyside discovered something extraordinary. He could remove a single cell from the early morula stage of the embryo—on about day 3, when the morula is eight cells large—and the embryo would still go on to create what appeared to be a healthy child. Since every cell in a morula is identical, this allowed him to put that single cell through scores of rigorous tests that could pinpoint chromosomal abnormalities or single-gene disorders in the embryo.

Verlinsky got on board immediately, publishing numerous papers on PGD, doing so well that he was able to establish his own clinic, the Reproductive Genetics Institute, in the early 1990s. In 2001, he published the first U.S. report of parents using PGD to give birth to an HLA-matched (tissue-matched) child so they could use the baby's cord blood stem cells to save their sick child, who had a lethal blood disorder—Franconi's anemia. The sick child, Molly Nash, was born without hip sockets or thumbs. She had holes in her heart and was deaf in her left ear. She had to be tube fed. She had multiple operations, but by age three her bone marrow was failing. She was due to die by the age of eight.

Because both Jack and Lisa Nash, Molly's parents, had one copy of the mutation (you must have two to get the disease), they had a 1 in 4 chance of giving birth to a second Franconi's child, if they conceived without PGD. So, as noted in the *Journal of the American Medical Association* (*JAMA*), Verlinsky sorted through the couple's embryos using PGD, and found embryos with the same tissue typing as their first child but with normal genes. The first two cycles ended in miscarriages. The second two produced embryos with the mutation. But the fifth cycle worked. Lisa Nash was pregnant. In the middle of the pregnancy, in March of 2000, Molly's cells were revealed to be leukemic. In June, Molly's condition worsened, and the Nashes were offered the option to deliver two months early to save Molly ASAP. They refused.

In August 2000, Adam was born. His cord blood stem cells were immediately given to his sister. Molly Nash was, as of the fall of 2004, 10 years old and in school. She still had Franconi's, but she was far healthier than she had ever been. The Nashes have since used PGD to give birth to a second healthy baby. Both Adam and the new baby, as of the fall of 2004, were healthy.

While this all generated huge controversy, Americans calmed down, as witnessed by a 2004 Johns Hopkins University survey showing that 61 percent approve of using PGD to select an embryo to benefit an ailing sibling, though 57 percent disapprove of using PGD to select embryos based on sex.

Not all of Verlinsky's projects have gone over so well with the public. In 2002, he reported in *JAMA* that he helped a woman with the early-onset Alzheimer's gene to have a child without the gene via PGD. This generated even more heat than his first report. Because of the gene, the mother, a geneticist, was soon likely to be unable to recognize that child. Furthermore, the mother had a second Alzheimer's-free child via PGD. Wrote Jennifer Foote Sweeny in *Salon* magazine, "Verlinsky, whose decision to facilitate the pregnancy was questioned in a short commentary in the same issue of *JAMA,* has defended the move. His responses to criticism, as reported in *The New York Times,* have the impudent ring of a kid who, after beating up a classmate, blurts out: 'It's a free country!' and moves on to his next victim. Asked if he had any reservations about treating the couple, Verlinsky said, 'It's totally up to the patient,' adding that single-parent families like this one would ultimately be increasingly common."

Still, many ended up agreeing with Verlinsky that it was not his business to judge. The woman was married, so the children have another parent, and the question, Would those children rather not have been born at all? is a powerful one that only those children can answer, many believed.

PGD has gone on to prompt controversies that may take a *very* long time to die down, however. In the mid-2000s, a rash of deaf couples used PGD to ensure they conceived deaf children. The deaf couples stated they believed deafness to be a way of life. Increasingly, parents have been trying to use IVF to choose the sex of their child. The list of strange requests is likely to grow.

Verlinsky says he refuses to help couples only seeking sex selection, although he gives them a full workup so they all end up knowing. He has set other limits for himself. (PGD, like so much of IVF, is unregulated by the U.S. government. Many believe PGD should not be over-regulated, but could use detailed guidelines like the U.K.'s.) He once rejected a couple who asked him to identify an embryo with Down syndrome, so they could give their existing Down-affected child a similar sibling. But in general, Verlinsky has just soldiered on, by 2003 doing more than 3,000 screenings for couples. And by 2004, Verlinsky's use of PGD to help parents save children came to seem almost the norm. There was little noise when he published a May 2004 *JAMA* article reporting on his success helping five of nine couples give birth to babies who were tissue matches for children who had either leukemia or lethal anemia.

"There is strong support for using these technologies when there is a health benefit, even when that benefit is for another person, but this support coexists with deep-seated worries about where all these new technologies may be taking us," Johns Hopkins University Genetics and Public Policy Center director Kathy Hudson said. There is nervousness for another reason: it is not until these PGD babies are in their forties or fifties that anyone will know for sure if the loss of one morula-stage cell matters.

But for those parents at risk of passing on genetic and chromosomal abnormalities to their children, it offers a seductive alterative. "In the future, there will be no IVF without PGD," is how Verlinsky has repeatedly put it, unbowed. "It's human nature. It's better to check."

Within a month of that last *JAMA* PGD publication, Verlinsky revealed just how unbowed he was. At a June 2004 conference he announced that, from spare embryos—10 of which were genetically defective—donated by couples in his clinic, he had created over 40 brand-new hES cell lines. With the exception of Harvard's Doug Melton, no single scientist had to that point created so many hES cell lines. By 2005, Verlinsky's stock would soar to over 120, all isolated with private funding.

By January 2005, he had opened up a private bank of those lines in the U.K. to sell to scientists worldwide. And in February 2005, he published a paper in *Reproductive Biomedicine Online* finding that cloned hES cells create genetically abnormal sperm and oocytes over 90 percent of the time. It was completely overlooked by the press. But it marked Verlinsky's as one of only two legitimate institutes in the United States (and only the third in the world, after—it seemed—the South Korea group) to openly clone human cells.

Furthermore, in late 2004, Verlinsky filed for two potentially important patents. One outlined his unique recipe for making hES cells. The other outlined a recipe for making hES-like cells out of adult cells by enucleating the hES cells and using nuclear transfer to fuse the two together. The resulting cells had the genome of the adult cell but were as proliferative as hES cells, he said. They were able to create all kinds of different adult cells, like hES cells. (The latter work would be published in January 2006.)

But it is Verlinsky's unique recipe for making hES cells that carries the most potential import. For his banked hES cells were derived from spare IVF-clinic embryos that were in the morula stage, and many from the exact same early morula stage at which PGD is conducted: the eight-cell stage. (Others came from the later morula stage, day 4.) Until Verlinsky, most hES cells had come from the day 5 blastocyst stage, which is made up of some 250 cells, of which about 30 are hES cells.

If true, Verlinsky's work indicated that, one day, couples may regularly ask their IVF doctor to remove one morula cell from each of their embryos for storage, to be turned into personalized hES cells for their children in the future, embryologist Peter Braude enthused to *Nature*. (Indeed, a *Nature* paper backing up Verlinsky would be published in October 2005.) The cells doctors have been removing to test for genetic stability may well have been essentially hES cells all along if Verlinsky is right.

This does not remove the controversy from the option. Each cell in early-stage morula embryos—unlike the hES cells most scientists derive from blastocyst-stage embryos—can go on to form a baby on its own, a process called "embryo-splitting." However, embryo-splitting has only been done up to the 48-cell morula-stage embryo. So when scientists refine this work to allow doctors to safely remove hES cells from later-stage embryos (in the later morula or blastocyst phase) without destroying them, women will be able to set aside controversy-free hES cells for their children.

Time will tell. One thing is clear. Someday, one hES cell may, as a matter of routine, be removed from the blastocyst-stage embryo right before it is implanted, eradicating controversy. "Single cell isolation for derivation of hES cells is a good strategy, and there are few reports of any problem when done for PGD," the NIH's Mahendra Rao notes. "I would simply add that lack of information does not allow us to predict how risky this procedure is for the developing embryo."

Adds ESI CEO Alan Colman: "It's possible and not possible. Can you imagine if every child had to go through that procedure? These are always going to be labor-intensive procedures. I think that could be possible, but for rich people. It's no different from cord blood. You don't find people in Africa laying down cord blood."

Would one solution be banking an hES cell for the public every time one is banked for a child? Says Colman: "Again, possible. But practically speaking, if you take a good blastocyst and disaggregate it to make an ES line, you'd probably have a two in five chance of getting a line. So if you are just taking one or two cells, you couldn't take many more without damaging the embryo. So the odds of getting an hES cell line in any one sampling would be quite low. It would be hit or miss in terms of each individual having a line laid down for them. So . . . you're talking about the future."

Indeed, scientists caution that even if/when this is pulled off, it may not represent a Final Answer for a long time. An uncountable number of hES cell lines must be studied before scientists fully understand them. And once parents have the option of saving hES cells for their children, many will want to keep them all, instead of giving any away.

But it all offers a tantalizing hint that the time may be coming when controversy over hES cell therapy will be limited to whether it works, just like any other form of medicine.

12

THE FALL OF SEOUL
AND THE RISE OF
SAN FRANCISCO

MINUTES BEFORE THE PIG IVF procedure commenced in the siz-zling malodorous Hongseong farm, the patient was sleeping peacefully in a bed of manure. Waking to the sight of several geneticists and clon-ers in masks and gloves scrambling toward her with outstretched arms to corner and sedate her, the pig began to scream. The screaming, to the ear of a city dweller who had been expecting a sound much closer to the word "oink," sounded far too alien—far too pterodactyl-like—to be ema-nating from something so pink and flabby and familiar. It inspired seven pregnant pigs trapped in a steel bicycle-rack-like structure beyond to add to the cacophany by repeatedly throwing themselves at their bars, making the sound of a small prison outbreak.

All told, it is a relief for everyone when Byeong Chun Lee is done injecting his 150 pig/human embryos and begins sewing up the incision with large graceful motions. He looks not unlike a puppeteer working some unseen puppet as he passes a curved needle resembling a gigantic eyelash down into his patient's flesh, high into the air, down again. When the sewing is done, plastic baggies on the pig's hooves are removed, a tag with the number 183 on it is stapled to her ear, and the teams burst out of the barn. They wash up in a shed and speed off with the grateful bonhomie of escaped Bonnies and Clydes.

Minutes later their SUVs—one containing Seoul National University (SNU) cloners, one containing Chungbuk University geneticists— peel into Hongseong's center, which comprises a few faceless cement buildings backed by rice paddies and sullen, low-lying, thickly vegetated mountains. They have come for a traditional Korean meal at the Cham Pork (True Pork) restaurant, as they have every week since 2003. From outside, the restaurant looks like a feed store. Its bathroom, in a shedlike structure next door, offers holes in the floor for toilets and a beautiful view of an unearthly green rice paddy—while the restaurant itself offers an expansive view of the dusty street. But it is lovely in the dining area, with its paper-and-wood sliding doors, its quietly padding husband-and-wife, cook-and-waiter team. Seconds after sitting on straw mats on the floor, the scientists are deluged with large and small bowls of kimchi (pickled vegetables), dried seaweed, rice, tea, and, of course, pork.

It is July 2005. The crew is between two more putative Woo Suk Hwang coups—his last two, as it will turn out, before disaster strikes.

Hwang has graced the pages of Korean schoolchildren's textbooks since he cloned the nation's first cows in 1999, says Lee, who runs Hwang's lab (with Hwang and a scientist named Sung Keun Kang). Kindergartens across South Korea have been switching from wooden to traditional metal chopsticks because of Hwang's claims that his success was due to same, Lee adds, seemingly somberly.

A geneticist chimes in, noting that his crew earlier took portable ultrasounds of the seven pregnant pigs. There were dozens of clones gestating inside each, although fewer than 10 would survive per pig. All were doing well, he reports.

Lee nods, then gently removes the metal chopsticks from the hand of a visiting Western writer, who has apparently been fumbling in a distracting fashion over a single grain of rice. The metal is, as advertised, extremely slippery. He hands over wooden ones. "No cloning for you, so you can just eat," he says, smiling.

On the drive back to Seoul, Lee notes that Hwang is not the only early bird in the lab. The entire staff gets in at 6:30 A.M. to make the daily meeting at 6:45 A.M. Then, from 7:00 to 8:30 A.M., there is an English lesson. "'This is an international lab,' Dr. Hwang always says. He gets mad if people speak Korean. It will help them most to understand

English, he believes." Hwang has remained largely untouched by all the publicity, but some things have changed, Lee notes. He used to travel with a rice cooker, rice, and kimchi to save money, but he stopped when the government gave him bodyguards. They needed more food than he could carry and it would be impolite not to bring food for all.

Lee makes a few phone calls on his Samsung cell, which the giant South Korean consumer electronics company gave several staffers. The phone has GPS navigation, a still camera, an MP3 player, and a video camera. Samsung, South Korea's biggest company, is continuing to exhibit a great deal of interest in Hwang's lab.

Lee flips his phone shut and notes with pride that two of the scientists at Genetic Savings and Clone, a controversial company in California that clones people's cats for $50,000 a pop, are from Hwang's lab. But when he is asked about dog cloning, he pauses. Again, it is July 2005. Dogs have not been cloned yet, according to the literature. No animal more intelligent and sentient than a cat has been cloned. And dogs share many diseases with humans. A cloned dog would be a big deal. "If we were working on that, we could not tell you before publication," Lee says carefully. When he is told that Hwang told the writer in the car a year ago that he was working on it, Lee acquiesces a bit, clearly curious about something. "How do you think people would react to a dog clone?" he asks. When the response is: "Emotionally, for and against," he falls silent. Hwang's lab is clearly still interested in the idea.

The SUV reaches Seoul two hours later. The city is a colorful jumble of Moscow-style apartment skyscrapers and squatting buildings. Merchandise, card tables, plastic bowls of circling eels, aquariums, and proprietors spill out onto sidewalks as armies of women in tiny heels totter up and down, clutching cell phones in one hand and flowered umbrellas for shade in the other. The cell-phone ring most popular here is, for some reason, the sound of tin cans trailing a wedding car. A bus passes. On it is a billboard of two stunningly sexy Western actors in the movie *The Island,* which is about stunningly sexy human clones kept on an island until their "originals" need their organs. The movie will tank in the United States, but boffo Korean sales will result in revenues over $21 million and make it a global winner. A devout Korean Christian who dislikes Hwang's work, yet likes him, had earlier pointed out a similar bus, laughing delightedly. Hwang himself is featured on the sides of buses.

From the outside, the building in which Hwang's labs reside looks bucolic. It sits at the end of one of a series of SNU streets lined in the middle with trees, including ginkgo bilobas, that are so lush that their branches are covered in leaves like the sleeves on a Cuban congo player's shirt. Hwang's building is made of gray cement, possesses friendly aquarium-colored windows, and stands on a wooded hill facing Mount Kwanak. Out front today, workers are taking apart a greenhouse. Tomorrow it will be gone, cleared out for the construction of a $26 million four-story lab to be built just for Hwang.

Inside, in an office shared by many staffers, two cats lollygag, one coiled in a rolling chair, the other sprawled on a windowsill. Freshly laundered shirts hang off hooks. Staffers work so many hours that many just sleep over, and others keep their pets here because they would never see them otherwise. All-nighters are not totally common in Korea, but they are not totally uncommon either. At Samsung, all pretense was junked long ago: the company has dorms.

One of Hwang's four paid Korean American interns notes that Hwang occasionally climbs mountains before work. Before a visiting writer can ask at just what ungodly hour Hwang could possibly fit this in, she is whisked off down the hall to his office. A bodyguard in suit and tie can often be seen in the hall outside this office, looking as though he feels silly.

Hwang is an expressive man. At a dinner honoring him in Washington, DC, two months from now (during which U.S. Senator Tom Harkin will announce that the room may have just been in the presence of a Nobel Prize winner), Hwang will laugh often, generally with his mouth open wide and his head thrown back, sunbursts of lines breaking out around his eyes. Nothing about his demeanor belies the fact that in a matter of weeks—November—his world will fall apart. He listens with his head cocked to the side like a sleepy cat, but occasionally pokes his head forward, as attentive as a butler, as if to say, "Don't worry, I was just reflecting on the profundity of your words," before retreating again. When he is flattered and embarrassed, he beams with his mouth shut and his eyes closed, rocking his head from one side to the other, as if flattering words are blows he must dodge. Several times he scrunches his face then splays it wide open, more facial aerobics apparently designed to say, "I am listening; I am thinking; I am listening again."

When he shakes old ladies' hands, he does it double-handed, holding on with a smile. With the discipline of a (vastly more animated) Queen's Guard, he looks directly into the eyes of whomever he is speaking with, never, ever over their shoulders. When he puts his hand on his partner Gerry Schatten's shoulder, he calls him, "My brother," and he will seem to mean it, passionately.

But here in the less public confines of his office yesterday he looked slightly older than usual. He seemed to have more white hair, which he had let grow slightly longish. He paced while he talked, looking at papers and computer screens, multitasking. Soon enough the reason for the distracted air would appear to become clear. He was about to follow up his second huge coup, which happened less than two months ago— and which in turn happened only 15 months after the first—with another big one next week.

That's what would appear to be the reason, anyway. This being Korea, land of surprises, the reality would turn out to be a little different.

.   .   .

"THIS IS a great experiment in democracy."

A few weeks earlier, on April 29, 2005, only a few days before South Korea's man of the hour would appear to score his second big coup, the United States's man of the hour sat for an interview. Irving Weissman was a top dog in his country for the same reason Hwang was top dog in Korea: he appeared to be between big coups. He had pulled off the first one in November, when California voted to spend $3 billion on stem cells. (Weissman was the major scientific force behind the historic move.) Another coup was on its way.

Weissman started speaking by phone with a writer even as he simultaneously juggled two other people in the room. Many in the stem cell world have hit Weissman's office at some point, so many know what his visitors were seeing: a man who looked like a lean Santa Claus bellowing articulately and happily about everything from the evils of Russian tyranny to the perils of NIH bureaucracy to the reason stem cells are prompting a paradigm shift—while sitting in what can only be described as a gargantuan shoe box stuffed with receipts. Piles of journals and academic papers tower so high on his desk it is possible to see his face over them, little more. There are piles under his desk, hiding his feet. There are piles on the floor leaning against the walls. And there likely *were* piles on the chairs earlier, before his visitors removed them to utilize at least some of the furniture in a more conventional fashion.

"Santa Claus–like" was how a former student in the lab of a Weissman colleague once described him. This was especially true, she said, when he was compared to stem cell pioneers Fred Gage of the Salk Institute and the California Institute of Technology's David Anderson, the other members of the trio that formed the fetal stem cell company, Stem Cells Inc., in the 1990s. Driving in a car on a winding road to a wine valley north of Melbourne, Australia, in October 2003, just after that nation's first national stem cell conference, the former student said, "Weissman is Santa Claus–like. Fred is more cerebral, although Irv is too. . . . A story: We went to a baseball game. Gage said his favorite book was some philosophy book. Irv said his favorite book was a cheap thriller. They turned to ask Dave the question, then turned back, not wanting to disturb him. In the middle of the ballgame, he was reading a textbook."

Weissman's choice of confidants and colleagues is perhaps telling, since the people closest to him are nothing like him in one way; just like him in another. Among them they all have different heads and wear different hats. Irv Weissman is a hydra-headed scientist and man.

For while he is Santa Claus–esque, having brought many academic gifts to the stem cell field, he is also a tough businessman who has driven the field hard at times. Weissman is generally credited, as noted before, with having prospectively identified in 1988, via surface markers, the mouse blood stem cell. Stem cells had been difficult to find for so long because, unlike mature adult cells, they have so few distinguishing markers on their surface. They are small, bald, immature, fairly featureless.

Weissman used antibodies that he knew marked *mature* blood cells until, after a laborious process of elimination, he was left with a single cell, which odds said had to be his hematopoietic stem cell (HSC). He swiftly characterized it (mostly by its lack of markers), patented it, then found the human equivalent in 1992 and patented that, too. (There is some debate over who definitively proved the cell was found, but Weissman generally is accorded the prize.) He built a company around the

HSC, Systemix, and promptly sold a 60 percent interest in it to Sandoz, Ltd., for $392 million. The first man to patent a human stem cell, he and his stockholders were also the stem cell field's first millionaires.

Irv Weissman had caused his first great hullabaloo.

"Next thing you know, someone will patent a zygote, and you won't be able to have a baby without a license," *Discover* quoted Memorial Sloan-Kettering hematologist David Golde as saying. Many scientists were not shy about expressing their horror that a human stem cell, or the process of isolating it, could be patented.

Another man—already a respected Stanford immunologist as Weissman was—might have taken his money, returned to his job, and resolved to hullabaloo no more. Not Weissman. In the mid 1990s, he, Gage, and Anderson built a virtual company around a human neural stem cell (NSC) patent. In 1997, they sold the company to a larger biotech called Cytotherapeutics. They sold it so persuasively to the company's leaders that *their* CEO was made CEO of the company that had bought *them*. Weissman was made scientific advisory board chairman.

Shortly after: more hullabaloos. Company investors became so besotted with the Weissman group's cells that they lost interest in the company's main product—encapsulation technology—and threw all their money behind stem cells. There were layoffs. By 2000, there was little of the old company left, but blood and neural stem cells still hadn't made any money. Weissman's crew took back their company in 2000, calling it Stem Cells Inc. "There was a lot of bad blood at the time, but that's to be expected," sighed Cytotherapeutics cofounder Michael Lysaght, who left the company in 1995.

Weissman helped set a precedent with this that many have angsted over since. Bush's restrictions have not posed the only continuing problems for the stem cell field. Patents have, some say. Some researchers persistently contend that the Wisconsin Alumni Research Foundation's (WARF's) late-1990s patenting of the hES cell has placed unreasonable limitations on their work. Among other things, pharmaceutical companies do not always ask for money, and for many rights to be signed away, in the exploratory stages, as WARF does. And Weissman's patented blood stem cell angered some in a field that had essentially been using kissing cousins of his cells since the 1950s.

Furthermore, where there are patents, there are lawyers, and Stem

Cells Inc. engaged in some expected legal bouts with at least one other NSC patent holder. The fights helped drive NeuralStem—the promising neural stem cell company described in the first chapter of this book—into a tight corner. As of the spring of 2006, NeuralStem will have survived many challenges. Still, the company, going through tough times anyway, gave up the lease on its labs. It does have some hugely promising preclinical animal trials going, and freezers of fetal neural stem cells. Success with animals could lead, in 2006, to a clinical trial for victims of paralysis after heart operations in 2006. Still, Neural-Stem is an example of how patents can slow research down.

On the other hand, patents can speed things up. Would any substantial U.S. funds have gone into fetal or hES cells at all in the early years without patents? The federal government, normally science's single biggest angel, put next to no money into the field.

Indeed, as of the spring of 2005 the only group advanced enough to be confidently awaiting FDA approval for a fetal or hES cell clinical trial was Weissman's Stem Cells Inc. (for Batten's disease). And Geron, which lost over $366 million since it was formed in 1990, would likely be next. The company was performing investigational new drug–enabling studies it hoped would lead to an FDA application for a 2006 hES cell trial for spinal cord injury patients, based on exciting work on once-paralyzed rodents. It is quite possible that without the patents, no trial would have happened for a long time. (Regardless, the cat was out of the bag. As of May 2005 alone, the U.S. Patent Office issued 410 patents with the phrase "stem cells" in the abstract, with 687 more pending.)

At any rate, for months Weissman had also been plotting the hullabaloo of his life: Proposition 71. He was joined by scientists, venture capitalists, and patient groups who ponied up 175 times more campaign money to fight for the initiative than those fighting against it: $35 million versus $200,000. The measure passed by a huge margin: 59 to 41 percent. It gave California, via the California Institute for Regenerative Medicine (CIRM), an annual stem cell budget for the next 10 years that was eight times bigger than the entire annual stem cell budget of the NIH—and 300 *times bigger* than the NIH's hES cell budget.

Essentially, it called for CIRM to spend $1 million a day for 10 years on stem cells.

As noted earlier, all this was followed by an unprecedented development: NIH officials publicly chastised Bush for his inadequate hES cell funding. Some interviewed for California jobs. Even NIH stem cell task force member Ron McKay met with the initiative's financial force— real-estate mogul Robert Klein—at the end of April, albeit just for a "chat." "It was quite amazing chatting to this guy," McKay said. "He was writing down numbers like $150 million, but if he puts in another $30 million then it could be $250 million. . . . It is all small change to him. It's quite impressive. This is just an amazing prospect, there is no doubt that California is making a statement."

The statement would be heard. By May 2005, two leading stem cell scientists would respond: Harvard ear stem cell scientist Stefan Heller and University of Michigan cancer stem cell scientist Michael Clarke. Both would commit to move to Stanford. A stem cell researcher at the Texas Heart Institute would note he lost candidates to California. On May 22, *The New York Times* would write: "Up and down the East Coast, stem cell researchers are feeling the tug of a powerful, invisible force. It is a wave of recruiting calls from institutions in California seeking to expand their research programs with help from Proposition 71." The NIH's James Battey would end up staying in DC. But the NIH's Arlene Chiu would head for CIRM, and months later NIH's Mahendra Rao would accept a job with California-based Invitrogen, to name a few.

"They're excited and they should be," said Weissman. In typical multitasking Weissman fashion, the adult stem and hES cell pioneer had become immersed in CIRM administrative details. He noted CIRM would not award grants the way the NIH does. It would be a fast process "based on past record and whatever it is you want to do, 1–2 pages, max." Judges would be tops in their fields, not the mix of the "top, and top of the bottom" that is the NIH way. (NIH review committees tend to comprise top specialists along with some chosen for geographic diversity or political reasons.) And the review panels would comprise 15 scientists and 9 patient advocates. The scientists would have the vote, but "the patient advocates will be there to remind us of what we are there for," Weissman said.

Weissman—only involved with the initiative now to the extent that he was applying for grants—was excited for many reasons. To him, Proposition 71 was a movement. He compared the U.S. scientific environment to that of Stalinist Russia's. "What's gone on in this country until now is the mirror image of what happened to Russia in the 1950s, with Lysenko," he said. (Trofim Lysenko was a scientist who had the ear of Stalin.) "Stalin used ideology, Lysenko's ideology, to ban Darwinian genetics and research and it lasted 50 years. . . . Of the two ways of doing things, the way of the typical American president, and the way of the typical Russian president, Bush chose the wrong president to emulate."

He continued: "The last few years have represented the first time in our history of medical research, as far as I know, that has had the government using political and religious ideology to trump merit review in science. We have never before said, 'You can't do this research because it offends us.' We've said, 'You can't do this research because it is dangerous.' Or, 'You can't do this research because it spreads infectious disease. . . .' We regulated recombinant DNA [genetic experimentation]. We didn't ban it. And look at how many lives have been saved, as a result."

The initiative was essentially launched in January 2002, a few months after Bush's hES cell edict. Weissman was presenting results of a National Academies of Science meeting on stem cells to Bush's bioethics committee. "I presented our results in a very objective fashion to the politically appointed presidential bioethics council, and at the end, I realized they didn't care about the science or the medicine. They wanted to know my personal religious beliefs; they wanted to know when I believed life began. I said that's not possible to know. You can have an argument about it, but what I'm telling you is, if you don't do this research, here are the lives, and here are the numbers of lives, that are going to be lost. They didn't care. That's when I started having strong objections to the whole process, to the way they do things."

When Weissman returned to California, "I didn't even have to talk about it. Within a couple of months, things just started happening. It had all been reported on and everyone was on it. The Bush administration had begun to infuriate families in California, especially parents of diabetic kids" (including the initiative's major financier, Klein, who has a diabetic child). "They were infuriated that the most promising set of methods that could lead to cures for our kids were being dropped for political or ideological or religious reasons. It was the parents who drove the process and got me, Paul Berg [Nobel Prize–winning geneticist] and Rusty [Fred] Gage involved in this."

Weissman was never sure Proposition 71 would win. "I worried all the way. I didn't think California would do it. I had misunderstood the politics of California; I had misunderstood the power of the Christian Right in California. . . . But it has turned out to be a wonderful experiment in democracy."

The scientists' experiment was instantly challenged by groups with pro-life ties, which sued, and some state lawmakers who wanted changes because, coming via a constitutional amendment, CIRM lay outside the state's sunshine laws for open meetings, among other things. While waiting for the challenges to be sorted out, CIRM found itself in financial limbo.

CIRM faced other headaches. As McKay noted, "They're obviously under pressure to get this off the ground." Many feared CIRM might move too fast to clinical trial, trying to meet expectations. "Already, because of basic science research on stem cells, our understanding of Parkinson's is improving rapidly. That may be one difficulty in a Proposition 71 world. One big advantage of the NIH is, in a way, that it is so remote from reality. That people can keep doing the basic science. I think that's tremendously important."

Still, McKay said, "We do need to build this capacity to let basic science into the clinic, so we can answer questions that can't be answered [in the lab]. They will be able to do that, in California, work on these problems in hospitals." He thinks California scientists should take the cells to hospitals in the near future. "Progress will be incremental. That's how it's going to work."

Furthermore, the United States has been too cautious about clinical trials, McKay felt. He pointed to Europe, which picked up the ball on some gene therapy trials (and hosted many of the first heart/bone marrow stem cell trials). "Those trials are working. They are not perfect. But a significant portion of the patients receive major benefit. Americans want Hollywood endings, but the way it works is you make incremental progress."

Whether CIRM worked might also depend on how interdisciplinary it could get. Controlling those potent cells requires knowledge of how the body does it. And as the body is interdisciplinary, so should the science be, McKay and others believed. Furthermore, because the way stem cells interact is "so complex, a lot of this is going to require some real technical stability to pull off." That is, the European model—where many scientists receive a salary and do not constantly seek grants—may be better for stem cells. Can highly skilled, experienced, interdisciplinary, long-term teams be built in California? "I think that is going to be *the* question. The other day at an NIH panel meeting, everyone was supposed to stand up and say something interesting at the end. Three other people said, 'We need to grow hES cells better.' I said, 'Our institutions may not be structured to support this kind of work right now.'" McKay laughed. "It went over like a lead balloon. . . . The NIH could get a little more creative."

Regardless, CIRM spurred competition from other states. In 2006, New York's legislature was due to vote on two multi-million-dollar stem cell bills. After Pennsylvania legislators worried their scientists would Go West, a $500 million stem cell bill was introduced there in 2005. By October 2005, Pennsylvania, New Jersey, and Delaware would be considering a $1 billion joint stem cell initiative, and bills or executive orders would be passed allowing expansive hES cell research in Massachusetts, New Jersey, Connecticut, and Illinois. Maryland would join them in 2006. Initiatives would be launched in Florida and Missouri.

The rest of the world was not taking California lying down, either. In April 2005, nations from Japan to Korea held the first meeting of a large alliance, the Asia-Pacific Stem Cell Network.

But the exhilarating fact remained that $1 million a day for 10 years was an enormous amount of money. "I think there's no doubt that, wherever I am working in the world in this field, I'd better get used to the idea that I'm going to be flying to L.A. and San Diego and San Francisco with reasonable frequency," said McKay. The bottom line might be that "even if things don't end up particularly well organized in California, they're going to get done. This field is going to flourish over the next five years in, and because of, California."

As if in response, within days of McKay's prediction, the United States's leading hES cell company, California-based Geron, would publish exciting news: scientists finally differentiated hES cells to blood cells that reconstituted a mouse immune system. The building blocks to create a chimeric immune system like the ones created at MGH that would accept hES cells—and thus someday perhaps solve the immune rejection problem associated with the cells—were laid down. "We are particularly excited by the developments," CEO Thomas Okarma would announce. "We are continuing to develop hematopoietic chimerism approaches as one means to limit the need for long-term immunosuppression for human embryonic stem-cell-derived grafts." A major hES cell technical hurdle would appear to have been cleared.

But because this was the ever-changing universe of the stem cell, a mere four weeks later, the field would appear to leap far beyond that hurdle.

* * *

IT ALL UNFURLED in a blindingly familiar way. On May 16, 2005, two legislators in the U.S. House of Representatives nearly came to blows when debating whether to vote on a new bill that would override the president's stem cell edict and let scientists use federal money to derive more hES cells from spare IVF-clinic embryos destined to be discarded. The bill had 201 cosponsors, only 17 votes shy of the 218 needed to win, and cosponsors believed they had a commitment for the remaining votes. The House decided to vote the next week, but not before Representative Rick Renzi (R-Arizona) had to be physically separated from Representative Mark Kirk (R-Illinois) at one point, *The Washington Post* reported.

Then, four days later, on Thursday, May 19, 15 months after the February 2004 South Korean cloning coup, the wires and the Internet begin to light up.

The South Koreans appeared to have done it again.

The barrage of stories began with another broken embargo. At 11:00 A.M., Reuters posted a story noting that Woo Suk Hwang's team had cloned 11 more hES cell lines and published the work in *Science*. There was a lull, then, at seconds past the official 2:00 P.M. embargo time, both UPI and the AP came out with similar stories. UPI reported that Hwang made the new announcement in London the day before, while visiting cloner Ian Wilmut. UPI added that Hwang's new cloned human cells had all been grown on human feeders, not mouse feeders—rendering them theoretically clinic-ready, unlike the presidential hES cells.

"It's unbelievably good news for patients," ACT's Robert Lanza told UPI. "It makes the technology accessible and practical to implement on a clinical scale." Calling the work "a major medical milestone," he added, "We've had hurdle after hurdle thrown at us in this country, both politically and financially. Unfortunately, you're going to see more and more of the major stem cell breakthroughs occurring overseas."

At about the same time *The New Scientist* posted a story headlined: "Cloned Human Embryos Deliver Tailored Stem Cells." This article noted that Hwang pulled it off with Gerald Schatten of the University of Pittsburgh, the same Schatten who had worried human cells could never be cloned. Astonishingly, the Koreans said an average of only 12 eggs per hES cell line was needed when the eggs came from women under 30. This meant a cloned hES cell line could be derived from *each egg donor*. "This paper has completely proven [adult] cell nuclear transfer as a technique in humans," Stephen Minger of King's College said. Eighteen women donated 185 eggs, of which 125 were from women under 30. "Hwang's laboratory is now way ahead of the field. . . . There is a good chance that the U.S. will be left behind as the situation on stem cell research there becomes more fragmented and incoherent."

After MSNBC, *Forbes*, Healthopedia, and CNN came out with stories that day, it was clear that even normally skeptical scientists were impressed. "This paper will be of major impact," MIT's Rudy Jaenisch told CNN. "The argument that it will not work in humans will not be tenable after this." Noting that many of the lines came from people with genetic disorders, which should let scientists see the progression of the diseases, neuroscientist Fred Gage of the Salk Institute for Biological Studies said: "Gigantic advance." *The Washington Post* quoted Johns Hopkins University human embryonic germ cell pioneer John Gearhart exalting: "They have increased the efficiency tenfold over what their paper was a year ago, and this is very important. It's kind of remarkable. It tells you how quickly things are moving." The *Post* also quoted Richard Doerflinger, deputy director of the Secretariat for Pro-Life Activities of the U.S. Conference of Catholic Bishops, as saying, "To say something was initially impossible but is now possible is not enough. We have to make moral decisions about whether we should do this."

By 11:30 A.M. on Friday, May 20, 2005, the story had appeared in some 300 publications; over 900 by 10:13 P.M. By noon on Sunday, the day before the U.S. House was scheduled to debate a bill to overturn Bush's hES cell edict, that number topped 1,000. President Bush threatened to veto the House bill, in the same breath denouncing Korean cloning even though it had nothing to do with the bill. Senator Arlen Specter threatened to counter Bush's veto by attaching an identical bill to an appropriations bill, adding: "The United States is being left farther behind every day, this morning by South Korea." Journalists debated it all on *This Week with George Stepanopolous*. *The New York Times* reminded readers that, even should the new hES cell bill pass, therapeutic cloning was still up in the air.

The same day, Seoul announced that Hwang needed help. He had been differentiating his cloned hES cells and trying them on animals, but he wanted to do more. Since 35 labs around the world offered aid, Seoul would establish a global consortium to work on getting cloned hES cells into the clinic.

Within weeks, the vortex of the stem cell universe had spun away from California—which had only just arrived there—back to South Korea again. Singapore stem cell researcher Vic Nurcombe would note, "The South Koreans have done this so fast that it beggars belief. It is so astonishingly fast." The fact that the crew retrieved so many lines from so few eggs, he said, "has taken my breath away. That's extraordinary. It's gobsocking. They have broken the dam. Who'd have thought it?"

Of course, a few days later, attention would swing back to Capitol Hill. On Tuesday, May 23, 2005, the House of Representatives fought in earnest, for over four hours, over whether to finally defy the Bush edict and allow federal funding for more hES cell lines derived from spare IVF-clinic embryos.

The debate ranged from emotional to wildly so, the tone set from the moment Sherrod Brown (D-Ohio) read with his right shoulder cocked forward, as if preparing to bound over the podium and wrestle the room. One pro-life, pro-hES-cell representative cried when he discussed a dying six-year-old girl who told him he was the only one who could help her. A pro-life, *anti*-hES-cell representative (Todd Akin, R-Missouri) cried when he read a story his daughter wrote about a depressed human clone. Two bills were debated, the first on cord blood stem cells. But most speeches focused on hES cells, and many were delivered at a volume just shy of a yell.

The two camps were organized. Many if not most of the anti-hES-cell troops pointed out that "we were all embryos once." Stunts abounded. One representative opened a thick book containing adult stem cell therapies—which numbered, the anti-hES-cell troops claimed constantly, 57—then opened a thin book containing hES cell therapies which numbered—as the anti-hES-cell troops also noted constantly— zero. Other anti-hES-cell representatives stood in front of garagantuan lists of the same therapies, with the same "nothing" listed under hES cells. That this had something to do with the fact that the adult stem cells in question, largely blood stem cells, had been around for 50 years, while hES cells had only been around for 3, the anti-hES-cell troops ignored. They also ignored the fact that, under Bush, more than $565 million *more* had been pumped into adult stem cells through the end of 2004, while only $55 million had been pumped, in total, into hES cells. A doctor in the anti-hES-cell camp also bizarrely claimed that the only place one could even find hES cell work was in "obscure rat and mouse journals," referring apparently to *Nature, Science,* and the *New England Journal of Medicine (NEJM).*

The anti-hES-cell camp repeatedly referred to what they called the "miraculous" ability of adult stem cells to form all functioning adult tissues just like hES cells, despite the fact that this had not been proven *in a single paper,* as the pro-hES-cell camp constantly pointed out to no avail. Pictures of Snowflake Children—children who were adopted as spare IVF-clinic embryos and came to visit Congress the day before— abounded. Most were not in the room: they were posing for more pictures with Bush. (While the Snowflake program is clearly a good thing, as many scientists have noted, it does not solve the spare IVF-clinic embryo problem: only 2 percent of embryos are put up for adoption.)

The combatants on the other side were not stunt-free, either. Pictures of sick children who could benefit from hES cells popped up behind countless pro-hES-cell warriors. Constant reference was made to the indelicate notion of turning "medical waste" into "medical miracles." After anti-hES-cell crusader Tom DeLay (R-Texas), who was under investigation for unethical behavior, made a speech about the immorality of hES cells and the "grisliness" of the bill, Representative Pete Stark (D-California) scathingly noted that he didn't need a lecture on "moral leadership" from DeLay. One pro-hES-cell representative constantly noted that federal dollars would not be used to procure hES cells under the new hES cell bill—neglecting to note, until the end of the debate, that she'd been parsing words: federal dollars would not be used to *procure* the cells, but they would be used for work *on* them. Representatives from both sides jumped to the podium unbidden when incensed. And to help make points both sides brought into the national spotlight the names of family members who were sick or who died.

But the bulk of the strange stunts unquestionably came from the anti-hES-cell side. Representative Michael Burgess (R-Texas), a doctor, played the washing-machine sound of a many-months'-old fetal heart into his microphone, without noting that neither hES cells, nor the five-day-old embryos from which they come, *have* hearts. The anti-hES-cell side constantly claimed hES cells cause tumors, never mentioning that this occurs mainly when they are undifferentiated, not differentiated— and never mentioning that *adult* stem cells are believed by many to be the cause of many cancers. A typical comment was made by Joseph Pitts (R-Pennsylvania), who offered the unappetizing and patently untrue statement that "the only thing" hES cells had created were "dead embryos and dead lab rats with tumors." Anti-hES-cell representatives constantly claimed "adult cells" could cure diabetes, never admitting they were referring not to stem cells, but to islet cells that cure nothing, just alleviate symptoms for a while.

An anti-hES-cell legislator flashed a picture of a woman who was supposedly cured of paralysis with adult stem cells. It was untrue: she merely had some sensation restored to her limbs. (She later wrote out-raged letters to the congressman.) Anti-hES-cell Representative Nancy Kaptur (D-Ohio) protested that there was something "all too convenient" about the fact that the South Korean paper came out just days before this debate. "I ask myself: which companies were behind it?" she yelled. (She added a claim that "the normal process was sub-verted" because, she said, there was never a subcommittee hearing on this bill, although numerous meetings on stem cells had been held ad nauseam over the past four years. "What are they hiding?" she asked dramatically.) Over and over anti-hES-cell representatives expressed their outrage at the notion of using federal tax dollars for hES cell research when "millions of Americans are pro-life," never acknowledg-ing that millions of Americans are against many things government does, like go to war, that they must pay for. Burt Stupak of Michigan, a democratic pro-lifer, offered his not-so-medical opinion that hES cells cause nothing but "tumors and death" and that a vote for the bill ensured "the cloning of a human baby"—even though the bill wasn't about cloning.

Anti-hES-cell Representative Steve King (R-Iowa) suggested that if the hES cell bill passed, a kind of "Nazi regime" experimentation might result. Anti-hES-cell soldier Roy Blunt (R-Missouri), majority whip, made the claim, "We are not whipping this vote." (That is, the anti-hES-cell troops were not doing any politicking.) "We feel this is a matter of real conscience." Over and over the anti-hES-cell troops claimed there was not a shred of evidence hES cells worked, despite the mounting numbers of studies indicating they, indeed, do.

Yet after 6:00 P.M., when the vote was finally taken, the hES cell bill passed 238–194 with 50 Republicans voting "yes," supporting the notion that since August 9, 2001, a rebellion against the Bush edict of that day had been building in the United States. First U.S. scientists rebelled, some leaving the country, others forming state and regional societies that aimed to create strength in numbers to move the science along. Then many states began defying the edict. Then members of the NIH began rebelling.

Now the Republican-led House of Representatives had defied the Bush edict, as well.

And, it shortly became clear, the Senate might not be far behind. The next day, House members presented to Senator Arlen Specter a copy of the legislation topped with a red bow. It was almost two years since Specter browbeat NIH officials for not pushing hard enough for a bill like this. Specter looked very different: he was bald, his trademark lamb hair having fallen out with chemo treatments he was undergoing for Hodgkin's lymphoma. He had become his own poster child for stem cells. But the moment was pure Specter. Instead of professing how moved he was that the moment he longed for had arrived, he was all blunt business. Bush had repeatedly threatened to veto any hES cell bill, and while the House bill did not pass with enough of a majority to override a Bush veto, Specter said, the Senate might pull it off. "If a veto threat is going to come from the White House, then the response from Congress is to override the veto, if we can. Last year we had a letter signed by some 58 senators, and we had about 20 more in the wings. I think if it really comes down to a showdown, we will have enough in the United States Senate to override a veto."

But the House did not. And that day, Senator Sam Brownback threatened that the minority of Republican senators against a similar bill in the Senate would stage a filibuster to keep it from passing.

Still, whatever vote was coming (in 2006—the 2005 Senate vote would be postponed) it would not affect the California initiative. And it would not impact South Korea, to whom international attention boomeranged back when it was reported that Hwang would receive an additional $1 million a year—added to the $2 million a year he was currently receiving—from his exhilarated government.

•   •   •

HWANG IN late July 2005 is fresh off what appears to be his second human cell cloning coup and about to jump into a third. But while he looks a bit frazzled today, he is as gracious as ever, scribbling his cell number on a card and handing it to a visitor while bowing, receiving the card of the visitor in both hands while bowing, and pretending to study it respectfully, as is the custom.

He is dressed in a white shirt and tie. His office is small and crammed with his secretary's desk, two couches, and his own desk, which is sandwiched between a window and a bookshelf. Landline and cell phones, his and hers, ring constantly. E-mails ding constantly. People quickly poke their heads in the door, then quickly disappear, constantly. One, SNU assistant professor Sung Keun Kang, the third member of the trio that heads the lab, stays.

There has been a small galaxy of luminaries cycling through lately, Hwang admits. Cloners Wilmut and Schatten are only two. He stretches his arm far down the back of the couch, relaxing a moment. University of Minnesota cloning expert Jose Cibelli. Curt Civin, editor of *Stem Cells*. Harvard hES cell researchers Doug Melton and Kevin Eggan. Memorial Sloan-Kettering Cancer Center hES cell researcher Lorenz Studer (to whom Hwang has sent both students and cloned stem cells) and blood stem cell pioneer Malcolm Moore of Rockefeller University. Roger Pedersen, who left the University of California at San

Francisco for the U.K. H. L. Trivedi, a kidney transplantation expert from India. UCSF hES cell researcher Linda Guitis will visit. Ron McKay, NIH stem cell task force member, may collaborate. Visiting this week is the Max Planck Institute's Hans Scholer, who made mouse oocytes from mouse ES cells.

Hwang shifts, snapping his hands together in his lap with his legs crossed, as contained now as earlier he was sprawled. He confirms that his lab clones 1,200 animal cells a day, "each student, 100 eggs a day." On a wall is what was clearly a gift—a *Nature* shot of him beaming in clean-room scrubs—framed. He looks like a pale blue astronaut (or a pale blue earless bunny) clowning for unseen kids. A plastic cow dangles from a cabinet.

Evan Snyder, a California Burnham Institute neural stem cell expert, may collaborate. Jonas Frisen, a Swedish neural stem cell pioneer, has visited and returns soon. Riken's Shin-Inchi Nishikawa has come several times. Hwang throws himself into a chair; checks his ringing cell phone; continues. Israel's Benjamin Reubinoff and Singapore's Alan Colman and Ariff Bongso hope to visit. . . .

"We have exchanged material transfer agreements [MTAs] with Harvard and Johns Hopkins," he says. "Also with a Hong Kong University team. We are making MTAs on a case-by-case basis."

Says Kang, "We have to make sure others' research matches ours."

Suddenly excited, Hwang confirms that the lab has already begun trying its cloned human cells on animal disease models. It turns out that, after the first coup in February 2004, he had volunteered to stop research while government officials cobbled together firm cloning laws, but he only did so from February to September. He leapt right back in that fall. The government, as usual, accommodated him.

He stands now, hands in pockets, as though still antsy about that lost six or so months. Says Kang, "For this recent paper, we came up with more than five improvements on the original protocol."

Hwang heads to a dry erase board and illustrates Kang's words. Between them they explain that the main improvement this time was the way the team extracted the hES cells: letting the embryos culture longer, then hatch naturally. "Hatching is easier." Most hES cell scientists mechanically dissociate hES cells from the inner cell mass. But Hwang's crew just placed a small slit in the zona pellucida (outer membrane) and, as the embryo expanded, its hES cells essentially gently extruded themselves, he says.

Hwang is pacing now, hands in pockets. "So we wait six or seven days, instead of the normal five. And we do not use enzymes." He sits, then leans forward, staring intently at his visitor.

Another call. Hwang gets it. Kang notes that they borrowed the idea of hatching from IVF, in which assisted hatching is done for many infertile couples. The zona pellucida can be very tough in older eggs, so IVF docs "induce a slit using a microneedle," Hwang says.

Someone pops his head in. A world leader in cellular signal transduction is outside, having just dropped by to say *anyeong haseyo* (hello). Hwang hangs up to step out and say *anyeong haseyo* back.

Kang says the crew expects differentiation work—proving the crew's cloned hES cells are functional in animals—will move swiftly. "We have no holidays in this lab," he says. "Many work until 9–10:00 P.M., others work until 11:00 P.M." In the after-hours, much of the work is talk: everyone discusses their experiments. Hwang, who has returned, buzzes again. His cell again. He answers, apologizes, and slips out of the room—again. It is 10:35 A.M.

A few minutes later, Hwang is back. "Would you like to see the labs now?"

He moves into the hall, expertly exchanges his shoes for rubber slippers outside doors leading to a cordoned-off hallway, shouts something to a dark figure moving behind blinds in a window in the hall, and waits patiently with a bemused look on his face as his foreign visitor hops about and nearly falls over trying to match his shoe-changing wizardry. On a screen, up pops what looks like a squarish piece of mica with a brown circle in the middle of it. "Nuclear transfer ES cells from an eight-year-old spinal cord injured boy," Hwang says.

There they are: what Hwang says are cloned hES cells. The colonies are in a strange squarish shape, Hwang explains, because he grew the cells on human feeders that have a slightly different quality from the mouse feeders most scientists use. (They are not always square.) These human feeder cells came from skin cells from this boy's own arm, to make sure the so-called cloned hES cells were totally happy. They are happy even after 46 passages, or over 100 doublings. Theoretically, the hES cells from this boy with spinal cord injury could provide cells for study to perhaps every spinal cord unit in the world, if they do not gain mutations. This is big partly because mature human cells cannot replicate, making it difficult to get enough diseased cells for study. But it is also big for genetic disease research. By bringing adult cells back to the embryonic state, one may see how certain diseases form and develop. Finally, this boy could well be the first to benefit from any therapies derived from his cells, since his body will never reject them. All the above, applies, of course, only if the cells are real clones. But Hwang's human cell cloning work has passed muster with one of the world's top journals *twice* now. Most of the hES cell field's top scientists have made pilgrimages to his door—as earlier they did to Israel. There seems no reason not to believe these are real clones.

Hwang whisks through another set of doors, buzzes through a lab, and enters a changing area. Within seconds, cued by, apparently, ESP, two students dressed in more light blue astronaut suits (head-to-toe one-piece scrubs) suddenly appear in the air shower, pop into the room, and dress the visitor in a similar suit. They are meticulous. One pinches the mask and puffs it out, then hooks it around the visitor's ears. The other pops the visitor's legs into the pants, offering a shoulder for balance. Hwang waits, having dressed in seconds on his own.

Next month, University of Toronto public health scientist Abdallah Daar will echo many when he notes that Hwang seems to be the kind of gifted scientist who, when he listens, can hear grass grow. "People revere him as a leader; he works incredibly hard; he is in there with his people all the time; he has a vision," Daar says. "But I think the guy is also just a great scientist. In this context, a great scientist is a guy who knows how to handle cells in a petri dish. In cell culture work there are days when you are tired and it won't work, and other days when you do it with love and care and passion and it works. There is a bit of art to it, the ability to change a little bit the conditions, to raise the temperature a bit, to add this molecule, to concoct this way, to make cell concentration a little higher or lower. . . . Hwang seems to have got it."

Hwang isn't working a petri dish right now, but it is certainly easy to see some of the other things he has down to an art form: the superman-fast changing abilities; the materialization of assistants to move fumbling visitors along ASAP, minimizing waste of time.

At last, into the air shower, which looks like a phone booth lined with air spigots. Then out (following a sign saying, Please Way Out) into the famous animal cloning room. It is huge. Seated at a long table like an antiseptic, outer-space Last Supper, are a dozen staffers in astronaut suits withdrawing oocytes from cow and pig ovaries that they pluck off pink piles on blue napkins scattered up and down the table. The pig ovaries look like chickpeas; the cow ovaries look like frankfurters. The staffers look as though they are taking miniature vacuums to the ovaries. They push syringes in gently, draw them out slightly. They do not talk. This helps keep the room as germ-free as possible, it is explained.

Hwang moves to one side of the lab, where good and bad embryos are separated in petri dishes. He moves to the other side, where embryos are culturing in incubators (20 hours for cows; 40 to 44 hours for pigs).

The next room offers the payoff: a handful of scientists cloning several cells an hour, their efforts beamed onto screens above them. There are 10—of the lab's total of 15—micromanipulators in this room. The three scientists here now are working on pig oocytes. A woman takes a long sharp pipette and slides it through the top of an egg and out the other side, as if stringing a bead on a necklace. She has done this near a part of the egg that contains a dark shadow, the polar body or nucleus. She gently squeezes two pipettes together, one still stoked through the top layer of the cell and one resting on the outside of the cell. The nucleus oozes out the hole, along with a tiny amount of cytoplasm, looking like a swarm of soapsuds. This method of gentle squeezing, while not devised by Hwang's team, was a key reason for its success, he has claimed. He makes a smaller slit than other scientists this way.

Another scientist has an enucleated egg attached to a docking pipette. He sends a pipette in. It is filled with a fibroblast. The pipette injects the fibroblast into a slit the scientist made in the egg earlier. To the side of the room is an electrofusion machine that will zap these embryos to life.

Hwang clutches the arm of a visitor who asks why he doesn't put only the nucleus of the adult cells in. "No! No! The whole cell! Why would you do that to a cell, cut out the nucleus?" he says dramatically. He smiles. "We compared. It is better this way, with large animals. The cell is not hurt."

"Good job," Hwang says to the scientist.

Back in Hwang's office, he disappears to greet a new visitor. Kang sits near a row of mementos from past visitors, which includes a red rhino with "Heinz Catsup" stamped on its back. He orders a lunch of Chinese fried rice and a black Korean seafood soup. As he eats, Kang explains there are 120 researchers including local collaborators at both SNU and Hanyang University. Forty are Hwang's core staffers. The lab will run off of $3 million annually for the next five years; the new building is costing $26 million. The building will have 30 micromanipulators. But things certainly seem to be moving fast without the new millions or the new building. The group has new cloning papers coming out soon, says Kang.

Everyone in the lab gets to "try everything" except hES cell work at this point (only 10 staffers work on human cells), Kang says. The inner core includes techies who took the jobs so they will be lined up to pounce on a student opening as soon as it occurs. "Our goal is to make everyone in this lab a valuable person, so it is a popular lab." The crew is collaborating with some top cell differentiation people, and some of the world's top gene manipulators, so they can correct genetic diseases in cloned cells. Thus most here are learning cloning, stem cell differentiation, *and* genetic techniques.

Such training is critical, he says. "Human oocytes are not easy to handle. You need to have full experience with animal, first. And they are sticky. If the pipette sticks to the oocytes, you have to adjust the location of the pipette. If you spend too much time on them, they will degenerate in vitro." To do it perfectly, "you need a year of practice, every day." And when you get away from it a while, you lose it, he says. "You have to practice again. Every day of practice is important."

The lab has gotten exotic from time to time. The team at one point tried to clone a mammoth but "it was very difficult to find intact cells." And before the human coup, they were able to make part-cow/part-human cloned blastocysts (enucleated cow eggs fused with human DNA), although they couldn't get hES cells from them.

They also have cloned transgenic cows that produce a human protein in their milk. "Cows are easy to clone, but they are subject to stress," Kang says. One human-protein cow died. So they do more pigs, which are subject to less stress, and can give birth to 10 to 15 fetuses at a time, perfect for cloning since it so often fails. The government plans to build them a "clean farm" to go with the clean after-birth facility. The crew is excited about the pigs, which are miniature and germ-free in addition to possessing genetic changes that will hopefully make them far less rejectable. "Different human genes are being transduced," he says. The pregnancy rate for the crews' cloned pigs is 30 percent; the delivery rate, 10 percent. The crew has implanted hundreds of embryos two times a week since 2003. The ones that live seem to have few abnormalities, Kang claims. If these figures are correct, Hwang has more than 100 cloned pigs—and scores of cloned part-human pigs—scampering about his antiseptic labs.

That first 2004 human cell cloning paper, Kang says, checking his watch and winding down, was a result of "pure donation, out of the pocket of a company head. No commitments. The second [human cell cloning] paper was funded by the government. Several companies have been interested in a more formal arrangement, but Dr. Hwang thinks you can't do whatever you want when supported by a company."

Is the lab's motto really "moving the heart of the sky"? Yes, he says laughing, although another way to put it is: "Do your best until the heart of God is moved by your effort, then He will help you."

And yes, Kang adds, like Korea's kindergarteners, some Western scientists out there are now using metal chopsticks to see if it will make them better cloners.

He grins.

•   •   •

THERE ARE many markets surrounding SNU Hospital. The markets offer mesmerizing arrays of just about everything: headless chickens racked up with one arm on their bellies like lines of headless pregnant women; rainbow displays of dead fish; bags of dried smelt; roots spilling wild beards over the side of barrels; baskets full of golden dried silkworms, treats for the kids. Seoul is so warm this week that fans are set up to blow cool air, not onto sweating customers, but onto sweating meats. Vending machines offering tiny cups filled with caffeine-free fruit drinks are scattered in every subway. Heat-loving *maemi* are so profligate their racket can drown out that of buses and cars.

SNU Hospital, which at this point seems likely to become the first in the world to give cloned human hES cells to patients, is a tall white skyscraper surrounded by underbrush screaming with *maemi*. High in the hospital, transplantation doctor Curie Ahn sits at her desk, jetlagged. She has just returned from visits to a New York blood bank and a Wisconsin IVF center. She was investigating how to run them because Hwang is about to open the first stem cell bank to offer cloned hES cells, she says.

She is jetlagged, but that doesn't stop her from tending to her parched visitor, ordering up two Cokes and a water. "I am sorry, we have no Diet Cokes," she says. "And we have no big Cokes like in America. But here are two little Cokes for you, OK?"

The office looks out over a gorgeous and immaculate green and red temple. "I know, I keep trying to get them to clean it up but they won't," she says smiling. She has a dry sense of humor, reddish hair, and laughs loudly and often.

Ahn got involved with Hwang because of the Confucian tradition of worshipping the body, which has left Koreans with an unusually severe shortage of organs and Ahn, as a transplantation doctor, with constant problems. Few pony up organs after death. "It is so terrible to see my patients suffer. So hard for them to get jobs when on dialysis. They suffer so much. Last year only 15 patients in all of Korea received heart/lung transplants. I went to Dr. Hwang after one of his pig papers. He was famous for cloned pigs. We decided to try to make transplantable pig organs. This was 2002."

Many of the pigs are being bred for their kidneys, Ahn says, although some are being bred for their hearts and islet cells. The human gene that Hwang's crew has been "knocking in" to their animals is complement regulating protein; the pig gene the crew plans to "knock out" is alpha-1-3-Galactosyltransferase. Pigs born with these human features grow organs that avoid the hyperacute rejection that occurs with normal animal transplants in humans, but "acute vascular rejection still seems to occur," Ahn says, "so we may not have perfected the approach yet. Still, it's amazing. We are testing on animals." Xenotransplantation has had its ups and downs through the years. Finding the right combo of "knock in" and "knock out" genes has been brutally difficult. But Hwang's team is following recipes similar to those of Harvard, the Mayo Clinic, and Revivicor, a company championed by transplantation pioneer Thomas Starzl of the University of Pittsburgh. There is a general sense that the right genes are finally being hunted down—and outside of Revivicor, no one possesses as many cloned transgenic pigs.

That night, Ahn takes her staff out to dinner. After kicking off their shoes, all sit on leather mats on the floor. There are barbecue grills in the middle of the table. "He is a special person," Ahn says of Hwang. "He is up at 4:00 A.M., doing his Tau Buddhism by 4:30 A.M., gets in at 6:00 A.M. We speak five times a week at 6:30 in the morning. It was very hard after that second paper. I was with him when he made the announcement in London. Everyone wants to talk to him; so many want to challenge him. We are just scientists, what do we know about ethics? The archbishop, religious extremists, animal rights people. . . . It was a tsunami after each of those papers."

She pauses as the waitress piles more strips of raw meat onto the barbecue.

"He has turned down many Big Pharmaceuticals. He wants it all to be not-for-profit. And he is humble still." She too tells the story of Hwang once bringing a rice cooker strapped to his side when he traveled to other countries. She has a different reason for why he no longer does it: "Now everyone is watching so he doesn't do it." But Hwang clings to many of his old habits, she says. "He still keeps the schedule very packed. He will shower at a former student's house in other countries so he does not have to rent a hotel room and can be back on the plane for sleep at night. He says every penny saved on these trips can go to one of his students' education in America. . . . I went to New York alone recently, to discuss the new stem cell bank we are starting. I was finally able to go to the Broadway, because Dr. Hwang was not with me, making me get back on the plane to sleep. I saw *Mamma Mia*. It was good."

She laughs.

Does Ahn think the South Korean government has what it takes to become a leader in the highly creative and volatile biotech field, especially given the stifling dictatorships in its not-so-distant past? She muses. "Park stole nothing, that was good," she says, of a dictator from the 1960s through the early 1980s. "But he did shoot his opponents." She laughs again. "And the next two dictators went to jail. . . ." The current president, while democratically elected, is believed by many to be lacking in some competence, she says. "But I think that is because this has been a one-party government for so long, the party in power has no idea how to govern. They will learn."

Indeed, University of Toronto political scientist Joseph Wong will note next month that he believes Korea's desire to become a biotech hub may actually have helped boost democracy here. He sees the imperative to go biotech—the imperative to fight China's ability to make everything cheaper by changing from an "industrial learning" to a "knowledge creation" economy—as having been so critical to Korea that it is changing in some key ways to meet biotech's uncertainties and demands. "In the 1980s and 1990s the government realized that competition from China meant they were going to have to become knowledge-based. The government started talking biotechnology and nanotechnology, but you can't really plan for something like that, no one person had all the right skills—in this new field no one even knows *today* what range of skills are needed. So they made the statement that they were emphasizing these technologies, and threw money on the table—and all of a sudden you have the Ministry of Commerce and Technology vying for it; the Ministry of Commerce and Industry vying; the Ministry of Health and Welfare . . ." The new "interministerial competition," combined with other powerful factors (like the 1997 financial crisis), has been forcing "the dismantling of the old-style coordinated developmental state."

Regardless, perhaps the quality that will best guide South Korea through the biotech era is its resilience, says Shin Yong Moon.

The heart of South Korea's stem cell operation, the Stem Cell Research Center (SCRC), is located below SNU Hospital. It is a pleasant wood-and-concrete building hidden behind a pleasant thruway lined with peddlers and a wall of traditional stone *tal* masks. The office of SCRC head Moon is almost militantly modest, filled with bare-necessity furniture and few photos. "I was three years old at the time of the Korean War," he says. "We had no rice, no school, no tables. All Koreans did something like begging. We were very happy that the American soldiers came. They built this building." He holds up his hand. "There were no crayons in school. I got crayons from the United States. Milk came from USAID."

Fifty years later South Korea, Moon notes, is home to two of the top consumer electronics and car companies in the world. It is the globe's most wired country in terms of broadband penetration. "When Samsung first invested in IT, nobody believed it would be successful here. Now: everywhere. My son can open a door or turn on the TV or a DVD or a gas range from far away, via computers. It is fantastic." And the SCRC has 40 hES cell lines—not including its cloned hES cell lines—which is far more than most nations (although Korea has not yet published evidence of all the lines). Some 1,942 people are involved in the SCRC's 60 projects: 30 principal investigators, over 600 PhDs, over 700 master's degrees, over 500 undergrads. The rest of the SCRC has not scored the coups that Hwang has. But this network, which allows far more innovative stem cell work, on far more hES cells, than the United States—and which is being run by a man who as a child depended on the United States for milk and crayons—is very well-positioned.

In many ways, biotech demands more from a nation than does IT: more multitasking, more creativity, and far, far more time, Wong will note a few days later. Thus, what Hwang has apparently done with cloning, generating such excitement that he has attained "rock star" status, is to "buy time. One of the gazillion dollar questions with respect to biotech is how much time governments and industries are going to be afforded in terms of enthusiasm"—a big problem given slowing economies and the fact that biotech is such a long-term proposition. "In many countries, when I talk to investors or government officials or firms, they are waiting for the biotech success story. They know it is the success story that is going to give them the time needed" for research. "Now South Korea has that success story, and that is great."

*   *   *

ON AUGUST 3, 2005, it happens again. Hwang and former naysayer Gerald Schatten introduce in *Nature* Snuppy, the first cloned dog. "South Korean Clones Dog, Causing Critics to Bark," reads the *Athens-Banner Herald* at 1:00 A.M. on August 4. "Give the Dog a Clone," chortles the U.K.'s *Sun*. *Businessweek*, *The Sydney Morning Herald*, *China Daily*, and *The Irish Examiner* all chime in. Writes *The New York Times* in a front page story on August 4: "South Korean researchers are reporting today that they have cloned what scientists deem the most difficult animal, the dog. The group worked for nearly three years, seven days a week, 365 days a year and used 1,095 eggs from 122 dogs before finally succeeding with the birth of a cloned male Afghan hound. The surrogate mother was a yellow Labrador retriever." The story notes that Genetic Savings and Clone of Sausalito, California, spent seven years and more than $19 million to clone a dog. But no go. Many stories note that dogs may be hardest to clone because the egg hits the fallopian tube in an immature state, and dogs only ovulate once or twice a year.

That night David Letterman and Jay Leno will duly note the advance. Leno will show laser beams emitting from the dog's eyes, an apparent reference to Godzilla; Letterman will say, "Biff, bring out Snuppy the world's first cloned dog" prompting Biff to appear with a rooster on a leash. By 8:05 A.M. on Friday, August 5, there will be 725 articles on the Web. *K9 Magazine* will be "appalled." *The Los Angeles Times* will call the move "monumentally unnecessary." It will be reported that Wilmut and Schatten are in Korea to work on a "secret experiment" that Schatten says will be a "milestone," and that they will have offices in Hwang's new digs. In its September, 75th-anniversary issue, *Fortune* will name Hwang one of "Ten to Watch" in the next 75 years, alongside Egypt president-in-training Gamal Mubarak and Google's founders.

In October, Hwang will launch the World Stem Cell Hub, based in South Korea, California, and England (and later, possibly, Spain, Sweden, and France). The goal: to generate, bank, study, and eventually try on patients 100 cloned hES cell lines a year worldwide and to teach human cell cloning techniques to scientists. When Hwang asks for skin donations to start things off at the beginning of November, a whopping 10,000 South Koreans fill out applications to have their cells cloned. Hwang's ambitious personal plan: to apply for permission to launch a small clinical trial of cloned hES cells for Parkinson's, and a small clinical trial of cloned hES cells for spinal cord injury in *both* the United States and Korea by the end of 2006.

"At the end of his talk, it was spectacular," says Dan Perry, head of Alliance for Aging Research, of a September 2005 dinner honoring Hwang as "Indispensable Person of the Year." "There was a spontaneous standing ovation, one of those things that just seem to come out of nowhere." Earlier that summer, when Hwang spoke at Baylor College, it was the same thing, Perry says. "He came into the room and it was electric. There were TV cameras all over the place and this huge entourage of Korean journalists following him." And there is more to come. "He has lots more papers in the works. Some very exciting results. He has already been transplanting his cloned cells into animals, attempting to repair damage of different kinds."

All this, and Hwang remains Hwang, Perry says. "At the September dinner I called him forward by saying that on three occasions within the last 18 months he has rocked the scientific world by bringing closer to patients universal repair kits made from their own cells. He talked in a halting, effective way about his 89-year-old mother who continues to live on a cow farm in Korea and all the diseases he hoped hES cells could help. Then Gerry Schatten talked about how Woo Suk's father died at the time of the Korean War and how Woo Suk, the youngest of six, was the only one to go to college because the family pooled their resources so he could have a chance. There were 320 people in a chandeliered ballroom at the Four Seasons in Washington, DC, full of health policy people who are usually table-hopping because they haven't seen each other over the summer. But as Woo Suk and Gerry spoke, there wasn't a fork hitting a plate."

Then it all falls apart. It does so in such a spectacular fashion, revolving around the biggest act of fraud in science history—in terms of the unprecedented number of participants and the extent of public interest—the nation and the field will be sent reeling. Eventually, it will cause scientific standards to be tightened globally. But it will also have dumbfounded scientists of all nationalities, who now understand overnight how much stem cells have come to mean to patients and politicians everywhere.

The unraveling of the Hwang myth starts out gently enough. On November 12, 2005, Schatten, Hwang's U.S. collaborator, issues a mysterious statement ending his 20-month collaboration with Hwang. The reason he gives is ethics: problems with egg donations. Schatten retreats. The next day, the mere 132 articles on the Web about Hwang are divided in focus between Schatten and Snuppy. The world is not stunned by the ramifications of Schatten's move yet.

It will be.

Shortly after, Hwang denies ethics violations but adds he is investigating Schatten's charge himself. Then he, too, mysteriously disappears. Some reporters note that Schatten was earlier associated with a scandal-ridden IVF clinic (although he was absolved of all wrongdoing); perhaps he is just gun-shy. Others surmise Schatten is jealous: after all, it was only after Hwang staffers helped him that he finally cloned a monkey embryo.

The South Korean government is almost shockingly unhelpful. On November 16 it announces, not that it is investigating the ethics charge, but that it has drafted a bill to give stricter oversight to future stem cell operations—and to give Hwang $15 million more. "We will render our utmost support to help stem cell research which hit hard times," explains an official.

On November 21, Sung Il Roh, head of the MizMedi Women's Hospital and IVF clinic, seems to explain Schatten's action. (Roh's MizMedi IVF clinic has been critical to Hwang. It has had bookend responsibilities. First, it supplied Hwang with human eggs on the front end. Then, after Hwang's SNU team was done making cloned embryos out of those eggs and patients' cells, Hwang gave them back to MizMedi, whose job next was to derive cloned personalized hES cells from the embryos.) Roh announced that, contrary to Hwang's assertions, 20 women were paid about $1,500 each to provide eggs for Hwang—an ethical no-no. He claims Hwang didn't know because Roh never told him, in order to protect him. The next day, the Korean Munhwa Broadcasting Corporation (MBC) TV network airs a report claiming that, actually, Hwang not only knew but paid for a whopping 600 eggs, and that at least one of his researchers provided eggs at no cost.

What happens next astounds many: *PD Diary*, the show running the MBC TV network report, receives 6,000 e-mailed complaints. Pictures of the producers' family members are posted and death threats issued. Over the next several weeks, 500,000 angry Korean e-mails will flood the station.

On November 24, Hwang resurfaces, admitting that he did find out—after *Nature* made that claim in May of 2004—that two of his researchers gave eggs, and that he lied thereafter to protect their privacy. He says he will resign from his world hub but stay in his lab. Since SNU finds he broke no laws—Korea only adopted laws against egg purchases in 2005—things look rosy. A poll indicates that more than 80 percent of Koreans support him. The Internet fan club "I Love Woo Suk" quickly chalks up 15,000 members. Hwang takes off for a countryside vacation, tears in his eyes, appearing every inch the noble defender of women. Many foreign institutes decide to continue working with him. SNU declines to accept his resignation. Pioneering Max Planck Institute stem cell biologist Davor Solter will note in early 2006 that the ethical violations could have ended up a "slap on the wrist problem."

Even so, the reactions that follow in South Korea shock the science world. On November 25, Hwang supporters start boycotting products advertised on *PD Diary*. On November 26, noting that 11 of 12 advertisers pulled out, *JoongAng Ilbo* writes, "Ordinary Koreans . . . seem to be rallying around the country's best-known scientist. Most of the action in this digital nation is on the Internet, where new support groups are being formed and more women are lining up to pledge eggs for his research use." So many hate e-mails flood the MBC TV network that the nation's president, Moo Hyun Roh, posts a Web letter telling his people to calm down. Women volunteering to donate eggs number 1,000—and counting. *JoongAng Ilbo* intones that Hwang's work "has mankind's entire future resting on it." Newspapers start talking about the "Hwang Woo Suk syndrome." On November 28, *The International Herald Tribune* adds that Hwang embodies "South Korea's dream of joining the ranks of countries that boast original technology, casting off the image of a country that copies goods developed first by countries like Japan."

Then, what should be a bomb. *PD Diary* producers announce in a news conference on December 1 that they tested some of Hwang's cloned hES cells and found that none of them were cloned—although some results were simply unreadable. *PD Diary* producer Han Hak Soo adds that it all began back in June 2005 (the month before the author of this book visited the Hwang team in Seoul). That month, *PD Diary* was alerted to the possibility of fraud by an insider (who will later turn out to be Young Joon Yoo, an appropriately named young, brand-new MD who was working toward a master's degree in Hwang's lab. The young Yoo was in charge of taking the Hwang team's cloned blastocysts, culturing them, and handing them off to MizMedi IVF clinic personnel who were, as noted, in charge of getting the hES cells out and sending them off for verifying DNA tests. Also as noted earlier, Hwang passed his cloned embryos to MizMedi this way because it had experience, having culled normal hES cells—one of which made the NIH list of presidential lines—from spare IVF-clinic embryos). On air, Soo adds that he has more witnesses.

On December 4, however, just hours before the MBC TV network is supposed to run its second program, this one to discuss fraud, it is revealed that two of Soo's witnesses talked to another network and claimed that MBC TV network staffers strong-armed them. The witnesses were former underlings at the all-important MizMedi IVF clinic: Sun Jong Kim and PhD student Jong Hyuk Park. (It will later be clear that these two, Kim and Park, led parallel lives: both spent time trying to get hES cells out of the cloned blastocysts handed to them by Hwang's SNU lab, and both landed in Schatten's lab in Pittsburgh.) The two underlings' message is that they were in Schatten's lab in October 2005 when the MBC TV network came calling, and that MBC bullied them into saying negative things about Hwang. It is also noted that the duo gave a written report to Schatten after the MBC TV network interview—which might explain why, within days, Schatten ended his partnership with Hwang.

The government's response to this is to offer military exemptions to Hwang's scientists. And the public is just plain furious. More protests follow, and the producers are forced to admit on air they did some strong-arming when getting statements from Hwang researchers. The second MBC TV show—which was supposed to introduce the fabrication charge— does not run.

Apparently emboldened, on December 5 the Hwang lab says it will offer no more cells for tests: "The cells will be naturally verified after several years when other scientists will reproduce the work," says Byeong Chun Lee, one of the triumvirate running Hwang's lab (and the one who regularly did the surgery on the cloned pig/humans). Lee's statement is backed up, bizarrely, by the Korean Federation of Science and Technology Societies: "No one—not Dr. Hwang, MBC, nor the public—should have to suffer anymore." Given that even the vaunted Wilmut redid some Dolly the sheep experiments for a skeptical public, this response is alarming to some onlookers. What is being hidden? How many people know—whatever it is? Yet shortly after, Korea's president simply asks Hwang to go back to work, as do 43 lawmakers. He asks this despite the fact that a company doing tests on Hwang's cells reported to his office on November 29 that the MBC

TV network's assertions seemed right: none of the cells it tested were clones.

The war continues. On December 4 at 11:30 at night, Hwang alerts *Science* to the fact that he has found that some of the pictures in his 2005 paper were duplicates. *Science* says it occurred by accident. But *Science* later notes that Hwang reported this only after an anonymous young South Korean scientist pointed it out on a Web site called BRIC (Biological Research Information Center). More than 200 posts follow in rapid succession, finding more duplicate shots. Another anonymous BRIC poster says there are also suspect DNA fingerprinting (illustration) replications. Yet on December 8, the astonishing result is that *PD Diary*—not just the canceled Hwang segment but the entire long-running show—is yanked off the air. It has received 20,000 angry postings. It has lost all its sponsors.

That appears to be that.

But the scientists' Internet postings turn out to be the winning salvos in this strange war. For, in response, SNU professors ask the university to look into Hwang's work. The University of Pittsburgh launches an investigation. *Nature* has called for one. *The New York Times* praises the young anonymous Korean scientists on the BRIC site. SNU finally orders a full-fledged investigation; Schatten asks that his name be taken off the *Science* paper; and somewhere in the middle of it all, Hwang lands in the hospital with exhaustion.

It is on December 16 that South Korea finally seems to have good reason to start seriously doubting its hero. On that day, Hwang asks *Science* to withdraw the 2005 article, saying that when it was published, his team only had eight stem cell lines on hand, not eleven as he had written in the paper. He is admitting he lied about three of the eleven lines—but claims he was also duped by underlings who replaced some of his cloned cells with IVF-clinic-derived cells. Two days later his lab is shut down, his computers confiscated, and SNU interviews of staffers begin. "I am deeply devastated," Hwang friend, stem cell researcher Hans Scholer of the Max Planck Institute, tells the press.

From there the story, as it unfolds in the Korean press, gets even more byzantine. On December 18 and 19 it is reported that Sun Jong Kim (the other half of the MizMedi duo who went on to Pittsburgh) said that Hwang *originally* made eight cloned hES cell lines—as Hwang contended in his confession—but that six were contaminated and died. Kim claims that Hwang then asked him to parlay photos of the two remaining lines into eleven— sure that he could make nine later and play catch up. Hwang lied, in other words, Kim claims, about *nine* lines, not three. "I think he was confident that he could produce more stem cells based on his experience of already creating six," Kim says. At this time it is also reported that both Schatten and the office of South Korea's president knew that the majority of the "cloned" stem cell lines had been contaminated back in January 2005, only about two months before the 2005 paper was submitted to *Science*, forcing the team to make new cloned hES cells. Since it takes at least three full months to verify that one truly has bona fide, "immortal" hES cells, all of the above should have been alarmed, many critics note.

The plot thickens. The third week in December it is reported that two stem cell papers from MizMedi, on which Hwang was *not* an author, are being retracted, and that Sun Jong Kim (half of the MizMedi/Pittsburgh duo) was a coauthor on *both*. (Shockingly, talks with editors and an April 10, 2006, PubMed search will reveal Kim cowrote *eight* non-Hwang MizMedi stem cell papers with photo or figure problems—mostly brazen duplications—six of which were retracted between November 2005 and April 10, 2006. Kim also cowrote the two soon-to-be-retracted Hwang papers. In fact, when "Kim SJ" and "embryonic stem" are fed to PubMed on April 10, the status of all Kim's stem cell papers will be: retracted, retracted, retracted, retracted, no problems, retracted, problematic (photo duplications), retracted, no problems, retracted, retracted, problematic (photo duplications). Kim's buddy Park will appear on five of those non-Hwang- (plus the two Hwang-) retracted papers, and another MizMedi boss, Hyun Soo Yoon, will appear on five of those retracted non-Hwang papers and one of the retracted Hwang papers. By contrast, in that period (by the end of April), Hwang's two hES cell papers will be his only ones retracted, and seven journals—and counting—will stand behind more than 25 of his other papers. Hwang is looking better by the minute.

Furthermore, soon after, Curie Ahn, the transplant doctor featured earlier in this chapter, resigns as Hwang's physician. She has begun doubting Hwang recently, she says. Yet it is reported that Ahn earlier flew to Pittsburgh to hand $10,000 to the father of whistle-blower

Kim—purportedly to help him with hospital expenses. (Kim was hospitalized after the MBC TV interview he gave in which he claimed Hwang forced him to fabricate the two human cell cloning papers.) It is also reported that Yoon, that second MizMedi boss with a string of retracted papers, gave $20,000 to Kim's father—and Ahn gave some $10,000 or more to Park, the other half of the MizMedi/Pittsburgh duo, who was also charged with getting hES cells from Hwang's cloned blastocysts, and also cowrote many retracted papers. An investigation is launched to see if the above actions constitute bribes. Watchers begin to seriously wonder just who was responsible for the Hwang fraud.

Revelations keep coming. On December 30, the Ministry of Science and Technology is accused of exerting pressure on SNU to delay announcing investigation results. That day the National Bioethics Committee says that Hwang's lab used 423 and 1,233 human eggs in his two papers, respectively. Hwang had claimed his team had used only 242 and 185 human eggs. The committee also finds that 16 egg donors landed in a hospital after showing symptoms of excessive ovulation. Donors may not have been given proper consent forms.

On January 1, 2006, *JoongAng Daily* reports that a genetic analyst on November 29, 2005, told the office of South Korea's president that not one of the Hwang cells it examined was a cloned ES cell—but the office did nothing. On January 3, MBC airs an interview with a researcher who says one Hwang staffer did some cloning using her own eggs.

Finally, on January 10, SNU's verdict: the 2004 and 2005 papers were *completely* fabricated. There are *no cloned hES cells*—at least, in existence as of January 10. Lines 2 and 3 are actually MizMedi stem cells taken from IVF-clinic embryos. And SNU reports that Hwang committed many ethics violations, getting letters of intent from subordinates to donate eggs and accompanying one to the hospital for egg retrieval. But Hwang did seem able to clone human embryos, SNU says. And Snuppy is real.

On January 12, *Science* retracts both cloned hES cell papers and Hwang apologizes—sort of. While admitting to some fabrication, he insists he thought he had *some* cloned ES cells. He was set up by MizMedi's Kim and others responsible for getting stem cells from his cloned blastocysts, he claims. "The researcher should have told me that he was not able to culture stem cells. I really don't know why he put

Korea to shame." He claims his team, led by successful animal cloner Lee, recently cloned wolves. He talks about his cloned part-human pigs. He says he is expert at making cloned human embryos—having generated 30 for the first paper, 71 for the second—if not hES cells, which he hadn't known. "We were crazy, crazy about work. . . . All I could see was whether I could make Korea stand in the center of the world."

That day, the unimpressed Seoul District Prosecutor's Office raids Hwang's home. It does so to look into Hwang's allegations, and to determine if his actions could be considered embezzlement of funds, which could lead to prison.

In January and February, the prosecutors raid scores of researcher homes and offices, collect countless computers and logbooks, and conduct countless interviews.

Prosecutors immediately discover that Sun Jong Kim, the MizMedi IVF-clinic underling who had claimed Hwang made him turn two cloned hES cells into eleven in photos—yet who is also a coauthor on at least eight retracted articles—has apparently deleted scores of Hwang lab files from his computer, as has a researcher in Hwang's SNU lab, Dae Gi Kwon. Finally, on February 6, a shocker: in direct opposition to SNU, prosecutors announce that Hwang may have been telling the truth when he claimed to be unaware he had *zero* cloned hES cells. Among other things, prosecutors say, he gave some to Memorial Sloan-Kettering in New York to be worked on—a stupid move if he knew they were fake.

Furthermore, on February 8 it is announced that prosecutors have tentatively concluded that the lines that were contaminated in January were deliberately ruined, since they were kept in different labs, yet all contracted the same virus: highly unlikely. And on February 16, only days after Schatten's University of Pittsburgh panel claimed he committed "scientific misbehavior" in not properly overseeing the work—but that he hadn't known about the fraud—Seoul prosecutors say not so fast, maybe he did. Kang, one of the three running Hwang's lab—the one who most often sits by Hwang, helping him with his English—testified that he overheard Hwang tell Schatten, the January two months before passing in the 2005 paper, that lines 2–7 had been contaminated and lines 4–7 had died—and that Schatten had said to write them up anyway.

In February, Hwang's lawyers announce they are asking for Hwang's right to patent his findings back from SNU—which is trying to cancel the application—and may sue Schatten for continuing to pursue his own human cell cloning patent. Also in February the nation's new science minister—who replaced the old, after the old minister was asked to step down, it is believed, because of Hwang—says he may give Hwang a second chance. Throughout this period, between 1,000 and 2,500 demonstrators gather every Saturday, cheering on Hwang.

By April, the story being pieced together by prosecutors appears to be this: Hwang's team at his SNU lab cloned many human embryos, possibly the 100-plus he claimed. But young MizMedi researchers were never able to turn the cloned embryos into stem cells—just as, indeed, some were bungling a lot of their normal IVF-clinic hES cell work, as well. Instead of fessing up—on many fronts, involving many papers—they covered up. And many who saw or suspected this shut up.

Prosecutors share with the press many details of their preliminary findings. They indicate they feel confident that some MizMedi staffers who worked on Hwang's cloned embryos were involved in fabrication. For example, Sun Jong Kim (the MizMedi/Pittsburgh staffer who talked under MBC TV pressure) was at one point in charge of taking stem cells from blastocysts and sending them for tests. Beyond the fact that Kim said he shot those falsified pictures at Hwang's command, hundreds of his lab files were destroyed on his computer. He was a coauthor on at least eight retracted papers. He never once, in all these years, sent an e-mail to anyone claiming he derived hES cells from cloned blastocysts. (Prosecutors plowed through thousands of e-mails.) Furthermore, Hwang named Kim as a likely fraudster.

Jong Hyuk Park, the other MizMedi/Pittsburgh underling involved in creating stem cells from the blastocysts—and in coauthoring several retracted papers—is also suspect, prosecutors believe as of early spring. He is the third and last staffer Hwang fingered. In a December 26 phone call with Hwang that Hwang secretly taped, Park blamed young SNU staffer (and original MBC whistle-blower) Yoo for problems—despite the fact that Yoo's involvement apparently stopped before stem cells were ever culled. Park was a coauthor on several of the retracted Yoon/Kim papers.

Yoon, an aforementioned MizMedi boss, has also been tagged by prosecutors as a prime suspect. He provided Kim with $20,000 just before Kim retracted his accusations. Furthermore, Hwang's "cloned" stem cells were always handed from the MizMedi clinic to the same person in the government for DNA analysis to "prove" cloning occurred, bypassing normal channels. That person was another Lee, Yang Han Lee at the National Institute of Science and Technology. Lee is friends with Yoon, having attended Hanyang University with him. Additionally, MizMedi chief Sung il Roh said he suspects Yoon. And Yoon (who ended up back at Hanyang) will be the first scientist involved in the scandal to, with Hwang, be ousted from the Korean Society for Molecular and Cellular Biology in March.

SNU researcher Dae Gi Kwon is also likely to be guilty of fabrication, prosecutors are reported to believe as of March. He too claimed Hwang forced him to fabricate. In his case, he said Hwang told him to take several samples of regular somatic (adult) cells from patients and split them for DNA analysis (pretending that one of the adult cells was a matching stem cell). His computer also contains hundreds of destroyed lab files. In November of 2005, he e-mailed Kim saying he could not make cloned hES cells.

Finally, there is Hwang. While prosecutors will continue to leak their belief that Hwang may not have known he had *zero* stem cells, he definitely lied—and several times. For in March *The Korea Times* reports he told prosecutors that he indeed ordered Dae Gi Kwon to manipulate lines 4 through 11 for DNA testing—his justification conceivably being he thought he at least *had* made the cells in the past. He lied about the eggs—how they were procured, how many, and from whom. He may have mismanaged millions. And he funneled money to suspect sources, from local politicians to the Karolinska Institute, where Nobel Prize winners are named (although Karolinska insists Hwang's $500,000 donation was aboveboard). The degree to which his U.S. partner Schatten was involved will remain vague in early April; however, it will be reported on April 11 that Schatten told prosecutors that Kang—steadfast Hwang friend—was correct: Hwang did tell Schatten in January that cells had died, and Schatten did say "go ahead," believing there were backup frozen vials of cloned hES cells. Some say this means Schatten did advise cutting corners, since there would not have been enough time to properly verify that the frozen cells were true hES cells between mid-January and mid-March, when the paper was submitted.

At last, on May 12, the prosecutors announce their investigation is over, and hand out their report and their indictments. The prosecutors, and a former Hwang associate interviewed for this book days later, offer big surprises. Foremost among them is the fact that Hwang-gate was apparently a virtual fabrication fest that ended up involving such a shocking number of active participants—at least 17—and lasted so long, one could be forgiven for thinking it offered only one lasting truth: when Hwang said his people were workaholics, he wasn't kidding. He and some of his colleagues apparently fabricated constantly, exhaustively, taking few breaks as they labored over those cloned cells that never were.

Prosecutors didn't know if the 2004 cell line was born of parthenogenesis (an unfertilized egg becoming embryolike on its own). They did determine, however, that Hwang and most of his team *genuinely believed* they'd made the world's first cloned hES cells. The gifted one here seemed to be Eul Soon Park, a young researcher in Hwang's SNU lab who developed Hwang's trademark gentle squeezing method. She made what seemed to be a good cloned blastocyst on February 18, 2003. (She would also turn out to be the underling who tried to clone her own eggs.) Young Joon Yoo, the young MD seeking an extra master's, was Hwang's team leader for cloned human stem cells. He was responsible for ensuring blastocysts made by the female Park were properly cultured and given to Jong Hyuk Park, part of the MizMedi duo discussed so often in the press. The male Park seemed to be the gifted one on the other end. He was chosen because of his supposed expertise getting hES cells out of IVF clinic embryos. Indeed, he was first author on MizMedi's—and Korea's—first official "normal" hES cell line paper, submitted to *Biology of Reproduction* in March 2003 and published in December 2003. (As of May 2006, it would also be one of the few *unretracted* MizMedi stem cell papers, although its first cell line, Miz-hES1, would prove astonishingly problematic and cause international problems, as will be seen.) The two Parks, between them, made what everyone instantly thought of as the world's first cloned hES cell line.

Then, in May 2003, the male Park gave that supposedly historic cell

line—called NT1, for "nuclear transfer number one"— to his sometime
MizMedi partner Sun Jong Kim for DNA fingerprinting tests to prove
cloning occurred. But Kim couldn't get DNA out. So—shockingly—
right off the bat, with no further ado, Woo Suk Hwang and Kang, the
quiet and loyal Hwang lab cohead who finished Hwang's sentences for
him, became fabricators, according to prosecutors. The two almost
instantly began ordering many people to falsify an astonishing amount
of data for that supposedly groundbreaking 2004 cloning paper.

First, according to Park and Kim, Hwang ordered Park to get DNA
from two somatic (adult) cells belonging to the person whose cells they
thought they cloned, so they could pass them off as an adult cell and its
matching cloned stem cell. (You can't tell a stem cell from a somatic cell
via most DNA tests.) Then he ordered those to be sent for a DNA fingerprinting test—which can't determine "stemness"—to genetic analyst
Yang Han Lee at the state center mentioned earlier. (Hwang denied
this; prosecutors believed Park and Kim.)

Then MizMedi research boss Yoon, with Park and Kang, tried to
make embryoid bodies (cell blobs from all main tissue groups grown in
petri dishes from true ES cells). They made one and sent it to Lee for
fingerprinting, but he couldn't get DNA out. So Yoon and Kang asked
Lee to fake it as Hwang had: by substituting DNA from that person's
normal somatic cells, not the "cloned" cells.

The fakery snowballed. After five attempts to get, for tests, a teratoma out of their cells—cell blobs of all tissue types created when true
ES cells are shot into immune-compromised mice—Hwang, Yoon, and
an underling made one. But again they couldn't get DNA out. So Lee
was asked to use the same somatic cell DNA data generated earlier to
"prove" the embryoid body's existence.

Most ironically, the team never even found out until after the scandal
broke that they had the wrong *somatic* cells to begin with. They were
working with the wrong woman's cells. They might have known this if
they hadn't instantly started fabricating.

Here the story gets extremely bizarre—and it is just the beginning.
Shortly after Park and Lee fabricated their tests, MizMedi underling
Kim began having trouble keeping that original "cloned" NT1 line propagating. (He was given the line to watch by Park, who stopped his fabricating work toward the end of 2003 to finish his PhD thesis.) So

supposedly without telling anyone, according to prosecutors, starting around December 2003, Kim mixed normal IVF-clinic hES cells from MizMedi's line Miz-hES1 into the "cloned" NT1 hES cell cultures. This made the "cloned" NT1 cells look as if they were proliferating like real stem cells. But it also meant that when he was asked, in February and September of 2004, to do more DNA fingerprinting on NT1 (to ensure stability), Kim had to fabricate yet again. That is, using the same fabricating trick his boss ordered him to perform, he sent around DNA from two sets of adult cells, not from any "cloned stem cells." This would also interfere with someone else's fabrication project, and actually impact regular hES cells labs worldwide—as will be seen.

In the meantime, a third wave of fakery was in the making. Also at the end of 2003, days before MizMedi's first official *normal* hES cell paper was to come out, MizMedi research boss Yoon discovered that the normal Miz-hES1 line—the very line Kim stole ES cells from to plop into the "cloned" Hwang NT1 hES cell line—was found to have accumulated genetic abnormalities. So Yoon in April 2004 called a meeting with Park, Kim, a researcher named Jeon Wuk Lee, and an (unnamed) Lee and told them Miz-hES1 had problems. This was big, not only because Kim was secretly passing off that normal hES cell line as the first cloned *Hwang* line. Miz-hES1 was the only Korean stem cell line to make the Bush cutoff, thus the only Korean line easily available to NIH-funded researchers around the world. MizMedi had received about $100,000 for it because it was presidential. Yoon told the group Miz-hES1 was being secretly replaced with Miz-hES5—and that they should tell researchers the new cells came from a frozen Miz-hES1 batch. Since Bush allowed no federal funding for cells made after August 2001—and since Miz-hES5 was made thereafter—the secret switch had global ramifications.

Back in the cloning world, the other fabricators were soldiering on. As noted above, Yoon and Hwang working together got a teratoma from NT1 around October 2003. Unsatisfied with that photo, Hwang simply ordered a photo of a teratoma of the Miz-hES1 cell line. Then rt-pcr tests on NT1, to find maternal and fraternal genes, failed. So Kang told SNU underling Jeon Hyon Yong to do rt-pcr on blastocysts from the NT1 donor, then he changed the label from rt-pcr "blastocysts" to rt-pcr "stem cells." And when the immunohistochemistry photo—in which NT1 cells were stained for certain qualities—didn't come out right, Hwang, Kang, and Park again just switched it for a Miz-hES cell photo.

According to prosecutors—and that former Hwang intimate—Kim repeatedly failed when charged with making more cloned hES cell lines for the next paper. So starting in September of 2004, he began smuggling into Hwang's lab hES cells from *many* normal MizMedi hES cell lines, passing them off as clones. He worked in the dark, using the excuse that, unlike other cells, he'd found that cloned cells seemed to like darkness. Some called him "God's Hands" and Hwang called him "teacher" as Kim sat hunched in the dark, bringing to life his colonies of fake clones.

Then, on January 9, 2005, a fungus hit, as the press noted. Fake cloned lines 4 to 7 died; fake cloned lines 2 and 3 survived. Hwang, believing the two surviving lines were real, and worried about being late to score patents for his country, ordered Kim to amplify data from the two to make it seem as if there were eleven (more lines than the seven Hwang thought they had earlier, because he liked the larger number). Kim ordered the same tests as before, fabricated in the same way—all based on two cloned hES cell lines that he'd never cloned in the first place.

Park by then had gone off to Pittsburgh to help Schatten (really) make the first cloned primate embryo, and Yoo had been fired by Hwang for being secretive about the "success" recipe for the first "cloned" cell, according to the former Hwang associate. So Kim (who replaced Park) and Dae Gi Kwon (who was mentioned so often by the press, and replaced Yoo) did most of the grueling falsification work on the 2005 cells at the behest of Hwang and Kang. Kwon this time was the one to pass only DNA from cell lines—not the telltale cells themselves—to transplant doctor Curie Ahn, who did HLA (tissue) matching. (Like stem cell network head Moon, Ahn was questioned and completely exonerated by prosecutors. The money she gave to Park and Kim in Pittsburgh she genuinely thought was for hospital and moving expenses—as is common when Korean students go abroad for a spell). Since Kwon only sent DNA, it was, again, difficult for anyone to know that, for each of two patients, he did not send a mature skin cell and a cloned ES cell. On February 2, 2005, Ahn therefore found that lines 2 and 3 seemed to be clones; and on March 22, that lines 4–11 were. The DNA fingerprinter found the same thing.

For the 2005 paper, Hwang got a bit footloose. He decided to substantially reduce the numbers of eggs that were actually used in a table he ordered Kwon to make with a single stroke of his pen. Other fabrications directed by Hwang and Kang included amplifying immunostaining, DNA fingerprinting, and karyoptyping results so it looked as if there were eleven lines, not two; making one teratoma look like seven; claiming in writing that all cells were made on human feeders, even though cloned NT4-NT11 cells were completely made up, and cloned NT2—which was actually of course an uncloned normal ES cell—was made on mouse feeders. They made no embryoid bodies this time, because on February 22, 2005, Hwang asked how long it took to make EBs. Kim said at least three to four weeks; Hwang said that would not do because he wanted to pass in the paper soon. He suggested that Kim use a slide of Miz-hES1, amplified ten times. Kim complied.

Hwang apparently genuinely didn't know until late 2005—around the start of the breaking news about the scandal—that his team had never made *any* cloned hES cells. But he ordered up an astonishing amount of fabrication on his own. In addition to research misconduct, Hwang is indicted for bioethics law violations and embezzlement. He allegedly misused $2.99 million in state funds and private gifts. He kept over 60 bank accounts under different names, and carried cash in bags between banks to avoid paper trails. He is not indicted for bribery. But if found guilty, he could go to jail for some 13 years.

Despite naming so many fraudsters, prosecutors end up handing out indictments to only five more people, since it is uncommon to criminalize scientists for fabrication alone. As it did for Moon and Ahn, the report goes out of its way to declare Hwang's key animal cloner, Byeong Chun Lee, innocent of scientific fraud. Indeed, the former Hwang associate notes that on December 6, when Lee and Ahn saw the BRIC postings, they became extremely worried at last. (Lee's face was "very pale," says the associate. And Ahn would, two days later, request the SNU investigation.)

Still, like Kang, Hwang's other main lab partner, Lee is indicted for fraudulently obtaining research funds. (In Lee's case, a former Hwang intimate says, this was about extra funds he appropriated for some poorly paid underlings. "He is a good person, very loyal," the associate says.)

MizMedi research boss Yoon is indicted for creating $50,000 in false receipts and embezzling research funds. Prosecutors urge the censuring of state geneticist Lee for accepting Yoon's money for the fingerprinting. Sang Sik Chang, another IVF-clinic head, is indicted for bioethics violations. Mentioned disapprovingly are MizMedi chief Sung Il Roh, for egg selling when it was frowned upon, and Yoon Young Hwang and Jung Hye Hwang, two Hanyang University doctors who the report says gave Hwang some ovaries after oophorectomies without getting authorization from patients. And Kim is indicted for destroying evidence and obstructing normal business operations at SNU.

Hwang's attorney says Hwang denies ordering false data for the 2004 paper, and denies embezzlement, claiming that he has made over $800,000 between his speaking engagements and articles. The Yonhap News Agency says Hwang is surprised and "tormented," and will fight the accusations at the trial.

But the scandal doesn't even end there. One week after the prosecutor's report is made public, NIH stem cell people are put on alert. The message: shipping of the NIH-funded stem cell line Miz-hES1 must go on hold because the line may have been replaced with one that can't receive NIH funding. NIH staffers scramble to get the prosecutor's report translated. That the Miz-hES1–to–Miz-hES5 switch may have global ramifications seems clear from a glance at SNU's final report, which said in March that some of the cells in Hwang's fake cloned NT1 line seemed to be Miz-hES5 cells, whereas prosecutors said Miz-hES1 cells were mixed in. This confusion could well be due to the fact that Yoon switched those two lines. Regardless, as of the end of May it will appear that work using Miz-hES1 lines must be halted and rethought worldwide.

The bottom line: the Woo Suk Hwang fraud is the biggest in science history in terms of the number of guilty parties. The vast majority of science frauds are committed by one person. The May 2006 prosecutor's report indicates that over 17 people were actively involved in this fraud. But this fraud is also biggest, some say, in terms of public interest. "Never before has a case of scientific misconduct been more widely reported in the popular press," wrote London School of Economics political scientist Herbert Gottweis and Riken translational researcher Robert Triendl in the February *Nature Biotechnology*.

* * *

HWANG-GATE WATCHERS offer any number of possible explanations for the fraud. In South Korea one can cheat more easily than in certain Western nations because it is not yet a democracy with an entrenched rule of law and separation of powers. In other words, some say this fraud occurred partly because it *could*. Korea became a democracy in name in 1987, but this incident's many witting and unwitting coconspirators, before and after the act, shows to some that various entities in government, education, and business are intertwined in a way reminiscent of the dictatorships of the past. The uncritical glorification, protection, and financial remuneration that Hwang received couldn't have happened in a true, established democracy, where check-and-balance-mechanisms, on many levels, can prevent unbridled cronyism, some contend. Gottweis and Triendl pointed to the fact that the president's science adviser was a Hwang-cloning-paper coauthor and that Hwang's lawyer doubled as head of the National Bioethics Committee *investigating* Hwang as examples of a lack of separation of powers in Korea. (Subsequently, both of those multitasking Hwang friends resigned their government posts.) Then there was the fact that Korea made up its bioethics rules *as*, instead of *before*, work proceeded, unlike the U.K. "Policy makers violated almost every existing rule for good governance in research," the duo noted. SNU president Un Chan Chung at one point claimed that, indeed, the blame belonged to many: "Most of us, in the name of national interests, exaggerated Dr. Hwang's stem cell research to make it a national aspiration." The government pressured Hwang's crew—then protected them in a fashion that was familiar to Koreans who lived through the dictatorships.

"The government policies supporting and financing Hwang's work, based on his scientific achievements, merged with nationalism and patriotism and created a quasi-fascist environment that suppressed criticism and the freedom to search for the truth," contended Korean University Asian studies expert Jang Jip Choi. Such problems have certainly dogged business in Korea. All of Korea's three big companies—which are run by *chaebols*, families that ruled during military dictatorships—have endured corruption scandals. Samsung and Hyundai have faced bribery charges. Daewoo has seen better days. "Some are comparing Hwang's woes to those of disgraced tycoon Woo Choong Kim, who cooked the books as he built a small textile-trading house called Daewoo into Korea's second-largest conglomerate before collapsing under a mountain of debt," *BusinessWeek* noted.

Then there is the chip on Korea's shoulder after, among other things, enduring 900 invasions in 2,000 years. A resulting pressure to succeed has led before to corners being cut in Korean academia, some say. In the first eight months of 2005, prosecutors penalized 61 professors and administrators, and in 2004, they penalized 23, "mostly for receiving bribes in exchange for granting tenure," wrote AP reporter Bo Mi Lim on February 12, 2006. "In one case last year, a university chancellor received $4 million from 42 candidates in exchange for appointing them as professors."

Many also suggest that the paternalistic Confucian culture at work in Korea—and Asia in general—played a role. The reason: it honors learning, but allows for little questioning of superiors, which may have left many underlings in the Hwang incident with little choice but to cheat at the command of various bosses. "The Hwang affair is a salutary reminder that scientific breakthroughs take more than just good brains, first rate facilities and generous government support," wrote *Financial Times* Asia commentator Guy de Jonquieres. "Just as important is a culture that encourages independence of mind, a spirit of challenge and healthy scepticism. Those attributes are not highly prized in Korea. Not only did Dr. Hwang's work escape rigorous scrutiny by colleagues, which could have exposed it as a hoax earlier; he himself was long immunised from criticism at home by the adulation of a fawning press, an adoring public and an admiring Government. . . . That is not to say that Asia's grand plans to advance the frontiers of science are all doomed to fail. Given the rapidly growing resources committed to them, statistically some are likely to succeed. But their chances of success would be improved if scientific discovery were free to proceed in an environment that was less authoritarian and deferential and more questioning and open to diversity of opinion." He went on to quote C. P. Snow, a Cambridge physicist: "'When you think of the long and gloomy history of man, you will find more hideous crimes have been committed in the name of obedience than have ever been committed in the name of rebellion.' Asian policy makers should take note."

Irv Weissman agrees with the above. "In the U.S. we don't have a hierarchical medical university system, so assistant professors work on their own. They don't work on a professor's projects—as they do in Asia or Europe—which gives the professor a lot of power. That is part of why such a problem could have come up." In Western labs, Weissman adds, there is an atmosphere of almost excessive criticism, as opposed to excessive deference. "If I give a seminar in my own lab, my fellows criticize me. They say, 'Bullshit.' And certainly my faculty colleagues criticize each other. We've learned to live this way and not take it personally."

But when all the extreme elements of Hwang-gate are factored in together, it is impossible to ignore the fact that the nature of the product played a huge role in the affair. Koreans are emotional, and unusually involved in their new democracy. Many issues, aided by the Web, have cranked up passions in recent years. However, a total of 20,000 patients had lined up, for a mere 10 slots, to have cells cloned. Even after Hwang confessed he fabricated some lines, close to 40 percent of Koreans continued to support him. Every weekend from January through March between 1,000 and 4,000 supporters held candlelit rallies—far more ralliers than normal in Korea. On February 4, a supporter even set himself on fire. In May 2006, even though Hwang had by then been terminated by SNU, Buddhists offered him a whopping $60 million to resume his work, and Hwang fans pledged to run pro-Hwang election candidates in over a dozen towns. Computer chips and cars made tiny South Korea a player on the world stage—yet no computer chip or car magnates were given their own stamps, appeared on the sides of buses, roused vast movements. The fact that Hwang's product might have saved lives—and not just some lives, but potentially an uncountable number of lives, afflicted by a theoretically endless number of diseases—played a massive role in Hwang-gate, many say.

"Hwang was not just a successful scientist," MBC TV network producer Han Hak Soo said in January. "He had become a Jesus figure, someone who said he could make the crippled walk again." Other Korean media reports have repeatedly referred to Hwang as Korea's economic "messiah." Some die-hard fans have become, says a former Hwang associate, "a kind of religious group."

Says Neuralstem CEO Richard Garr: "I do think there is something special/different about stem cell technology, and it is evident in the international coverage his 'work' received. I think that with 'chip' technology, or information technology, there is a feeling that each new 'discovery' is inevitable, and in fact, incremental, regardless of how really important it is. History seems to bear this out. . . . [But] stem cells are magic. They are the 'end' thing, the ultimate cure. To repair and replace diseased tissue is kind of a human dream, it promises hope for the hopeless, and a kind of immortality of the body and mind. It goes to our most basic and probably hardwired fears. You don't get that 'connection' with other technologies, even if they are equally as important or far reaching in their impact on our lives." Australian hES cell pioneer Alan Trounson agrees: "The research work grabbed the imagination of the world—the United Nations, and considerable attention from societies and conferences." The astonishing Hwang-gate is therefore "somehow related to the 'magic' of the possible therapies that may evolve in this area."

Indeed, some note, while medical products in general can elicit a more emotional response than other kinds of products, many people instinctively grasp the extraordinary, historic potential—*potential*—of stem cells in particular. Whether adult or embryonic, they are related to the cells of our birth, and just about everyone finds birth a miracle. (How did we come from a single cell? Who knows?) So the notion that scientists could co-opt the power of such a cell to build a new arm, a new eyeball, a new head is instinctively grasped and instinctively believed, rightly or wrongly. Stem cells gave messianic flair, not only to Hwang but—by virtue of its wholehearted embrace—to the long-suffering South Korea.

*   *   *

REGARDLESS, THE UPSHOT is that by 2006 the pendulum will have swung firmly away from Seoul and back to San Francisco. This may not be a permanent move. There are some 1,942 people involved in Korea's official stem cell network. Korea is one of the few countries that continually make and improve their own normal hES cell lines culled from

IVF-clinics. (Although not, obviously, hES cells from cloned embryos. But Hwang's human cloning approach may yet prove to work with variation. Indeed, many scientists expect this, since his papers made so much sense, following animal cloning so closely.) Furthermore, much of Hwang's animal cloning work remains unchallenged. In March 2006, two separate groups will confirm in *Nature* that Snuppy is real. Additionally, says Trounson, "Before all this, I knew Hwang for ten years from his cattle and pig cloning work. He led a very good team." Hwang's reputation was so good in animal cloning that Trounson had accepted Hwang's request to critique his first human cell cloning paper before it was published. Furthermore, a Hwang student was on the Texas team that cloned the first cat, and the Hwang team helped Schatten do what he failed to do for years alone: clone monkey embryos. (The journal in which that paper was published would reconfirm its validity in April.) Finally, after the prosecutor's report, on May 24, 2006, *Chosun Ilbo* will report that the Korean government will spend $420 million on stem cell and cloning work focusing, at first, on animals over the next decade, since three of five of Korea's best life science technologies come from Hwang, and those all involve cloning and/or animals. Even if Korea totally avoids human cells, half a billion is a hefty sum—and "they have pretty skilled people," notes Rao. There is little doubt: Hwang's animal cloning work can't be dismissed.

And animal cloning matters, not just because it allows prize animal genomes to be preserved—and, theoretically, extinct animals to be revived. Cloning is the most efficient way to introduce a genetic change into the entire germline of most animals. It is also the best way to introduce multiple genetic changes into animals. One fascinating reason for the latter: as scientists try to insert new genes into cells, the cells continue to divide and wear out. Once a gene is successfully inserted, the cell is very often old, near death. But one can clone it to "rejuvenate it," make it young again, as Hematech CEO James Robl notes—and thus ready to accept another new gene. Using this technique, Robl has created what one scientist calls "the $100 million cow." Enough new human immune system genes have been introduced into this cow that it can muster up human antibodies in response to the challenge of all kinds of viruses. It is a living factory for human antibodies that Robl collects, and will eventually sell to patients to fight any number of diseases.

Then there is Revivicor CEO David Ayares, who has also been making transgenic pigs via cloning—if in a more rigorous, controlled, published fashion than Hwang. Ayares recently signed a contract with a major company to sell his part-human pig tissue for, among other things, the creation of long-lasting replacement heart valves—a gigantic global market. Hwang is not alone in his belief that cloning may let scientists meet a midrange goal of regenerative medicine: save millions of lives with part-human organs until we figure out how to make 100 percent human organs out of stem cells.

However, as of early 2006, Koreans will be considered "radioactive" in the regenerative medicine field, Burnham Institute stem cell scientist Evan Snyder will note. The affair has given ammunition to some hES cell opponents who claim the entire field must be as unscrupulous as Hwang. And some scientists say that so many in Korea were involved in the affair, officials on many levels must demonstrate that all kinds of ethics rules and laws have been shored up before Western doors will swing open as freely as before. "A lot of people who got involved with Hwang are now wishing to Christ they didn't," notes one prominent U.S. researcher. It will be a while before trust will be fully regained. Furthermore, even before Hwang-gate erupted, in July 2005, Korean stem cell network head Shin Yong Moon declined to admit he believed the nation held a firm lead. "I understand why California is watching us," he said. "But we are watching California. And Big Pharma is watching everyone."

The reason: $3 billion is $3 billion—and the United States is the United States. Despite the fact that lawsuits may tie up CIRM's $3 billion until March 2007 at the earliest, California keeps making tracks in the sand—and scientists keep making tracks to it. In the fall of 2005, the FDA green-lit Weissman's Stem Cells Inc. request for a Batten's disease clinical trial. Its launch in 2006 will represent the world's first legitimate trial of purified fetal or embryonic stem cells. (Weissman's are fetal.) As noted, the world's second such trial (this one of hES cells) is likely to be launched by another California company: Geron. When in January 2006, NIH staffer Mahendra Rao moved to the California-based stem cell company Invitrogen, he was up-front about the reason: CIRM (i.e., California's stem cell money and relaxed stem cell laws). That same month—for the same reason—it was announced that Martin

Pera, chief of the embryonic stem cell unit of the Australian Stem Cell Centre, was moving to head the new Institute for Stem Cell and Regenerative Medicine at the University of Southern California, Los Angeles—and that the COO of the Australian center, Dianna Devore, was moving to San Diego to join a commercial stem cell venture. It was also announced that $25 million was donated to USC for a stem cell center. The next month, the ACT cloning company moved its headquarters from Massachusetts to California. In April, CIRM won the first round of lawsuits, scored large loans, and said it was making therapeutic cloning a priority. In May, $16 million was donated to UCSF's stem cell program.

Furthermore, because of CIRM, Stanford has acquired funding to hire three more embryonic stem cell scientists in 2006—one of whom will focus on human cell cloning. This will bring the number of scientists working on stem cells at Stanford to over 100. "We remain confident," reported Weissman in February of 2006.

By late 2006 (when this book goes to press), things will certainly not be proceeding brilliantly everywhere. A *Politics and the Life Sciences* report will have verified the fears of many U.S. scientists: in 2004, the United States wrote only 30 percent of all ES cell publications, compared to 51 percent for five other biotechnologies including DNA microarrays, polymerase chain reaction, yeast two-hybrid screening, green fluorescent protein expression tagging, and RNA interference. Unsurprisingly, there has been some chaos in the Middle East. Leadership changes and political clashes at Misr University have caused problems for Ismail Barrada's nascent stem cell lab in Egypt. As of April 2006 it will be uncertain in whose hands the lab will survive—and with what cells. As of September 2006, outside of the work of hES cell pioneer Ariff Bongso, Singapore will still have put out few hES cell papers despite millions spent. And the U.K., after failing to drum up significant funding, lost the second—and as of March 2006, only other—man to ever clone a human embryo, Miodrag Stojkovic. In September 2005 he announced he was leaving the Center for Life in Newcastle for the Prince Felipe Research Center in Valencia, Spain, because support is better there.

Still, many say "just wait" when it comes to the U.K., where over 10 labs have been diligently working on hES cells for years longer than

most nations, and which boasts some true stem cell stars including
Wilmut, the University of Sheffield's Peter Andrews, and Cambridge
University's Austin Smith. In December 2005, the U.K. government
committed $177 million more to stem cell research for a two-year
period, and in February 2006, it launched a stem cell alliance with
India. "I know the U.K. will be a player," says McKay. In the Middle
East, Barrada's lab was hardly junked, but fought over. Furthermore, a
major Egyptian vaccine company, Vacsera, began talking to him about
launching a stem cell lab. And the Kuwaiti government, which earmarked $300 million for scientific research including stem cells, sent
feelers to Barrada and companies in California and South Korea. An
IVF clinic in Egypt intends to move from cord blood to hES cells. Other
new adult stem cell labs started in 2005 in Egypt and Dubai. Iran's
Royan Institute has been turning its published hES cell line into
insulin-producing cells—and in the fall of 2005 announced that it had
produced more hES cell lines. It is also exploring bone marrow stem
cells for heart patients. It is cloning sheep. In February 2006, Iran president Mahmoud Ahmadinejad used stem cells in a weapons speech to
demonstrate the nation's strength: "The young people have attained
technology on nuclear power and stem cells with their bare hands." In
late 2005, the director of the film *Syriana* noted that a Hezbollah
leader—after essentially kidnapping him—started a political chat with
him by bragging that the terrorist organization was the first to issue a
fatwa allowing therapeutic cloning.

Singapore is still Singapore. The new prime minister is the son of
the founding prime minister. An analyst notes that the "creativity" initiatives have local students groaning that "now they are forcing us to be
creative." No top U.S. stem cell people joined Singapore's workforce
for two years after its lavish 2003 conference. But it did attract Dundee
University's David Lane, a top cancer researcher, for a two-year stint.
Johns Hopkins has expanded its presence in, and Duke University
launched a partnership with, the nation. There are other U.S. collaborations, including one between Weissman and a Singapore team on
cancer stem cells. Weissman will lose two top cancer researchers to
Singapore in early 2006 because the delay in CIRM funding has stalled
the arrival of non-Bush hES cell lines to Stanford. The glittering core
talent previously assembled—including Bongso, Nurcombe, Edison

Liu, Bing Lim, Alan Colman, Larry Stanton, and Patrick Tan—should result in some spectacular work, many believe. And finally, while Singapore may be a place of limited personal freedoms, it is a place of "scientific freedom," Weissman stated pointedly in February 2006. As Western scientists slowly start accumulating on its shores, "Singapore is going to have to change to adopt even more of a Western lifestyle. It is going to be a great place to do science."

Overall, globally, there is much to watch. Israel keeps flooding international journals with hES cell papers. At Singapore National University, there is an adult bone marrow stem cell clinical trial for torn cartilage. At Johns Hopkins and Chicago's Northwestern Memorial Hospital, there are adult bone marrow stem cell trials for lupus. In the fall of 2006, a spinal cord injury trial involving a form of adult neural stem cell was due to be launched at the National Hospital, Queens Square, led by Geoffrey Raisman. By 2006, Case Western Reserve University Hospital in Ohio; Helsinki University in Helsinki, Finland; Hurkisondas Nurrotumdas Hospital in Mumbai, India; the Hunter Medical Research Institute in Newcastle, Australia; and the University of Rochester in New York—among others—had all added themselves to the long list of centers launching bone marrow stem cell trials for heart disease. The Texas Heart Institute received FDA approval to launch a second clinical trial, this one with a different subset of bone marrow cells. At Duke and Indiana universities, there are adult bone marrow stem cell clinical trials for peripheral artery disease. Osiris, of Baltimore, Maryland, is running adult bone marrow stem cell trials for both heart disease and torn meniscus (knee cartilage). At Fudan University, Harvard graduate Jianhong Zhu is still giving adult neural stem cells to brain trauma patients. Excitingly, as of fall 2006, most of the MGH kidney/bone marrow transplant patients were still immunosuppressive-free, including Derek Besenfelder and Janet McCourt (a feat still as yet unpublished in a medical journal). That program continues apace. Stanford University's Samuel Strober wowed peers with his September 2005 *NEJM* report that he used an approach similar in philosophy to MGH's when he administered allogeneic stem cells to blood cancers in 27 patients—and only saw graft-versus-host disease in 2.

Across the world, scientists are racing to do what Hwang failed to do: clone those first human cells. Northwestern University has been trying bone marrow stem cells on patients with rheumatoid arthritis. Several centers worldwide are seeing success with corneal epithelial stem cells and blindness, based on work done by Michele De Luca of the Veneto Eye Bank Foundation. The first company dedicated solely to cancer stem cells has been built: OncoMed Pharmaceuticals. Countries as diverse as the Czech Republic, Turkey, and Spain have all created their own hES cell lines. Thailand has pledged $50 million to stem cell work; Vietnam is interested. Yann Barrandon of the Ecole Polytechnique Federale de Lausanne in Switzerland has isolated hair stem cells, and Harvard's Kaylene Simpson has isolated breast stem cells. The University of Texas plans to give bone marrow stem cells to kids with traumatic brain injuries in late 2006. The list of creative new therapies and approaches grows daily. Limitations of adult stem cell therapy, as outlined in this book, continue to apply. But the knowledge being gained is invaluable and lives are being extended—and extended again.

The bottom line: things are moving. Despite all the caterwauling and controversy, all the posturing, prevaricating, fabricating, and outright fear, stem cells have been teaching the world a "gobsocking" number of revolutionary notions about the plasticity of the human body and the malleability of life, scientists say.

The world has learned that the brain can regenerate itself. The world has learned that one person's immune system can be co-opted to fight another person's cancer and two peoples' immune systems can be grown in the buried laboratory of a single patient's bone marrow. The world has learned that there is a hierarchy among cancer cells that can mimic the hierarchy of the tissue system of the body, and may have been pointing at the single cancer cell whose eradication is "necessary and sufficient" for a cure, as Weissman has put it, all along. The world has learned that many if not *all* the tissue systems of the body possess a hierarchy that, all along, has been pointing straight at the single normal cells that may be necessary and sufficient to lead to cures for many diseases. The world has learned that more animals than was thought are not born with a fixed supply of eggs; that young blood can rejuvenate old muscle; that the clock in old human cells can be turned back to the embryonic state—then cranked forward again. The world has learned that a single, wildly robust embryonic stem cell can not only morph into all the mature cells of the body but also produce more cells than exist in all the human bodies of the world put together. The world has learned that someday, IVF babies may be born with their own tailored, uncontroversial supply of hES cells already banked away.

The world has accumulated knowledge about the "enormous" potential of stem cells with "remarkable speed," as Solter has noted. Knowledge may well accumulate faster with time. Ignorance will undoubtedly fester and spread just as rapidly. Regulations established to best utilize the knowledge and temper the ignorance will sometimes be too restrictive, sometimes too loose. But the stage has been set, in the labs and government halls of several nations, many researchers say. In the future, no matter which group of scientists, states, or nations pulls into the lead or falls behind in the area of regenerative medicine, there will probably always be another group just behind it, to pick up the slack and carry on. Too many people, from too many different walks of life, have become besotted with the notion that the secret to curing the body has essentially been hiding within the body, all along. "The technology will survive despite any obstacle," says Barrada. The river of science is finding its way.

Notes

Unless otherwise noted, all quotations are from author interviews.

PROLOGUE

Interviews: Ismail Barrada, scientific founder and head, Stem Cell Research Center, Misr University for Science & Technology, 6th of October City, Egypt, October 16–20, 2003, in Egypt; Karl Skorecki, director, Rappaport Family Institute for Research in the Medical Sciences, Haifa, Israel, October 24, 2003, in Israel; Benjamin Reubinoff, director, Hadassah University Human Embryonic Stem Cell Research Center, Jerusalem, Israel, October 22, 2003, in Israel, and August 24, 2003, by phone; Nissim Benvenisty, head, Department of Genetics, Hebrew University in Jerusalem, Israel, October 26, 2003, in Israel, and August 2003 by phone; Shimon Slavin, head, Department of Bone Marrow Transplantation, Hadassah University, Jerusalem, Israel, October 22 and 26, 2003, in Israel, and August 2003 by phone; Lior Gepstein, stem cell researcher, Rappaport Family Institute, Haifa, Israel, October 7, 2004, by phone; Yair Reisner, stem cell researcher, Weizmann Institute, Rehovot, Israel, August 21, 2003, by phone; Irving Weissman, director, Stanford University Institute for Cancer/Stem Cell Biology, April 29, 2005, February 6, 2006, and February 22, 2006, by phone; Woo Suk Hwang, former POSCO Chair Professor, Seoul National University, July 21, 2005, in South Korea; Curie Ahn, Transplantation Medicine doctor, Seoul National University Hospital, July 22, 2005, in South Korea; Sung Keun Kang, College of Veterinary Medicine, Seoul National University, July 21, 2005, in South Korea; Byeong Chun Lee, College of Veterinary Medicine, Seoul National University, July 23, 2005, in South Korea; Lars Arhlund-Ricter, director, Unit of Molecular Embryology, Karolinska Institute of Stockholm, Sweden, August 9, 2004, at the Jackson Laboratory on Mount Desert Island, ME; David Ayares, CEO, Revivicor, Blacksburg, VA, February 10, 2006, by phone.

3 uncannily aligned: "The Stargazers of Ancient Egypt," *BBC World Services,* November 17, 2000 (www.bbc.co.uk); K. Spence, "Ancient Egyptian Chronology and the Astronomical Orientation of the Pyramids," *Nature* 408 (November 16, 2000): 320–24.

4 enduring . . . recede: Alexander Stille, *The Future of the Past* (New York: Picador, 2003): 7. 32.

4 "primordial" . . . man: "Life in Ancient Egypt," Carnegie Museum of Natural History, www.carnegiemuseums.org, as of October 13, 2005.

4 "Houses of Eternity": "Egypt's Pyramids: Houses of Eternity," *National Geographic* video, 1978.

4 "Barrada's e-mails": Author interview with Ismail Barrada, October 16, 2003.

5 Names of Israelis: G. Vogel, "In the Mideast, Pushing Back the Stem Cell Frontier," *Science* 295 (March 8, 2002): 1818–20.

5 first truly successful systemic gene therapy trial: L. Thompson, "Human Gene Therapy Harsh Lessons, High Hopes," *FDA Consumer Magazine,* September–October 2000; personal communication (e-mail) from Human Gene Therapy Program codirector Fulvio Mavilio, Instituto Scientifico, Hospital San Raffaele, Milan, Italy, June 12, 2005; M. Cavazzana-Calvo et al., "The Future of Gene Therapy: Balancing the Risks and Benefits of Clinical Trials," *Nature* 427 (February 2004): 779–81.

5 4,000 patients treated: M. Jullig, "Gene Therapy in Orthopaedic Surgery: The Current Status," *ANZ Journal of Surgery* 74 (January–February 2004): 46–54.

5 cured a seven-month-old: A. Aiuti, "Correction of ADA-SCID by Stem Cell Gene Therapy Combined with Nonmyeloablative Conditioning," *Science* 296 (June 28, 2002): 2410–13; J. Steinberg, *Canadian Jewish News,* July 11, 2002.

5 The child is now three: Author interview with Shimon Slavin in Israel, October 22, 2003.

6 suicide bombing . . . two weeks ago: "Seaside Establishment in Port City of Haifa Was Owned by and Catered to Both Groups," Associated Press, October 5, 2003.

6 meant for his Rambam: "She Ate, Paid Her Check, Then Slaughtered 21 Israelis," *New York Post,* October 28, 2003; "Palestinian Poll Reveals 3 in 4 Support Haifa Bombing—Investigators: Female Bomber Ate First Then Killed," *ICEJ News,* October 21, 2003.

7 Knesset member: "We Speak Out Against the Attempted Murder of MK Issam Makhoul," petition written by Ittijah (Union of Arab Community Based Associations) on November 14, 2003; D. Ratner, "MK's Wife Unhurt After Bomb Explodes Under Her Car," *Haaretz,* October 25, 2003; "Car of Legislator in Israel Attacked," Associated Press, October 24, 2003.

7 100 more bombs: C. Burns, Upsurge in Violence in Israel, *CNN,* December 26, 2003.

7 gowned Byeong Chun Lee: Author at the Hongseong farm watching operation, July 23, 2005.

8 Days earlier: In-house CD-ROM of birth in Woo Suk Hwang's Seoul National University transgenic pig lab, summer of 2005, given to author July 22, 2005.

11 newt extract: C. J. McGann, "Mammalian Myotube Dedifferentiation Induced by Newt Regeneration Extract," *Proceedings of the National Academy of Sciences* 98 (November 20, 2001): 13699–704; H. Pearson, "The Regeneration Gap," *Nature* 414 (November 22, 2001): 388–90; J. P. Brockes and A. Kumar, "Appendage Regeneration in Adult Vertebrates and Implications for Regenerative Medicine," *Science* 310 (December 23, 2005): 1919–23.

11 dearth of needed cell types: Personal communication (e-mails) from Jeremy Brockes, University College London, February 10–13, 2006.

11 Don Oberdorfer: Don Oberdorfer, *The Two Koreas: A Contemporary History* (New York: Basic Books, 2001): 445.

11 one publicly listed: J. Wong, "South Korean Biotechnology—a Rising Industrial and Scientific Powerhouse," *Nature Biotechnology* 22 (December 2004): DC42–47.

11 creating *all-human*: W. S. Hwang, "Patient-Specific Embryonic Stem Cells Derived from Human SCNT Blastocysts," *Science* 308 (June 17, 2005): 1777–83; W. S. Hwang, "Evidence of a Pluripotent Huan Embryonic Stem Cell Line Derived from a Cloned Blastocyst," *Science* 303 (March 12, 2004): 1669-74.

11 "astonishing": Author interview with Vic Nurcombe, Singapore Institute of Molecular and Cell Biology researcher, May 24, 2005.

11 end of 2005: "Panel Finds Hwang Deliberately Fabricated Results," *Chosun Ilbo*, December 23, 2005; D. Normile, "Korean University Will Investigate Paper," *ScienceNOW Daily News*, December 13, 2005.

11 biggest act of fraud: Horace Freeland Judson, *The Great Betrayal: Fraud in Science* (Orlando, FL: Harcourt, 2004): 139; H. Gottweis, "South Korean Policy Failure and the Hwang Debacle," *Nature Biotechnology* 24 (February 2006): 141–43.

12 stem cell papers: D. Kennedy, "Editorial Retraction," *Science* 311 (January 20, 2006): 335.

12 will not be among: "Seoul Prosecutor's Group 3 Report: Woo Suk Hwang Case," May 12, 2006 (in Korean): 62.

12 will hold up . . . largely pig and cow cloning: H.G. Parker, "Molecular Genetics: DNA Analysis of a Putative Dog Clone," *Nature* 440 (March 9, 2006): E1–2; Seoul National University Investigation Committee, "Molecular Genetics, Verification That Snuppy Is a Clone," *Nature* 440 (March 9, 2006): E2–3; personal communication (e-mail) from Ralph Gwatkin, editor, *Molecular Reproduction and Development*, April 4, 2006; personal communication (e-mail) from John Kastelic, coeditor in chief, *Theriogenology*, April 4, 2006; personal communication (e-mail) from Judith Jansen, managing editor, *Biology of Reproduction*, April 5, 2006; personal communication (e-mail) from Robb Krumlauf, editor in chief, *Developmental Biology*, April 7, 2006; personal communication (e-mail) from R. J. Pauson, editor, *Fertility & Sterility*, April 7, 2006; personal communication (e-mail) from D. K. C. Cooper, editor, *Xenotransplantation*, April 10, 2006.

12 David Ayares: Author interview with David Ayares, February 10, 2006.

12 continues elsewhere: M. Stojkovic, "Derivation of a Human Blastocyst After Heterologous Nuclear Transfer to Donated Oocytes," *Reproductive Biomedicine Online* 11 (August 2005): 226–31; C. Hall, "USCF Resumes Human Embryo Stem Cell Work, Scientists Hope to Generate Lines by Cloning Donated Eggs," *San Francisco Chronicle*, May 6, 2006; G. Vogel, "Picking Up the Pieces After Hwang," *Science* 312 (April 28, 2006): 516–17.

12 with the cash: J. Ioannidis, "Materializing Research Promises: Opportunities, Priorities and Conflicts in Translational Medicine," *Journal of Translational Medicine* 2 (2004).

13 found himself scooped: K. Hubner, "Derivation of Oocytes from Mouse Embryonic Stem Cells," *Science* 300 (May 23, 2003): 1251–56; F. Flam, "Ingenious Genetics: New Ways to Create Medically Promising Stem Cells May Raise Fewer Moral and Ethical Objections," *Philadelphia Inquirer,* May 2, 2004.

13 world's first sperm: N. Geijsen, "Derivation of Embryonic Germ Cells and Male Gametes From Embryonic Stem Cells," *Nature* 427 (January 8, 2004): 148–54.

13 two females: T. Kono, "Birth of Parthenogenetic Mice That Can Develop to Adulthood," *Nature* 428 (April 22, 2004): 860–64.

CHAPTER 1

Interviews: Ron McKay, Stem Cell Task Force member, National Institutes of Health, April 28 and April 29, 2005, by phone and April 8, 1999, by phone; Irving Weissman, director, Stanford University Institute for Cancer/Stem Cells, February 7, 2001, in Palo Alto, CA, and April 29, 2005, by phone; Evan Snyder, director, Stem Cell and Regeneration program, Burnham Institute, formerly of Harvard University, March 19 and March 20, 1999, by phone, and March 30, 1999, in Boston, MA; Douglas Kondziolka, neurosurgeon, University of Pittsburgh, at the University of Pittsburgh, PA, March 9, 1999; Sylvia Elam, stroke patient from Scottsdale, AZ, and her husband, Ira, at the University of Pittsburgh, March 9, 1999, and October 31, 1999, by phone; Gary Snable, former CEO of Layton Bioscience, March 1999, by phone; Virginia Lee, University of Pennsylvania, at the university, March 1999; Mahendra Rao, Stem Cell Research Task Force member, NIH, at the NIH, July 25, 2003; Richard Garr, CEO, NeuralStem, College Park, MD, February 23, 2001 (in MD), July 24, 2003 (in DC), September 22, 2002, and February 27, 2003, in e-mails; Ronald Numbers, University of Wisconsin, Department of Medical History, June 6, 2005, by phone; Nathaniel Comfort, Department of History, Johns Hopkins University, June 13, 2005, by phone.

18 Founded by Roger Williams: "Unitarianism in America, Roger Williams 1603–1683," www.harvardsquarelibrary.org, accessed October 10, 2005; C. Haynes, "Freedom of Religion Must Be for Everyone: Inside the First Amendment," www.firstamendmentcenter.org, November 1, 1998.

18 Providence became: "Providence," 514 Broadway Web site, Rhode Island School of Design Museum, http://tirocchi.stg.brown.edu/514/story/providence_1.html, accessed October 12, 2005; "History and Facts, America's Rennaissance City, Three and One Half Centuries at a Glance," www.providenceri.com/history/centuries.1.html, accessed October 11, 2005.

18 Pittsburgh, Pennsylvania: Pittsburgh History Series Teacher's Guide, www.wqed.org, accessed October 11, 2005; K. Zangrilli, "Tours Become Lesson in Living History," Pittsburgh History and Landmarks Foundation www.phlf.org, accessed October 11, 2005; E. Dyer, "Arthursville Abolitionists Ran Underground Railroad through Pittsburgh," *The Pittsburgh Post-Gazette,* www.postgazette.com, accessed Oct 10, 2005; "Civil Rights in Pittsburgh," www.freedomcorner.org, accessed October 11, 2005; North Side History, Carnegie Library of Pittsburgh, www.clpgh.org/exhibit/neighborhoods/northside/nor_n4.html, accessed October 10, 2005; "Pittsburgh, Pennsylvania," Microsoft Encarta Encyclopedia 2000, www.thecity ofpittsburgh.com, accessed October 12, 2005; "Downtown: Abraham Lincoln,"

Carnegie Library of Pittsburgh, www.clpgh.org/exhibit/neighborhoods/downtown/down_n121.html, accessed October 11, 2005; "North Side: Our Gay Village: Millionaire's Road," *Pittsburgh Gazette Times,* May 15, 1927.

19 For the first time: Author interview with Ronald Numbers, June 6, 2005; author interview with Nathaniel Comfort, June 13, 2005.

19 "The last thing": H. W. Brands, "Founders Chic: Our Reverence for the Fathers Has Gotten Out of Hand," *The Atlantic Monthly,* September 2003.

20 when the brain goes: "The Uniform Determination of Death Act," The National Conference of Commissioners on Uniform State Laws, www.nccusl.org/nccusl/uniformact_summaries/uniformacts-s-udoda.asp; M. Luckham, "Brain Stem Death and Organ Transplantation: Should the Donor Be Anaesthetised?" www.sprconsilio.com/braindead.html; Council of Europe, Explanatory Report to the Additional Protocol to the Convention on Human Rights and Biomedicine Concerning Transplantation of Organs and Tissues of Human Origin, ETS No. 186.

20 The U.K., among other: G. Bahadur, "The Moral Status of the Embryo: The Human Embryo in the U.K. Human Fertilisation and Embryology (Research Purposes) Regulation 2001 Debate," *Reproductive Biomedicine Online* 7 (July–August 2003): 12–16; Article 3 of the UK Human Fertilisation and Embryology Act (London: Her Majesty's Stationery Office, 1990).

21 400,000 spare in vitro fertilization: The RAND-SART Survey of 430 US-assisted reproductive technology facilities, Rand Institute and the Society for Reproductive Technology, 2003; D. I. Hoffman et al., "Cryopreserved Embryos in the United States and Their Availability for Research," *Fertility and Sterility* 79 (May 2003): 1063–69.

21 made zero sense: S. Hall, *Merchants of Immortality: Chasing the Dream of Human Life Extension* (Boston, MA: Houghton Mifflin Company 2003): 294–313; A. Parson, *The Proteus Effect* (Washington, DC: The Joseph Henry Press, 2004): 177.

21 the government had spent: A. Robeznieks, "The Politics of Progress: How to Continue Stem Cell Research Despite Limitations," *AMNews,* American Medical Association, August 9, 2004; Online *Newshour with Jim Lehrer,* "Stem Cell Research," August 10, 2004.

21 "has the potential": NIH Director's Statement on Research Using Stem Cells—Statement of Harold Varmus, MD, Director, National Institutes of Health, Department of Health and Human Services, before the Senate Appropriations Subcommittee on Labor, Health and Human Services, Education and Related Agencies, December 2, 1998.

22 "heal itself": Author interview with William Haseltine, then CEO of Human Genome Sciences, December 4, 2000, at the First Annual Conference on Regenerative Medicine in Washington, DC.

22 "Magic seeds": Author interview with Evan Snyder, then Harvard University researcher, March 19, 1999.

22 "You can increase": Author interview with Irving Weissman, February 7, 2001.

23 average life span . . . "circumvent aging": L. Hayflick, *How and Why We Age* (New York: Ballantine Books, 1996): 66, 84–87; L. Hayflick, "The Future of Aging," *Nature* 408 (November 9, 2000): 267–69.

23 it was announced: J.A. Thomson, "Embryonic Stem Cell Lines Derived from Human Blastocysts," *Science* 282 (November 6, 1998): 1145–47.

23 Five months after: Author interviews with Douglas Kondziolka and Sylvia Elam before, during, and after her surgery, at University of Pittsburgh, March 9 and 10, 1999.

23 first 12: N. Boyce, "Transplant of Cultured Cells May Pave Way for Stroke Treatment," *The New Scientist,* July 11, 1998: 23.

23 *politically* controversial: ibid.

24 they were brewed: Author interview with Gary Snable, April 9, 1999; personal communication (e-mails) from Peter Andrews, University of Sheffield stem cell researcher, March 30 and March 31, 1999.

24 decades earlier: L. J. Kleinsmith and G. B. Pierce, "Multipotentiality of Single Embryonal Carcinoma Cells," *Cancer Research* 24 (1964): 1544–51.

24 through a mouse: Author interview with Peter Andrews of the University of Sheffield, in Bar Harbour, ME, August 2004.

24 brain-neuron-like: Personal communication (e-mail) from Peter Andrews of the University of Sheffield, April 19, 1999; P. W. Andrews, "Retinoic Acid Induces Neuronal Differentiation of a Cloned Human Embryonal Carcinoma Cell Line in Vitro," *Developmental Biology* 103 (June 1984): 285–93; S. R. Kleppner, "Transplanted Human Neurons Derived from a Teratocarcinoma Cell Line (Ntera-2) Mature, Integrate, and Survive for Over 1 Year in the Nude Mouse Brain," *The Journal of Comparative Neurology* 357 (1995): 618–32.

24 no supply of: Author interview with Sam Weiss of the University of Calgary, March 1999; author interview with Richard Garr, February 23, 2001.

24 The very existence of: P. S. Eriksson, "Neurogenesis in the Adult Human Hippocampus," *Nature Medicine* 4 (November 1998): 1313–17; A. J. S. Rayl, "Research Turns Another 'Fact' into Myth," *The Scientist* 13 (February 15, 1999).

24 So Layton's: J. Q. Trojanowski, "Neurons Derived from a Human Teratocarcinoma Cell Line Establish Molecular and Structural Polarity Following Transplantation into the Rodent Brain," *Experimental Neurology* 122 (1993) 283–294.

24 made a claim: Author interview with Gary Snable, April 9, 1999.

24 nearly mature neurons: Layton Bioscience Inc, Web site as of July 25, 1998, "Lead Product Technologies." ("LBS neurons exhibit the properties of site-specific differentiation. That is, they are sufficiently plastic so that they continue to differentiate after implantation.")

24 "normal" stem cell: Personal communication (e-mail) from Cesario Borlongan, NIH, National Institute on Drug Abuse, April 2, 1999; K. Fackelmann, "Stroke Rescue: Can Cells Injected into the Brain Reverse Paralysis?" *Science News Online,* August 22, 1998; author interview with Ron McKay, April 8, 1999 ("So there are people who are putting cells which are known to be transformed into people at the moment, Layton. I just don't understand that. . . . My thought is it would be nice to see hard data there. . . . That's just a mess. . . . I've grafted many, many neural cells which have grown for long periods in culture and not seen tumors. But I've also grown cells that have been grown for short times which can form tumors").

24 60 chromosomes: Personal communication (e-mail) from Gary Snable, April 14, 1999; personal communication (e-mail) from Peter Andrews, stem cell researcher at the University of Sheffield, U.K., March 30, 1999.

24 reversed or blocked: Personal communication (e-mail) from the following: Gene Johnson, Washington University Medical School, November 23, 1999; Aviva Tolkovsky, Department of Biochemistry, Cambridge University, U.K., November 23,

1999; Derek Raghavan, current chairman, Cleveland Clinic Taussig Cancer Center, November 23, 1999; and Richard Momparler, Department of Pharmacology, University of Montreal, Canada, November 23, 1999; S. R. Kleppner, "Transplanted Human Neurons Derived from a Teratocarcinoma Cell Line (Ntera-2) Mature, Integrate, and Survive for Over 1 Year in the Nude Mouse Brain," *The Journal of Comparative Neurology* 357 (1995): 618–32.

24 only one paper: C. V. Borlongan, "Transplantation of Cryopreserved Human Embryonal Carcinoma-Derived Neurons (NT2N cells) Promotes Functional Recovery in Ischemic Rats," *Experimental Neurology* 149 (February 1998): 289–93.

25 two more: C. V. Borlongan, "Cerebral Ischemia and CNS Transplantation: Differential Effects of Grafted Fetal Rat Striatal Cells and Human Neurons Derived from a Clonal Cell Line," *NeuroReport* 9 (November 16, 1998): 3703–9; S. Saporta, "Neural Transplantation of Human Neuroteratocarcinoma (hNT) Neurons into Ischemic Rats. A Quantitative Dose-Response Analysis of Cell Survival and Behavioral Recovery," *Neuroscience* 91 (1999): 519–25.

25 passive avoidance: Personal communication (e-mail) from Clive Svendsen, then at the University of Cambridge Centre for Brain Repair, November 19, 1999.

26 out of her wheelchair: Author interview with Ira and Sylvia Elam, Scottsdale, AZ, October 31, 1999.

26 even greater significance: Author interview with Ron McKay, April 8, 1999.

27 While the child was: Author interview with Alan Flake, University of Pennsylvania, at Roger Williams Medical Center, April 8, 2003; A. Flake et al., "Treatment of X-Linked Severe Combined Immunodeficiency by In Utero Transplantation of Paternal Bone Marrow," *New England Journal of Medicine* 335 (1996): 1806–10.

28 her cells divide more slowly: Author interview with Moryama Reyes, University of Minnesota, at Roger Williams Medical Center, April 8, 2003; Author interview with Victor Nurcombe, Stem Cell and Tissue Repair Lab, Institute of Molecular and Cell Biology, Singapore, in Singapore, October 31, 2003.

29 "political correctness": Author interview with Peter Quesenberry, Roger Williams Medical Center, Research Department Chair April 8, 2003.

29 heated up: A. Regalado, "'Supercell' Controversy Sets off a Scientists' Civil War," *The Wall Street Journal,* June 21, 2002, B1.

29 she had found an hES-like: Y. Jiang, "Pluripotency of Mesenchymal Stem Cells Derived from Adult Marrow," *Nature* 418 (July 4, 2002): 41–49; S. P. Westphal, "Ultimate Stem Cell Discovered," *The New Scientist,* January 23, 2002.

29 number of different: D. S. Krause, "Multi-Organ, Multi-Lineage Engraftment by a Single Bone Marrow-Derived Stem Cell," *Cell* 105 (May 4, 2001): 369–77; *Associated Press,* "Scientists Tout Alternatives to Stem Cells at US Senate," June 12, 2003; C. N. Shen, "Transdifferentiation of Pancreas to Liver," *Mechanisms of Development* 120 (January 2003): 107–16; D. Orlic, "Stem Cells for Myocardial Regeneration," *Circulation Research* 91 (December 13, 2002): 1092–1102.

29 The claim of Reyes: Y. Jiang, "Pluripotency of Mesenchymal Stem Cells Derived from Adult Marrow," *Nature* 418 (July 4, 2002): 41–49.

29 However, many adult stem cell researchers: D. J. Anderson, "Can Stem Cells Cross Lineage Boundaries?" *Nature Medicine* 7 (April 2001): 393–95; R. F. Castro, "Failure of Bone Marrow Cells to Transdifferentiate into Neural Cells in Vivo," *Science* 297 (August 23, 2002): 1299; H. Pearson, "Stem Cells Take Knocks," *Nature News Service*

(September 2, 2002); A Wagers, "Little Evidence for Developmental Plasticity of Adult Hematopoietic Stem Cells," *Science* 297 (September 27, 2002): 2256–59.

29 two 2003 *Nature* papers were: X. Wang, "Cell Fusion Is the Principal Source of Bone-Marrow-Derived Hepatocytes," *Nature* 422 (April 24, 2003): 897–901 (e-pub March 30, 2003); G. Vassilopoulos, "Transplanted Bone Marrow Regenerates Liver by Cell Fusion," *Nature* 422 (April 24, 2003): 901–4 (e-pub March 30, 2003).

29 this might well: A. Medvinsky, "Stem Cells: Fusion Brings Down Barriers," *Nature* 422 (April 24, 2003): 823–25.

29 "she'd never been": A. Regalado, "'Supercell' Controversy Sets off a Scientists' Civil War," *The Wall Street Journal* June 21, 2002: B1.

31 Weissman paper: A. J. Wagers, "Little Evidence for Developmental Plasticity of Hematopoietic Stem Cells," *Science* 297 (September 27, 2002): 2256–59.

32 *create* hES cells: J. E. Robinet, "Pittsburgh Scientists Vie for Piece of the Stem Cell Research Pie," *Pittsburgh Business Times*, December 13, 2002; Pennsylvania Consolidated Statutes, Title 18, 3216, 3203.

33 Linda Lester: Author interview with Linda Lester, endocrinologist and stem cell scientist, Oregon National Primate Research Center, April 26, 2004.

33 15 Senate hearings: Senate floor statement by U.S. Senator Tom Harkin (D-Iowa) on the passing of Christopher Reeve on October 11, 2004.

34 Lindazo Cheng: L. Cheng, "Human Adult Marrow Cells Support Prolonged Expansion of Human Embryonic Stem Cells in Culture," *Stem Cells* 21 (2003): 131–42.

34 generally believed: Federal Funding for Stem Cell Research—Hearing Before the Senate Appropriations Subcommittee on Labor, Health and Human Services, Education, and Related Agencies, May 22, 2003. Specter said at the meeting that the information that all NIH-funded hES cells were grown on mouse feeders was given to him by "the Secretary of HHS, [Thomas] Thompson; Dr. Allen Spiegel, the director of the National Institute on Diabetes, Digestive and Kidney Diseases . . . by Dr. James Battey, Chairman of the NIH Stem Cell Task Force, all from NIH, and from other scientists as well: Dr. Roger Pedersen from Cambridge, Dr. George Daley from MIT. There is a considerable body of additional authority for the proposition that the existing cell lines, stem cell lines, are contaminated with mouse feeders."

34 "I'm concerned": Ibid. A partial, fairly accurate transcript of the proceedings can be found at www.access.gpo.gov/congress/senate.

37 "revolution": R. McKay, "Stem Cells–Hype and Hope," *Nature* 406 (July 27, 2000): 361–64.

37 "wimp": Author interview with Ron McKay, April 29, 2005; author attended meeting.

38 "my book": Arlene Chiu and Mahendra S. Rao, *Human Embryonic Stem Cells* (Totowa, NJ: Humana Press, 2003).

38 so broad: Author interview with Mahendra Rao, July 25, 2003; author interview with Victor Nurcombe, October 31, 2003; M. Wadman, "Licensing Fees Slow Advance of Stem Cells," *Nature* 435 (May 19, 2005): 272–73.

39 March 2006: J. F. Loring and C. Campbell, "Intellectual Property and Human Embryonic Stem Cell Research," *Science* 311 (March 24, 2006): 1716–17.

39 multipotent mouse neural stem cell: A. Kalyani, "Neuroepithelial Stem Cells from the Embryonic Spinal Cord: Isolation, Characterization, and Clonal Analysis," *Developmental Biology* 186 (June 15, 1997): 202–23.

39 Sam Weiss: B. A. Reynolds et al., "Generation of Neurons and Astrocytes from Isolated Cells of the Adult Mammalian Central Nervous System," *Science* 255 (March 27, 1992): 1707–10. Author interview with Angelo Vescovi, codirector of Research of the Institute for Stem Cell Research, Hospital San Raffacle, Milan, Italy, March 1999. "Before 1992 there was nothing there. We didn't know there were stem cells in the adult brain."

40 lost their ability: Author interview with Mahendra Rao, July 25, 2003; author interview with Richard Garr, July 24, 2003; United States Patent Application 20020064873, Kind Code A1, Renji Yang and Karl Johe, May 30, 2002, "Stable Neural Stem Cell Lines" ("With the previous culture conditions, it had been difficult to expand CNS stem cells beyond about 30 cell-doublings at which point a majority of the cells have lost their capacity for neuronal differentiation"); personal communication (e-mail) from Richard Garr, September 20, 2004 ("Everyone that has been able to produce dopaminergic cells in vitro, has seen the phenotype disappear within 2 months of transplantation"); J. H. Kim, "Dopamine Neurons Derived from Embryonic Stem Cells Function in an Animal Model of Parkinson's Disease," *Nature* 418 (July 4, 2002): 50–56.

42 "living things": Author interview with CEO Thomas Okarma, Geron, Menlo Park, CA, in CA, February 12, 2001.

43 problems getting backing: Author interview with Richard Garr, February 23, 2001; author interview with Irv Weissman, February 7, 2001.

45 by 2002: Personal communication (e-mail) from Richard Garr, September 22, 2002.

45 by July 2003: Personal communication (e-mail) from Richard Garr, February 26, 2003; author interview with Richard Garr, July 24, 2003

46 "To gain access": E-mail from Mahendra Rao, May 17, 2005.

47 "great experiment": Author interview with Irv Weissman, April 29, 2005.

47 "breathtaking": Author interview with Ron McKay, April 29, 2005.

CHAPTER 2

Interviews: Ismail Barrada, scientific founder and head, Stem Cell Research Center, Misr University for Science and Technology, 6th of October City, Egypt, June 10 and June 11, 2003, in Washington, DC, and October 16–20, 2003, in Egypt; the late Souad A. Kafafi, then chancellor, Misr University for Science and Technology, October 19, 2003, in Egypt; Mahmoud Naguib, then president, Misr University for Science and Technology, October 19, 2003, in Egypt.

48 "contradiction": Author interview with Ismail Barrada, June 11, 2003.

49 launderers: D. Eggen, "U.S. Ties Hijackers' Money to Al Qaeda," *The Washington Post,* October 7, 2001, A1; B. Loeterman, "Inside the Terror Network," *Frontline,* Transcript Program 2009, January 17, 2002.

49 Their men: Academy for Educational Development, "United Arabic Emirates Case Study: Women and Information Technology in the UAE," http://projects.aed.org /techequity/UAE.htm, accessed October 11, 2005. see also L. Wright, "The Kingdom of Silence," *The New Yorker,* January 5, 2004.

49 to drive, or to leave: Ibid.

49 custom dictates: U.S. Department of State, "Country Reports on Human Rights Practices 2004: United Arab Emirates," released February 28, 2005.

49 able to vote: Ibid.; U.S. Department of State, "Country Reports on Human Rights Practices 2004: Saudi Arabia," released February 28, 2005; Lori B. Andrews, *The Clone Age: Adventures in the New World of Reproductive Technology* (New York: Henry Holt, 1999): 5.

49 marry: Ibid.

49 board a bus: L. Wright, op cit.

49 imprisoned for sharing: U.S. Department of State, "Country Reports on Human Rights Practices 2004: Saudi Arabia," released February 28, 2005.

51 "new oil": "Stem Cells Arrive in Saudi Arabia," *Science* 296 (June 28, 2002).

51 $100 million: "A Risky Experiment in Biotechnology," *The Financial Times,* June 25, 2002.

51 *fatwas:* "Muslim Scientists to Enter Stem Cell Field," *Bioventure View,* June 25, 2002; Z. Amanullah, "Does Islam Offer a Way Out of the Stem-Cell Debate?" www.alt.muslim, June 17, 2002.

51 holy greenlight: Personal communication (e-mail) from Ismail Barrada, April 14, 2006.

51 some 4 percent of Saudis: T. Friend, "Saudis Take Lead on Stem Cell Cloning," *USA Today,* July 8, 2002.

51 isolated an hES cell line: H. Baharvand, "Establishment and In Vitro Differentiation of a New Embryonic Stem Cell Line from Human Blastocyst," *Differentiation* 72 (June 2004): 224–29; L. Walters, "Human Embryonic Stem Cell Research: An Intercultural Perspective," *Kennedy Institute of Ethics Journal* 14 (March 2004): 3–8; "Producing Heart Cells by using Embryonic Stem Cells," *Ente Khab,* July 15, 2003: 8.

51 weapons in a war: "Iran Achieves Stem Cells Technology," www.khamenei.de, September 3, 2003; T. Reichhardt, "Religion and Science: Studies of Faith," *Nature,* December 8, 2004; "Iran, 10th Country to Produce Embryonic Human Stem Cell," *IRNA,* September 3, 2003.

52 Jordan is rumored: Author interview with Ismail Barrada, October 17, 2003.

52 Severino Antinori: "Human Cloning Project Claims Progress," *Gulf News,* March 4, 2002.

52 eccentric: Stephen Hall, *Merchants of Immortality* (New York: Houghton Mifflin, 2003): 277–82.

52 in Egypt . . . it can: Marcia C. Inhorn, *Local Babies, Global Science: Gender, Religion, and In Vitro Fertilization in Egypt* (New York: Routledge, 2003): 1–98; Lori B. Andrews, *The Clone Age: Adventures in the New World of Reproductive Technology* (New York: Henry Holt, 1999): 8.

52 "There are boats": Author interview with Ismail Barrada, June 11, 2003.

52 Nasser's ability: Noah Feldman, *After Jihad* (New York: Farrar, Straus and Giroux, 2003): 162–73; Fareed Zakaria, *The Future of Freedom: Illiberal Democracy at Home and Abroad* (New York: W. W. Norton, 2003): 119–59.

53 indicated to many: Author interview with Ismail Barrada, October 19, 2003.

53 some of its streets: "'Bulls-eye' Say Egyptians as They Celebrate Anti-US Attacks," AFP, September 11, 2001; "Around the World, US security Is Tightened," CNN, September 11–12, 2001; D. Pipes, "A Middle East Party," *The Jerusalem Post,* September 14, 2001.

53 men led by: M. Weaver, "Pharoahs in Waiting," *Atlantic Monthly*, October 2003.

53 Egypt isn't . . . days of Nasser: Ibid; B. Lewis, "The Revolt of Islam: When Did the Conflict with the West Begin, and How Could It End?" *The New Yorker*, November 19, 2001; B. Lewis, "What Went Wrong?" *The Atlantic Monthly*, January 2002.

53 656 scientists: *The Pulse: The ISSCR Newsletter*, July 1, 2003.

54 hundreds of years: "Science and Technology: Historic Innovation, Modern Solutions, Cutting Edge Science in the Middle East," PBS, 2002 (which notes that, in the seventeenth century, "the Ottomans and their Muslim contemporaries in Mughal India and the Persian Safavid Empire ceased to support scientific research and innovations"); N. Ahmed, *Book Review–Intellectual Achievements of Muslims* Islamic Research Foundation International, www.irfi.org, accessed October 8, 2005.

54 Zewail: "Egypt Honors Zewail," www.sis.gov.eg/zewail/html/egypt.htm, accessed October 8, 2005; "Egypt Harvests Nobel Prize for Third Time," *Egypt Magazine*, issue 20, Spring 2000.

54 own wife: M. Abdel-Malek, "Lubna Abdel-Aziz: The Eyes Have It," *Al-Ahram Weekly*, December 30, 1998; R. Woffenden, "The Reluctant Movie Star," *The Cairo Times*, November 16–22, 2000; R. Dowell, "Lubna Abdel Aziz: Egypt's Beloved Star Returns," *Enigma Magazine*, www.enigma-mag.com/interview12.htm, accessed October 14, 2005.

54 Egypt is like . . . than its neighbors: Noah Feldman, *After Jihad: American and the Struggle for Islamic Democracy* (New York: Farrar, Straus and Giroux, 2003): 162–73; Fareed Zakaria, *The Future of Freedom: Illiberal Democracy at Home and Abroad* (New York: W. W. Norton, 2003): 119–59; B. Lewis, "The Revolt of Islam: When Did the Conflict with the West Begin, and How Could it End?" *The New Yorker*, November 19, 2001; B. Lewis, "What Went Wrong?" *The Atlantic Monthly*, January 2002.

56 In 1989: "Montrealer Goes to Israel for Experimental Cancer Treatment," *Canadian Jewish News*, August 30, 2001; C. Poynton, "Minigrafty-Transplant Hope," *Lymphoma Winter Newsletter*, issue 41, Spring 2000.

56 giving her transplants: Personal communication (e-mail) from Human Gene Therapy Program codirector Fulvio Mavilio, Instituto Scientifico, Hospital San Raffaele, Milan, Italy, June 12, 2005, M. Cavazzana-Calvo et al., "The Future of Gene Therapy: Balancing the Risks and Benefits of Clinical Trials," *Nature* 427 (February 2004): 779–81.

56 aside from clandestine: Michael B. Oren, *Six Days of War: June 1967 and the Making of the Modern Middle East* (New York: Oxford University Press, 2002); Thomas L. Friedman, *Longitudes and Attitudes: The World in the Age of Terrorism* (New York: Anchor Books, 2002).

56 *Gulf Today*: "Only 38% Will Vote Bush: Survey," *The Gulf Today*, October 16, 2003.

56 *Khaleej Times*: M. Galadari, "American Veto Surprises All," *Khaleej Times*, October 16, 2003.

56 All day: "Mubarak: UN Should Play Decisive Role in Restoring Authority to Iraqis," *The Egyptian Gazette*, October 17, 2003.

57 celebrating . . . all month: "October War Anniversary, Three Decades of Liberty, Peace and Development," *Egypt Magazine*, October 2003; G. Nikrumah, "Three Decades on: Bittersweet Memories on the 30th Anniversary of the October 1973 War Were Splashed on the Pages of the Egyptian Press," *Al-Ahram Weekly*, October 9–15, 2003.

57 "reading my e-mails": Author interview with Ismail Barrada, October 16, 2003.
59 science in the way: Alexander Stille, *The Future of the Past* (New York: Picador, 2003); "Egypt's Pyramids: Houses of Eternity," *National Geographic* video.
59 this month: "German Team Finds Secrets of Mummies' Preservation," *China Daily,* October 23, 2003.
61 6th of October City: L. Seif and W. Wali, "Sixth of October City," *Business Today—Egypt,* June 2001.
61 largest private: "Message from the Chancellor," Misr University for Science and Technology catalog; S. Shehab, "First Year German U," *Al-Ahram Weekly,* October 2–8, 2003; F. Richter, "Higher Thinking," *American Chamber of Commerce in Egypt, Business Monthly,* May 2005.
62 later publish: H. Baharvand, op. cit.
66 "The Egyptian people . . . Pharonic": Author interview with Souad Kafafi.

CHAPTER 3

Interviews: Benjamin Reubinoff, director, Hadassah University Human Embryonic Stem Cell Research Center, Jerusalem, Israel, October 22, 2003, in Israel, and August 24, 2003, by phone; Ariff Bongso, Singapore National University, October 31, 2003, in Singapore, and September 25, 2003, and May 3, 3005, by phone; Alan Trounson, Monash Institute of Reproduction and Development, Monash, Australia, in Monash, October 13, 2003; Martin Pera, director, Australian Stem Cell Centre's embryonic stem cell program, Monash, Australia, in Monash, October 14, 2003; Karl Skorecki, director, Rappaport Family Institute for Research in the Medical Sciences, Haifa, Israel, October 24, 2003, in Haifa; Nissim Benvenisty, head, Department of Genetics, Hebrew University in Jerusalem, Israel, October 26, 2003, in Israel, and August 2003, by phone; Lior Gepstein, stem cell researcher, Rappaport Family Institute, Haifa, Israel, October 7, 2004, by phone; Dan Kaufman, Stem Cell Institute, University of Minnesota, Minneapolis, MN (formerly in the University of Wisconsin lab of James Thomson), November 3, 2002, in MN; Douglas Melton, co-director, Harvard University Stem Cell Institute, Cambridge, MA, June 17, 2004, by phone.
70 *Science*: G. Vogel, "In the MidEast, Pushing Back the Stem Cell Frontier," *Science* 295 (March 8, 2002): 1818–20.
70 intifada: J. Babbin, "Fencing out Terror: Peace through Security," *The National Review,* November 25, 2004.
71 travelers' warning: "Israel, the West Bank and Gaza Travel Warning," U.S. Department of State, October 20, 2003 ("The Department of State warns U.S. citizens to defer travel to Israel, the West Bank and Gaza"), expatexchange.com.
71 Gaza: "US Convoy in Gaza Bombed; Three Americans Killed," *PBS Online Newshour,* October 15, 2003.
72 Between 1996: "Science in Israel: 1996–2000," ISI Essential Science Indicators, December 31, 2001.
73 Teva Pharmaceuticals: Teva Pharmaceutical Industries Ltd, www.tevapharm.com; "The History of Teva Pharmaceutical Industries," www.tevapharm.com/about/history.asp, accessed October 14, 2005.
73 first two: J. A. Thomson, "Embryonic Stem Cell Lines Derived from Human Blas-

tocysts," *Science* 282 (November 6, 1998): 1145–47; B. Reubinoff, "Embryonic Stem Cell Lines from Human Blastocysts: Somatic Differentiation In Vitro," *Nature Biotechnology* 18 (April 2000): 399–404.

74 British scientist: M. J. Evans, "Establishment in Culture of Pluripotential Cells from Mouse Embryos," *Nature* 292 (July 9, 1981): 154–56; G. R. Martin, "Isolation of a Pluripotent Cell Line from Early Mouse Embryos Cultured in Medium Conditioned by Teratocarcinoma Stem Cells," *Proceedings of the National Academy of Sciences* 78 (December 1981): 7634–38.

75 Palmiter created: C. M. Smith, "Technical Knockout: Gene-Targeting Strategies Provide an Avenue for Studying Gene Function," *The Scientist* 14 (July 24, 2000); R. D. Palmiter, "Dramatic Growth of Mice that Develop from Eggs Microinjected with Metallothionein-Growth Hormone Fusion Genes," *Nature* 300 (December 16, 1982): 611–15.

76 boosting the success: Agency for Science, Technology and Research, Singapore: www.astar.edu.sg/astar/exploit/action/pressreleasedetails.do?id=2925937c4eAg, accessed October 14, 2005; A. Bongso, "Improved Pregnancy Rate after Transfer of Embryos Grown in Human Fallopian Tubal Cell Coculture," *Fertility and Sterility* 58 (September 1992): 569–74.

76 those five-day-old cells: Author interview with Doug Melton, June 17, 2004.

76 submitted his paper: A. Bongso, "Isolation and Culture of Inner Cell Mass Cells from Human Blastocysts," *Human Reproduction* 9 (November 1994): 2110–17.

77 specialist Alan Trounson: Author interview with Alan Trounson, October 13, 2003.

78 Back in 1994: Author interviews with Ariff Bongso, October 31, 2003, and May 3, 2005.

81 four out of five: G. Vogel, "In the MidEast, Pushing Back the Stem Cell Frontier," *Science* 295 (March 8, 2002): 1818–20.

81 Since 1995 . . . before publishing: Author interview with Karl Skorecki, October 24, 2003; Stephen Hall, *Merchants of Immortality: Chasing the Dream of Human Life Extension* (Boston: Houghton Mifflin: 2003): 159–63.

82 team came out with: B. Reubinoff, "Embryonic Stem Cell Lines from Human Blastocysts: Somatic Differentiation In Vitro," *Nature Biotechnology* 18 (April 2000): 399–404.

82 two Israelis: J. Itskovitz-Eldor, "Differentiation of Human Embryonic Stem Cells into Embryoid Bodies Compromising the Three Embryonic Germ Layers," *Molecular Medicine* 6 (February 2000): 88–95.

83 *Journal of Translational Medicine*: J. Ioannidis, "Materializing Research Promises: Opportunities, Priorities and Conflicts in Translational Medicine," *Journal of Translational Medicine* 2 (January 31, 2004).

83 lentiviral vectors: M. Gropp, "Stable Genetic Modification of Human Embryonic Stem Cells by Lentiviral Vectors," *Molecular Therapy* 7 (February 2003): 281–87.

83 In France: M. Cavazzana-Calvo, "Gene Therapy of Human Severe Combined Immunodeficiency (SCID)-X1 disease," *Science* 288 (April 28, 2000): 669–72; E. Marshall, "Second Child in French Trial Is Found to Have Leukemia," *Science* 229 (January 13, 2003): 320; J. Kaiser, "Panel Urges Limits on X-SCID Trials," *Science* 307 (March 11, 2005).

84 Just in case: M. Schuldiner, "Selective Ablation of Human Embryonic Stem Cells Expressing a 'Suicide' Gene," *Stem Cells* 21 (2003): 257–65.

84 not arouse: M. Drukker, "Characterization of the Expression of MHC Proteins in Human Embryonic Stem Cells," *Proceedings of the National Academy of Sciences* 99 (July 23, 2002): 9864–69.

84 chick embryo: R. S. Goldstein, "Integration and Differentiation of Human Embryonic Stem Cells Transplanted to the Chick Embryo," *Developmental Dynamics* 225 (September 2002): 80–86.

85 October of 2000: M. Schuldiner, "Effects of Eight Growth Factors on the Differentiation of Cells Derived from Human Embryonic Stem Cells," *Proceedings of the National Academy of Sciences* 97 (October 10, 2000): 11307–12.

85 November of 2000: M. Amit, "Clonally Derived Human Embryonic Stem Cell Lines Maintain Pluripotency and Proliferative Potential for Prolonged Periods of Culture," *Developmental Biology* 227 (November 2000): 271–78.

85 And in 2001: S. Assady, "Insulin Production by Human Embryonic Stem Cells," *Diabetes* 50 (August 2001): 1691–97; I. Kehat, "Human Embryonic Stem Cells Can Differentiate into Myocytes with Structural and Functional Properties of Cardiomyocytes," *Journal of Clinical Investigation* 108 (August 2001): 407–14; B. Reubinoff, "Effective Cryopreservation of Human Embryonic Stem Cells by the Open Pulled Straw Vitrification Method," *Human Reproduction* 16 (October 2001): 2187–94.

85 first to describe creating true neurons: M. Schuldiner, "Induced Neuronal Differentiation of Human Embryonic Stem Cells," *Brain Research* 913 (September 21, 2001): 201–5.

85 hES cells into primitive blood cells: D. S. Kaufman, "Hematopoietic Colony-forming Cells Derived from Human Embryonic Stem Cell," *Proceedings of the National Academy of Sciences* 98 (September 11, 2001): 10716–21.

86 telomerase: M. Tzukerman, "Identification of a Novel Transcription Factor Binding Element Involved in the Regulation by Differentiation of the Human Telomerase (hTERT) Promoter," *Molecular Biology of the Cell* 11 (December 2000): 4381–91.

86 voltage: M. K. Carpenter, "Enrichment of Neurons and Neural Precursors from Human Embryonic Stem Cells," *Experimental Neurology* 172 (December 2001): 383–97.

86 pop neural cells: B. E. Reubinoff, "Neural Progenitors from Human Embryonic Stem Cells," *Nature Biotechnology* 19 (December 2001): 1134–40; S. C. Zhang, "In Vitro Differentiation of Transplantable Neural Precursors from Human Embryonic Stem Cells," *Nature Biotechnology* 19 (December 2001): 1129–33.

86 made human veins: S. Levenberg, "Endothelial Cells Derived from Human Embryonic Stem Cells," *Proceedings of the National Academy of Sciences* 99 (April 2, 2002): 4391–96.

86 heart tissue in a dish: I. Kehat, "High-Resolution Electrophysiological Assessment of Human Embryonic Stem Cell-Derived Cardiomyocytes: A Novel In Vitro Model for the Study of Conduction," *Circulation Research* 91 (October 18, 2002): 659–61.

86 Geron also got cultures: C. Xu, "Characterization and Enrichment of Cardiomyocytes Derived from Human Embryonic Stem Cells," *Circulation Research* 91 (September 20, 2002): 501–8.

86 cells that form placenta: R. H. Xu, "BMP4 Initiates Human Embryonic Stem Cell Differentiation to Trophoblast," *Nature Biotechnology* 20 (December 2002): 1261–64.

86 "stemness" genes: M. Ramalho-Santos, "'Stemness': Transcriptional Profiling of Embryonic and Adult Stem Cells," *Science* 298 (October 18, 2002): 597–600; N. B. Ivanova, "A Stem Cell Molecular Signature," *Science* 298 (October 18, 2002): 601–4.
86 create vessels: S. Gerecht-Nir, "Human Embryonic Stem Cells as an In Vitro Model for Human Vascular Development and the Induction of Vascular Differentiation," *Laboratory Investigation* 83 (December 2003): 1811–20.
86 hES cells on a scaffold: S. Levenberg, "Differentiation of Human Embryonic Stem Cells on Three-Dimensional Polymer Scaffolds," *Proceedings of the National Academy of Sciences* 200 (October 28, 2003): 12741–46.
86 make liver cells: L. Rambhatla, "Generation of Hepatocyte-like Cells from Human Embryonic Stem Cells," *Cell Transplantation* 12 (2003): 1–11.
86 Pumilio-2 gene: F. L. Moore, "Human Pumilio-2 Is Expressed in Embryonic Stem Cells and Germ Cells and Interacts with DAZ (Deleted in Azoospermia) and DAZ-like Proteins," *Proceedings of the National Academy of Sciences* 100 (January 2003): 538–43.
87 genes in mouse: N. Sato, "Molecular Signature of Human Embryonic Stem Cells and Its Comparison with the Mouse," *Developmental Biology* 260 (August 15, 2003): 404–13.
87 skin cells: H. Green, "Marker Succession During the Development of Keratinocytes from Cultured Human Embryonic Stem Cells," *Proceedings of the National Academy of Sciences,* 100 (December 23, 2003): 15625–30.
87 tumors: J. M. Sperger, "Gene Expression Patterns in Human Embryonic Stem Cells and Human Pluripotent Germ Cell Tumors," *Proceedings of the National Academy of Sciences,* 100 (November 11, 2003): 13350–55.
88 second scientist in the world: T. Ben-Hur, "Transplantation of Human Embryonic Stem Cell-Derived Neural Progenitors Improves Behavioral Deficit in Parkinsonian Rats," *Stem Cells* 22 (2004): 1246–55.
88 first scientist in the world: I. Kehat, "Electromechanical Integration of Cardiomyocytes Derived from Human Embryonic Stem Cells," *Nature Biotechnology* 22 (October 2004): 1282–89.

CHAPTER 4

Interviews: Shimon Slavin, head, Department of Bone Marrow Transplantation, Hadassah University, Jerusalem, Israel, October 22, 2003 in Israel, and October 26 2003 in Israel, and August 2003, by phone; Dan Kaufman, Stem Cell Institute, University of Minnesota, Minneapolis, MN (formerly in the University of Wisconsin lab of James Thomson), November 3, 2003, in MN; John Dick, May 10, 2005; Richard Childs, National Heart Lung and Blood Institute, NIH, Bethesda, MD, March 4, 2004, by phone, and February 17, 2005, in MD; Stephen Rosenberg, chief, National Cancer Institute Surgery Branch, Bethesda, MD, several interviews in MD and by phone, January–March 1998; Karl Skorecki, director, Rappaport Family Institute for Research in the Medical Sciences, Haifa, Israel, October 24, 2003, in Israel; Lior Gepstein, stem cell researcher, Rappaport Family Institute, Haifa, Israel, October 7, 2004, by phone.
90 "Take two stem": Author interview with Dan Kaufman, November 3, 2003.

90 50,000 bone marrow transplants: J. M. Goldman, "The International Bone Marrow Transplant Registry," *International Journal of Hematology* 76 (August 2002): 393–97.

90 first truly successful: See note for p. 5.

91 humans in 1956: Medhunters, *Transplant Timeline,* www.medhunters.com/articles/transplantTimelineBMT.html, accessed October 14, 2005.

91 the Greeks: Douglas Starr, *Blood: An Epic History of Medicine and Commerce* (New York: Knopf, 1998): 5–22.

92 430 BC: M. E. Weksler, "The Immune System, Amyloid-Peptide, and Alzheimer's Disease," *Immunological Reviews* 205 (June 2005): 244.

92 Medawar: E. Simpson, "Reminiscences of Sir Peter Medawar: In Hope of Antigen-Specific Transplantation Tolerance," *American Journal of Transplantation* 4 (December 2004): 1937–40.

92 Owen: R. D. Owen, "Immunogenetic Consequences, of Vascular Anastomoses Between Bovine Twins," *Science* 102 (1945): 400; I. L. Weissman, "The Road Ended up at Stem Cells," *Immunological Reviews* 185 (2002): 159–174.

93 As a resident . . . wild again: Several author interviews with Steven A. Rosenberg, December 1997–March 1998; Steven A. Rosenberg, *The Transformed Cell: Unlocking the Mysteries of Cancer* (New York: Avon Books, 1992).

94 Canadian researchers: J. E. Till, "A Direct Measurement of the Radiation Sensitivity of Normal Mouse Bone Marrow Cells," *Radiation Research* 14 (1961): 213–22.

95 *Science,* in 1976: S. Slavin, "Long-Term Survival of Skin Allografts in Mice Treated with Fractionated Total Lymphoid Irradiation," *Science* 193 (September 1976): 1252–54.

95 in the 1970s: P. L. Weiden, "Antileukemic Effect of Graft-versus-Host Disease in Human Recipients of Allogeneic-Marrow Grafts," *New England Journal of Medicine* 300 (May 1979): 1068–73; P. L. Weiden, "Antileukemic Effect of Chronic Graft-versus-Host Disease: Contribution to Improved Survival after Allogeneic Marrow Transplantation," *New England Journal of Medicine* 304 (June 18, 1981): 1529–33.

95 University of Wisconsin: M. M. Horowitz, "Graft-versus-Leukemia Reactions after Bone Marrow Transplantation," *Blood* 75 (February 1990): 555–62.

96 widely agreed: Author interview with Richard Childs, March 4, 2004; H. J. Kolb, "Donor Leukocyte Transfusions for Treatment of Recurrent Chronic Myelogenous Leukemia in Marrow Transplant Patients," *Blood* 76 (December 1990): 2462–65.

96 "He was cured": S. Slavin, "Allogeneic Cell Therapy: The Treatment of Choice for All Hematologic Malignancies Relapsing Post BMT," *Blood* 87 (May 1996): 4011–13; S. Slavin, "Cellular-Mediated Immunotherapy of Leukemia in Conjunction with Autologous and Allogeneic Bone Marrow Transplantation in Experimental Animals and Man," *Blood* 72 (1988, suppl.): 407a.

97 Fuchs: L. Luznik, "Donor Lymphocyte Infusions to Treat Hematologic Malignancies in Relapse after Allogeneic Blood or Marrow Transplantation," *Cancer Control* 9 (March 2002): 123–37.

97 Rainer Storb: "F. Baron, "Hematopoietic Cell Transplantation: Five Decades of Progress," *Arch Medical Research* 34 (November–December 2003): 528–44.

98 Rosenberg et al.: M. E. Dudley, "Cancer Regression and Autoimmunity in Patients after Clonal Repopulation with Antitumor Lymphocytes," *Science* 298 (October 25, 2002): 850–54.

98 1984 and 1997: Author interview with Richard Childs, March 4, 2004; S. Morecki, "Induction of Graft vs. Tumor Effect in a Murine Model of Mammary Adenocarcinoma," *International Journal of Cancer* 71 (March 1997): 59–63; M. Moskovitch, "Anti-Tumor Effects of Allogeneic Bone Marrow Transplantation in (NZB X NZW)F1 hybrids with Spontaneous Lymphosarcoma," *Journal of Immunology* 132 (February 1984): 997–1000.

98 Richard Childs: R. W. Childs, "Successful Treatment of Metastatic Renal Cell Carcinoma with a Nonmyeloablative Allogeneic Peripheral-Blood Progenitor-Cell Transplant: Evidence for a Graft-versus-Tumor Effect," *Journal of Clinical Oncology* 17 (July 1999): 2044–49; R. Childs, "Regression of Metastatic Renal-Cell Carcinoma after Nonmyeloablative Allogeneic Peripheral-Blood Stem-Cell Transplantation," *New England Journal of Medicine* 343 (September 14, 2000): 750–58.

99 making headway: L. Luznik, "Successful Therapy of Metastic Cancer Using Tumor Vaccines in Mixed Allogeneic Bone Marrow Chimeras," *Blood* 101 (February 15, 2003): 1645–52.

99 Slavin, once again, performed: Author Interview with Richard Childs, March 4, 2004; H. Y. Knobler, "Tolerance to Donor-type Skin in the Recipient of a Bone Marrow Allograft. Treatment of Skin Ulcers in Chronic Graft-versus-Host Disease with Skin Grafts from the Bone Marrow Donor," *Transplantation* 40 (August 1985): 223–25.

99 Massachusetts General Hospital: T. R. Spitzer, "Combined Histocompatibility Leukocyte Antigen-matched Donor Bone Marrow and Renal Transplantation for Multiple Myeloma with End Stage Renal Disease: The Induction of Allograft Tolerance through Mixed Lymphohematopoietic Chimerism," *Transplantation* 68 (August 27, 1999): 480–84.

100 In 1991: Steven A. Rosenberg, *The Transformed Cell: Unlocking the Mysteries of Cancer* (New York: Avon Books, 1992): 298—29.

100 4,000 patients: See note for p. 5.

101 to tamp: "Report from the ASGT Ad Hoc Committee on Retroviral-Mediated Gene Transfer to Hematopoietic Stem Cells, April 2003.

101 The trial was: Personal communication (e-mail) from Fulvio Mavilio, codirector, Human Gene Therapy program Instituto Scientifico, Hospital San Raffaele, Milan, Italy, June 12, 2005.

104 Skorecki . . . "in culture": Author interview with Karl Skorecki, October 24, 2003; Stephen Hall, *Merchants of Immortality: Chasing the Dream of Human Life Extension* (Boston: Houghton Mifflin Company, 2003): 159–63.

105 Since, even though . . . "living history": Author interview with Karl Skorecki, October 24, 2003.

106 Gepstein: Author interview with Lior Gepstein, October 7, 2004.

107 hanging by the neck: G. Myre, "Palestinian Group Executes 2 Suspected of Aiding Israel," *The New York Times,* October 24, 2003.

108 Yet to be . . . "will last": Author interview with Nissim Benvenisty, October 26, 2003.

110 Outside Benvenisty's . . . petri dish of cells: Author interview with Maya Schuldina.

111 "Oh, and one day . . . part of it": Author interview with Shimon Slavin, October 26, 2003.

CHAPTER 5

Interviews: Ariff Bongso, Department of Obstetrics and Gynecology, National University of Singapore, October 31, 2003 in Singapore, and September 25, 2003, by phone, and May 3, 2005, by phone; Victor Nurcombe, Stem Cell and Tissue Repair Lab, Institute of Molecular and Cell Biology, Singapore, October 31, 2003, in Singapore, and October 12, 2003, in Melbourne, Australia, and May 24, 2005, by phone; Bing Lim, senior group leader, Genome Institute of Singapore (GIS), Singapore, November 3, 2003, in Singapore; Douglas Melton, cofounder, Harvard University Stem Cell Institute, June 17, 2004, by phone; Alan Colman, CEO, ES Cell International, November 4, 2003, in Singapore, and October 11, 2003, in Melbourne, Australia, and May 4, 2005, by phone; Sai Kiang Lim, group leader, Stem Cell and Developmental Biology, GIS, Singapore, November 3, 2003, in Singapore, and May 27, 2005, by phone; Hwai Loong Kong, executive director, Biomedical Research Council, A*Star, Singapore, November 4, 2003, in Singapore; Eng Hin Lee, head, Tissue Engineering, National University of Singapore, November 1, 2003, in Singapore; James Goh, Department of Orthopedic Surgery, National University of Singapore, November 1, 2003, in Singapore; Philip Yeo, chairman, A*Star, Singapore, October 31, 2003, in Singapore; Ron McKay, NIH Stem Cell Task Force member, Bethesda, MD, April 28 and 29, 2005, by phone; Irving Weissman, director, Stanford University Institute for Cancer/Stem Cell Biology, October 29, 2003, in Singapore.

113 against the law: "Five Things We Hate About Singapore Taxis," *Lonely Planet: Singapore* (Victoria, Australia: Lonely Planet Publications, July 2003): 61.

113 $4 billion: D. Normile, "Can Money Turn Singapore into a Biotech Juggernaut?" *Science* 297 (August 30, 2002): 1470–73.

114 cameras: D. Baum, "The Ultimate Jam Session," *Wired,* November 2001.

114 last tiger: *Lonely Planet: Singapore*: 66.

115 peanut shells: Ibid., 135.

115 $1,000 ticket: JoAnn Craig, *Culture Shock: A Guide to Customs and Etiquette, Singapore* (Portland, OR: Graphic Arts Center Publishing, 2001): 34–38.

115 "I always wanted": Author interview with Vic Nurcombe, October 31, 2003, and May 24, 2005.

116 1993 *Science* paper: V. Nurcombe, "Developmental Regulation of Neural Response to FGF-1 and FGF-2 by Heparan Sulfate Proteoglycan," *Science* 260 (April 2, 1993): 103–6.

118 $1 million: Author interview with Ariff Bongso, October 31, 2003.

120 butt or candy: "The Litter of the Law in Singapore," Reuters www.singapore-window.org/sw01/01030672.htm, March 6, 2001.

120 gum: "The Son Rises: Singapore," *The Economist,* July 22, 2004 ("The authorities are not yet ready to allow Singaporeans complete freedom to chew gum either: they have lifted a ban on imports, but only of the sugar-free sort, only to be sold in pharmacies and only to purchasers who register with the pharmacist first"); J. Gampell, "Singapore: Where Having Fun Is Now O.K.," *The New York Times,* April 24, 2005 ("It's still illegal to chew gum in Singapore [except, of course, for the "therapeutic," prescription version]").

120 one-party: "The Son Rises: Singapore" *The Economist,* July 22, 2004; D. Baum, "The Ultimate Jam Session," *Wired,* November 2001.

120 strict libel: "Singapore: Country Reports on Human Rights Practices 2002," U.S. Department of State, March 31, 2003.

120 "libel suits": "The Son Rises: Singapore," *The Economist,* July 22, 2004; Author interview with Suddharta Dubay, November 3, 2003; "Annual Reports 2005, Singapore," Amnesty International, web.amnesty.org/report2005/sgp-summary-eng.

120 "won" by the government: E. Ellis, "Singapore Authorities Use Libel Laws to Silence Critics," *The Australian,* September 26, 2002.

120 never be upgraded: "The Son Rises: Singapore," *The Economist,* July 22, 2004; Author interview with Suddharta Dubay, November 3, 2003.

120 "lack of access to": "Singapore, Report 2005," Reporters Without Borders, www.rsf.org/article.php3?id_article=13440&Valider=OK.

120 execution rates: "Annual Reports 2005, Singapore," Amnesty International, web.amnesty.org/report2005/sgp-summary-eng; European Union Annual Report on Human Rights, 2002: 87. Member states of the EU resolved in a 1998 policy paper on capital punishment to "work towards universal abolition of the death penalty."

120 Chen: L. Yong, "Student Forced a Shut Down Blog Following Libel Threat," freespeech.civiblog.org, May 4, 2005; J. H. Chen, "Apology," www.scs.uiuc.edu/~chen6/blog; "State Science Agency Warns Off Weblog Critic," *Index on Censorship,* April 5, 2005.

120 "benign dictatorship": D. Baum, op. cit., W. Safire, "Bloomberg News Humbled," *The New York Times,* August 29, 2002 ("It has to do with my old pal Lee Kuan Yew, who prefers to be called 'senior minister' rather than dictator of Singapore, and whose family members have been doing exceedingly well lately"); N. Ferguson, "Overdoing Democracy," *The New York Times,* April 13, 2003 ("the benign despotism of Lee Kuan Yew in Singapore . . .").

121 $20.3 million: S. C. Polad, "Embryonic Stem Cell Funding: California Here I Come?" *Kennedy Institute of Ethics Journal* 14 (December 2004): 407–9.

122 Friedman: Thomas Friedman, *The Lexus and the Olive Tree* (New York. Anchor, 2000): 229.

123 Lee Kuan Yew: "The Son Rises: Singapore," *The Economist,* July 22, 2004 ("Such policies, of course, helped propel Singapore from third world to first in a generation. The economy is still going strong: in the second quarter, it posted growth of almost 12%, albeit compared with a low base during the same period last year, when the region was hit by an outbreak of the SARS respiratory disease").

123 per capita income: D. Normile, op. cit.

123 China's growing ability: C. Chandler, "Coping with China: As China Becomes the Workshop of the World, Where Does That Leave the Rest of Asia?" *Fortune,* January 20, 2003.

123 LED signs: D. Baum, op. cit.

125 no islet stem cell: Y. Dor, "Adult Pancreatic Beta-Cells as Formed by Self-Duplication Rath The Stem Cell Differentiation," *Nature* 429 (May 6, 2004): 41–46.

126 will be unhappy: Author interview with Linda Lester, Oregon National Primate Research Center, April 26, 2004.

128 "shocked": X. Wang, "Kinetics of Liver Repopulation After Bone Marrow Transplantation," *The American Journal of Pathology* 161 (2002): 565–74; L. M. Eisenberg, "Stem Cell Plasticity, Cell Fusion, and Transdifferentiation," *Birth Defects Research* 69 (August 25, 2003): 209–18.

128 2001 *Nature*: D. Orlic, "Bone Marrow Cells Regenerate Infarcted Myocardium," *Nature* 410 (April 5, 2001): 701–5.

128 *Nature* that will claim: L. B. Balsam, "Haematopoietic Stem Cells Adopt Mature Haematopoietic Fates in Ischaemic Myocardium," *Nature* 428 (April 8, 2004): 668–73.

129 *Nature* editorial: "No Consensus on Stem Cells," *Nature* 587 (April 8, 2004).

129 *Times* reporter: G. Kolata, "Promise, in Search of Results; Stem Cell Science Gets Limelight; Now It Needs a Cure," *The New York Times*, August 24, 2004.

129 grousing: A. Regalado, "'Supercell' Controversy Sets Off a Scientists' Civil War," *The Wall Street Journal* June 21, 2002: B1.

130 Melton: Author interview with Douglas Melton, June 17, 2005.

130 McKay: Author interview with Ron McKay, April 28, 2005.

131 Weissman: T. Bearden, "Extended Interview: Irving Weissman," *PBS Online Newshour Transcript*, July 2005.

132 backup: D. Normile, op. cit.

133 "for libel": "Attacks on the Press 2003: Report on Singapore," Committee to Protect Journalists, March 11, 2004; W. Safire, "Essay; The Dictator Speaks," *The New York Times*, February 15, 1999; S. Mydans, "Singapore Journal; Soapbox Orators Stretch the Limits of Democracy," *The New York Times*, September 2, 2000. ("Speakers here will be subject, just like everybody else, to the Internal Security Act, under which they can be jailed without trial as threats to national security, and to the laws against libel and slander that have been vigorously employed in Singapore to harass and bankrupt opposition politicians").

133 co-opted . . . "in the body": Jean D. Peduzzi-Nelson, University of Alabama at Birmingham, Testimony, Senate Commerce Subcommittee on Science, Technology, and Space, July 14, 2004; A. Regalado, "'Supercell' Controversy Sets Off a Scientists' Civil War," *The Wall Street Journal*, June 21, 2002: B1. (Immediately after he hears of Verfaillie's work, Sen. Sam Brownback says, according to the *WSJ*: "I am heartened to know that scientific research has proven, once again, that destructive human embryo research and human cloning are unnecessary.")

134 Droge: Author interview with Peter Droge, associate professor, Nanyang Technological University, Singapore, October 30, 2003.

137 "Outstanding conference": Author interview with Ariff Bongso, October 31, 2003.

140 Censorship: D. Normile, op. cit.; D. Baum, op. cit.

141 disdain: D. Normile, op. cit.

141 Yeo: Author interview with Philip Yeo, October 31, 2003.

142 one party . . . sued . . . Civil rights: "The Son Rises: Singapore," *The Economist*, July 22, 2004 ("But many Singaporeans assume that Mr Lee père struck a deal with Mr Goh, engineering his rise to power in exchange for a promise that he would eventually give way to Mr Lee fils. At any rate, Mr Goh made the younger Mr Lee his deputy from the moment he succeeded the elder Mr Lee as prime minister, in 1990. The new man has never had any rivals. The members of the ruling People's Action Party [PAP], which controls all but two seats in parliament, selected him unanimously as prime minister, as did the party's executive committee and the cabinet. The new Mr Lee appears to see nothing wrong with this sort of arrangement. He fiercely defends the PAP's more underhand election tactics, such as threatening to put districts that vote for the opposition at the bottom of the list for public spending. He claims to want a more vigorous public debate, but promises to "demolish" any critic who undermines the government's standing. The threat is real: in recent years, one opposition figure found himself bankrupt after losing a defamation suit, while another wound up in jail"); D. Baum, "The Ultimate Jam Session," *Wired,* November 2001; European Commission Report, "The EU's Relations With Singapore," October 2004 ("The last national elections took place on November 2001. The People's Action Party that has governed Singapore for the last 42 years won 75.3% of all the votes, capturing 82 out of 84 seats in the Parliament. The opposition was not represented in 11 out of 15 Group representation Constituencies for the election campaign").

142 relatives: "The Son Rises: Singapore," *The Economist,* July 22, 2004 ("Mr Lee senior, now 80 years old, still sits in the cabinet [as senior minister, rather than prime minister], and supervises the Government Investment Corporation [GIC], which manages Singapore's foreign reserves. Meanwhile, Mr Lee junior's wife, Ho Ching, runs Temasek, a government holding company that owns stakes in Singapore's biggest firms, while his brother, Lee Hsien Yang, runs Singapore Telecommunications, the biggest local firm of all. Fans of the family argue that this concentration of power stems simply from its members' remarkable talent, not their connections. They claim the younger Mr Lee's rapid ascent through the ranks of the army to become a brigadier-general by the age of 32 rested purely on merit, as did his promotion to the post of deputy prime minister after only six years in politics").

153 "contempt": James Fallows, *Looking at the Sun: The Rise of the New East Asian Economic and Political System* (New York: Vintage 1995): 324.

153 Daar: Author interview with Abdallah Daar, McLaughlin Center for Molecular Medicine, University of Toronto, August 20, 2005.

154 Mombaerts: Author interview with Peter Mombaerts, Laboratory of Developmental Biology and Neurogenetics, Rockefeller University, October 10, 2005.

154 Andrews: Author interview with Peter Andrews, Department of Biomedical Science, University of Sheffield, October 11, 2005.

CHAPTER 6

Interviews: Thomas Murray, president, The Hastings Center, October 7, 2003, by phone; Rudy Jaenisch, MIT stem cell researcher, September 25, 2003, by phone; Jerry Yang, head, Center for Regenerative Biology, Storrs, CT, August 6, 2003, by phone, and September 15–16 in CT, and March 30, 2005, in CT and April 4, 2005, by phone; Shaorong Gao, Assistant Professor, Center for Regenerative Biology, Storrs, CT, March 30, 2005, in CT and April 11, 2005, by phone; Norio Nakatsuji, Kyoto University, Institute for Frontier Medical Sciences, Kyoto, Japan, March 26, 2004, in Kyoto; John McDonald, Kimmel Cancer Center, Johns Hopkins University, Baltimore, MD, October 25, 2004, by phone; Arthur Caplan, director, University of Pennsylvania, Center for Bioethics, November 2003, by phone; Mahendra Rao, Stem Cell Research Task Force member, NIH, at the NIH, July 25, 2003, and February 12, 2004, by phone; Woo Suk Hwang, Veterinary Medicine Teaching Hospital, Seoul National University, Seoul, South Korea, in New York, June 4, 2004, and February 12, 2004, by phone; Shin Yong Moon, chief, South Korean Stem Cell Center, February

12, 2004, by phone and June 3, 2004, in New York; James Battey, chief, NIH Stem Cell Task Force, February 26, 2004, by phone; Shin-ichi Nishikawa, group director, Riken Center for Developmental Biology, Kobe, Japan, in Kobe, March 26, 2004; Teru Wakayama, team leader, Laboratory for Genetic Reprogramming, Riken Center for Developmental Biology, March 26, 2004, in Kobe; Stephen Minger, director, King's College Stem Cell Biology Laboratory, London, England, March 24, 3005, by phone; Akira Iritani, chairman, Genetic Engineering, Kinki University, March 25, 2004 in Wakayama City; Anthony Perry, Riken, Mammalian Molecular Embryology, March 29, 2004 in Kobe; Rudy Jaenisch, MIT, September 25, 2003, by phone; Peter Mombaerts, Laboratory of Developmental Biology and Neurogenetics, Rockfeller University, October 10, 2005, in New York.

156 behaviorally different: A. F. Savage, "Behavioral Observations of Adolescent Holstein Heifers Cloned from Adult Somatic Cells," *Theriogenology* 60 (October 1, 2003): 1097–100.

157 *The Scientist*: R. Lewis, "The Clone Reimagined: Nuclear Reprogramming Remains a Major Black Box in Somatic Cell Nuclear Transfer," *The Scientist* 19 (April 25, 2005): 13.

158 "combination activation": X. Yang, "Nuclear Transfer in Cattle: Effect of Nuclear Donor Cells, Cytoplast Age, Co-culture, and Embryo Transfer," *Molecular Reproduction Development* 35 (May 1993): 29–36.

158 Willadsen: S. M. Willadsen, "Nuclear Transplantation in Sheep Embryos," *Nature* 320 (March 6, 1986): 63–65.

159 mid 1990s . . . Dolly: Ian Wilmut et al., *The Second Creation: Dolly and the Age of Biological Control* (New York: Farrar, Straus & Giroux, 2000); I. Wilmut, "Viable Offspring Derived from Fetal and Adult Mammalian Cells," *Nature* 385 (February 27, 1997): 810–13.

159 'When Ian": Author interview with Alan Trounson, October 13, 2003.

159 Yang cloned: University of Connecticut Animal Science, Faculty Member Profile, Xiangzhong (Jerry) Yang, www.canr.uconn.edu/ansci/faculty/jxy.htm; author interview with Jerry Yang, March 30, 2005.

160 Nakatsuji: Author interview with Norio Nakatsuji, Kyoto University, March 26, 2004.

161 $30 to $35 billion: R. Weiss, "Japanese Clone 8 Calves from Cow," *The Washington Post,* December 9, 1998: A01.

161 both lungs: W. Hathaway, "Scientist Working Against the Odds," *The Hartford Courant,* September 12, 2005.

162 born in 1957 . . . "meal": W. Hathaway, "Ambassador of Cloning: UConn's Yang and a Duplicated Friend: It's All in the DNA," *The Hartford Courant,* September 23, 2000: A1.

162 organizations: China Bridges International, www.ia.uconn.edu/china.html.

163 Yang, of all people: Author interview with Jerry Yang, March 30, 2005; "Success Stories; Dr. Xiangzhong (Jerry) Yang," documentary, RTHK, Hong Kong TV; W. Hathaway, "Scientist Working Against the Odds," *The Hartford Courant,* September 12, 2005.

163 South Korean researchers: W. S. Hwang, "Evidence of a Pluripotent Human Embryonic Stem Cell Line Derived from a Cloned Blastocyst," *Science* 303 (March 12, 2004): 1669–74 (E-pub February 12, 2004).

163 Gerry Schatten: C. Simerly, "Molecular Correlates of Primate Nuclear Transfer Failures," *Science* 300 (April 11, 2003): 297.

164 "biologically impossible" . . . embryonic cells: Ian Wilmut et al., *The Second Creation*, op. cit.; Gina Kolata, *Clone* (New York: Morrow, 1998).

164 John McDonald: Author interview with John McDonald, October 25, 2004.

165 Rudolf Jaenisch: Author interview with Rudy Jaenisch, September 25, 2003.

165 66 of the world's: The InterAcademy Panel on International Issues Statement on Human Cloning, www.interacademies.net/iap/iaphome.nsf/(weblinks)/WWWW-5RHFLT, September 22, 2003.

166 40 and 120 days: M. Revel et al., "Report of the Bioethics Advisory Committee of The Israel Academy of Sciences and Humanities," August 2001: 14; U.S. National Bioethics Advisory Committee Report, "Ethical Issues in Human Stem Cell Research, Vol. III: Religious Perspectives," Washington, DC: September 1999.

167 Many major: UNESCO: Universal Declaration on the Human Genome and Human Rights Paris 1997; World Health Organization, "Ethical, Scientific and Social Implications of Cloning in Human Health," Geneva, 1998; Council of Europe, "Additional Protocol to the Convention for the Protection of Human Rights and Dignity of the Human Being with Regard to the Application of Biology and Medicine, on the Prohibition of Cloning Human Beings," Paris, 1998.

167 against it: Webcast of the proceedings of the UN General Assembly's Legal Committee vote on cloning, November 6, 2003, www.un.org/law/cloning/.

167 Dr. Thomas Murray: Author interview with Thomas Murray, October 7, 2003.

167 "wholeheartedly": Webcast of the proceedings of the UN General Assembly's Legal Committee vote on cloning, November 6, 2003, www.un.org/law/cloning/.

168 All three voted: Press Report, Fifty-eighth General Assembly, Sixth Committee, 11th and 12th Meetings (A.M. and P.M.), October 21, 2003.

168 a human: Ibid.

168 use therapeutic cloning: Ibid.

168 same point: Ibid.

169 only to destroy it: M. Revel et al., op. cit., pp. 19, 20.

169 limited human therapeutic cloning: X. Bosch, "Spain to Allow Human Embryo Research. Cabinet Approves Law Reforms to Allow Use of Frozen Embryos to Establish Stem Cell Lines," *The Scientist*, July 31, 2003.

169 Brazil's UN delegation: Ad Hoc Committee on an International Convention against the Reproductive Cloning of Human Beings, www.un.org/law/cloning, accessed December 2003, October 2005.

169 would be allowed: L. Massarani, "Brazil Edges Toward Therapeutic Cloning," ScienceDevelopment.Net, June 24, 2002; UN Sixth Committee—II Session of the Ad Hoc Committee on an International Convention against Reproductive Cloning of Human Beings, Statement by the Brazilian Delegation, New York, September 24, 2002.

169 "scientific research": Press Report, Fifty-eighth General Assembly, Sixth Committee, 11th and 12th Meetings (A.M. and P.M.), October 21, 2003.

170 degenerative brain: Ibid.

170 ban only: Ad Hoc Committee on an International Convention against the Reproductive Cloning of Human Beings, op. cit.

170 "human entity": U.S. National Bioethics Advisory Committee Report, "Ethical

Issues in Human Stem Cell Research, Volume III: Religious Perspectives," September 1999.

170 like Islam: M. Revel et al., op. cit., 13.

170 into effect: European Commission, Directorate General, "Survey on Opinions from National Ethics Committees or Similar Bodies, Public Debate and National Legislation in Relation to Human Embryonic Stem Cell Research and Use," September 2003.

171 more liberal: Ibid.; Center for Genetics and Society, National Policies Governing New Technologies of Human Genetic Modification: A Preliminary Survey, September 2003.

171 as was Singapore: Ad Hoc Committee on an International Convention Against the Reproductive Cloning of Human Beings, op. cit.

171 Gandhi: Statement by Dr. M. Gandhi, Counsellor on the Elaboration of a Mandate to a Convention Against the Reproductive Cloning of Human Beings on September 24, 2002," http://secint04.un.org/india/ind637.pdf, accessed October 24, 2005

171 human cloning alone: Center for Genetics and Society, "National Policies Governing New Technologies of Human Genetic Modification: A Preliminary Survey," September 2003.

171 Caplan: Author interview with Arthur Caplan, November 23, 2003.

172 ethically sound: European Commission, Directorate General, "Survey on Opinions from National Ethics Committees or Similar Bodies, Public Debate and National Legislation in Relation to Human Embryonic Stem Cell Research and Use," September 2003.

173 Trounson: Author interview with Alan Trounson, October 13, 2003.

173 Mahendra Rao: Author interview with Mahendra Rao, June 25, 2003.

174 *Nature's* news section: D. Cyranoski, "Koreans Rustle Up Mad Cow Resistant Cows," *Nature* 426 (December 18, 2003).

174 pigs: H. G. Hong, "Seoul National Geneticists Lead Advance," *JoongAng Daily,* February 12, 2004.

174 "electrified": K. Tae-gyu, "Professor Confident of Commercializing Mad Cow Disease-Free Clones," *The Korea Times,* December 11, 2003.

175 broken earlier: I. Ho-jun, "JoongAng Hurts Korea's Science Reputation," *Chosun Ilbo,* February 12, 2004.

177 Still, as the days passed . . . everyone they saw: Author interview with Woo Suk Hwang June 4, 2004, and February 12, 2004; author interviews with Shin Yong Moon, February 12, 2004, and June 3, 2004; C. Dreifus, "A Conversation with Woo Suk Hwang and Shin Yong Moon: 2 Friends, 242 Eggs and a Breakthrough," *The New York Times,* February 17, 2004; B. Demick, "South Korea a Fertile Field for Research into Cloning," *The Los Angeles Times,* February 17, 2004.

178 *Science* paper: W. S. Hwang, op. cit.

178 Mahendra Rao: Author interview with Mahendra Rao, February 12, 2003.

179 Franco Marincola: Personal communication (e-mail) from Franco Marincola, director of immunogenetics, NIH, February 2004.

179 James Battey: Author interview with James Battey, February 26, 2004.

180 Shin-ichi Nishikawa: Author interview with Shin-ichi Nishikawa, March 26, 2004.

180 Teru Wakayama: Author interview with Teru Wakayama, March 26, 2004.

180 Norico Nakatsuji: Author interview with Norio Nakatsuji, March 26, 2004.

180 end of February . . . vote delay: Author interview with Shin Yong Moon, June 3, 2004; S. Sung-sik, "Korea Sees Big Future in Bio Research," *Joongang Daily,* February 26, 2004; K. Tae-gyun, "Cloning Expert Hwang Woo-suk: One of a Kind," *The Korea Herald,* February 20, 2004; "Ministry Considers Creating Embryo Research Complex," *The Korea Times,* February 27, 2004.

181 Lancet: H. J. Kang, "Effects of Intracoronary Infusion of Peripheral Blood Stem-Cells Mobilised with Granulocyte-Colony Stimulating Factor on Left Ventricular Systolic Function and Restenosis after Coronary Stenting in Myocardial infarction: The MAGIC Cell Randomised Clinical Trial," *The Lancet* 393 (March 6, 2004): 751–56.

181 Hwang received: Author interview with Woo-Suk Hwang, June 3, 2004.

181 Frost & Sullivan: "Competitive Positioning Strategies for South Korean Biotech Companies," Frost & Sullivan, November 6, 2003, can be ordered from www.docu mus.com and many other sites.

182 In March 2004 . . . "It's crazy": J. Gillis et al., "NIH: Few Stem Cell Colonies Likely Available for Research of Approved Lines, Many Are Failing," *The Washington Post,* March 3, 2004.

182 anonymous egg: D. Normile, "Research Ethics. South Korean Cloning Team Denies Improprieties," *Science* 304 (May 14, 2004): 945.

183 Bernie Siegel: Author interview with Bernie Siegel, head of GPI, June 2, 2004.

184 hurt their reputation: D. Cyanoski, "Stem-Cell Research: Crunch Time for Korea's Cloners," *Nature* 429 (May 6, 2004): 12–14; D. Cyranoski, "Korea's Stem-Cell Stars Dogged by Suspicion of Ethical Breach," *Nature* 429 (May 6, 2004): 3.

184 John Wagner: Author interview with John Wagner, University of Minnesota, June 2, 2003.

185 In the lobby: Author interview with Woo Suk Hwang, June 4, 2004.

185 McDonald: Author interview with John McDonald, October 25, 2004.

187 The apology: Correction, *Nature* 429 (June 10, 2004).

190 turn down: Y. Wonsup, "Stem Cell Researcher Under Police Protection," *The Korea Times,* September 15, 2004.

190 Nature Biotechnology: J. Wong, "South Korean Biotechnology—A Rising Industrial and Scientific Powerhouse," *Nature Biotechnology* 22 (December 2004): DC 42–47.

190 bodyguard: Y. Wonsup, op. cit.

190 "Hwang is": Author interview with Stephen Minger, March 24, 2005.

190 Schatten: "Korean Cloner Is the Envy of Science: Pioneering Researcher Has Support of Government," Associated Press, April 27, 2005.

191 McDonald: Author interview with John McDonald, October 25, 2004.

191 breed corruption: Bruce Cumings, *Korea's Place in the Sun: A Modern History* (New York: W. W. Norton, 2005): 398.

191 "'could be described'": Michael Breen, *The Koreans: Who They Are, What They Want, Where Their Future Lies* (New York: Thomas Dunne, 2004): 145–46.

191 "Unlike Western": Ibid.

192 Eckert: Carter J. Eckert et al., *Korea Old and New: A History* (Seoul: Ilchokak 1990): 406.

192 900 times: Dan Oberdorfer, *The Two Koreas: A Contemporary History* (New York: Basic 2001): 3.

192 harnessed: Eckert, op. cit., 408.

192 Confucianism's reverence: Eckert, op. cit., 410.

192 level of education: Breen, op. cit., 64.

192 college entrance: Breen, op. cit., 66–67.

192 Daar: Author interview with Abdallah Daar, McLaughlin Center for Molecular Medicine, University of Toronto, August 20, 2005.

192 Faiz Kermani: F. Kermani, "A Viewpoint on South Korean Biotech," *DDT* 10 (May 2005): 685–88.

192 subsequent developments . . . business, and government: Cumings, op. cit., 396–403.

192 International Monetary Fund: Breen, op. cit., 160.

193 Fallows: James Fallows, *Looking at the Sun: The Rise of the New Economic and Political System* (New York: Vintage 1995): 394.

193 Changes have included: J. Wong, "From Learning to Creating: Biotechnology and the Postindustrial Developmental State in Korea," *Journal of East Asian Studies* 4(2004): 491–517.

193 "fractious but effective": Oberdorfer, op. cit., 1.

193 Wong: Wong, op. cit., 491–517.

194 Russian scientists . . . endangered/extinct species: Richard Stone, *Mammoth: The Resurrection of a Stone Age Giant* (Cambridge: Perseus, 2001); *"Land of the Mammoth,"* Discovery Channel DVD; J. Ryall, "Japanese Professor Plans Pleistocene Park," *The Vladivostok News,* August 15, 2002; T. Radford, "Japanese Scientists Take First Step Towards Cloning Ice Age Beast," *The Guardian,* July 16, 2003.

195 Akira Iritani: Author interview with Akira Iritani, March 25, 2004.

195 Kazufumi Goto: K. Goto, "Fertilization of Bovine Oocytes by the Injection of Immobilised, Killer Spermatozoa," *Veterinary Row* 127 (November 24, 1990): 517–200.

199 "impossible" to clone: Yang speech, The First New England Symposium on Regenerative Biology and Medicine, September 15, 2003; A. Regalado, "Could a Skin Cell Someday Replace a Sperm or an Egg?" *The Wall Street Journal,* October 17, 2002.

200 both Buddhism and Shintoism: K. Bowman, "Culture, Brain Death and Transplantation," *Progress in Transplantation* 13 (September 2003): 211–17.

200 rules for hES: S. Harris, "Asian Pragmatism," *EMBO Reports* 3 (2002): 816–17.

201 The offices outside: Author interview with Teru Wakayama, March 26, 2004.

202 "magic" hands: E. Pennisi, "Clones: A Hard Act to Follow," *Science* 288 (June 9, 2000): 1722–27.

204 *Nature*: T. Wakayama, "Full-Term Development of Mice From Enucleated Oocytes Injected with Cumulus Cell Nuclei," *Nature* 394 (July 23, 1998): 369–74.

204 generally known: H. Alton, "Mouse-cloning Researcher Sues UH," *The Honolulu Star Bulletin,* July 29, 1999; A. Salkever, "Who Owns the Clues?" Salon.com, August 16, 1999.

204 infamous premature paper . . . into closing: Michael West, *The Immortal Cell: One Scientist's Quest to Solve the Mystery of Human Aging* (New York: Doubleday, 2003); E. Check, "Biotech Firm's Accounts Scrutinized," *Nature* 417 (May 23, 2002): 370; J. Fox, "Human Cloning Claim Renews Debate," *Nature Biotechnology* 20 (January 2002): 4–5.

206 Perry looks satisfied: Author interview with Anthony Perry, March 29, 2004.

CHAPTER 7

Interviews: Kotaro Yoshimura, Department of Plastic Surgery, University of Tokyo, Tokyo, Japan, March 24, 2004, in Tokyo, and April 25, 2005, by e-mail; John Dick, director, Program of Stem Cell Biology, University of Toronto, May 10, 2005, by phone; Michael Andreef, Oncology, MD Anderson Cancer Center, Houston, TX, January 14, 2005, in Houston; Ron McKay, NIH Stem Cell Task Force member, Bethesda, MD, April 8, 1999, by phone, and April 28 and 29, 2005, by phone; Evan Snyder, then Harvard University neuroscientist and neural stem cell pioneer, March 30, 1999, in Boston; G. Barry Pierce, former pathologist, University of Colorado, August 1999, by phone; Stewart Sell, research physician, Wadsworth Center, NYS Department of Health, Albany, NY, August 1999, by phone; Ralph Steinman, head, Laboratory of Cellular Physiology and Immunology, Rockefeller University July 1999, by e-mail; Ethan Dmitrovsky, Thoracic Oncology, Dartmouth College, December 9, 1999, by phone; Peter Andrews, March 30 and 31, 1999, by e-mail; Ann Hamburger, Oncology, University of Maryland, College Park, MD, December 20, 2004, by phone; Philip Amrein, Oncology, Dana Farber Hospital, Boston, MA, December 17, 2004, by phone; George Demetri, MGH, Boston, MA, July 8, 2004, by phone; Drew Pardoll, Oncology, Johns Hopkins University, August 10, 2004, by phone; Samuel Waxman, Oncology, Mt. Sinai Hospital, New York, NY, August 4, 2004, in NY; Craig Jordan, Oncology, University of Rochester, Rochester, NY, October 2004, by phone; Jeffrey Rosen, Program in Developmental Biology, Baylor College of Medicine, Houston, TX, January 11, 2004, in Houston; Diane Howard, Oncology, University of Kentucky, January 3, 2005, by phone; Sam Weiss, University of Calgary, neural stem cell pioneer, March 1999, by phone; Mahendra Rao, NIH Stem Cell Task Force member, Bethesda, MD, April 22, 2005, by phone; Ariff Bongso, hES cell pioneer, Singapore National University, Singapore, May 3, 2005, by phone; Alan Colman, CEO, ESI, Singapore, May 4, 2005, by phone.

212 Wayne Morrison: A. Salleh, "Tissue Engineers Grow New Breasts," ABC Science Online, August 15, 2003.

212 MIT: C. Kuperwasser, "Reconstruction of Functionally Normal and Malignant Human Breast Tissues in Mice," *Proceedings of the National Academy of Sciences* 101 (April 6, 2004): 4966–71 (e-pub March 29, 2004).

214 Marc Hedrick: D. A. De Ugarte, "Comparison of Multi-Lineage Cells from Human Adipose Tissue and Bone Marrow," *Cells Tissues Organs* 174 (2003): 101–9.

214 Yoshimura: Personal communication (e-mail) from Kotaro Yoshimura, April 25, 2005.

215 each other's work: Author interview with Evan Snyder, March 30, 1999 ("Neural stem cells are responsive to EGF. One of the characteristics of a tumor is mutant or upregulated EGF. So you can argue that if you select cells that are exceptionally responsive to EGF you're actually selecting for cells that are likely to become tumors"); author interview with Ron McKay, April 8, 1999 (Cancer cells and stem cells "are the same kind of beastie. You need to know the subtle differences between them so you're riding them, and not the other way around").

215 For there. . . . gone awry: T. Reya, "Stem Cells, Cancer, and Cancer Stem Cells," *Nature* 414 (November 1, 2001): 105–11; M. F. Clarke, "Epigenetic Regulation of Normal and Cancer Stem Cells," *Annals of the New York Academy of Science* 1044

(June 2005): 90–93; M. Kucia, "Trafficking of Normal Stem Cells and Metastasis of Cancer Stem Cells Involve Similar Mechanisms: Pivotal Role of the SDF-1-CXCR4 Axis," *Stem Cells* (May 11, 2005); M. Dean, "Tumour Stem Cells and Drug Resistance," *Nature Reviews Cancer* 5 (April 2005): 275–84; Al Hajj, "Self-Renewal and Solid Tumor Stem Cells," *Oncogene* 23 (September 20, 2004): 7274–82; D.T. Scadden, "Cancer Stem Cells Refined," *Nature Immunology* 5 (July 2004): 701–3; D. R. Bell, "Stem Cells, Aging, and Cancer: Inevitabilities and Outcomes," *Oncogene* 23 (September 20, 2004): 7290–96.

216 "'caricatures'": Author interview with Barry Pierce, August 1999; author interview with Stewart Sell, August 1999; S. Sell, "Stem Cell Origin of Cancer and Differentiation Therapy," *Critical Reviews of Oncology and Hematology* 51 (July 2004): 1–28.

216 normal healthy: M. F. Clarke, "Epigenetic Regulation of Normal and Cancer Stem Cells," *Annals of the New York Academy of Science* 1044 (June 2005): 90–3; D. Pearton, "Sharpening the Focus; The Business of Epithelial Cell Biology. An Interview with Chris Potten," *International Journal of Developmental Biology* 48 (2004): 193–96; A. Z. Rizvi, "Epithelial Stem Cells and Their Niche: There's No Place Like Home," *Stem Cells* 23 (February 2005): 150–65; author interview with John Dick, May 10, 2005.

217 Jeffrey Rosen: F. Behbod, "Will Cancer Stem Cells Provide New Therapeutic Targets?" *Carcinogenesis* 26 (2004): 703–711.

217 can form cancers: Y. Li, "Stem/Progenitor Cells in Mouse Mammary Gland Development and Breast Cancer," *Journal of Mammary Gland Biological Neoplasia* 10 (January 2005): 17–24; S. A. Bapat, "Stem and Progenitor-like Cells Contribute to the Aggressive Behavior of Human Epithelial Ovarian Cancer," *Cancer Research* 65 (April 15, 2005): 3025–29.

218 For decades: Personal communication with Johns Hopkins University oncologist Burt Vogelstein, October 1999.

218 equally able to go: Author interview with John Dick, May 10, 2005 ("The other model says there is an equal probability of any cell going into the cell cycle, which would give every cell tumor initiating capacity").

219 McKay: Author interview with Ron McKay, April 8, 1999.

221 Goldberg: R. Goldberg, "Reagan's Medical Revolution," *The Washington Times*, June 8, 2004.

222 Gepstein: I. Kehat, "Electromechanical Integration of Cardiomyocytes Derived from Human Embryonic Stem Cells," *Nature Biotechnology* 22 (October 2004): 1282–89.

222 cured Parkinson's: T. Ben-Hur, "Transplantation of Human Embryonic Stem Cell-Derived Neural Progenitors Improves Behavioral Deficit in Parkinsonian rats," *Stem Cells* 22 (2004): 1246–55.

223 Anna Kenney . . . MSCs to them: ISSCR 2nd Annual Meeting, Final Program, June 10–13, 2004.

225 support the whole: G. Tabatabai, "Lessons from the Bone Marrow: How Malignant Glioma Cells Attract Adult Haematopoietic Progenitor Cells," *Brain* June 9, 2005; E. I. Fomechnko, "Stem Cells and Brain Cancer," *Experimental Cell Research* 306 (June 10, 2005): 323–29; A. Nakamizo, "Human Bone Marrow Derived Mesenchymal Stem Cells in the Treatment of Gliomas," *Cancer Research* 65 (April 15, 2005): 3307–18.

225 Snyder: Author interview with Evan Snyder, March 30, 1999; S. Yip, "Neural Stem Cell Biology May Be Well Suited for Improving Brain Tumor Therapy," *Cancer Journal* 9 (May-June 2003): 189–204.

226 Amrein: Author interview with Philip Amrein, December 17, 2004.

227 Jordan: Author interview with Craig Jordan, October 2004.

227 Dick: Author interview with John Dick, May 10, 2005.

229 AMN107: "Increasing Benefit Seen in Novel Drug AMN107 That Treats Gleevec Resistance," *Medical Research News,* April 23, 2005.

229 Gary Gilliland: B. J. Huntly, "MOZ-TIF2, but Not BCR-ABL, Confers Properties of Leukemic Stem Cells to Committed Murine Hematopoietic Progenitors," *Cancer Cell* 6 (December 2004): 587–96.

230 In fact. . . . until the 1990s: S. Sell, "Stem Cell Origin of Cancer and Differentiation Therapy," *Critical Reviews of Oncology and Hematology* 51 (July 2004): 1–28; author interview with Jeffrey Rosen, January 11, 2005; F. Behbod, "Will Cancer Stem Cells Provide New Therapeutic Targets?" *Carcinogenesis* 26 (2004): 703–711; W. B. Frye, "Julius Friedrich Cohnheim," *Clinical Cardiology* 25 (2002): 575–77.

233 weird cancer: Ann Parson, *The Proteus Effect* (Washington, DC: Joseph Henry, 2004): 42–45.

233 single cell: L. J. Kleinsmith, "Multipotentiality of Single Embryonal Carcinoma Cells," *Cancer Research* 24 (October 1964): 1544–51.

233 Pierce noted: Author interview with Barry Pierce, August 1999.

233 The two confirmed: D. Solter, Extrauterine Growth of Mouse Egg-Cylinders Results in Malignant Teratoma," *Nature* 227 (August 1, 1970): 503–4.

234 Beatrice Mintz: B. Mintz, "Normal Genetically Mosaic Mice Produced from Malignant Teratocarcinoma Cells," *Proceedings of the National Academy of Sciences* 72 (September 1975): 3585–89.

234 SSEA-1: D. Solter, "Monoclonal Antibody Defining a Stage-Specific Mouse Embryonic Antigen (SSEA 1)," *Proceedings of the National Academy of Sciences* 75 (November 1978): 5565–69.

234 mouse ES cell: M. J. Evans, "Establishment in Culture of Pluripotential Cells from Mouse Embryos," *Nature* 292 (July 9, 1981): 154–56; G. R. Martin, "Isolation of a Pluripotent Cell Line from Early Mouse Embryos Cultured in Medium Conditioned by Teratocarcinoma Stem Cells," *Proceedings of the National Academy of Sciences* 78 (December 1981): 7634–38.

235 Jeffrey Rosen: Author interview with Jeffrey Rosen, January 11, 2005.

235 Sid Salmon: Author interview with Craig Jordan, October 2004; Author interview with Ann Hamburger, December 20, 2004; A. W. Hamburger, "Primary Bioassay of Human Tumor Stem Cells," *Science* 197 (July 29, 1977): 461–63.

236 Canadian group: J. E. Till, "A Direct Measurement of the Radiation Sensitivity of Normal Mouse Bone Marrow Cells," *Radiation Research* 14 (1961): 213–22; A. Becker, "Cytological Demonstration of the Clonal Nature of Spleen Colonies Derived from Transplanted Mouse Marrow Cells," *Nature* 197 (1963): 452–54.

236 Leo Sachs: L. Degos, "Differentiation Therapy of Human Leukemia," paper presented at the International Symposium on Predictive Oncology and Intervention Strategies, Paris, France, February 9–12, 2002; M. Paran, "In Vitro Induction of Granulocyte Differentiation in Hematopoietic Cells from Leukemic and Non-Leukemic Patients," *Proceedings of the National Academy of Sciences* 67 (November

15, 1970): 1542–49; Z. X. Shen, "All-Trans Retinoic Acid/As203 Combination Yields a High Quality Remission and Survival in Newly Diagnosed Acute Promyelocytic Leukemia," *Proceedings of the National Academy of Sciences* 101 (April 13, 2004): 5328–35.

237 In the mid 1980s . . . mature adult blood cell: L. Degos, "The History of Acute Promyelocytic Leukemia," *British Journal of Haematology* 122 (August 2003): 539–53.

238 made vice president: D. Bradley, "Biography of Zhu Chen," *Proceedings of the National Academy of Sciences* 101 (April 13, 2004): 5325–27.

238 Chen reported: Z. X. Shen, "All-Trans Retinoic Acid/As203 Combination Yields a High Quality Remission and Survival in Newly Diagnosed Acute Promyelocytic Leukemia," *Proceedings of the National Academy of Sciences* 101 (April 13, 2004): 5328–35.

238 Rosen: Author interview with Jeffrey Rosen, January 11, 2005.

239 Weiss: Author interview with Sam Weiss, March 1999.

240 McKay: Author interview with Ron McKay, April 28 and 29, 2005.

240 Weissman's suggestion: T. Reya, "Stem Cells, Cancer, and Cancer Stem Cells," *Nature* 414 (November 1, 2001): 105–11.

241 Clarke: M. Al-Hajj, "Prospective Identification of Tumorgenic Breast Cancer Cells," *Proceedings of the National Academy of Sciences* 100 (April 1, 2003): 3547–49.

241 Dirks: S. K. Singh, "Identification of a Cancer Stem Cell in Human Brain Tumors," *Cancer Research* 63 (September 15, 2003): 5821–28.

241 Kornblum: H. D. Hemmati, "Concerns Stem Cells Can Arise from Pediatric Brain Tumors," *Proceedings of the National Academy of Sciences* 100 (December 9, 2003): 15178–83.

241 Hedgehog: S. S. Karhadkar et al., "Hedgehog Signalling in Prostate Regeneration, Neoplasia and Metastasis," *Nature* 431 (October 7, 2004): 707–12; P. Sanchez et al., "Inhibition of Prostate Cancer Proliferation by Interference with Sonic Hedgehog-GLI1 Signaling," *Proceedings of the National Academy of Sciences* 101 (August 24, 2004): 12561–66.

244 in culture: D. Rubio, "Spontaneous Human Adult Stem Cell Transformation," *Cancer Research* 65 (April 15, 2005): 3035–39; C. Blackstock, "Stem Cells Cancer Fear," *The Guardian,* April 21, 2005.

244 Harvard neuroscientist . . . "cellular events": C. Fox, "Making Neurons: Newly Published Recipes Direct Neural Stem Cell Research," *The Scientist* 14 (September 18, 2000): 1.

245 Rao: Author interview with Mahendra Rao, April 22, 2005.

245 Colman: J. J. Buzzard, "Karyotype of Human ES Cells During Extended Culture," *Nature Biotechnology* 22 (April 2004): 381–82, and P. W. Andrews's response 382; author interview with Alan Colman, May 4, 2005.

245 Bongso: Author interview with Ariff Bongso, May 3, 2005.

246 Thomson: "Workshop on Current Protocols in Stem Cell Biology," Jackson Lab, August 13, 2005.

246 Okarma: Author interview with Geron CEO Thomas Okarma, February 12, 2001.

246 Cytori Therapeutics; "Stem Cells from Fat Tested to Regrow Breast Tissue," Reuters, April 20, 2006.

247 patient: Author interview with Yoshimura patient, New York, April 26, 2005.

CHAPTER 8

Interviews: Thomas Spitzer, director, Bone Marrow Transplant Program, MGH, July 7, 2004, by phone, and July 12, 2004, and February 15, 2005, in Boston; Megan Sykes, chief, Bone Marrow Transplantation Section, MGH, July 1, 2004, by phone, and March 17, 2005, by phone; ESI CEO Alan Colman, Singapore, May 4, 2005, by phone; Irving Weissman, head, Stanford University Institute for Cancer/Stem Cell Biology, Palo Alto, CA, February 7, 2001, in Palo Alto; Shimon Slavin, Department of Bone Marrow Transplantation and Cancer Immunotherapy, Hadassah Medical Center, Jerusalem, Israel, October 22 and October 26, 2003, in Israel, and August 2003, by phone; Richard Childs, National Heart, Lung, and Blood Institute, NIH, Bethesda, MD, February 17, 2005, in MD, and March 4, 2004, by phone; Janet McCourt, MGH kidney transplant patient, Milton, MA, February 15, 2005, in Milton; Ephraim Fuchs, Division of Hematopoiesis, Johns Hopkins University, Baltimore, MD, July 20, 2004, by phone; Derek Besenfelder, MGH kidney transplant patient, Massachusetts resident, February 16, 2005, at MGH.

251 Instead of . . . Weissman says: Author interview with Irv Weissman, February 7, 2001; author interview with Shimon Slavin, August 2003; author interview with Alan Colman, May 4, 2005.

255 Diana Bianchi: D. W. Bianchi, "Male Fetal Progenitor Cells Persist in Maternal Blood for as Long as 27 Years Postpartum," *Proceedings of the National Academy of Sciences* 93 (January 23, 1996): 705–8; K. Khosrotehrani, "Multi-Lineage Potential of Fetal Cells in Maternal Tissue: A Legacy in Reverse," *Journal of Cell Science* 118 (April 15, 2005): 1559–63.

257 The world's first . . . rejection problem with stem cells: Thomas E. Starzl, *The Puzzle People: Memoirs of a Transplant Surgeon* (Pittsburgh: University of Pittsburgh Press, 2003); F. Delmonico, "Interview with Dr Joseph Murray," *American Journal of Transplantation* 2 (October 2002): 803–6; E. Simpson, "Reminiscences of Sir Peter Medawar: In Hope of Antigen-Specific Transplantation Tolerance," *American Journal of Transplantation* 4 (December 2004): 1937–40; D. Hatch, "Kidney Transplantation," emedicine.com, November 23, 2004; A. B. Cosimi, "Use of Monoclonal Antibodies to T-cell Subsets for Immunologic Monitoring and Treatment in Recipients of Renal Allografts," *New England Journal of Medicine* 305 (August 1981): 308–14.

262 The year 1967 . . . day after day, with drugs: Thomas E. Starzl, *The Puzzle People,* op. cit.; R. Calne, "The History and Development of Organ Transplantation: Biology and Rejection," *Baillieres Clinical Gastroenterology* 8 (September 1994): 389–97; D. Hatch, op. cit.

266 It was in 1991: M. H. Sayegh, "Immunologic Tolerance to Renal Allografts after Bone Marrow Transplants from the Same Donors," *Annals of Internal Medicine* 114 (June 1, 1991): 954–95.

266 MGH's Megan Sykes . . . MGH approach on its patients: Author interviews with Megan Sykes, March 17, 2005, and July 1, 2004; author interviews with Shimon Slavin, October 22 and 26, 2003, and August 2003; author interviews with Thomas Spitzer, July 7 and 12, 2004, and February 15, 2005; T. Fehr, "Tolerance Induction in Clinical Transplantation," *Transplant Immunology* 13 (2004): 117–30; T. Kawai, "Long-term Outcome and Alloantibody Production in a Non-Myeloablative Regimen for Induction of Renal Allograft Tolerance," *Transplantation* 68 (December 15, 1999):

1767–75; Y. Sharabi, "Mixed Chimerism and Permanent Specific Transplantation Tolerance Induced by a Nonlethal Preparative Regimen," *The Journal of Experimental Medicine* 169 (February 1989): 493–502; S. Delis, "Donor Bone Marrow Transplantation Chimerism and Tolerance," *Transplant Immunology* 13 (2004): 105–15; T. R. Spitzer, "Combined Histocompatibility Leukocyte Antigen-matched Donor Bone Marrow and Renal Transplantation for Multiple Myeloma with End Stage Renal Disease: The Induction of Allograft Tolerance Through Mixed Lymphohematopoietic Chimerism," *Transplantation* 68 (August 27, 1999): 480–4; S. Strober, "Approaches to Transplantation Tolerance in Humans," *Transplantation* 77 (March 27, 2004): 932–96.

272 Janet McCourt: Author interview with Janet McCourt, February 15, 2005.

277 "strongest immune response": Author interview with Megan Sykes, July 1, 2004.

277 same way: Author interview with Richard Childs, March 4, 2004.

279 solid tumors: R. Childs, "Regression of Metastatic Renal-Cell Carcinoma after Nonmyeloablative Allogeneic Peripheral-Blood Stem-Cell Transplantation," *New England Journal of Medicine* 343 (September 14, 2000): 750–8.

279 to write reviews: R. Childs, "Nonmyeloablative Blood Stem Cell Transplantation as Adoptive Allogeneic Immunotherapy for Metastatic Renal Cell Carcinoma," *Critical Reviews in Immunology* 21 (2001): 191–203; R. Childs, "Allogeneic Stem Cell Transplantation for Renal Cell Carcinoma," *Current Opinions in Urology* 11 (September 2001): 495–502; R. Childs, "Advances in Allogeneic Stem Cell Transplantation: Directing Graft-versus-Leukemia at Solid Tumors," *Cancer Journal* 8 (January–February 2002): 2–11.

279 while Sykes: Author interviews with Megan Sykes, July 1, 2004, and March 17, 2005.

279 Childs likes to put it: Author interview with Richard Childs, February 17, 2005.

279 40 percent: "Multiple Myeloma and Other Plasma Cell Neoplasms (PDQ): Treatment," Health Professional Version, National Cancer Institute, April 12, 2005.

280 stop-gap: L. Luznik, "Donor Lymphocyte Infusions to Treat Hematologic Malignancies in Relapse After Allogeneic Blood or Marrow Transplantation," *Cancer Control* 9 (March/April 2002): 123–37.

280 And in 1990: M. M. Horowitz, "Graft-versus-Leukemia Reactions after Bone Marrow Transplantation," *Blood* 75 (February 1, 1990): 555–62.

280 "These results": H. J. Kolb, "Donor Leukocyte Transfusions for Treatment of Recurrent Chronic Myelogenous Leukemia in Marrow Transplant Patients," *Blood* 76 (December 15, 1990): 2462–65.

281 the inflammation caused: Author interviews with Megan Sykes, July 1, 2004, and March 17, 2005; T. Wekerle, "Induction of Tolerance," *Surgery* 135 (April 2004): 359–64.

282 85 percent: F. Baron, "Hematopoietic Cell Transplantation: Five Decades of Progress," *Arch Medical Research* 34 (November–December 2003): 528–44.

283 those delayed: M. Y. Mapara, "Donor Lymphocyte Infusions Mediate Superior Graft-versus-Leukemia Effects in Mixed Compared to Fully Allogeneic Chimeras: A Critical Role for Host Antigen-Presenting Cells," *Blood* 100 (September 1, 2002): 1903–9.

284 tumors much better: Author interview with Ephraim Fuchs, July 20, 2004; L. Luznik, "Successful Therapy of Metastatic Cancer Using Tumor Vaccines in Mixed Allogeneic Bone Marrow Chimeras," *Blood* 101 (February 15, 2003): 1645–52.

284 "wake up": W. D. Shlomchik, "Prevention of Graft Versus Host Disease by Inactivation of Host Antigen-Presenting Cells," *Science* 285 (July 16, 1999): 412–15; B. R. Dey, "Anti-Tumour Response Despite Loss of Donor Chimaerism in Patients Treated with Non-Myeloablative Conditioning and Allogeneic Stem Cell Transplantation," *British Journal of Haematology* 128 (February 2005): 351–59.

284 *no* immune systems: L. Luznik, "Successful Therapy of Metastatic Cancer Using Tumor Vaccines in Mixed Allogeneic Bone Marrow Chimeras," *Blood* 101 (February 15, 2003): 1645–52.

288 most promising approaches: Ibid.

288 Drew Pardoll: Y. Cui, "Immunotherapy of Established Tumors Using Bone Marrow Transplantation with Antigen Gene-Modified Hematopoietic Stem Cells," *Nature Medicine* 9 (July 2003): 952–58.

290 Derek Besenfelder: Author interview with Derek Besenfelder, February 16, 2005.

292 Vacanti brothers: K. Ogawa, "The Generation of Functionally Differentiated, Three-Dimensional Hepatic Tissue from Two-Dimensional Sheets of Progenitor Small Hepatocytes and Nonparenchymal Cells," *Transplantation* 77 (June 27, 2004): 1783–89.

292 overlooked *Nature* paper: I. M. Conboy, "Rejuvenation of Aged Progenitor Cells by Exposure to a Young Systemic Environment," *Nature* 433 (February 17, 2005): 760–4.

293 *criminalize* all embryo creation: S. LeBlanc. "Researchers say Romney stem cell plan world criminalized their work," Associated Press, February 10, 2005.

295 Medawar: R. J. Duquesnoy, "Early History of Transplantation Immunology," Transplant Pathology Internet Services, www.tpis.upmc.edu/tpis/immuno/_www Hist_part1.htm, accessed October 24, 2005.

CHAPTER 9

Interviews: Ruth Pavelko, Texas Heart Institute patient, January 13–14, 2005, in Houston, TX, and April 4, 2005, and June 13, 2005, by phone; Lior Gepstein, stem cell researcher, Rappaport Family Institute for Research in the Medical Sciences, Haifa, Israel, October 7, 2004, by phone; Douglas Losordo, chief, Division of Cardiovascular Research, Caritas St. Elizabeth Medical Center, December 15, 2004, by phone; Richard Cannon, principle investigator of the cardiovascular branch, NIH National Institute of Heart, Lung, and Blood, January 19, 2005, by phone; Edward Yeh, chair, Department of Cardiology, MD Anderson Cancer Center, Houston, TX, January 11, 2005, in Houston; Elizabeth McNally, Associate Professor of Medicine, University of Chicago, Chicago, IL, January 3, 2005, by phone; Takayuki Asahara, team leader, Laboratory for Stem Cell Translational Research, Riken Center for Developmental Biology, Kobe, Japan, July 7, 2004, by phone; Masataka Sata, Assistant Professor, University of Tokyo, Tokyo, Japan, March 23, 2004, in Tokyo; Loren Field, Department of Pathology, Indiana University, Indianapolis, IN, May 2005, by phone; Silviu Itescu, Assistant Professor of Clinical Medicine, Columbia University, New York, NY, and founder, Mesoblast Ltd, Melbourne, Australia, February 9, 2005, and July 3, 2005, by phone; Emerson Perin, director, New Cardiovascular Interventional Technology, Texas Heart Institute, Houston, TX, January 12–14, 2005 (in TX), October 28 2004 (by phone), and January 27, 2005, February 2005, and March 4, 2005 (by phone), and June 13, 2005 (e-mail); Guilherme Silva, cardiology fellow, Texas Heart Institute, January 13, 2005, in TX; Andrew Pavelko, Ruth's husband, January 13–14, 2005, in TX; Amit Patel, Department of Cardiology, University of Pittsburgh, January 28, 2005 (by phone) and June 27, 2005, (by e-mail); Franca Angeli, cardiology fellow, Texas Heart Institute, January 12–14, 2005; Jeffrey Wilson, director, Cell Processing Laboratory, Center for Cell and Gene Therapy, Baylor College of Medicine, Houston, TX, in Houston, January 13, 2005.

296 unique view: Author interview with Emerson Perin, March 4, 2005; R. Kornowski, "Left Ventricular Mapping and Myocardial Revascularization," *Current Interventional Cardiology Reports* 1 (1999): 117–26; G. Van Langenhove, "Nonfluoroscopic Endoventricular Electromechanical Three-Dimensional Mapping—Current Status and Future Perspectives," *Japanese Circulation Journal* 65 (August 2001): 695–701.

297 "Here we" . . . varied degrees: Author interviews with Emerson Perin, January 27, 2005; Ruth Pavelko, January 13 and 14, 2005; and Guilherme Silva, January 13, 2005.

298 more scar: Author interviews with Emerson Perin, January 27, 2005; and Amit Patel, January 28, 2005; Howard J. Leonhardt, "Myoblast Transplantation for Heart Repair: A Review of the State of the Field, June 9, 2005," EuroPCRonline.com; B. Bittira, "In Vitro Preprogramming of Marrow Stromal Cells for Myocardial Regeneration," *Annals of Thoracic Surgery* 74 (October 2002): 1154–59; J. S. Wang, "The Coronary Delivery of Marrow Stromal Cells for Myocardial Regeneration: Pathophysiologic and Therapeutic Implications," *Journal of Thoracic Cardiovascular Surgery* 122 (October 2001): 699–705.

298 Ruth will die: Author interview with Andrew Pavelko, January 14, 2005.

299 French cardiologists: J. Couzin, "Renovating the Heart," *Science* 304 (April 19, 2004): 192–94.

299 Orlic: D. Orlic, "Bone Marrow Cells Regenerate Infarcted Myocardium," *Nature* 410 (April 2001): 701–5.

299 sans arrhythmias: J. Couzin, "Renovating the Heart," *Science* 304 (April 9, 2004): 192–94.

299 explosion of phase 1: Ibid.

300 add to atherosclerosis: Author interviews with Perin, October 28, 2004, January 12–14, 2005; Author interview with Amit Patel, January 28, 2005; author interview with Richard Cannon January 19, 2005; M. Sata, "Hematopoietic Stem Cells Differentiate into Vascular Cells that Participate in the Pathogenesis of Atherosclerosis," *Nature Medicine* 8 (April 2002): 403–9; I. I. Masaaki, "Transplant Graft Vasculopathy," *Circulation* 108 (2003): 3056–58; N. M. Caplice, "Smooth Muscle Cells in Human Coronary Atherosclerosis Can Originate from Cells Administered at Marrow Transplantation," *Proceedings of the National Academy of Sciences* 100 (April 15, 2003): 4754–50; Q. Xu, "Role of Stem Cells in Atherosclerosis," *Archives des Maladies du Coeur et des Vaisseaux* 98 (June 2005): 672–76; author interview with Silviu Itescu, July 3, 2005.

302 European papers: E. C. Perin, "Transendocardial, Autologous Bone Marrow Cell Transplantation for Severe, Chronic Ischemic Heart Failure," *Circulation* May 13, 2003: 2294–302.

306 generally believed: www.isscr.org (as of October 24, 2005).

308 January 2005: G. V. Silva, "Mesenchymal Stem Cells Differentiate into an Endothelial Phenotype, Enhance Vascular Density, and Improve Heart Function in a Canine Chronic Ischemia Model," *Circulation* 111 (January 18, 2005): 150–55.

309 Takayuki Asahara: Author interview with Takayuki Asahara, July 7, 2004.

312 distinguished yet maverick scientist . . . through the years: Robert Cooke, *Dr. Folkman's War: Angiogenesis and the Struggle to Defeat Cancer* (New York, Random House, 2001).

313 *Science*: T. Asahara, "Isolation of Putative Progenitor Endothelial Cells for Angiogenesis," *Science* 275 (February 14, 1997): 964–79; Q. Xu, "Role of Stem Cells in Atherlosclerosis," *Archives des Maladies du Coeur et des Vaisseaux* 98 (June 2005): 672–76.

320 Class 10,000 . . . Ruth's heart: Author interview with Jeffrey Wilson, January 13, 2005.

328 *"Nature Medicine"*: A. Saiura, "Circulating Smooth Muscle Progenitor Cells Contribute to Atherosclerosis," *Nature Medicine* 7 (April 2001): 382–83.

329 *Circulation Research*: K. Tanaka, "Diverse Contribution of Bone Marrow Cells To neointimal Hyperplasia after Mechanical Vascular Injuries," *Circulation Research* 93 (October 17, 2003): 783–90.

330 *100 percent*: Y. Hu, "Endothelial Replacement and Angiogenesis in Arteriosclerotic Lesions of Allografts Are Contributed by Circulating Progenitor Cells," *Circulation* 108 (December 23, 2003): 3122–27.

330 Hu would publish: Y. Hu, "Abundant Progenitor Cells in the Adventitia Contribute to Atherosclerosis of Vein Grafts in ApoE-Deficient Mice," *Journal of Clinical Investigation* 113 (May 2004): 1258–65

331 angiostatin: K. S. Moulton, "Inhibition of Plaque Neovascularization Reduces Macrophage Accumulation and Progression of Advanced Atherosclerosis," *Proceedings of the National Academy of Sciences* 100 (April 15, 2003): 4736–41.

332 Cannon: Author interview with Richard Cannon, January 19, 2005.

334 This is unproven: Personal communication with Harvard University stem cell expert Doug Melton, October 4, 2005.

334 Johns Hopkins's Joshua Hare: www.hopkinsmedicine.org/Press_releases/2005/03_25_05.html; Author interview with Silviu Itescu, July 3, 2005.

335 atherosclerosis: M. Ii, "Transplant graft Vasculopathy: A Dark Side of Bone Marrow Stem Cells?" *Circulation* 108 (December 23, 2003): 3056–58.

335 Weissman: L. B. Balsam, "Haematopoietic Stem Cells Adopt Mature Haematopoietic Fates in Ischaemic Myocardium," *Nature* 428 (April 8, 2004): 668–73.

335 McNally: Author interview with Elizabeth McNally, January 3, 2005.

335 Field: Author interview with Loren Field, May 2005; C. E. Murry, "Haematopoietic Stem Cells Do Not Transdifferentiate into Cardiac Myocytes in Myocardial Infarcts," *Nature* 428 (April 8, 2004): 664–68.

CHAPTER 10

Interviews: John McDonald, stem cell researcher, Johns Hopkins University, Baltimore, MD, October 25, 2004, by phone; Rudy Jaenisch, MIT, Boston, MA, September 25, 2003, by phone; Shaorong Gao, Assistant Professor, Center for Regenerative Biology, Storrs, CT, March 30, 2005, in Storrs, and April 11, 2005, by phone; Wise Young, December 15, 2004, February 2005, and March 2005 by phone; Geoffrey Raisman, Spinal Repair Unit, University College London, London, ENG, October 26, 2004, by phone; Alex Zhang, Beijing Institute of Geriatrics, Xuanwu Hospital, January 3, 2005, by phone; Jerry Yang, head, Center for Regenerative Biology, Storrs, CT, August 6, 2003, by phone and September 15–16 in CT, and March 30, 2005 in CT and April 4, 2005, by phone; Bruce Lahn, University of Chicago, geneticist, October 2004; Andy Peng, head, Sun Yat Sen University, Stem Cell Biology and Tissue Engineering, December 12, 2004, by phone; Stephen Minger, director, King's College Stem Cell Biology Laboratory, London, England, March 24, 2005, by phone; Sherry and Ricky Ashmore, Keansburg, NJ, June 2001, in Asbury Park, NJ; David Mehnert, writer and former researcher, June–November 2000, (by phone); Peter Andrews, Professor of Biomedical Science, University of Sheffield, October 5, 2005 (by phone).

338 Hong Yun Huang . . . every foreigner's operation: J. Guest, "Rapid Recovery of Segmental Neurological Function in a Tetraplegic Patient Following Transplantation of Fetal Olfactory Bulb-Derived Cells," *Spinal Cord* (September 6, 2005); D. Cyranoski, "Paper Chase," *Nature* 437 (October 6, 2005): 810–11; H. F. Judson, "The Problematic Dr. Huang Hongyun," *Technology Review,* January 2005.

340 avert its eyes: C. Cookson, "Generous Staffing and Permissive Laws Aid Asia's Largest Stem Cell Effort," *Scientific American,* June 25, 2005; J. Hepeng, "China Beckons to Clinical Trial Sponsors," *Nature Biotechnology* 768 (June 30, 2005).

343 supply labs: "Landmark Stem Cell Law Caps Year of Progress in Connecticut," *PR Newswire,* June 19, 2005.

343 publicly threatened: "UConn Said Close to Creating Human Embryonic Cells," NewsMax.com, March 26, 2005.

344 first to: X. C. Tian, "Meat and Milk Compositions of Bovine Clones," *Proceedings of the National Academy of Sciences* 102 (May 3, 2005): 6261–66 (e-pub April 13, 2005).

344 Yang rushes into his office: Author interview with Jerry Yang, March 30, 2005.

345 job offer: Ibid.

345 Lin Song Li: *Global Watch Mission Report: Stem Cell Mission to China, Singapore and South Korea,* September 2004 (www.global.watchonline.com/online-pdfs/36206mR.pdf).

346 other cloning coup: L. Wang, "Generation and Characterization of Pluripotent Stem Cells from Cloned Bovine Embryos," *Biological Reproduction* 73 (July 2005): 149–55.

346 Zhu Chen: op cit: *Global Watch Mission Report;* D. Bradley, "Biography of Zhu Chen *Proceedings of the National Academy of Sciences* 101 (April 13, 2004): 5325–27.

347 traditionally credits: Ian Wilmut et al., *The Second Creation: Dolly and the Age of Biological Control* (New York: Farrar, Straus and Giroux, 2000).

347 *Science*: O. Normile, "Asia Jockeys for Stem Cell Lead," *Science* 307 (February 4, 2005): 660–64.

347 *Nature Biotechnology*: L. Zhenzhen, "Health Biotechnology in China—Reawakening of a Giant," *Nature Biotechnology* 22 (2004): DC13–DC18.

347 *Nature*: M. M. Poo, "Cultural Reflections," *Nature* 428 (March 11, 2004): 204–5.

348 Glenn Rice: A. McCook, "When Science Switches Shores," *The Scientist* 19 (March 28, 2005).

349 Wu told *Nature Biotechnology*: L. Zhenzhen, "Health Biotechnology in China—Reawakening of a Giant," *Nature Biotechnology* 22 (2004): DC13–DC18.

349 yet another: K. Chien, "The New Silk Road," *Nature* 428 (March 11, 2004): 208–9.

349 Bruce Lahn: Author interview with Bruce Lahn, October 2004.

349 Andy Peng: Author interview with Andy Peng, December 12, 2004.

350 Sir George Radda: Author interview with George Radda, October 28, 2003.

350 *Nature* article: X. Yang, "An Embryonic Nation," *Nature* 428 (March 11, 2004): 210–2.

351 Stephen Minger: Author interview with Stephen Minger, March 24, 2005.

351 courting Yang: Author interview with Jerry Yang, March 30, 2005.

351 20 staffers: Global Watch Mission Report: Stem Cell Mission to China, Singapore, and South Korea, September 2004, op. cit.

352 This included . . . associated company: Author interview with Andy Peng, December 12, 2004; "China Seeks Extended Collaboration with EU on Stem Cell Research," *CellNews*, January 26, 2005.

352 Guang Xiu Lu: Author interviews with Jerry Yang, March 30, 2005, and April 4, 2005; K. Leggett, "China Stem-Cell Research Leaps Ahead," *The Wall Street Journal*, March 6, 2002; Reproductive and Genetic Hospital of CITIC-Xiangya Web site, www.hn-ivf.cn/en/main.asp (as of April 23, 2006).

353 Alex Zhang: Author interview with Alex Zhang, January 3, 2005.

356 Other labs: Author interview with Andy Peng, December 12, 2004.

357 Wise Young's vast network: Author interview with Wise Young, February 2005.

357 "His breath smells": Author interview with Sherry and Ricky Ashmore, June 2001.

360 1,500 transplants . . . by 2005: Emcell, www.emcell.com/en/about/index.html, October 6, 2005; PubMed search as of October 6, 2005; G. Cook, "Absence of Data on Clinic's Therapies Provokes Skepticism," *The Boston Globe*, September 26, 2004.

360 An ALS association: "Emcell Review," ALS Therapy Development Foundation, www.als.net, June 2002.

361 in the early 1990s . . . among other disorders: F. Fleck, "Russians Start Human Fetal Tissue Transplant Operations," *Reuters News Service*, March 21, 1993; M. Wallace, "Joint Venture: Russia and Discredited California Plastic Surgeon Set Up Fetal Cell Institute in Moscow," CBS's *60 Minutes*, April 11, 1993; V. Gloger, "Die Miss-Brauchten Foten," *Der Stern*, February 4, 1993.

361 Well over 200: William Freed, *Neural Transplantation* (London: MIT Press): 201.

361 Isaacson: Personal communication with Ole Isaacson, Harvard University, December 5, 1999.

361 not reliable: Ibid.; Author interview with Peter Andrews, October 5, 2005.

362 license to practice: M. Wallace, "Joint Venture," CBS's *60 Minutes*, op. cit.

362 before the 1990s: V. I. Shumakov, "Transplantation of Cultures of Human Fetal Pancreatic Islet Cells to Diabetes Melitus Patients," *Klin Med 61* (February 1983): 46–51; M. Micheda, "Quo Vadis?" Fetal Tissue Transplantation," *Journal of Hematotherapy* 5 (1996): 185–88.

362 negative Western press: See note above for p. 361 (in the early 1990s . . .)

362 But bizarrely . . . international journal: NeuroVita Clinic site, www.neurovita.ru/publish/eng_article02.html (as of November 16, 2005); "A. S. Bruhovetsky's Fetal Stem Cell Treatment of Chronic Spinal Cord Injury in Moscow," CareCure Online Forums, August 3, 2001–August 30, 2004, www.carecure.org/forum/archive/index .php/t-39642.html (as of October 2005); "Moscow Stem Cell Research Clinic," CareCure Forums, June 17, 2005–July 8, 2005, www.carecure.org/forum/showthread .php?t=923 (as of October 2005); Description, "Cell Transplantation Surgery and Human Organ Bioengineering," a joint private/public venture, www.brainhealing .com, 2000–2001.

362 Gennady Sukhikh: M. A. Aleksandrova, "Transplantation of Cultured Neural Cells from Human Fetuses into the Brain of Rats Exposed to Acute Hypoxia," *Bulletin of Experimental Biology and Medicine* 137 (March 2004): 262–65 and "Behavior of Human Neural Progenitor Cells Transplanted to Rat Brain," *Developmental Brain Research* 134 (2002): 143–48.

362 peddling human fetal cells: A. Loshak, "Moscow: Stem Cell Capitol of the World," *Moscow News,* March 14, 2005.

362 Rabinovich: S. S. Rabinovich, "Transplantation Treatment of Spinal Cord Injury Patients," *Biomedical Pharmacotherapeutics* 57 (November 2003): 428–33; Care Cure Web site, www.carecure.org (as of November 16, 2005).

363 Victor Seledtsov: Personal communications with Victor Seledtsov of the Institute of Clinical Immunology, Novosibirsk, Siberia, November 2004.

363 Wise Young: Author interview with Wise Young, December 15, 2004.

363 *unregulated* clinics . . . in business anyway: M. Danilova, "Stem Cell Treatment Arrives, But at a Price," Associated Press, March 11, 2005; N. Titova, "Moscow Beauty Salons Are Offering Bogus Stem-Cell Treatments for Wrinkles, Grey Hair and Other So-Called Ailments," *Newsweek International,* November 8, 2004.

364 Ukrainian clinics: A. Zarembo, "Outside the US, Businesses Run with Unproved Stem Cell Therapies," *Los Angeles Times,* February 20, 2005.

364 Medra . . . stem cell Web sites: Author interviews with David Mehnert, Former researcher aiding Kristina Friedman, June–November 2000; Rader clinic brochures from November 1997 to 2001; Initial Rader clinic research materials and documents, 1997–2000; A. Zarembo, "Outside the US, Businesses Run with Unproved Stem Cell Therapies," *The Los Angeles Times,* February 20, 2005; Rader's Web site, Medra.com; L. Mecoy, "Stem Cells, Hopes Lure Many Abroad," *The Sacramento Bee,* January 9, 2005.

365 "conspiracy": A. Zarembo, "Outside the US, Businesses Run with Unproved Stem Cell Therapies," *The Los Angeles Times,* February 20, 2005.

365 Year after year . . . respected journals: C. Cookson, "The Strange Tale of Ukrainian Stem Cell Experts, American Investors and Caribbean Tourism," *The Financial Times,* November 10, 2004; A. Zarembo, "Outside the US, Businesses Run with Unproved Stem Cell therapies," *The Los Angeles Times,* February 20, 2005.

366 "In 2004, the U.S.": Author interviews with Wise Young, December 15, 2004, February and March 2005.

369 Hong Yun Huang: H. Freeland Judson, "The Problematical Dr. Huang Hongyun," *Technology Review,* January 2005; D. McElroy, "Doctor Attacked Over" 'Miracle Cure' Based on Aborted Foetuses," *News Telegraph,* May 12, 2004.

371 John McDonald . . . "that data": Author interview with John McDonald, October 25, 2004.

372 lawsuits: P. Elias, "Lawsuits Filed to Invalidate California's $3 Billion Stem Cell Institute," Associated Press, February 24, 2005.

372 Senator Sam Brownback: Office of Legislative Policy and Analysis, Legislative Updates, Cloning.

CHAPTER 11

Interviews: Mary Herbert, scientific director, Newcastle Centre for Life, Newcastle University, Newcastle Upon Tyne, UK, April 27, 2005, by phone; Evelyn Telfer, Institute of Cell and Molecular Biology, University of Edinburgh, Edinburgh, Scotland, May 13, 2005, by phone; Thomas Toth, director, MGH Vincent Obstetrics and Gynecology In Vitro Fertilization Unit, September 13, 2004, by phone; Jonathan Tilly, director, Vincent Center for Reproductive Biology, Massachusetts General Hospital, August 24, 2004, and October 28, 2005, by phone; Richard Anderson, Centre for Reproductive Biology, Edinburgh, Scotland, May 12, 2005, by phone; Roger Gosden, Center for Reproductive Medicine and Fertility, Cornell Weill Medical College, May 13, 2005, by phone; Mahendra Rao, NIH Stem Cell Task Force member, April 22, 2005, by phone; Alan Colman, CEO of ESI May 4, 2005.

375 expensive: "Reproductive Health Services, Mandatory Insurance Coverage for In Vitro Fertilization Leads to Greater Access of Services But Does Not Improve Pregnancy Rates, Study Says," Kaisernetwork.org, August 29, 2002 ("David Guzick of the University of Rochester School of Medicine notes that the average cost of IVF treatment is $40,000."); K. Johnson, "Infertility Treatments Exemplify How Therapy Costs Vary From Plan to Plan by Region," *Managed Healthcare Executive,* June 1, 2005 ("a total treatment cost averaging about $36,000").

375 few states: "Reproductive Health Services, Mandatory Insurance Coverage for In Vitro Fertilization Leads to Greater Access of Services but Does Not Improve Pregnancy Rates, Study Says," Kaisernetwork.org, August 29, 2002.

375 14 states: "Facing Infertility," *FDA Consumer Magazine, November–December 2004.*

375 Society for Assisted Reproductive Technology: The National Fertility Directory, "The Costs of IVF," www.SART.org.

376 immoral: Staff conversations, Center for Reproductive Medicine and Fertility, Weill Medical College of Cornell University, New York, NY, July 30, 2005.

376 bear a child with someone: Staff conservations, Columbia University's Center for Women's Reproductive Care, New York, NY; J. Egan, "Wanted: A Few Good Sperm," *The New York Times Magazine*, March 19, 2006.

376 a few patients: Staff conservations, Columbia University's Center for Women's Reproductive Care, New York, NY.

376 often *aren't* approved: Author conversation with Weill Medical College geneticist, Fred Gilbert, August 31, 2004.

377 CDC: www.cdc.gov.

377 Rao: Author interview with Mahendra Rao, April 22, 2005; J. Thomson, "Funds of Human Embryo Research in the U.S.," *Nature Biotechnology"* (April 1, 1999): 312.

378 die naturally: C. Packer, "Why Menopause?" *Natural History,* 1998.

378 Tilly: Author interview with Jonathan Tilly, August 24, 2004.

378 Gosden: Roger Gosden, *Cheating Time: Science Sex and Aging* (New York: W. H. Freeman, 1996): 264.

378 1.8 percent of Fortune 500: The 2005 Fortune 500, Fortune.com.

378 since 1982: Women and Education Statistics, U.S. Census Bureau, http://spe cials.about.com/zxfcp0.htm?gs=womensissues&u=womensissues.about.com/library/ blwomeneducationstats.htm; National Center for Policy Analysis, "The College Gender Gap," *Daily Policy Digest*," June 25, 2002.

378 law school: Women's Bar Association of Massachusetts, "WBA Part-Time Report," www.womensbar.org.

378 most must work: M. Warner, "Economic Development Strategies to Promote Quality Child Care," Department of City and Regional Planning, Ithaca, NY, 2004.

378 78 cents: U.S. Department of Labor 2003 study; CNN.com (www.cnn.com /2005/US/Careers/06/23/women.salary), last accessed October 23, 2005.

378 been dropping: K. Hansen, "Ten Powerful Career Strategies for Women," www.quintcareers.com/women_career_strategies.html, last accessed October 23, 2005.

378 70 percent of the poor: "Women in Politics: Beyond Numbers," http://archive .idea.int/women/parl/ch2c.htm, last accessed October 23, 2005.

379 never recover: A. Honebrick, "Treatment of Menopausal Symptoms Post Women's Health Initiative: Refinement of Existing Treatments and Development of New Therapies," *Expert Opinions on Emerging Technologies* 10 (August 2005): 619–41.

379 Heart disease: "Interview: Raymond Woosley," *Frontline* transcript, PBS, www.pbs.org, last accessed October 25, 2005.

379 cashing in . . . after 54: K. Campbell, "Men Decide It's Never Too Late to Have Kids," *The Christian Science Monitor,* September 1, 2004.

379 The events that led: Author interview with Jonathan Tilly, August 24, 2004.

380 "I lived it": R. Mishra, "Defying Dogma," *The Boston Globe,* July 26, 2004.

380 repeated to the press: K. Powell, "Age Is No Barrier," *Nature* 432 (November 4, 2004): 40–42.

381 Telfer: E. E. Telfer, "Germline Stem Cells in the Postnatal Mammalian Ovary: A Phenomenon of Prosimian Primates and Mice?" *Reproductive Biology and Endocrinology* 2 (May 18, 2004): 24.

381 Spradling: A. C. Spradling, "Stem Cells: More Like a Man," *Nature* 428 (March 11, 2004): 133–34.

381 Damewood: M. Damewood, "ASRM Comments on Research Showing Germline Stem Cells in Female Mice," ASRM, March 10, 2004.

381 "Textbook rewrite?": J. Couzin, "Textbook Rewrite? Adult Mammals May Produce Eggs after All," *Science* 303 (March 12, 2004): 1593.

381 "Astonishing": E. Croager, "Egg-Citing Fertility Finding," *Nature Reviews Molecular Cell Biology* 5 (April 2004): online highlights, www.nature.com.

381 Ng: B. C. Heng, "'Waste' Follicular Aspirate from Fertility Treatment—a Potential Source of Human Germline Stem Cells?" *Stem Cells and Development* 14 (February 2005): 11–14.

381 *Nature*: J. Johnson, "Germline Stem Cells and Follicular Renewal in the Postnatal Mammalian Ovary," *Nature* 428 (March 11, 2004): 145–50.

382 In 1999: G. J. Perez, "Prolongation of Ovarian Lifespan into Advanced Chronological Age by Bax-Deficiency," *Nature Genetics* 21 (February 1999): 200–3.

382 called S1P: Y. Morita, "Oocyte Apoptosis Is Suppressed by Disruption of the Acid Sphingomyelinase Gene or by Sphingosine-1-phosphate Therapy," *Nature Medicine* 6 (October 2000): 1109–14.

382 cigarettes kill: T. Matikainen, "Aromatic Hydrocarbon Receptor-Driven Bax Gene Expression is Required for Premature Ovarian Failure Caused by Biohazardous Environmental Chemicals," *Nature Genetics* 28 (August 2001): 355–60.

382 pups: F. Paris, "Sphingosine-1-Phosphate Preserves Fertility in Irradiated Female Mice without Propagating Genomic Damage in Offspring," *Nature Medicine* 8 (September 2002): 901–2.

382 ceramide: G. I. Perez, "A Central Role for Ceramide in the Age Related Acceleration of Apoptosis in the Female Germline," *The FASEB Journal* 19 (May 2005): 850–52.

384 DMBA: T. Matikainen, op. cit., 498.

385 might explain: K. Powell, "Age Is No Barrier," *Nature* 432 (November 4, 2004): 40–42.

385 paper came out: K. Oktay, "Embryo Development after Heterotopic Transplantation of Cryopreserved Ovarian Tissue," *The Lancet* 363 (March 13, 2004): 837–40.

386 Zuckerman: S. Zuckerman, "The Number of Oocytes in the Mature Ovary," *Recent Progress in Hormone Research* 6 (1951): 63–108.

388 Van Eck: G. Eck, "Neo-Ovogenesis in the Adult Monkey," *The Anatomical Record* 125 (1956): 207–24.

391 Jacques Cohen: J. A. Barritt, "Mitochondria in Human Offspring Derived from Ooplasmic Transplantation," *Human Reproduction* 16 (March 2001): 513–16.

392 the FDA demanded: H. Firfer, "How Far Will Couples Go to Conceive?" CNN.com, accessed June 17, 2004.

392 Jamie Grifo: D. Grady, "Pregnancy Created Using Egg Nucleus of Infertile Woman," *The New York Times,* October 14, 2003.

392 Mary Herbert: Author interview with Mary Herbert, April 27, 2005.

394 hardly all kudos . . . a pregnancy: Author interviews with Mary Herbert, April 27, 2005; Evelyn Telfer, May 13, 2005; Richard Anderson, May 12, 2005; and Roger Gosden, May 13, 2005.

394 July 2005: J. Johnson, "Oocyte Generation in Adult Mammalian Ovaries by Putative Germ Cells in Bone Marrow and Peripheral Blood," *Cell* 122 (July 29, 2005): 303–15.

394 "concerns": E. Telfer, "On Regenerating the Ovary and Generating Controversy," *Cell* 122 (September 23, 2005): 821–22.

394 written response: J. Johnson, "Setting the Record Straight on Data Supporting Postnatal Oogenesis in Female Mammals," *Cell Cycle* 4 (November 2005): e36–e42.

397 "from 1921": R. Pearl, "Studies on the Physiology of Reproduction in the Domestic Fowl," *Journal of Experimental Zoology* 34 (1921): 101–18.

397 "hundreds of examples": J. Johnson, "Setting the Record . . . ," op. cit.

399 Antonin Bukovsky: A. Bukovsky, "Oogenesis in Cultures Derived from Adult Human Ovaries," *Reproductive Biological Endocrinology,* 3 (May 5, 2005): 17.

399 proven he was right: Author interview with Angelo Vescovi, codirector of research of the Institute for Stem Cell Research, Hospital San Raffaele, Milan, Italy, March 1999.

399 Macklis: Author interview with Jeffrey Macklis, Harvard University neuroscientist, September 6, 2000.

399 University of Auckland: M. A. Curtis, "A Histochemical and Immunohistochemical Analysis of the Subependymal Layer in the Normal and Huntington's Disease Brain," *Journal of Chemistry and Neuroanatomy* 30 (July 2005): 55–66.

399 "artificial ova creation": "Researchers Create Artificial Ova from Stem Cells—Pro-lifers Warn Against Embryo Farming." Lifesitenews.com, May 11, 2005, accessed May 2005.

400 Scholer: J. Kehler, "Generating Oocytes and Sperm from Embryonic Stem Cells," *Seminars in Reproductive Medicine* 23 (August 2005): 222–23.

400 Julang: J. Couzin, "Stem Cells: Another Route to Oocytes?" *Science* 309 (September 23, 2005): 1983

400 over 120: http://stemride.com/products.htm, last accessed October 31, 2005.

400 Verlinsky's journey . . . Chicago: Chicago Jewish Community Online, HIAS celebrates 90th anniversary, JUF News and Community Affairs, www.juf.org.

400 $9 million: J. Laidman, "Harnessing the Power of Creation: Advances Spark Birth of Hope, Controversy," *The Toledo Blade,* April 20, 2005.

401 Handyside: A. H. Handyside, "Pregnancies from Biopsied Human Preimplantation Embryos Sexed by Y-specific DNA Amplification," *Nature* 344 (April 19, 1990): 768–70.

401 Franconi's anemia: Y. Verlinsky, "Preimplantation Diagnosis for Franconi anemia combined with HLA Matching," *Journal of the American Medical Association* 285 (June 27, 2001): 3130–33.

401 both Jack . . . were healthy: J. Laidman, "Harnessing the Power of Creation: Advances Spark Birth of Hope, Controversy," *The Toledo Blade,* April 20, 2005; Ronald Bailey, *Liberation Biology: The Scientific and Moral Case for the Biotech Revolution* (Amherst, NY: Prometheus Books, 2005): 102–3.

402 Alzheimer's gene: Y. Verlinsky, "Preimplantation Diagnosis for Early-Onset Alzheimer Disease Caused by V717L Mutation," *Journal of the American Medical Association* 287 (February 27, 2002): 1018–21.

402 *Salon*: J. Sweeney, "A Cruel Choice: A Woman Decides to Have a Child Knowing That She's about to Descend into Dementia. That's Morally Indefensible," *Salon,* March 1, 2002.

402 deaf couples used: C. Dennis, "Genetics: Deaf by Design," *Nature* 431 (October 21, 2004): 894–96.

403 Down syndrome: M. Healy, "Embryo Diagnosis Stirs Controversy," *The Los Angeles Times,* July 29, 2003.

403 Kathy Hudson: "Genetic Testing of Embryos to Pick 'Savior Sibling' OK with Most Americans," *Medical News Today,* May 4, 2004.

403 "In the future": M. Healy, "Embryo Diagnosis Stirs Controversy," *The Los Angeles Times,* July 29, 2003.

404 February 2005: V. Galat, "Cytogenetic Analysis of Human Somatic Cell Haploidization," *Reproductive Biomedicine Online* 10 (February 2005): 199–204.

404 Peter Braude: H. Pearson, "Early Embryos Fuel Hopes for Shortcut to Stem-Cell Creation," *Nature* 22 (November 2004).

404 backing up: Y. Ching, "Embryonic and Extraembryonic Stem Cell Lines Derived from Single Mouse Blastomeres," *Nature* (October 16, 2005; e-pub ahead of print, www.nature.com).

404 essentially hES cells: N. Strelchenko, "Momla-Derived Human Embryonic Stem Cells," *Reproductive Biomedicine Online* 9 (December 2004): 623–29.

404 up to the 48-cell: Author interview with Mahendra Rao, April 22, 2005.
405 Rao: Ibid.
405 Colman: Author interview with Alan Colman, May 4, 2005.

CHAPTER 12

Interviews: Irving Weissman, head, Stanford University Institute for Cancer/Stem Cell Biology, Palo Alto, CA, April 29, 2005, by phone, February 19 and 22, 2006, by phone, and February 7, 2001, in Palo Alto; Ron McKay, NIH Stem Cell Task Force member, Bethesda, MD, July 31, 2003, by phone, and April 28–29, 2005, by phone; Vic Nurcombe, Stem Cell and Tissue Repair Lab, Institute of Molecular and Cell Biology, Singapore, October 31, 2003, in Singapore, and May 24, 2005, by phone; NIH Stem Cell Task Force member Mahendra Rao, Baltimore, MD, July 25, 2003, in MD, and April 22, 2005, by phone; Douglas Melton, cofounder, Harvard Stem Cell Institute, Cambridge, MA, June 17, 2004, by phone; Richard Garr, CEO, Neural-Stem, College Park, MD, February 3, 2001, in MD, and June 23, July 5, and July 7, 2005, by e-mail; Ismail Barrada, scientific founder and head, Stem Cell Research Center, Misr University for Science & Technology, 6th of October City, Egypt, June 22, June 24, and June 28, 2005, by e-mail; Stephen Minger, director, King's College Stem Cell Biology Laboratory, London, England, March 24, 2005, by phone; Michael Lysaght, director, Center for Biomedical Engineering at Brown University and cofounder of Cytotherapeutics, Providence, RI, March 2001, by phone; John Dick, director, Program of Stem Cell Biology, University of Toronto, Toronto, Canada, May 10, 2005, by phone; Shimon Slavin, director, Department of Bone Marrow Transplantation, Hadassah University, Jerusalem, Israel, July 9, 2005, by phone; Woo Suk Hwang, Seoul National University, Seoul, South Korea, July 21, 2005, in Seoul; Curie Ann, Seoul National University Hospital, July 22, 2005, in Seoul; Byeong Chun Lee, Seoul National University, July 23, 2005, in Seoul, Sung Keun Kang, Seoul National University, July 21, 2005, in Seoul; Shin Yong Moon, Stem Cell Research Center, Seoul, South Korea, July 21, 2005, in Seoul; Joseph Wong, University of Toronto, November 4, 2005 by phone; Abdallah Daar, University of Toronto, August 20, 2005, by phone; Dan Perry, Alliance for Aging Research, September 16, 2005, by phone; Michael Warner, Biotechnology Industry Organization, Chief of Policy, Washington, DC, August 22, 2005, by phone; Greg Horowitt, Global Connect, San Diego, CA, August 20, 2005, by phone; Yann Barrandon, Ecole Polytechnique Federale de Lausanne, Lausanne, Switzerland, October 10, 2005, by phone; David Ayares, CEO, Revivicor, Blacksburg, VA, February 10, 2006, by phone; Thomas Starzl, Starzl Institute, University of Pittsburgh, February 9, 2006, by phone; Davor Solter, Max Planck Institute of Immunobiology, Freiberg, Germany, February 9, 2006, by phone; James Robl, CEO, Hematech, February 6, 2006, by phone; Alan Trounson, scientific director, Monash IVF, Melbourne, Australia, February 14, 2006, by phone; Hyun Soo Yoon, former director, MizMedi Medical Research Center, current professor at Hanyang University, July 20, 2005, in Seoul; Tariq Hussain, author of *The Diamond Dilemma—Shaping Korea for the 21st Century*, April 21, 2006, by phone.
407 first cows: A. Mandavilli, "Profile: Woo-Suk Hwang," *Nature Medicine* 11 (May 2005): 464.

408 boffo Korean sales: G. Snyder, "Hollywood & New World Order," *Variety,* September 26, 2005.

411 in 1988: G. J. Spangrude, "Purification and Characterization of Mouse Hematopoietic Stem Cells," *Science* 241 (July 1, 1988): 58–62.

411 patented it . . . first millionaires: P. Radetsky, "The Mother of All Blood Cells," *Discover* 16 (March 1995); M. Freudenheim, "Sandoz Buys 60% Stake in Systemix," *The New York Times,* December 17, 1991.

411 in 1992: C. M. Baum, "Isolation of a Candidate Human Hematopoietic Stem-Cell Population, *Proceedings of the National Academy of Sciences* 89 (April 1, 1992): 2804–8.

412 virtual company: Author interview with Irv Weissman, February 7, 2001.

412 In 1997 . . . board chairman: "Cytotherapeutics Acquires Stem Cell Product Pipeline," *Biotech Patent News* 11 (August 1, 1997).

412 CEO: Author interview with Michael Lysaght, March 2001.

412 Shortly after . . . Stem Cells Inc.: W. Donovan, "A Tough Capsule to Take," *The Providence Journal Bulletin,* October 3, 1999; "Recent Changes in Stock Listings," *The Wall Street Journal,* May 25, 2000; "Stemcells, Inc. Reports Third Quarter Financial Results," *Business Wire,* November 14, 2000; "Cytotherapeutics Plans Layoffs," *The Boston Globe,* July 10, 1999.

412 Lysaght: Author interview with Michael Lysaght, March 2001.

412 unreasonable limitations: M. Wadman, "Licensing Fees Slow Advance of Stem Cells," *Nature* 435 (May 19, 2005): 272–73; S. G. Stolberg, "Patent on Human Stem Cell Puts U.S. Officials in Bind," *The New York Times,* August 17, 2001; author interview with Victor Nurcombe, October 31, 2003; author interview with Mahendra Rao, July 25, 2003.

412 there are lawyers . . . slow research down: Personal communication (e-mail) from Richard Garr, June 23, July 5, and July 7, 2005.

413 Batten's disease: G. Vogel, "Ready or Not: Human ES Cells Head Toward the Clinic," *Science* 308 (June 10, 2005); "Stock Report for Week Ending Feb 11, Mixed Bag," www.stemnews.com/archives/000293.html.

413 Geron: K. Philipkoski, "Race to Human Stem Cell Trials," Wired News, April 19, 2005.

414 chastised Bush: C. Holden, "Restiveness Grows at NIH Over Bush Research Restrictions," *Science* 308 (April 15, 2005): 334–335.

414 Some interviewed: Author interview with Mahendra Rao, April 22, 2005.

414 McKay: Author interview with Ron McKay, April 28, 2005.

414 lost candidates: J. Harkinson, "Stem Cell Restrictions Could Send Texas Medical Center Researchers Fleeing to California," *The Houston Press,* May 19, 2005.

414 Rao: Personal communication (e-mail) from Mahendra Rao, September 11, 2005.

414 "They're excited": Author interview with Irv Weissman, April 29, 2005.

415 diabetic child: D. Duncan, "Biotech and Creativity," *San Francisco Chronicle,* October 3, 2004.

416 As McKay noted . . . "little more creative": Author interviews with Ron McKay, April 28 and 29, 2005.

417 "I think there's no doubt": Author interview with Ron McKay, April 29, 2005.

417 exciting news: L. Wang, "Hematopoietic Development for Human Embryonic Stem Cell Lines," *Experimental Hematology* 33 (September 2005): 987–91.

418 came to blows: C. Connolly, "House Bill to Ease Stem Cell Curbs Gains Momentum," *Washington Post,* May 19, 2005.

418 done it again: W. S. Hwang, "Patient-Specific Embryonic Stem Cells Derived from Human SCNT Blastocysts," *Science* 308 (June 17, 2005): 1777–83 (e-pub May 19, 2005).

418 broken embargo: Personal communication (e-mail) from MSNBC writer Alan Boyle, May 20, 2005.

419 *New Scientist*: R. Hooper, "Cloned Human Embryos Deliver Tailored Stem Cells," *The New Scientist,* May 19, 2005.

419 *Washington Post*: R. Weiss, "Koreans Say They Cloned Embryos for Stem Cells," *The Washington Post,* May 20, 2005.

420 "being left farther": P. Baker, "Bush Vows He'll Veto Stem Cell Measure," *The Washington Post,* May 21, 2005.

420 *New York Times*: "A Surprising Leap on Cloning," *The New York Times,* May 22, 2005.

420 global consortium: "Consortium to Help Hwang's Research," *KBS Global,* May 22, 2005.

420 Nurcombe: Author interview with Vic Nurcombe, May 24, 2005.

420 House of Representatives fought: U.S. House of Representatives debate on bills HR 810 and HR 2520, May 25, 2005.

421 Snowflake Children . . . for adoption: W. Saletan, "Leave No Embryo Behind: The Coming War over In Vitro Fertilization," *Slate,* June 3, 2005; Snowflakes Embryo Adoption Program Web site: www.nightlight.org.

422 cure nothing: G. Vogel, "Still Waiting Their Turn," *Science* 308 (June 10, 2005): 1536–37.

423 the Senate . . . keep it from passing: C. Holden, "Embryonic Stem Cells: Spotlight Shifts to Senate after Historic House Vote," *Science* 308 (June 3, 2005): 1388–89; S. G. Stolberg, "Sponsor of Stem Cell Bill Says Senate Could Override a Veto," *The New York Times,* May 25, 2005.

424 it was reported: "South Korea to Provide Extra $1M for Stem Cell Researcher," Associated Press, May 25, 2005.

434 On August 3, 2005: B. C. Lee, "Dogs Cloned from Adult Somatic Cells," *Nature* 436 (August 4, 2005): 641.

436 unprecedented number: Horace Freeland Judson, *The Great Betrayal: Fraud in Science* (Orlando, FL: Harcourt, 2004): 139; H. Gottweis, "South Korean Policy Failure and the Hwang Debacle," *Nature Biotechnology* 24 (February 2006): 141–43.

436 tightened globally: "Scientists Agree Global Stem Cell Guidance," Press Association, February 24, 2006; N. Wade, "A New Fraud Busting Test for Photos," *International Herald Tribune,* January 25, 2006; S. Dolbee, "Egg Donors Would Be Reimbursed, Not Paid," *The San Diego Union Tribune,* February 1, 2006; "Female Research Team Members to Be Banned from Donating Eggs," Japan Economic Newswire, March 6, 2006; "Getting Beyond Hwang," *Nature Biotechnology* 24 (January 1, 2006): 1.

436 mysterious statement . . . donations: J. Duffield, "University of Pittsburgh Researcher Ends Collaboration with South Korean Stem Cell Program," University of Pittsburgh press statement, November 12, 2005.

436 denies ethics violations: T. G. Kim, "In House Egg Donation Not Unethical,"

The Korea Times, November 15, 2005; T. G. Park, "Hwang Knew of Staff's Egg Donations," *JoongAng Ilbo*, November 24, 2005.

436 investigating: K. C. Kim, "Hwang, Bewildered, Promises Answers Soon," *JoongAng Ilbo*, November 17, 2005.

436 scandal-ridden: P. Elias, "U.S. Cloning Researcher Tainted by Scandal," Associated Press, December 16, 2005; R. Weiss, "S. Korea Crisis May Affect U.S. Debate," *The Washington Post*, November 20, 2005; P. D. Iglauer, "Hwang's 40 Foot Gorilla," *The Korea Times*, December 19, 2005. (Quote from the *Post*: "Ten years ago, revelations about criminal practices at a University of California fertility program led investigators to Schatten, who was then at the University of Wisconsin. He had an arrangement to obtain eggs from the clinic in Irvine, Calif., where, it turned out, doctors were impregnating women with embryos made from other women's eggs and distributing excess eggs to researchers without institutional approval. One Irvine doctor was eventually convicted on federal charges, and two others fled the country to avoid prosecution. Schatten was cleared of any wrongdoing.")

437 jealous: Y. J. Oh, "Don't Turn Hwang into Dr. Moreau," *The Korea Times*, November 20, 2005; S. H. Choe, "Scientist Faked Stem Cell Study," *The International Herald Tribune*, December 15, 2005.

437 Hwang staffers helped: T. G. Kim, "Schatten's Break With Hwang: Interest or Principle?" *The Korea Times*, November 21, 2005; "MBC Charges Stem Cell Research A Sham," *Chosun Ilbo*, November 28, 2005.

437 unhelpful . . . an official: S. Y. Hwang, "Seoul to Enact Stem Cell Research Law Next Year," *The Korea Herald*, November 16, 2005.

437 November 21 . . . Schatten's action: "Hwang Woo-Suk," *Wikipedia*, http://en.wikipedia.org/wiki/Hwang_Woo-Suk (as of March 9, 2006).

437 Sung Il . . . 600 eggs: Ibid *Wikipedia;* "Hwang Paid 20 Women for Donating Ova," *Chosun Ilbo*, November 21, 2005.

437 eggs at no cost . . . 6,000 e-mailed: "Hwang Supporters Condemn MBC Investigative Report," *Chosun Ilbo*, November 23, 2005.

437 pictures of the . . . death threats: J. Brooke, "Korean Cloning Pioneer Is Popular Despite Ethical Lapses," *The New York Times*, November 28, 2005.

437 500,000 angry: B. Demick, "South Korean Cloning Scandal Takes Toll on Whistleblowers," *The Los Angeles Times*, February 14, 2006.

437 lied thereafter: "Suspicion Mounting Again: Hwang Woo Suk Ought to Cooperate for DNA Tests of Stem Cells," *The Korea Times,* December 1, 2005; "*Time* Talks to the Controversial South Korean Cloning Pioneer," *Time* online, December 5, 2005.

437 Korea only . . . 15,000 members: J. Y. Kwon, "Nation debates Hwang's ethics," *The Korea Herald*, November 11, 2005. (Hwang's Internet fan club can be found at http://cafe.daum.net/ilovews.)

437 vacation: T. G. Kim, "Cloning Pioneer Stays Away from Research," *The Korea Times*, November 28, 2005.

438 foreign institutes: C. Holden, "Korean Cloner Admits Lying about Oocyte Donations," *Science* 310 (December 2, 2005): 1402–3.

438 Davor Solter: Author interview, Davor Solter, February 9, 2006.

438 boycotting products: S. J. Jung, "Public Rallies Behind Fallen Scientist," *Dong AIlbo*, November 26, 2005.

438 11 out of 12 advertisers: D. S. Seo, "Roh Decries Public Backlash at Hwang Report," *The Korea Times*, November 27, 2005.

438 *JoongAng Ilbo*: D. H. Wohn, "Cloner Drops Out of Public View," *JoongAng Ilbo*, November 26, 2005.

438 calm down: J. Brooke, "Clone-Gate Scoop Boomerangs on South Korean News Show," *International Herald Tribune*, December 19, 2005.

438 number 1,000: "Number of Aspirants for Egg Donation Rapidly Soaring," *The Korea Times*, November 27, 2005; S. H. Choe, "S. Koreans Support Disgraced Scientist," *International Herald Tribune*, November 28, 2005.

438 "syndrome" . . . "like Japan": S. H. Choe, "S. Koreans Support Disgraced Scientist," *International Herald Tribune*, November 28, 2005.

438 producers announce: "Getting Beyond Hwang," *Nature Biotechnology* 24 (January 2006): 1; J. Herskovitz, "S. Korea TV Documentary to Question Stem Cell Study," Reuters, December 2, 2005; "A Credible Investigation," *JoongAng Ilbo,* December 4, 2005.

438 *PD Diary* producer . . . by an insider: J. S. Cho, "MBC Proposes a Joint Probe with Hwang," *The Korea Times*, December 2, 2005.

438 possibility of fraud: J. Brooke, "Clone-Gate Scoop Boomerangs on South Korean News Show," *International Herald Tribune*, December 19, 2005.

438 turn out to be: B. Demick, "South Korean Cloning Scandal Takes Toll on Whistle-Blowers," *The Los Angeles Times*, Feburary 16, 2006; J. Y. Kwon, "SNU Professors Play Dumb in Stem Cell Probe," *The Korea Herald*, January 31, 2006; "Hwang Probe Turns to Cell Contamination," *Chosun Ilbo*, January 18, 2006.

438 brand-new MD . . . . handing them off: "Prosecution Summons Six Junior Stem Cell Researchers," J. Y. Kwon, *The Korea Herald*, January 16, 2006.

438 who were, as noted, in charge . . . from spare IVF-clinic: Author interview with Woo Suk Hwang, June 4, 2004.

439 more witnesses: J. S. Cho, "MBC Proposes a Joint Probe With Hwang," *The Korea Times*, December 2, 2005.

439 On December 4 . . . Jong Hyuk Park: "MBC Apologizes for Bullying in Hwang Exposé," *Chosun Ilbo*, December 4, 2005; "Apology from MBC Is Not Enough," *Dong A-Ilbo*, December 6, 2005.

439 It will later . . . by Hwang's SNU lab: D. H. Wohn, "Hwang Saga Conjures Up More Queries," *JoongAng Ilbo*, January 23, 2006.

439 The two underlings' message . . . bullied them: "Hwang's Team Seeks to Regain Its Credibility," *JoongAng Daily*, December 6, 2005; "Hwang Staff Bullied into Smearing Stem Cell Research," *Chosun Ilbo*, December 4, 2005; "Korean Broadcaster Apologises over Cloning Allegations," *The New Scientist*, December 5, 2005.

439 gave a written . . . partnership: "Hwang Staff Bullied into Smearing Stem Cell Research," *Chosun Ilbo*, December 4, 2005; "MBC Apologizes for Unethical Reporting," *JoongAng Daily*, December 5, 2005.

439 military exemptions: T. G. Kim, "Cloning Pioneer Stays Away from Research," *The Korea Times*, November 28, 2005.

439 More protests follow: "S. Korean Cloning Pioneer May Soon Return to Work," *AFX News Limited*, December 4, 2005.

439 producers are forced . . . not run: "MBC Apologizes for Bullying in Hwang

Exposé," *Chosun Ilbo*, December 4, 2005; "Integrity of Stem Cell Research Lies in Follow-Up Studies," *Yonhap News*, December 5, 2005; "Apology From MBC Is Not Enough," *Dong A-Ilbo*, December 6, 2005.

439 no more cells: "Hwang's Team Won't Retest Its Stem Cells for Validity," *The Korea Herald*, December 5, 2005.

439 Lee's statement is backed up: "Hwang's Team Seeks to Regain Its Credibility," *JoongAng Daily*, December 6, 2005.

439 the vaunted Wilmut: E. Check, "Stem Cell Scientist Asks for Retraction," *Nature*, December 14, 2005; "In a Reversal, U.S. Colleague Questions Hwang's Research," *JoongAng Ilbo*, December 15, 2005.

439 43 lawmakers: "Wang Woo-Suk," *Wikipedia*, op. cit.

440 On December 4 . . . replications: "Technical Question Raised on Korean Stem Cell Report," Associated Press, December 6, 2005; D. Normile, "Korean University Will Investigate Cloning Paper," *ScienceNOW Daily News*, December 12, 2005.

440 "More than 200 posts": S. Chong, "How Young Korean Researchers Helped Unearth a Scandal," *Science* 311 (January 6, 2006): 22–25.

440 yanked off the air: J. Brooke, "Clone-Gate Scoop Boomerangs on South Korean News Show," *International Herald Tribune*, December 19, 2005.

440 all its sponsors: Ibid.

440 winning salvos: D. Normile, "Korean University Will Investigate Cloning Paper," *ScienceNOW Daily News*, December 12, 2005.

440 SNU professors . . . .Pittsburgh launches: P. Elias, "Korea's Cloning Research Comes Under Investigation," Associated Press, December 14, 2005; M. Lemonick, "The Rise and Fall of the Cloning King," *Time*, January 1, 2006.

440 *New York Times* praises: N. Wade, "Korean Scientist Said to Admit Fabrication in a Cloning Study," *The New York Times*, December 16, 2005.

440 SNU finally . . . exhaustion: H. J. Jin, "Schatten Part Responsible: *Science* Won't Drop Name from Paper," *The Korea Herald*, December 16, 2006.

440 in the middle of it all: "Prof Hwang Keeps Mum over Scandal," *Yonhap*, December 15, 2005.

440 South Korea finally seems . . . IVF-clinic-derived cells: "What Went Wrong in the Hwang Affair?" *Chosun Ilbo*, December 16, 2005; "Cell-Making Process Can Be Repeated," *The Korea Times*, December 16, 2005 (text of Hwang's speech).

440 lab is shut down: D. H. Wohn, "Scientist Admits Copying Photos," *JoongAng Daily*, December 19, 2005; "Stem Cell Researcher under Fire," Associated Press, December 19, 2005; "Professor Hwang's Lab Is Closed," *Dong A-Ilbo*, December 20, 2005.

440 computers confiscated: J. Herskovitz, "South Korea Panel Seizes Stem Scientist's Computer," Reuters, December 19, 2005.

440 SNU interviews: P. H. Wohn, "Scientist Admits Copying Photos," *JoonAng Daily*, December 19, 2005.

440 "deeply devastated": P. Reber, "Cloning Pioneer Hwang Wants to Retract *Science* Article," *The Bangkok Post*, December 17, 2005.

440 On December 18 and 19 . . . and died: S. Bhattacharya, "Cloning Pioneer Asks to Retract Landmark Study," *The New Scientist*, December 16, 2005; "Hwang Grilled as SNU Inquiry Gets Under Way," *Chosun Ilbo*, December 18, 2006.

441 Kim claims . . . "creating six": D. H. Wohn, "Scientist Admits Copying Photos,"

JoongAng Daily, December 19, 2005; Y. S. Kim, "SNU Starts Probe into Hwang's Research," *The Korea Times*, December 28, 2005.

441 At this time . . . many critics note: "Hwang Grilled as SNU Inquiry Gets Unde Way," *Chosun Ilbo*, December 18, 2006; S. Y. Lee, "Roh Aide Faces Heat in Stem Cell Row," *The Korea Herald*, December 19, 2005.

441 The plot thickens . . . one of the retracted Hwang papers: "Fresh Mixup Casts Doubt on Cloning Pioneer's Research," *Chosun Ilbo*, December 15, 2005; N. Wade, "Scientist Faked Stem Cell Study, Associate Says," *The New York Times*, December 15, 2005; "Hwang Achievements Succumb to the Domino Effect," *Chosun Ilbo*, December 19, 2005; E. Reich, "Cloning Crisis Goes from Bad to Worse," *The New Scientist*, December 20, 2005; "Retraction of Publication," *Biology of Reproduction* 74 (March 2006): 611; "Editorial Retraction," *Stem Cells Express*, April 2, 2006; "Retraction Notice to 'Contribution of the PI3K/Akt/PKB Signal Pathway to Maintenance of Self-Renewal in Human Embryonic Stem Cells,'" *FEBS Letters* 580 (February 20, 2006): 1529; "Editorial Retraction," *Molecules and Cells* 21 (February 2006): 166 (this notice states that three MizMedi papers were retracted); www.ncbi.nlm.nih .gov/entrez/query.fcgi?CMD=search&DB=pubmed (as of April 10, 2006).

441 his only ones retracted: Personal communication (e-mail) from Jennifer Couzins, *Science* writer, April 10, 2006; PubMed search April 9 and 10, 2006.

441 seven journals—and counting: H. G. Parker, "Molecular Genetics: DNA Analysis of a Putative Dog Clone," *Nature* 440 (March 9, 2006): E1–2; Seoul National University Investigation Committee, "Molecular Genetics, Verification That Snuppy is a Clone," *Nature* 440 (March 9, 2006): E2–3; personal communication (e-mail) from Ralph Gwatkin, editor, *Molecular Reproduction and Development*, April 4, 2006; personal communication (e-mail) from John Kastelic, coeditor in chief, *Theriogenology*, April 4, 2006; personal communication (e-mail) from Judith Jansen, managing editor, *Biology of Reproduction*, April 5, 2006; personal communication (e-mail) from Robb Krumlauf, editor in chief, *Developmental Biology*, April 7, 2006; April 7, 2006; personal communication (e-mail) from R. J. Paulson, editor, *Fertility & Sterility*, April 17, 2006; personal communication (e-mail) from D. K. C. Cooper, editor, *Xenotransplantation*, April 10, 2006.

441 doubting Hwang: D. W. Wohn, "Hwang Allies Continue to Back Away," *JoongAng Daily*, December 16, 2005.

441 to hand $10,000 to the father . . . $20,000 to Kim's father: "Hwang Associates Gave Key Witness $30,000," *Chosun Ilbo*, December 27, 2005; J. S. Cho, "DNA Report Supports Hwang's Claim," *The Korea Times*, December 27, 2005; "Fresh Success Claims Add Confusion in Hwang Scandal," *Chosun Ilbo*, December 28, 2005.

442 string of retracted papers: See note above for "The plot thickens . . . one of the retracted Hwang papers," in which all of the retracted papers are cited. The 2004 *Science* paper and the *Biology of Reproduction* paper are the only two non-Yoon retracted papers.

442 gave some $10,000 or more to Park, the other half . . . constitute bribes: "Cash Went to Researcher in the U.S.," *Chosun Ilbo*, December 28, 2005; "Patient-Specific Stem Cells Do Not Exist," *Chosun Ilbo*, December 29, 2005.

442 National Bioethics committee . . . consent forms: J.K. Min, "Stem Cell Researcher Used More Eggs Than Reported," *OhMyNews International*, December 30, 2005.

442 using her own eggs: T.G. Kim, "Hwang Researcher Forced to Donate Eggs," *The*

Korea Times, January 3, 2006; H. J. Jin, "Ethical Issues Resurface Over Hwang," *The Korea Herald*, January 3, 2006.

442 SNU's verdict . . . Snuppy is real: Investigation Committee Report (ICR), Seoul National University, January 10, 2006, www.useoul.edu/sk_board/boards/sk_news_read.jsp?board=11769&p_tid=61982&id=26092 (as of March 16, 2006).

442 On January 12, *Science*: D. Kennedy, "Editorial Retraction," online in *Science Express* on January 12, 2006.

442 While admitting: "Hwang Apologizes but Insists He Was Duped," *Chosun Ilbo*, January 12, 200.

442 set up by MizMedi's Kim . . . "Korea to shame": H. J. Jin, "Hwang Demands Prosecution to Probe Alleged Fraud," *The Korea Herald*, December 22, 2005; South Korea's Hwang Blames Researchers for Fake Stem Cell Work," *Bloomberg.com*, January 12, 2006; H. J. Jin, "Hwang Accuses Research Partner of Fabricating Stem Cell Data," *The Korea Herald*, January 13, 2006.

443 cloned wolves . . . making cloned human embryos: "Hwang Apologizes but Insists He Was Duped," *Chosun Ilbo*, January 12, 2006

443 if not hES cells . . . "center of the world": "Koreans Blinded to Truth about Claims on Stem Cells," *Newsbytes*, January 12, 2006.

443 raids Hwang's home: "Hwang Apologizes but Insists He Was Duped," *Chosun Ilbo*, January 12, 2006; "Hwang Woo-Suk," *Wikipedia*, op. cit.

443 look into Hwang's allegations: "Prosecutors Raid Hwang's Home," *Chosun Ilbo*, January 12, 2006.

443 embezzlement of funds: S. J. Jung, "SNU: Hwang Got Road Advisory Funding," *Dong A-Ilbo*, January 14, 2006; T. H. Kim, "Hwang's Collaborators Summoned," *The Korea Times*, January 23, 2006; J. Herskovitz, "South Korea Widens Probe of Tainted Scientists," Reuters, February 1, 2006.

443 In January and February: "S. Korean Prosecutors Widen Probe of Scientist Hwang," Reuters, January 13, 2006; J. Y. Kwon, "Prosecutors Raid Hwang's Home," *The Korea Herald*, January 13, 2006.

443 at least eight: "Fresh Mixup Casts Doubt on Cloning Pioneer's Research," *Chosun Ilbo*, December 15, 2005; N. Wade, "Scientist Faked Stem Cell Study, Associate Says," *The New York Times*, December 15, 2005; "Hwang Achievements Succumb to the Domino Effect," *Chosun Ilbo*, December 19, 2005; E. Reich, "Cloning Crisis Goes from Bad to Worse," *The New Scientist*, December 20, 2005. (See also note above to "The plot thickens . . . one of the retracted Hwang papers.")

443 deleted scores . . . Dae Gi Kwon: D. H. Wohn, "Hwang Saga Conjures Up More Queries," *JoongAng Ilbo*, January 23, 2006; "Hwang Stem Cells Include None from Clones," *Chosun Ilbo*, January 25, 2006; T. H. Kim, "Prosecutors Look into How Hwang Got Ova," *The Korea Times*, January 26, 2006.

443 claimed to be unaware: T. H. Lee, "Prosecutors Say Hwang Was Unaware," *Dong A-Ilbo*, February 6, 2006.

443 deliberately ruined . . . highly unlikely: "Cell Lines Sabotaged, Prosecutors Suspect," *JoongAng Daily*, February 8, 2006.

443 February 16 . . . hadn't known: N. Wade, "University Panel Faults Cloning Coauthor," *The New York Times*, February 11, 2006.

443 not so fast . . . up anyway: J. Y. Kwon, "Schatten May Have Been Involved in Fabrication," *The Korea Herald*, February 16, 2006; T. G. Kim, "Hwang Aide Says Schatten Knew All," *The Korea Times*, Feburary 16, 2006.

444 sue Schatten: "Hwang to Sue US Collaborator Over Stem Cell Patent," *Chosun Ilbo*, February 13, 2006.

444 asked to step down: H. Gottweis, "South Korean Policy Failure and the Hwang Debacle," *Nature Biotechnology* 24 (February 2006): 141–43.

444 second chance: T. G. Kim, "New Science Minister Wants to Give Hwang Second Chance," *The Korea Times*, February 10, 2006.

444 researchers were never . . . shut up: T. H. Lee, "Stem Cell Tampering Claims Continue," *Dong A-Ilbo*, January 24, 2006; T. H. Lee, "Hwang's Assistant Forged Stem Cell Data," *Dong A-Ilbo*, January 27, 2006; J. Y. Yoon, "Co-Researcher of Hwang Faked Stem Cells Alone," *The Korea Times*, April 3, 2006. (See also note above to "The plot thickens . . . one of the retracted Hwang papers.")

444 in charge . . . for tests: D. H. Wohn, "Hwang Saga Conjures Up More Queries," January 23, 2006; T. G. Kim, "Hwang Blames Research Partner for Data Fabrication," *The Korea Times*, January 12, 2006.

444 eight retracted: H. J. Jin, "Hwang Accuses Research Partner of Fabricating Stem Cell Data," *The Korea Herald*, January 13, 2006; see note above to "The plot thickens . . . one of the retracted Hwang papers."

444 He never once . . . thousands of e-mails: K. K. Jin, "Investigators Secure Stem Cell E-Mails," *Donga.com*, January 16, 2006; J. Y. Kwon, "Prosecution Steps Up Hwang Probe," *The Korea Herald*, January 17, 2006; "Stem Cell Fraud Search Continues," *Dong A-Ilbo*, January 17, 2006.

444 Hwang named Kim: T. G. Kim, "Hwang Blames Research Partner for Data Fabrication," *The Korea Times*, January 12, 2006.

444 creating stem cells: J. Y. Kwon, "Prosecution Summons Six Junior Stem Cell Researchers," *The Korea Herald*, January 16, 2006.

444 coauthoring several: See note above to "The plot thickens . . . one of the retracted Hwang papers."

444 staffer Hwang fingered: "Hwang Apologizes but Insists He Was Duped," *Chosun Ilbo*, January 12, 2006; J. Y. Kwon, "Prosecution Summons Six Junior Stem Cell Researchers," *The Korea Herald*, January 16, 2006; "Cash Went to Researcher in the U.S.," *Chosun Ilbo*, December 28, 2005; "Patient-Specific Stem Cells Do Not Exist," *Chosun Ilbo*, December 29, 2005.

444 In a December . . . ever culled: J. Y. Kwon, "Prosecution Summons Six Junior Stem Cell Researchers," *The Korea Herald*, January 16, 2006.

444 Yoon, an aforementioned: D. H. Wohn, "Hwang Saga Conjures Up More Queries," *JoongAng Daily*, January 23, 2006; "Hwang Targets Suspected of Coordinating Stories," *Chosun Ilbo*, January 13, 2006; T. G. Kim, "Scientists Oust Hwang from Fellowship," *The Korea Times*, March 13, 2006.

445 provided Kim with $20,000: "Seoul Prosecutor's Group 3 Report: Woo Suk Hwang Case," May 12, 2006 (in Korean): 124.

445 Hwang's "cloned" . . . Hanyang University with him: D. H. Wohn, "Hwang Saga Conjures Up More Queries," *JoongAng Daily*, January 23, 2006.

445 Additionally, Mizmedi chief: "Hwang Targets Suspected of Coordinating Stories," *Chosun Ilbo*, January 13, 2006.

445 Yoon (who ended up back at Hanyang): T. G. Kim, "Scientists Oust Hwang from Fellowship," *The Korea Times*, March 13, 2006.

445 also likely to be guilty . . . e-mailed Kim: J. Y. Kwon, "Researcher says Hwang Ordered Fabrication," *The Korea Herald*, Feburary 1, 2006; T .H. Kim, "Hwang Ordered Manipulation of Lab Samples: Researcher," *The Korea Times*, January 31, 2006; "Hwang Ordered Faking of Stem Cell DNA Tests," *Chosun Ilbo*, January 31, 2006; T. H. Kim, "Prosecutors Look into How Hwang Got Ova," *The Korea Times*, January 26, 2006; H. S. Lee, "Hwang Probe Focus Shifts to Mice DNA," *The Korea Times*, February 1, 2006

445 not have known he had *zero*: B. J. Moon, "Prosecutors Still Unsure of Hwang's Culpability," *JoongAng Ilbo*, March 8, 2006; T. H. Lee, "Prosecutors Say Hwang Was Unaware," *Dong A-Ilbo*, February 6, 2006; J. Y. Yoon, "Co-Researcher of Hwang Faked Stem Cells Alone," *The Korea Times*, April 3, 2006.

445 indeed ordered: K. Rahn, "Disgraced Prof. Hwang Admits Data Fabrication," *The Korea Times*, March 6, 2006.

445 lied about the eggs: "The Hwang Case, Researcher Did Not Give Her Eggs Freely," Asia News/Agencies, February 3, 2006; M. J. Ser, "2nd Panel Cites Hwang's Team for Ethics Lapses," *JoongAng Daily*, January 14, 2006.

445 mismanaged millions: "Audit Board Says Hwang Misused Funds," *Chosun Ilbo*, February 6, 2006; D. S. Seo, "Hwang Misappropriated $7 Mil. In Research Funds," *The Korea Times*, February 6, 2006.

445 Karolinska insists: T. G. Kim, "Sweden Denies Hwang's Lobbying," *The Korea Times*, February 28, 2006.

445 Schatten told prosecutors: H. J. Jin, "Schatten Denies Any Part in Hwang Fraud," *The Korea Herald*, April 11, 2006; K. Rahr, "Schatten Denies Involvement in Fabrication," *The Korea Times*, April 11, 2006.

445 enough time: Off-record talks with scientists who say three months is needed to prove hES cell creation; W. S. Hwang, "Patient Specific Embryonic Stem Cells Derived from SCNT Blastocysts," *Science* 308 (June 17, 2005): 1777–83; C. Holden, "Schatten: Pitt Panel Finds 'Misbehavior' but Not Misconduct," *Science* 311 (February 17, 2006) 928–29.

446 their report: "Seoul Prosecutor's Group 3 Report: Woo Suk Hwang Case," May 12, 2006 (in Korean).

446 fabricated constantly: Ibid.; D. Y. Wohn, "Korean Cloning Scandal: Prosecutors Allege Elaborate Deception and Missing Funds," *Science* 312 (May 19, 2006): 980–81.

446 *genuinely believed*: "Prosecutor's Report," op. cit.; Wohn, "Korean Cloning Scandal," *Science*, op. cit.

446 Eul Soon Park . . . supposed expertise: "Prosecutor's Report": 47–61, 110, 143.

446 first author: J. H. Park, "Establishment and Maintenance of Human Embryonic Stem Cells on STO, a Permanently Growing Cell Line," *Biology of Reproduction* 69 (December 2003): 2007–14.

446 few *unretracted*: PubMed search, May 25, 2006.

446 The two Parks: "Prosecutor's Report," op. cit.: 47–61 and 143.

447 became fabricators: Ibid.: 52 and 143; Wohn, "Korean Cloning Scandal," *Science*, op. cit.

447 Hwang ordered . . . to fake: "Prosecutor's Report," op. cit.: 46, 47, 52, and 53.

447 fakery snowballed: Ibid.: 120.

448 Back in the . . . Miz-hES cell photo: Ibid.: 47–61 and 143.

449 Kim repeatedly failed . . . Kim complied: Ibid.: 14–16, 62–78, and 144.

450 he ordered up . . . 13 years: Ibid. (see especially charts on pp. 143 and 144); Wohn, "Korean Cloning Scandal," *Science*, op. cit.; J. S. Chang, "S. Korean Cloning Scientist Hwang Is Indicted," Associated Press, May 12, 2006.

450 innocent of scientific fraud . . . fraudulently obtaining research funds: "Prosecutor's Report": 62; Wohn, "Korean Cloning Scandal," *Science*, op. cit.

451 false receipts and embezzling: "Prosecutor's Report": 118; Wohn, "Korean Cloning Scandal," *Science*, op. cit.

451 Mentioned disapprovingly . . . from patients: "Prosecutor's Report": 112.

451 Kim is indicted . . . at SNU: Ibid.: 10–11; Wohn, "Korean Cloning Scandal," *Science*, op. cit.

451 2004 paper . . . $800,000: Wohn, "Korean Cloning Scandal," *Science*, op. cit.

451 Yonhap News Agency: J. S. Chang, "S. Korean Cloning Scientist Hwang Is Indicted," Associated Press, May 12, 2005.

451 SNU's final report: Seoul National University Investigation Committee, Summary of Final Report on Professor Hwang Woo-suk's Research," March 23, 2006.

451 number of guilty parties: Horace Freeland Judson, *The Great Betrayal: Fraud in Science* (Orlando, FL: Harcourt, 2004): 139.

451 is also biggest: H. Gottweis, "South Korean Policy Failure and the Hwang Debacle," *Nature Biotechnology* 24 (February 2006): 141–43.

452 Herbert Gottweis . . . Robert Triendl: Ibid.

452 both of those multitasking Hwang friends resigned: D. S. Seo, "Hwang Misappropriated $7 Million in Research Funds," *The Korea Times*, February 6, 2006; "Korea and Beyond," *Genetic Crossroads*, January 27, 2006.

452 Un Chan Chung: A. Faiola, "Koreans 'Blinded' to Truth about Claims on Stem Cells," *The Washington Post*, January 13, 2006.

452 "The government policies" . . . Jang Jip Choi: T. H. Kim, "Hwang Debacle Seen as 'Regression of Democracy,'" *The Korea Times*, January 12, 2006.

452 Korea's three big companies . . . bribery charges: J. Jettel, "Scandals Won't Erode Investors' Confidence," *The Korea Times*, April 2, 2006; T. H. Kim, "Chaebol Face Increasing Calls for Better Corporate Governance," *The Korea Times*, March 30, 2006; "Korea Needs Outstanding Leadership" *The Korea Herald*, April 10 2006.

453 Daewoo: "Daewoo Corruption Scandal Widens," *BBC News*, February 2, 2001; M. Hennock, "Daewoo Runaway Hopes for Leniency," *BBC News*, June 14, 2005.

453 *BusinessWeek*: M. Ihlwan, "The Cloning Crisis Clouding Korea," *BusinessWeek*, December 19, 2005.

453 Korean academia: G. Poupore, "Breeding Culture of Dishonesty," *The Korea Times*, January 22, 2006.

453 In the first . . . "them as professors": B. M. Lim, "A Culture of Obedience Is Faulted," Associated Press, February 12, 2006

453 Confucian culture: "In S. Korea, 'No' Gets Failing Grade: Questions Loom for Seniority-Based Culture after Stem Cell Scandal," Associated Press, January 7, 2006; "Korea Needs Outstanding Leadership," *The Korea Herald*, April 10 2006.

453 "The Hwang Affair" . . . "take note": G. de Jonquieres, "A Korean Science Lesson for the Whole of Asia," *Financial Times*, January 31, 2006.

454 Irv Weissman: Author interview with Irving Weissman, Feburary 19, 2006.

454 Koreans are emotional: Personal communication from Joesph Wong, Feburary 20, 2006; Michael Breen, *The Koreans: Who They Are, What They Want, Where Their Future Lies* (New York: St. Martin's, 2004).

454 Hwang confessed: K. Rahn, "Disgraced Prof. Hwang Admits Data Fabrication," *The Korea Times*, March 6, 2006.

454 candlelit rallies: S. W. Park, "Poems, Flags and Cheers for Disgraced Scientist," *JoongAng Ilbo*, January 23, 2006; "Backers Rally on Eve of Hwang's Summons," *JoongAng Daily*, March 2, 2006.

454 set himself on fire: B. Demick," South Korean Cloning Scandal Takes Toll On Whistle-Blowers," *The Los Angeles Times*, February 14, 2006; "S. Korean Commits Suicide over Hwang Scientist Case," Reuters, February 4, 2006.

454 $60 million: "Buddhists to Offer More Than 60 Million Dollars to Former Cloning Pioneer," *Asia News*, May 9, 2006.

454 Hwang fans: "Hwang's Supporters," *The Korea Herald*, May 11, 2006.

454 "a Jesus figure":"Koreans Blinded to Truth about Claims on Stem Cells," *Newsbytes*, January 12, 2006.

455 Richard Garr: Personal communication (e-mail) from Richard Garr, February 19, 2006.

455 Alan Trounson: Personal communication (e-mail) from Alan Trounson, March 7, 2006.

455 1,942 people: Author interview with Hyun Soo Yoon, Hanyang University stem cell researcher , July 20, 2005, in Seoul.

456 Snuppy is real: "Some Vindication for S. Korean Scientist?" United Press International, March 12, 2006; H. G. Parker, "Molecular Genetics: DNA Analysis of a Putative Dog Clone," *Nature 440* (March 9, 2006): E1–2; Seoul National University Investigation Committee, "Molecular Genetics, Verification That Snuppy Is a Clone," *Nature* 440 (March 9, 2006): E2–3.

456 Trounson: Author interview with Alan Trounson, February 14, 2006.

456 Hwang student: Author interview with Woo Suk Hwang, June 4, 2004.

456 reconfirm its validity: T. G. Kim, "New Science Minister Wants to Give Hwang Second Chance," *The Korea Times*, February 10, 2006; H. Gottweis, "South Korean Policy Failure and the Hwang Debacle," *Nature Biotechnology* 24 (February 2006): 141–43; personal communications (e-mail) from Ralph Gwatkin, editor in chief, *Molecular Reproduction and Development*, April 4, 2006; personal communications (e-mail) from John Kastelic, coeditor in chief, *Theriogenology*, April 4, 2006; personal communications (e-mail) from Robb Krumlauf, editor in chief, *Developmental Biology*, April 7, 2006; personal communications (e-mail) from Judith Jansen, managing editor, *Biology of Reproduction*, April 5, 2006; personal communications (e-mail) from R. J. Paulson, editor, *Fertility & Sterility*, April 17, 2006; personal communications (e-mail) from D. K. C. Cooper, editor, *Xenotransplantation*, April 10, 2006.

456 *Chosun Ilbo*: "Government Recognized Use of Hwang-Developed Technologies," *Chosun Ilbo*, May 24, 2006.

456 can't be dismissed: See note above for "reconfirm its validity."

456 entire germline: Author interview with David Ayares, February 10, 2006.

456 James Robl: Author interview with James Robl, Feburary 6, 2006.

456 "the $100 million cow": I. Oransky, "Cloning for Profit," *The Scientist* 19 (January 31, 2005): 41.

457 David Ayares: Author interview with David Ayares, Feburary 10, 2006.

457 "radioactive": S. Beardsley, "Down in Flames," *Scientific American*, February 20, 2006.

457 has given ammunition: "Ethics and Fraud," *Nature* 439 (January 12, 2006): 117–18.

457 Shin Yong Moon: Author interview with Shin Yong Moon, July 21, 2005.

457 FDA green-lit: P. Elias, "FDA Approves First Brain Stem Cell Transplant," Associated Press, October 20, 2005.

457 same month . . . Los Angeles: "$25M Donation Will Fund Stem Cell Research Center," *California Healthline*, February 27, 2006.

458 Diana Devore: "Stem Cell Gold," *Science* 311 (January 20, 2006): 333.

458 $25 million: "$25M Donation Will Fund Stem Cell Research Center," *California Healthline*, February 27, 2006.

458 ACT cloning: Press release at www.advancedcell.com/press-release/advanced-cell-technology-formally-moves-headquarters-and-opens-research-facility-in-california.

458 CIRM won: "Judge Clears Way for Stem Cell Research," *Stanford Daily*, April 26, 2006.

458 therapeutic cloning a priority: C. Hall, "UCSF Resumes Human Embryo Stem Cell Work; Scientists Hope to Generate Lines by Cloning Donated Eggs," *San Francisco Chronicle*," May 6, 2005.

458 $16 million: "College Gets $16 Million for Stem Cell Center," Associated Press, May 11, 2006.

458 Stanford has acquired . . . 2006: Author interview with Irv Weissman, February 22, 2006.

458 verified the fears: E. Merg, "U.S. Lags in Stem Cell Work, Study Finds," *The Daily Princetonian*, September 26, 2005.

458 Leadership changes . . . with what cells: Personal communications (e-mails) from Ismail Barrada, April 7–8, 2006.

458 Miodrag Stojkovic: M. Stojovic et al., "Derivation of a Human Blastocyst after Heterologous Nuclear Transfer to Donated Oocytes," *Reproductive Biomedicine Online* 11 (August 2005): 226–31.

459 committed $177 million: "Blair Lures Stem Cell Talent to U.K. as Bush Ban Stalls Science," *Bloomberg.com*, February 1, 2006.

459 stem cell alliance with India: "India, Britian, to Collaborate on Stem Cell Research," *newKerala.com*, February 22, 2006.

459 fought over . . . sent feelers: Personal communications (e-mails) from Ismail Barrada, November 11, 2005, January 2006, and April 14, 2006.

459 cord blood . . . started in 2005 in Egypt: C. Dabu, "Stem Cell Science Stirs Debate in Muslim World Too," *The Christian Science Moniter*, June 22, 2005.

459 Royan Institute: See Royal Institute Web site at www.royaninstitute.org/ (as of April 12, 2006).

459 Iran president: "Iran May Reconsider Nuclear NTP Membership, " Reuters, February 11, 2006.

459 David Lane: "Movers," *Nature* 431 (October 14, 2004): 882.

459 lose two top cancer researchers . . . "great place to do science": Author interviews with Irv Weissman, February 19, 2006.

460 torn cartilage: Author interview with James Goh, Department of Orthopedic Surgery, National University of Singapore, November 1, 2003.

460 lupus: "NIH Launches Study of Hematopoietic Stem Cell Transplantation for Severe, Treatment-Resistant Lupus," NIH/National Institute of Arthritis and Musculoskeletal and Skin Diseases, May 18, 2004.

460 Geoffrey Raisman: J. Cornwell, "The Miracle Worker," *The Sunday Times Magazine*, April 9, 2006.

460 Texas Heart Institute: "Aldagen, Texas Heart Institute Begin Study of Stem Cell Therapy for Heart Failure Patients," *Pharmacy Choice*, March 27, 2006.

460 peripheral artery disease: J. Adlersberg, "Stem Cells Fix Damaged Leg Arteries," WABC, July 5, 2005.

460 Osiris: "Osiris Cleared by FDA to Begin Stem Cell Trial for Knee Repair; The Announcement Marks the Third Adult Stem Cell Product That Osiris Has in Human Clinical Trials," *Business Wire*, April 1, 2005.

460 Fudan University: Global Watch Mission Report: South Korean Mission to China, Singapore, and South Korea, September 2004.

460 MGH kidney/bone marrow transplant patients: Personal communications (e-mails) with Thomas Spitzer, MGH immunologist, March 26 and 27, 2006.

460 scientists are racing: "Scientists in US To Try Human Cloning Koreans Faked," *Bloomberg.com*, April 13, 2006.

461 rheumatoid arthritis: "Stem Cells Put Woman's Arthritis in Remission," Reuters Health, www.reutershealth.com, September 25, 2004.

461 Michele De Luca: Author interview with Yann Barrandon, stem cell researcher, Ecole Polytechnique Federal de Lausanne.

461 OncoMed Pharmaceuticals: See OncoMed Pharmaceuticals Web site at www.oncomed.com (as of April 14, 2006).

461 Thailand . . . Vietnam: D. Cyranoski, "Far East Lays Plans to be Stem-Cell Hot Spot," *Nature* 438 (November 10, 2005): 135.

461 Yann Barandon: Author interview with Yann Barrandon, Ecole Polytechnique Federale de Lausanne stem cell researcher.

461 Kaylene Simpson: W. Cromie, "Complete Breast Is Grown from a Single Stem Cell," *Harvard University Gazette*, February 9, 2006.

461 "necessary and sufficient": T. Reya et al., "Stem Cells, Cancer and Cancer Stem Cells," Nature 414 (November 1, 2001): 105–11.

462 "remarkable speed": D. Solter, "From Teratocarcinomas to Embryonic Stem Cells and Beyond: A History of Embryonic Stem Cell Research," *Nature Reviews Genetics* 7 (April 2006): 319–27.

462 "The technology will survive": Personal communication (e-mail) from Ismail Barrada, June 22, 2005.

Index

117 impanted blastocyst